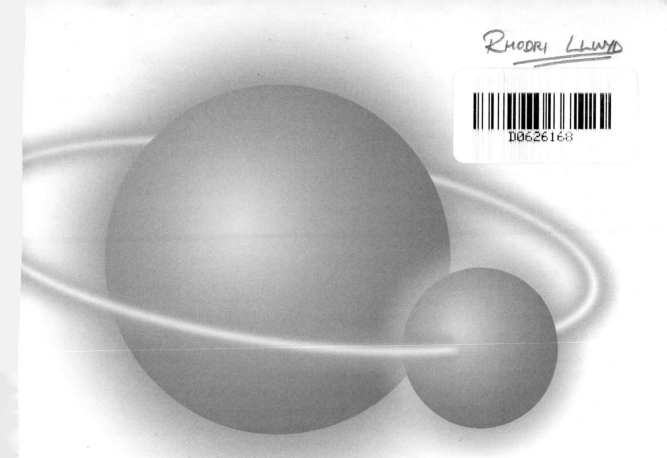

Calculations for A-level
Physics

T.L. Lowe

J.F. Rounce

FOURTH EDITION

First published in 1987 by:
Stanley Thornes (Publishers) Ltd

Fourth edition published in 2002 by:
Nelson Thornes Ltd
Delta Place
27 Bath Road
CHELTENHAM
GL53 7TH
United Kingdom

08 09 / 10 9 8 7

A catalogue record for this book is available from the British
Library

ISBN 978 0 7487 6748 9

Illustrations by TechSet Ltd and Oxford Designers and
Illustrators
Page make-up by TechSet Ltd

Printed and bound in Slovenia by DELO tiskarna
by arrangement with Korotan-Ljubljana

Contents

Preface

As a result of the continuing popularity of this book the fourth edition has all the main features of the earlier editions. Some changes have been made to match alterations to the A-level specifications of the national Awarding Bodies, and up-to-date examination questions have replaced the older ones.

As in the past, A-level students will find *Calculations for A-level Physics* a valuable aid to their studies. It is designed to help both those students who are already achieving success in their physics calculations work and those who have less experience or little confidence.

Relevant theory is described for each topic and this is accompanied by worked examples. These are followed by practice exercises for that topic. At the end of each chapter is a series of examination exercises for which hints and answers are provided. These are designed to widen the experience and improve the confidence of all students.

Explanation of the techniques needed for handling physics calculations continues to be an important part of the book. Chapter 1 shows how calculation questions are approached and the mathematics needed for A-level physics is described in Chapter 2. These two chapters in particular will help those students who find it difficult to approach calculations work.

Astronomy, Medical Physics and Rotational Dynamics are topics which may be offered as options, by certain Awarding Bodies, and chapters are devoted to these because of their special calculations.

An important feature of this book is the inclusion of recent examination questions. These questions come from the papers of a whole range of A-level Awarding Bodies and, together with similar questions, form the final exercise for each topic covered. The final chapter comprises examination questions on topics throughout the book and especially includes long questions that involve more than one topic.

The authors are grateful to the Awarding Bodies that have allowed their questions to be used and extend their thanks to the following:

Assessment and Qualifications Alliance (AQA)
Northern Ireland Council for the Curriculum Examinations and Assessment (CCEA)
University of London Examinations (Edexcel)
Oxford, Cambridge and RSA Examinations (OCR)
Welsh Joint Education Committee (WJEC)

The abbreviations shown alongside the Awarding Bodies are those used with the questions in this book to indicate their origin. Other terms used are 'part' to indicate that only part of the question has been used, 'Spec' for questions from Special level papers and 'Nuff' refers to the Nuffield examinations of OCR.

The answers given and the associated working and hints are solely the responsibility of the authors, and the Awarding Bodies are in no way responsible for these.

T L Lowe
J F Rounce

How to use this book

For many students it is the calculations of A-level physics that make the subject difficult. You may be asking yourself one or more of the following questions:

What mathematical skills do I require?
Why do I get stuck so often?
What are the rules for presenting the calculation?
How can I practise calculations with help available whenever I need it?

A good teacher can help with these difficulties but the benefit of *Calculations for A-level Physics* is that you can refer to it whenever you need help.

This book contains a vast number of calculations exercises. Even if you are already coping well with calculations you will find plenty of interesting and informative problems to work on. In each chapter there are exercises for practice followed by examination questions, most of which have featured in recent GCE A-level examinations.

The chapters and exercises may be worked through in order but many students will not want to use the book in this way. The authors suggest that you first read the introductory chapters (1 to 3) before exercises are attempted. The exercises may be attempted in any order but you should remember that any question may assume knowledge from earlier chapters. Answers to all questions are given towards the end of the book for you to check your work. Obtaining correct answers is rewarding and builds confidence, which is important for any future calculation work. Hints are provided for examination questions (page 291 on) to help you if you get stuck. Always try hard for a few minutes to solve a problem before resorting to hints.

The topics covered here should satisfy almost all of the calculations needs for an A-level student. What is important for you is to decide which topics are not required for the examination that you intend to sit. These topics can be omitted and, in particular, *you need to get a copy of the relevant specification for your examination*. Ask yourself, for example, if you need to study optional topics such as Astronomy, Medical Physics or Rotational Dynamics. If you do, certain chapters have been allocated specifically for these popular options.

Have a look at the list of contents and note the Miscellaneous Questions chapter, which contains a mixture of past A-level examination questions that will serve well for your revision when you have succeeded with other exercises. Turn also to Tables I and II (page 319) to see what useful information is given there and don't forget the index at the back of the book. This will be valuable when you need to find the meaning of a word or an explanation of a concept.

Section A
Basic ideas

1
How to approach a calculation

In this chapter a question of the kind found in A-level physics examination papers is answered. The question concerns an electric circuit calculation and, although the physics for such calculations is not discussed until Chapter 20 is reached, you can learn a lot from the question and its answer. The answer shows how you can present a calculation so that it is easily understood by an examiner or by other people. The comments which follow the calculation will prepare you for answering all the other kinds of calculations you can meet.

Example 1 – an A-level question

A 6.0 V, 3.0 W light bulb is connected in series with a resistor and a battery as in Fig. 1.1. The battery's EMF is 6.0 V and its internal resistance is 4.0 Ω.

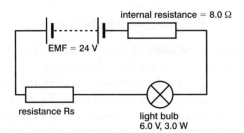

Fig. 1.1   Circuit diagram for worked example 1

The resistance R_s is chosen so that the bulb is run under the conditions for which it was designed, i.e. at a power of 3.0 W.

Calculate:
(a) the current that flows through the bulb under the design conditions
(b) the resistance of the bulb under these conditions
(c) the total resistance of the circuit
(d) the value of the resistance R_s.

Answer

(a) Power $P = V \times I$
$$\therefore \quad 3.0 = 6.0 \times I$$
$$\therefore \quad I = \frac{3.0}{6.0} = 0.50 \, \text{A}$$

(b) Resistance $R = \dfrac{V}{I} = \dfrac{6.0}{0.50} = 12 \, \Omega$

(c) Total resistance $= \dfrac{EMF}{I} = \dfrac{12}{0.50} = 24 \, \Omega$

(d) Resistance R_s = total resistance – bulb resistance
$$\qquad\qquad\qquad\qquad\qquad - \text{internal resistance}$$
$$= 24 - 12 - 8 = 4 \, \Omega$$

Reading the question

Don't be surprised if you have to read a question a few times to understand it. At first you discover which branch of physics is concerned, perhaps electricity or mechanics. You ask yourself whether you have met a similar question before. 'Do I recall one or more formulae likely to fit the question?' 'What have I got to work out?'

Read the question until it makes sense

The question in our worked example should make you think of a light bulb that would normally be used with 6.0 volts across it and would produce 3.0 joules of heat and light energy per second, i.e. its power output is 3.0 watts. The formulae that might come into your mind are $P = V \times I$, $P = V^2/R$ and $P = I^2 R$, where P is the power of the bulb, V the voltage across it, I the current through it and R its resistance.

Diagrams

A diagram contains information that can be seen at a glance and it can be easier to work from than the lengthy wording of a question. If a diagram is not provided with a question or asked for, a quick sketch may be worthwhile. Note that the question in our example would have been complete without the diagram.

Diagrams can be very helpful

For diagrams you will need to be familiar with some symbols such as those for the battery, light bulb and resistors that are shown in Fig. 1.1. The lines drawn with arrows on them to represent forces are another example.

Symbols for electric circuits are listed on page 319.

Symbols for quantities and units

You will find in this and other physics books that there is a well-established set of abbreviations for most physics quantities. F for force and R for resistance are examples. You soon become familiar with them. V is used for potential difference (or voltage) and I is used for electric current ('Intensité de courant' in French). Greek symbols are often used; the symbol λ (pronounced lambda) for wavelength is an example.

Using upper case and lower case letters can distinguish between two similar quantities, e.g. masses M and m in the formula $F = \dfrac{GMm}{r^2}$ in which M could be the mass of the earth and m the mass of a satellite pulled towards the earth with a force F. Subscripts serve the same purpose,

e.g. m_1 and m_2 for two masses. R_s was used for the series resistance in our worked example (1). In Example 3 of Chapter 4 abbreviations H_A and H_B are used for two horizontal forces, similarly V_A and V_B for two vertical forces.

For some quantities many different symbols are used. You will find d, s, x, r and other letters used for distances.

Whatever letters you decide upon you must state what they are being used for.

State what your symbols stand for

You have much less choice with units. The international agreement known by the name of 'Système International' (SI) fixes the units you MUST use and the accepted abbreviations for them. Examples are the ampere for current and its abbreviation A, the metre (m) for measuring distances (lengths) and kilogram (kg) for a mass. A list of the SI units you will meet is given in Table I on page 319. Note that the SI unit for resistance measurements is the ohm and its abbreviation is Ω (the Greek letter omega). We have already had an example: $8.0\,\Omega$ for the internal resistance of the battery.

It is customary to write units in the singular. So we see 3 metre rather than 3 metres. You should adopt this practice in your calculations. Otherwise 'metres' might be mistaken for 'metre s' (interpreted as 'metre second'). However, in text, the plural is acceptable because it makes more comfortable reading. For example '3 joules of heat' was mentioned above. When abbreviations are used 3 joule is written as 3 J and we certainly do not put an s after the J here. The J s (joule second) applies to quite a different quantity.

Write units in the singular in all calculations 3 volt or 3 V, not 3 volts

All the formulae you learn should work with SI units and all formulae used in this book work with SI units.

Formulae work with SI units

Multiples of SI units such as the kilometre (km, a thousand times a metre) and megawatt (a million times a watt) may be used when stating the size of a large quantity. Submultiples of the SI units such as the milliampere (a thousandth of an ampere) or a centimetre (one hundredth of a metre) may be convenient for describing small quantities.

1 kilovolt = 1000 × 1 volt = 1000 volt,

or 1 kV = 1000 V

and

1 milliampere = 1 ampere/1000

or $\dfrac{1}{1000}$ ampere, or 1 mA = $\dfrac{1}{1000}$ A

A list of the multiples and submultiples that you can use is shown in Table II on page 319.

A symbol placed before a unit to make it into a multiple or submultiple is a 'prefix' and it is fitted to the front of the unit with no gap. A gap must be used when a unit is made up of other units. Thus a newton metre is written as N m. A gap between the m and s in 'm s' causes the unit to read as 'metre second' whereas 'ms' denotes 'millisecond', the 'm' for milli being a prefix.

2 ms means 2 millisecond, but 2 m s means 2 metre second

If the values you put into a formula are measurements in SI units the answer you will get is in SI units. In the worked example the figures given (the data) are all for measurements in whole SI units, namely volts, watts and ohms. Otherwise it is advisable to convert a value given as a multiple or submultiple into a value in whole SI units, as explained in Chapter 3.

You may decide to do a calculation using a multiple or submultiple of a unit but you have to be very sure of what you are doing and of what units will apply to the answer you get for your calculation. It is safer to work in whole SI units.

Whatever units are used you must show the units in your working. For every item you work out you must show the unit. So in our worked example, at the end of the calculation for part (a) of our answer we see the symbol A for ampere. It can be very cumbersome to insert the unit for every quantity every time it is used. So in part (d) of the calculation you see '= 24 − 12 − 8' without any Ω signs, but the unit is shown with the final resistance of 4 Ω.

Show the unit with each quantity calculated

Getting an answer

When you have read the question, or even as you are reading it, your aim should be to rewrite the question's information, perhaps first as a labelled diagram and then as one or more equations. In

our example the diagram is already there so we move on to choosing an equation for part (a) of the answer.

Equations you might think of that relate the current (I) we want to calculate to the voltage (V), power (P) and resistance (R) are $R = V/I$ and $P = VI$, and you may know $P = V^2/R$ and $P = I^2 R$.

Consider relevant formulae

Note that we usually leave out multiplying signs and write $P = VI$ instead of $P = V \times I$ as long as no confusion results. When values are inserted for V and I the $\times$ must be used. Otherwise a product like 240×2 would become 2402 instead of 480.

Multiplying signs between symbols can be left out

Regarding the formula $R = V/I$, you can rearrange it to get $V = IR$ or $I = V/R$. This rearranging is called 'transposition' and the rules for this are explained in Chapter 2. Similarly we can get $I = P/V$ as a second formula for I. You might wonder which of these formulae for current is suitable for our part (a) answer. The one to choose is $I = P/V$ or $P = VI$ where I is the current through the bulb, P is the power of the bulb and V is the voltage across the bulb.

So the answer is $I = \dfrac{3.0}{6.0} = 0.50$ A.

Select an appropriate formula

The formula $I = V/R$ would be suitable if V were the voltage across the bulb and R its resistance, but we don't know the value of this resistance. There could be a temptation to put into this formula whatever resistance value is available, namely the 8.0 Ω internal resistance of the battery and this would be quite wrong.

For any formula remember the conditions under which it works

So for part (a) we use $P = VI$ or $I = P/V$.

Rearranging the $P = VI$ formula for P to get the formula $I = P/V$ is very easy and quick to do. Formulae needed in some questions can be more tedious to transpose and, if the values that are to be put into the equation are not too complicated, it is best to start by entering the data in the formula you remember. Transposition is delayed until the calculation has achieved some simplification. In our worked example the equations are simple, as are the values to go into them, so there is little to choose between entering data first or rearranging equations first.

When an equation is complicated consider entering values before rearranging the equation

Write your answer as a series of equations. These may be linked by words of explanation and the symbol '∴' which stands for 'therefore' or 'it follows that' is particularly useful. It was used in part (a) of our answer. In place of several equations one equation can often be continued through a number of steps, as in part (b), where we see $R = V/I = 6.0/0.50 = 12\,\Omega$ instead of

$$R = V/I$$
$$\therefore \quad R = 6.0/0.5$$
$$\therefore \quad R = 12\,\Omega.$$

Write a calculation as a series of equations

Using your calculator

It is assumed that you have an electronic calculator which has keys for sin, cos and tan (for use with angles) and for the log of a number. These keys and others on such a 'scientific' calculator are essential for A-level physics calculations. Some calculators conform to the 'VPAM' specifications and display not just the last number you have entered or an answer but show all the values and the operations (such as adding and multiplying) that you have keyed in. The answer is then displayed as well when you press the equals key.

So to work out 6.0/0.50 in part (b) of the worked example, you remember that 6.0/0.50 is the same as $6.0 \div 0.50$ and key in $6.0 \div 0.50 =$ and the calculator display is exactly as shown in Fig. 1.2 or is the same except that the 1.200^{01} answer is replaced by 12. or by 12.0000 and the number of noughts (zeros) may be different. The differences are the result of the calculator having a number of different 'modes,' i.e. ways of working. The mode we want to use is the 'scientific mode' and it is this mode that gives the display shown in Fig. 1.2.

Advice given in this book for calculator use will apply to the Casio *fx*-83WA calculator. The procedures for switching on this calculator,

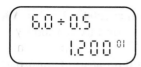

Fig. 1.2   A calculation displayed on a VPAM calculator

selecting the 'scientific mode' and clearing the screen for another calculation are described in Chapter 2.

From the answer displayed as 1.200^{01} you get the expected answer of 12 by multiplying the 1.200 by 10. If the two small figures were 02 you would multiply by 10 a second time, 03 a third time to give 1200. When the small figures are 11, for example, multiplying by 10 eleven times would be inconvenient and this is one reason for keeping the 1.200 and, as explained in Chapter 2, we then write 1.200×10^1 instead of 1.200^{01}.

'I'm stuck'

How often do you hear these despairing words when a calculation question is tried? Even when you have a good knowledge of physics and the appropriate maths you can get stuck.

You might then read the question again and ask yourself:

- Does the question fit what I have been trying to do?
- Is there a diagram I could draw?
- Have I pictured the situation described by the question or have I had in my mind a circuit without a battery or other voltage supply?
- Have I missed an equation that is needed? Perhaps it is in the list provided with the exam paper.
- Are there words in the question that I have disregarded, perhaps 'in series' in our example? If a diagram had not been provided it would have been essential to appreciate that 'in series' meant that the circuit components formed a single loop.

A word like 'series' can make a lot of difference to a calculation. It is a key word. Similar key words often met are 'smooth', 'slowly' and 'steady'. In mechanics questions a 'smooth surface' is one that is so smooth that it cannot provide any force parallel to its surface. So a smooth floor can push upwards and prevent a person falling but cannot provide a force to stop sliding. An object 'raised slowly' means it rises so slowly that it has no kinetic energy and gains only gravitational potential energy. A 'steady speed' means no change in speed.

If at first you don't succeed . . .

Good luck

We all make silly mistakes sometimes, so never get too disappointed. Rough checks are mentioned in Chapter 2 and these will minimise errors. Hurrying encourages errors, of course. Leaving a question and returning to it can waste time but may give you a fresh view of a problem and lead to a successful answer. So take care and good luck with your calculations!

2
Essential mathematics

Expressions and equations

An *expression* is a combination of numbers and symbols. Simple examples are the sum $3 + 2$, the difference $3 - 2$ and the product 3×2.

In any expression the order of multiplication or adding is not important,

e.g. $\mathbf{3 \times 2 = 3 \times 2}$ **and** $\mathbf{2 + 7 = 7 + 2}$.

The order of subtraction DOES matter, e.g. $3 - 2$ is not the same as $2 - 3$.

Alphabetical symbols are used to represent numbers either for convenience or because the number is not yet known.

An equation shows that two expressions have equal size or value, e.g. $3 + 2 = 4 + 1$ or $x = 7$.

A quantity is a number or a measurement. A measurement is a number times a *unit*, e.g. 3 times a metre or 3 metre. (Note that units used in calculations are written in the singular.)

Abbreviations are used for units, e.g. m for metre. The 'Système International' (SI) specifies the symbols to be used for units. Units are discussed in Chapter 3.

A formula is an equation which shows how a quantity on the left may be calculated by inserting values of quantities on the right, e.g. area = length × width or $A = L \times w$.

Note that a × sign is usually omitted if no confusion will result, e.g. $A = Lw$, and

$\mathbf{3a}$ **means** $\mathbf{3 \times a}$.

Fractions

A half is obtained by sharing one (equally) between two or dividing 1 by 2. For a half we write $1 \div 2$ or $1/2$ or $\frac{1}{2}$. Half is part of a whole one, so it is a fraction.

3 divided by 6 also equals a half, so that $\frac{3}{6} = \frac{1}{2}$. This illustrates that multiplying or dividing the top (the numerator) AND the bottom (the denominator) of a fraction by the same number does not change its value, e.g. $\frac{1}{2} = \frac{1 \times 3}{2 \times 3} = \frac{3}{6}$.

A number multiplying a fraction multiplies just the numerator, and a number dividing a fraction multiplies the denominator, e.g.

$$\frac{2}{9} \times 4 = \frac{8}{9} \qquad \frac{1}{2} \div 2 = \frac{1}{2 \times 2} = \frac{1}{4}$$

Note that a fraction 'of' a number means the fraction 'times' the number, e.g. 'a quarter of 3' means $\frac{1}{4} \times 3$ which is $\frac{3}{4}$.

'of' means 'times'

To multiply a fraction by a fraction the numerators are multiplied and the denominators are multiplied, e.g.

$$\frac{9}{10} \times \frac{2}{3} = \frac{18}{30} \left(or \ \frac{6}{10} \right)$$

Simplifying an expression means to rewrite it with smaller numbers or fewer numbers, e.g. $\frac{18}{30}$ above was simplified to $\frac{6}{10}$ and also equals $\frac{3}{5}$.

Reducing two numbers in an expression, as in the 18/30 fraction above, is called 'cancelling'. So too is the removal of two numbers, as in $\frac{9.3 \times 5}{9.7 \times 5}$, which simplifies to $\frac{9.3 \times 1}{9.7 \times 1}$ which equals $\frac{9.3}{9.7}$.

Cancelling in equations is discussed later. Cancelling units features in Chapter 3.

Simplifying may be useful when a fraction has been divided by a fraction, e.g.

$$\frac{1/3}{1/2} = \frac{2 \times 1/3}{2 \times 1/2} = \frac{2/3}{1} = \frac{2}{3}$$

The reciprocal of a fraction is obtained by turning the faction upside down. So the reciprocal of 2/3 is 3/2. The reciprocal of a simple number e.g. of 7 (which can be written as 7/1) is 1/7.

The reciprocal of 2/3 is 3/2 and of 7 is 1/7

Percentage

'per' means 'for each' and 'cent' denotes 100, so 50 percent (written as 50%) means 50 for each 100 or 50 out of each hundred, i.e. a fraction 50/100 or a half. So $p\%$ means a fraction, $\frac{p}{100}$, $p\%$ of y means $\frac{p}{100} \times y$, and a fraction $\frac{a}{b} = \frac{p}{100}$, so that

$$p = \frac{a}{b} \times 100 \qquad (2.1)$$

These facts are best recalled by remembering that 50% means $\frac{1}{2}$ or $\frac{50}{100}$.

Using brackets

$a \times (b + c)$ or $a(b + c)$ can be rewritten as $ab + ac$. So $3(5 + 2)$, for example, means 3 times the sum of $5 + 2$. The multiplication or other action applied to a bracketed expression applies to everything within the brackets. So $3(5 + 2)$ (which is 3×7 or 21) = 3×5 plus 3×2 or $15 + 6$.

$$a(b + c) = ab + ac \qquad (2.2)$$

Numbers that are multiplying are called factors and in the expression $ab + ac$ the a is a factor of both ab and ac. It is 'common' to both. Taking out a common factor is the reverse of the process described above. $ab + ac$ becomes $a(b + c)$.

In a fraction such as $\frac{8 + 4x}{2}$ the dividing line shows that 2 divides both the 8 and the $4x$ and the effect

is the same as $\frac{8}{2} + \frac{4x}{2}$ which equals $4 + 2x$. But $8 + \frac{4x}{2}$ equals $8 + 2x$.

When two bracketed expressions multiply the rule is

$$(a + b)(c + d) = ac + ad + bc + bd \qquad (2.3)$$

You can test this rule with simple numbers.

Working with + and − signs

If multiplying brackets contain differences we need rules concerning the effect of a − sign before a number, i.e. a negative number. The rule is that two negative numbers give a positive product (− times − gives +), − times + gives −, and + times + gives +. (Note here that a number with no + or − before it is regarded as +.) So using equation 2.3 the expression $(3 − 2)(7 − 3)$ could be written as $21 − 9 − 14 + 6$.

A similar rule applies to sums and differences. For example, the heat in joules required to warm a kilogram of water from temperature $T_1°$ to $10°$ is $4200(10 − T_1)$ and for $T_1 = 9°$ equals 4200, but if $T_1 = −9°$ we have $4200(10 − −9)$ and the $−−$ must give + to give a $19°$ temperature rise.

The rules are

$$−− \text{ gives } + \quad − + \text{ or } + − \text{ gives } − \quad + + \text{ gives } +$$

Your calculator

For A-level physics calculations you need an electronic calculator, and it should be a scientific one so that it will handle, for example, the powers, logarithms, sines and cosines explained later in this chapter. This book describes the use of the Casio *fx*-83WA calculator. Other calculators are similar.

Your *fx*-83WA is switched on by pressing the AC/ ON key.

The calculator has a number of different ways of working, i.e. different modes, and pressing the MODE key (near top right of the keypad) three times will show you the choice of modes.

Now use the MODE key again and press 1 when 'Comp' is displayed to select that, then 1 when 'Deg' is displayed to select degrees for angles (discussed later), and finally 2 when 'Sci' is displayed to get scientific mode (soon to be explained). In response to your selecting scientific mode you are asked to enter a number; you should enter 4. As a result the four-figure answer of 4.000 is obtained in the following calculation.

Now you can test the calculator by pressing the AC/ON key to clear the screen, then entering, for example, $8 \times 0.5 =$. Your 8×0.5 to be calculated is shown on the screen and the answer is displayed on the right as 4.000, a four-figure answer.

Now try $7 \times 3 =$. Your answer is 2.100^{01}. You expected 21 or 21.00? Well, you are using scientific mode, which will be very useful. Just multiply the 2.100 by 10, i.e. move the decimal point one place to the right. Do this only once as the 1 in the small 01 on the right indicates. You now have 21.00. (2.100^{02} would indicate $2.100 \times 10 \times 10$.)

Dividing, adding and subtracting are achieved in the same way but using the $\div$, $+$ and $-$ keys.

Some care is needed with dividing.

Consider $\dfrac{9.1}{3.5 \times 4.7}$. This is the same as $\dfrac{9.1}{3.5}$ DIVIDED by 4.7, as explained above. So the calculator entry should be $9.1 \div 3.5 \div 4.7$.

An example of another difficulty is $\dfrac{9.34}{2.11 + 3.79}$, where you could unintentionally get the answer for $\dfrac{9.34}{2.11} + 3.79$. The simplest procedure is to use the calculator for $2.11 + 3.79$ to get 5.90, clear the calculator, then use $9.34 \div 5.90$ to get 1.583. Alternatively, if you know how to use it, the calculator's memory can help.

Brackets are handled by the calculator just as you would expect. For example, entering $6(2 + 3) =$ gives the answer 3.000^{01} meaning 30.00 or 30.

Simple rules for handling equations

If the whole of one side of an equation is multiplied, divided, added to or reduced by any number, then the equation will remain true if the same is done to the other side.

Examples are

- $x + 2 = 5$ gives $x = 5 - 2$ by subtracting 2 from each side, i.e. $x = 3$
- $4x = 8$ gives $x = 2$ when each side is divided by 4
- $6x = 4$ or $x = 4/6$ can be written as $3x = 2$ or $x = 2/3$
- $\dfrac{x}{2} = 7$ becomes $x = 14$ by multiplying both sides by 2.

Example 1

Calculate the time for which an electric heater must be run to produce 7200 joules of heat if the potential difference across the heater is 12 volts and the current flowing is 2.5 amperes.

Answer

The formula usually learnt is 'heat produced in joules $= VIt$' where V is the potential difference in volts, I is the current in amperes and t is the time in seconds.

$\therefore \quad 7200 = 12 \times 2.5 \times t$

$\therefore \quad 7200 = 30 \times t$

Dividing both sides by 30 (or moving the 30 to the left-hand side where it will divide) we get $\dfrac{7200}{30} = t$ which can be rewritten as $t = \dfrac{7200}{30}$.

$\therefore \quad t = 240$ second or 4 minute

(The $\therefore$ symbol denotes the word 'therefore' or 'it follows that.')

Cancelling in an equation

Simplifying or removing a pair of numbers in an equation is called cancelling. In the equation $3(2x + 3) = 3(x + 5)$ the threes cancel when both sides of the equation are divided by 3. If Example 1 above had been $7200 = 12 \times 2 \times t$ and you noticed that 12 divides nicely into 72 you might have divided both sides of the equation by 12 (it would still be true). You would get $600 = 2t$ so that $t = 300$ s.

Simplifying $3.1x + 5 = 4.4 + 5$ to $3.1x = 4.4$ is also an example of cancelling.

Solving an equation with two unknowns

Solving means discovering the value of a quantity.

Suppose a rectangular block of material measures 2.0 m by 3.0 m by 1.0 m and has a mass of 15 000 kg and we want to calculate its density.

The formula for density is $\rho = \dfrac{mass}{volume}$ or $\dfrac{M}{V}$, so that $\rho = \dfrac{15\,000}{V}$

This equation contains two unknown quantities, namely ρ and V, and as it stands cannot give a value for ρ. Further information is needed, another equation.

We have the formula for V which is $V = length \times width \times depth$.

$$V = 2.0 \times 3.0 \times 1.0 = 6.0 \text{ cubic metre}$$

This value for V can be substituted in the formula for ρ.

$$\therefore \quad \rho = \frac{15\,000}{6.0}$$

$$\therefore \quad \rho = 2500 \text{ kilogram per cubic metre}$$

An equation with two unknowns has been solved for d by having a second equation that provides a value for V to be substituted in the first equation.

Proportionality

Two quantities, say x and y, are proportional if doubling x causes y to double and tripling x triples y, etc. We then write $x \propto y$ and the equation $x = ky$ must be obeyed, the k being a constant (unaffected by the values of x and y). Also of course $y \propto x$.

$$\text{If } x \propto y \text{ then } x = ky \qquad (2.4)$$

If x changes from x_1 to x_2 causing y to change from y_1 to y_2 then, if x and y are proportional,

$$\frac{x_1}{y_1} = \frac{x_2}{y_2} (= k) \qquad (2.5)$$

If doubling x halves y, or vice versa, we have 'inverse proportionality' and $x = k/y$ or $x \propto k/y$ or $x \propto 1/y$. Consequently

$$\text{if } x \propto 1/y \text{ then } \frac{x_1}{x_2} = \frac{y_2}{y_1} \text{ or } x_1 y_1 = x_2 y_2 \qquad (2.6)$$

The term 'ratio' refers to a comparison of two quantities and is usually expressed as a fraction, e.g. the ratio of a 3 metre length to (i.e. 'compared with') a 2 metre length is 3 to 2, or 3:2 or 3/2. Two quantities that are proportional are in constant ratio. For $y = kx$ the ratio $\dfrac{y}{x} = k$.

Exponents

a^2 means $a \times a$, a^3 means $a \times a \times a$, etc., so that $10^2 = 100$ and $10^3 = 1000$, etc. The small superscript numbers are called exponents or indices or powers. The number below an index is the base and the base and exponent together can be called a power.

When two numbers with exponents are multiplied their exponents add. So $10^2 \times 10^3$ ($= 10 \times 10 \times 10 \times 10 \times 10$) $= 10^5$.

When two numbers divide their exponents subtract so that $10^3/10^2 = 10^{3-2} = 10^1$ or simply 10.

$$a^b \times a^c = a^{(b+c)} \qquad (2.7)$$

$$a^b/a^c = a^{(b-c)} \qquad (2.8)$$

Note that $10^0 = 1$, e.g. $10^2/10^2 = 10^{2-2}$ or 10^0 but clearly equals 1.

The reciprocal of a number with an exponent is obtained by putting a $-$ sign before the exponent. For example, the reciprocal of 10^2 ($= 1/10^2 = 10^0/10^2 = 10^{0-2}$) $= 10^{-2}$.

$$a^0 = 1 \text{ and } a^{-1} = \frac{1}{a}$$

Another useful fact is

$$(a^x)^y = a^{xy} \qquad (2.9)$$

For example, $(10^3)^2 = 10^6$.

9

A root of a number when multiplied by itself, perhaps more than once, gives the number concerned, e.g. the fourth root of 16 (written as $\sqrt[4]{16}$) is 2 because $2 \times 2 \times 2 \times 2$ or $2^4 = 16$, or $\sqrt[4]{16} \times \sqrt[4]{16} \times \sqrt[4]{16} \times \sqrt[4]{16} = 16$.

A second root is called a square root and a third root is called a cube root. The 2 at the front of a square root is usually omitted.

For example,

$$\sqrt{9} = 3 \text{ and } \sqrt[3]{8} = 2$$

When you consider that $x^{0.5} \times x^{0.5} = x^1$ or x and also $\sqrt{x} \times \sqrt{x} = x$ you can see the rule that

$$x^{1/2} = \sqrt{x}$$

Similarly $\sqrt[n]{x} = x^{1/n}$, e.g. $\sqrt[4]{16} = 16^{1/4} = 16^{0.25}$, and your calculator (see below)will tell you that this equals 2.

Note that

$$\sqrt[n]{ab} = \sqrt[n]{a} \times \sqrt[n]{b} \qquad (2.10)$$

so that $\sqrt{4 \times 9} = \sqrt{4} \times \sqrt{9} = 2 \times 3 = 6$ and, even more usefully,

$$\sqrt{4 \times 10^6} = \sqrt{4} \times \sqrt{10^6} = 2 \times 10^3$$

The exponential function

This is a number close to 2.718, which is always denoted by e and which has the property that a graph of $y = e^x$ has, at any point on it, a slope equal to the value of y for that point. This is illustrated in Fig. 2.1a.

This property is also possessed by $y = e^{-x}$ and by $y = e^{-(x-c)}$ except that the slope is negative (see Fig. 2.1b) so $y = -$slope. This last equation can be written as

$$y = e^c \times e^{-x} \text{ or } y = y_0 e^{-x}$$

where y_0 is the value of y when $x = 0$.

This relationship applies, for example, to radioactive decay in Chapter 28.

Exponential graphs are discussed in Chapter 30.

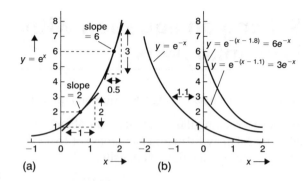

Fig. 2.1　Exponential graphs involving e

e can be very useful as the base for logarithms, as explained later in this chapter.

Powers of ten and standard form

A large number like $1\,000\,000\,000$ is more conveniently written as 10^9, i.e. as a power of 10. Similarly $0.000\,01 (= 1/100\,000) = 10^{-5}$ $0.007 = 7 \times 10^{-3}$ and $70\,000 = 7 \times 10^4$.

Writing a number with a power of 10 overcomes a difficulty. For example $70\,000$ implies a number known to be exactly $70\,000$, not even $70\,001$. Very few physics measurements would be so precise. There would be some experimental error so that perhaps only the 7 and first two zeros are reliably known. We can then write 7.00×10^4. This number is said to be in 'standard form' because it is written with one digit only (the 7) in front of the decimal point and shows the appropriate number of digits after the decimal point as well as the correct power of 10.

In standard form write 3456 as 3.456×10^3

The scientific mode on your calculator gives answers in standard form. An answer displayed as $2.100^{\,01}$, as mentioned earlier, is of course using the small figures to show the power of 10 and should be read as 2.100×10^1. Similarly $1.234^{\,12}$ means 1.234×10^{12}.

Logarithms

If $10^L = x$ then L is called the logarithm of x or, more exactly, the logarithm to the base 10 of x.

$$10^{\log x} = x$$

For example $10^3 = 1000$ so that log (to the base 10) of 1000 is 3.

We write $\log_{10} 1000 = 3$.

If the base is not specified then we assume it to be 10, so that log 2 is taken to mean $\log_{10} 2$.

The exponential function e is quite often used as the base for logarithms and $\log_e$ is described as the natural logarithm and is denoted by ln. So $\log_e 7.388$ or $\ln 7.388$ is 2 because $e^2 = 7.388$.

Since $x = 10^{\log x}$ and $e^{2.3} = 10$ we have $x = (e^{2.3})^{\log x} = e^{2.3 \log x}$, which means that

$$\ln x = 2.3 \log x \text{ (a very useful rule)} \quad (2.11)$$

Other useful rules for handling logarithms are

$$\log ab = \log a + \log b \quad (2.12)$$
$$\log \frac{a}{b} = \log a - \log b$$

$$\text{and} \quad \log a^b = b \log a \quad (2.13)$$

so $\log(1000 \times 100) = \log 1000 + \log 100$
$$= 3 + 2 = 5$$
and $\log 100^3 = 3 \times \log 100 = 3 \times 2 = 6$.

Powers and logs on your calculator

To obtain the square root of a number on your *fx-83WA* calculator you use the $\sqrt{}$ key. Entering AC $\sqrt{}$ 4 = produces the answer 2. If you had an answer of 4 displayed after some calculation (or press 4 and =) you could then press the $\sqrt{}$ key and get 2.

For 5^2 or 5^3 enter the 5 first, press the = key and then use x^2 or x^3.

The x^y key allows a number like 2.1^5 to be calculated using the keys 2.1 x^y 5 =, which gives 4.084^{01} or 40.84.

A most important key is the one marked EXP. Its effect is '$\times$ 10 to the power of,' so that 4 EXP 2 = gives an answer of 4×10^2 or 400, and 4.000^{02} is displayed.

Experimental errors

When a length is measured with a metre rule the reading is taken of the nearest marking above or below the length. This means that a measurement of say 23.3 centimetre may be too high or too low by an amount (e.g. 0.05 cm) corresponding to half the spacing of the markings. So the possible error is + or −0.05 cm and we record the measurement as 23.3 ± 0.05 cm.

We can also express the possible error as a percentage of the measurement. The 0.05 as a % of 23.3 is $\frac{0.05}{23.3} \times 100\%$ *or* 0.2146%, but 0.2 is near enough for indicating error, so we have $23.3 \pm 0.2\%$.

The possible error in a measurement may also be indicated by simply limiting the number of digits used for the recorded value. 53.3 will be regarded as having a $\pm$ possible error that would take the right hand 3 digit halfway up towards 4 or down towards 2. This is the same as ± 0.05 in our example.

The 53.3 is described as comprising three 'significant figures,' the 5, 3 and 3.

A 0 in front of 53.3 would serve no purpose, the 0 at the front of 0.4 serves to emphasize the presence of the decimal point and the 00 in the number 0.004 acts as a spacer to show that the 4 means 4 thousandths, these zeros all being examples of figures that are NOT significant. Thus 0.004 53 has three significant figures, 0.004 530 has four.

When it was recommended that you set your calculator to scientific mode (Sci) and follow this by keying in a number 4 you were choosing answers to be limited to four significant figures (sig figs).

In your calculations you must not give an answer that suggests a very inappropriate accuracy. For a simple rule never give an answer to an accuracy better than that of the least accurate quantity used in the calculation, i.e. no more sig figs than the least accurate value used in the calculation. This usually means that you will shorten your final answers to two sig figs. During your calculation shorten any longer numbers to four sig figs and limit your calculator to four sig fig answers.

Note that a number like 2.371 is reduced to two sig figs as 2.4 rather than 2.3, because 2.4 is closer to the 2.371. We have 'rounded' up. For a number like 3.65, which is half way between 3.6 and 3.7, the practice is to round up, i.e. write 3.7 for the two sig fig value.

Calculus notation

An increase in a quantity e.g. in time t, can be denoted by Δt (pronounced 'delta t') and, in the branch of mathematics called calculus, a very small increase in t is denoted by δt (also 'delta t').

The change δt, may be associated with a change in a quantity e.g. displacement x and this change δx will necessarily be very small because of the smallness of δt. δx divided by δt, i.e. $\delta x/\delta t$, then tells us the velocity (the rate of change of displacement).

(To specify a velocity at a precise time t and not an average value at around the time t we need to have δx for a time δt with δt infinitely small. δx

will of course be infinitely small too but $\delta x/\delta t$ will tell us the velocity at exactly time t. The value of $\delta x/\delta t$ when δt is infinitely small (approaching zero) is written as $\delta x/\delta t$ $\delta t \rightarrow 0$ or more briefly as dx/dt ('dee x by dee t').

For most purposes a physics student need not (but should) distinguish between $\delta x/\delta t$ and dx/dt.

So dx/dt denotes rate of change of x with change of t. If it so happens that $dx/dt = $ constant (a velocity for example might be constant) then $dx/dt = $ any distance/time taken.

Some rules of geometry

Angles may be measured either in degrees (one revolution is 360 degrees (360°)) or in radians (rad), whose size is such that 2π rad equals one revolution. Some useful facts about angles are shown in Fig. 2.2.

Pythagoras' theorem

In a right-angled triangle (Fig. 2.3) the longest side (the hypotenuse) has a length c related to the lengths a and b by

$$c^2 = a^2 + b^2 \qquad (2.14)$$

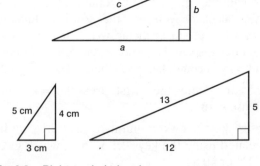

Fig. 2.3   Right-angled triangles

Well-known examples of right-angled triangles are the 3, 4, 5 and 5, 12, 13 triangles shown in Fig 2.3.

Isosceles and equilateral triangles

The isosceles triangle has two sides of equal length and so two of the angles are equal (Fig 2.4a).

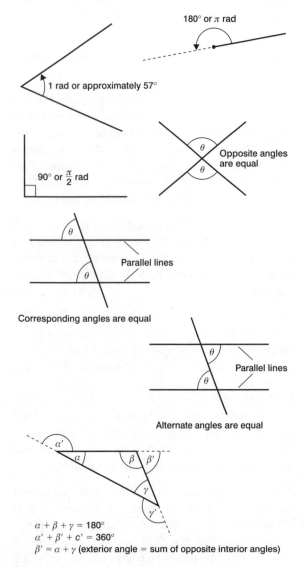

$\alpha + \beta + \gamma = 180°$
$\alpha' + \beta' + c' = 360°$
$\beta' = \alpha + \gamma$ (exterior angle = sum of opposite interior angles)

Fig. 2.2   Useful information concerning angles

An equilateral triangle has three sides of equal length and each angle equals 60° (Fig. 2.4b).

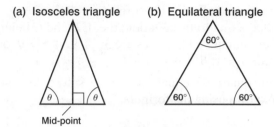

Fig. 2.4   Isosceles and equilateral triangles

Some properties of circles, discs and spheres

The circumference of a circle is $2\pi r$ (where r is its radius) or π times the diameter. The value of π is 3.142 or 22/7. A disc's area is πr^2. For spheres volume $= 4\pi r^3/3$ and surface area $= 4\pi r^2$.

As shown in Fig. 2.5a, an angle of 1 radian subtends, at any radius r, an arc equal to a fraction $1/2\pi$ of the circumference, i.e. it subtends an arc of length r.

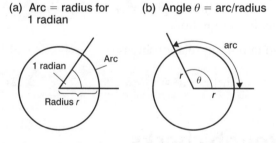

Fig. 2.5   Using radians

The size of any angle in radians equals the arc it subtends divided by the radius (Fig. 2.5b).

$$\theta = \frac{\text{arc}}{r} \qquad (2.15)$$

Trigonometrical ratios

The size of any angle θ can be specified by imagining it to be part of a right-angled triangle and then describing the resulting shape of the triangle as shown in Fig. 2.6. For example when $\theta = 60°$, the ratio of the *adjacent side* to the *hypotenuse* is $\frac{1}{2}$. So b/c, which we call cosine θ, is 0.5.

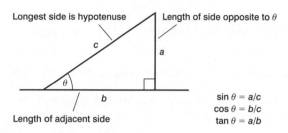

Fig. 2.6   Trigonometrical ratios

The most useful ratios are

> $\text{sine } \theta \ (\text{or sin } \theta) = a/c \qquad (2.16)$
> $\text{cosine } \theta \ (\text{or cos } \theta) = b/c \qquad (2.17)$
> $\text{tangent } \theta \ (\text{or tan } \theta) = a/b \qquad (2.18)$

For a given θ (in degrees or radians) we can get $\sin\theta$, $\cos\theta$, etc. using suitable electronic calculators or tables and similarly can deduce θ from any given trigonometrical ratio.

Small angles

For a small angle (θ about 5° or less), $\tan\theta \approx \sin \approx \theta$ in radians and $\cos\theta \simeq 1$, to better than 1%. ('$\simeq$' denotes 'approximately equals'.)

Large angles

For $\theta = 90°$, $\sin\theta = 1$, $\cos\theta = 0$ and $\tan\theta = \infty$. For $\theta > 90°$, we can still use $\sin\theta$, $\cos\theta$, etc. if we apply suitable rules as illustrated in Fig. 2.7a.

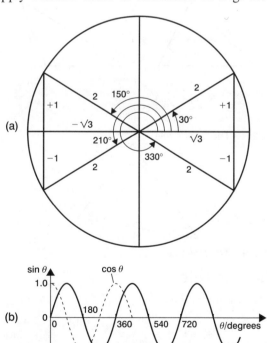

Fig. 2.7   Trigonometrical ratios for large angles

This shows that a negative sign must be given to an opposite side that is below the horizontal axis and to an adjacent side if it is to the left, e.g. $\tan 210 = \dfrac{-1}{-\sqrt{3}} = 0.58$ (same as $\tan 30$). The relationship between $\sin \theta$ and θ is shown as a graph in Fig. 2.7b.

Trig ratios on your calculator

Your fx-83WA calculator has keys for sin, cos and tan. These are used in the way described earlier for the $\sqrt{\ }$ key.

The SHIFT key (top left of the keypad) allows you to obtain a function indicated above the key. The sin key has $\sin^{-1}$ marked above it and so it is this key to use when, for example, you want to find the angle whose sine is 0.5. Apart from having to use the SHIFT key the procedure is as for obtaining a trig ratio. So AC SHIFT $\sin^{-1} 0.5 =$ or AC $0.5 = $ SHIFT $\sin^{-1} =$ will give the answer $30°$.

The cosine rule

This rule is an extension of the rule (or theorem) of Pythagoras and applies to a triangle of any shape. It relates the lengths a, b and c of the triangle's sides (see Fig. 2.8).

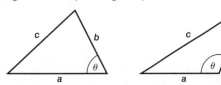

$$c^2 = a^2 + b^2 - 2ab \cos \theta$$

Fig. 2.8   The cosine rule

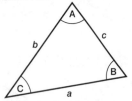

Fig. 2.9   The sine rule

The sine rule

This rule states that

$$\frac{a}{\sin A} = \frac{b}{\sin B} = \frac{c}{\sin C} \qquad (2.20)$$

(see Fig. 2.9).

Quadratic equations

An equation having the form $Ax^2 + Bx + C = 0$ is called a quadratic equation, the A, B and C being fixed numbers. Examples are $3x^2 + 9x + 5 = 0$, or (with $A = 1, B = 0, C = -4$), $x^2 - 4 = 0$.

Any quadratic equation can be solved for the unknown using the formula

$$x = \frac{-B \pm \sqrt{B^2 - 4AC}}{2A} \qquad (2.21)$$

so that for $3x^2 + 9x + 5 = 0$ we get $\dfrac{-9 \pm \sqrt{9^2 - 4 \times 3 \times 5}}{2 \times 3}$ which simplifies to $x = \dfrac{-9 \pm \sqrt{21}}{6} or -1.5 \pm 0.7638$.

So there are two possible answers, namely $-1.5 + 0.7638$ and $-1.5 - 0.7638$, i.e $= -0.7362$ and -2.2638.

In the case of $x^2 - 4 = 0$, which means $x^2 = 4$, the above formula is not needed. $x = 2$ or -2.

Note that multiplying brackets in the way explained earlier, if applied to an expression like $(x + 2)(x + 3)$ will give $x^2 + 5x + 6$, which is a quadratic equation.

An important relationship is

$$(x + a)(x - a) = x^2 - a^2 \qquad (2.22)$$

Rough checks

It is easy to make a mistake in a calculation, by pressing the wrong key on a calculator for example. If you wanted to add 3.132 to 0.8401 but accidentally pressed the $\div$ key instead of the $+$ key you would get 3.728. But you can see at a glance that the answer should be more than 3.9. If you have a rough idea of the answer you expect you can eliminate mistakes.

For the expression $\dfrac{5.923}{2.202 \times 1.461}$ you would expect an answer not much different from $\dfrac{6}{2 \times 1.5}$ or $\dfrac{6}{3}$ which is 2. When your calculator gives you 1.841 you believe it. If it gives you 3.930 you've made a mistake (you've used a $\times$ sign instead of a second $\div$).

3
Units and dimensions

Measuring a quantity

When a length is measured as 7 feet it means 7 times the length of a foot. What is measured (i.e. the quantity) consists of a number (7) multiplied by the chosen unit (foot, metre, etc.).

Fundamental and derived quantities

Several quantities, like mass, length, time, temperature, are called fundamental or base quantities while others are derived from these. One example of a derived quantity is a velocity which is a length divided by a time.

SI units

The SI system was mentioned in Chapter 1.

The system uses seven base units including the kilogram (kg), metre (m), and second (s), and all other SI units are derived from these: e.g. metre per second for velocity.

Dimensions

Regardless of the units employed a velocity is always a length divided by a time and a force is always a mass multiplied by a length and divided by time squared as seen from $F = ma$ (Equation 5.5, Chapter 5) or $F = mv^2/r$ (Equation 8.4, Chapter 8). We write

$$[\text{Force}] = \text{mass} \times \text{length/time}^2 \text{ or}$$
$$\text{mass} \times \text{length} \times \text{time}^{-2}$$

The multiplying quantities (mass, length and time here) are the 'dimensions' of the derived quantity (force in the example used here). So the dimensions of a quantity are the base quantities from which it is made up in the same way that the dimensions of a box would be length, width and depth of the box.

Square brackets are used to indicate 'the dimensions of' and the symbols M, L and T are used to denote mass, length and time when we are dealing with dimensions. Thus the dimensions of a force are M, L and T^{-2} and we can write

$$[F] = MLT^{-2}$$

(An identity sign $\equiv$ may be used in place of the equals sign here because the equality is true under all circumstances, not just for particular values of the quantities concerned.)

To decide the dimensions of a quantity a definition or formula for it is usually required. As an example for a pressure P the formula $P = $ force/area could be used. It is advisable to be familiar with the dimensions of force (MLT^{-2}) and then $[P] = MLT^{-2}/L^2 = ML^{-1}T^{-2}$

Some quantities are dimensionless, i.e. their dimensions are zero. They are simply numbers, perhaps ratios of similar quantities. An angle is an example (an angle in radians equalling an arc divided by radius, i.e. a length divided by a length giving L^0). The symbols Q, I, θ may be used for the dimensions of charge, current and temperature.

Important properties of dimensions

'Three pints plus two pints equals five pints' is always true but 'three pints plus two kilograms equals ...' is meaningless in an equation since all of the terms in an equation must have the same dimensions, i.e. each must be the same kind of quantity. This fact can be useful for checking equations.

The dimensions of a unit must be the same as those of the quantity to which it applies. So in place of

$$[F] = MLT^{-2}$$

we can write

$$[\text{newton}] = [\text{kilogram}] \, [\text{metre}] \, [\text{second}]^{-2}$$

or

$$[\text{N}] = [\text{kg}] \, [\text{m}] \, [\text{s}]^{-2}$$

and $\text{kg} \, \text{m} \, \text{s}^{-2}$ is a suitable unit for any force.

(In fact the definition of the newton means that one newton corresponds to ONE kilogram and $\text{m} \, \text{s}^{-2}$. So $1 \, \text{N} = 1 \, \text{kg} \, \text{m} \, \text{s}^{-2}$.)

Example 1

What are the dimensions of (a) force, (b) moment, (c) work, (d) pressure?

Method

We need to relate each of these quantities to quantities whose dimensions are known.

(a)   Force = Mass × Acceleration

$$[F] = M \times \frac{L}{T^2} \quad \text{or} \quad MLT^{-2}$$

(b)   Moment = Force × Perpendicular distance

$$\therefore [\text{Moment}] = MLT^{-2} \times L \quad \text{or} \quad ML^2T^{-2}$$

(c) $[\text{Work}] = [\text{Force} \times \text{Distance}] = MLT^{-2} \times L$
$$\text{or} \quad ML^2T^{-2}$$

(d) $\text{Pressure} = \dfrac{\text{Force}}{\text{Area}}$

$$[p] = \frac{MLT^{-2}}{L^2} \quad \text{or} \quad ML^{-1}T^{-2}$$

Answers

(a) MLT^{-2} (b) ML^2T^{-2}
(c) ML^2T^{-2} (d) $ML^{-1}T^{-2}$.

Example 2

Which one of the following has different dimensions from the others?

A stress × strain   B stress/strain
C pressure   D potential energy per unit volume
E torque   [O & C 94]

Method

Answer **A** mentions stress which is defined as force per unit area while strain has no dimensions because it is the ratio of increase in length to original length. So **A** has the dimensions of stress (F/A or MLT^{-2}/L^2 or $ML^{-1}T^{-2}$).

Answer **B** clearly has the same dimensions so we are looking for an answer whose dimensions are not $ML^{-1}T^{-2}$.

Answer **C** is also a force divided by an area by definition of pressure.

Now for **D** we look for a definition or formula concerning potential energy.

$PE = mgh$ or   PE = work done in lifting = weight × height may be useful.

mgh has dimensions $M(LT^{-2})L$ and, more easily, weight × height has dimensions $(MLT^{-2})L$. Dividing either expression by volume (L^3) we get $ML^{-1}T^{-2}$.

So for **D** the dimensions are also $ML^{-1}T^{-2}$ and the different answer must be **E** (where torque = force × distance, see Chapter 33, page 274, and has dimensions $(MLT^{-2})L$ or ML^2T^{-2} confirming our answer).

Answer

E.

Example 3

Which of the following units could be used for capacitance?

A $\text{kg} \, \text{m}^2 \, \text{s}^{-1} \, \text{C}$ B $\text{kg} \, \text{m}^2 \, \text{s}^{-2} \, \text{C}^2$
C $\text{kg}^{-1} \, \text{m}^{-1} \, \text{C}^2$ D $\text{kg}^{-1} \, \text{m}^{-2} \, \text{s}^2 \, \text{C}^2$
E $\text{kg}^{-1} \, \text{m}^{-2} \, \text{s}^3 \, \text{C}$

Method

Some relationships that might be useful are

$$\text{capacitance } C = Q/V$$

and $\frac{1}{2}CV^2$ = work done (or energy stored): see Chapter 22.

Neither of these formulae gives an immediate answer because the volt for V does not appear in the answers suggested.

Now $V = \dfrac{\text{work done}}{\text{charge moved}} = \dfrac{\text{force} \times \text{distance}}{\text{charge}}$

and a suitable unit for V (using C for coulomb now) is

$$\frac{(\text{kg} \, \text{m} \, \text{s}^{-2}) \, \text{m}}{\text{C}} \quad \text{or} \quad \text{kg} \, \text{m}^2 \, \text{s}^{-2} \, \text{C}^{-1}$$

For capacitance we get $\text{C}/\text{kg} \, \text{m}^2 \, \text{s}^{-2} \, \text{C}^{-1}$ or $\text{C}^2 \, \text{kg}^{-1} \, \text{m}^{-2} \, \text{s}^2$

Answer

D.

Checking equations and units

All terms in an equation must have the same dimensions, i.e. it is homogeneous. This can be useful for checking the correctness of an equation. For example the lens equation

$$\frac{1}{u} + \frac{1}{v} = \frac{1}{f}$$

(see Chapter 15) might, by mistake, be written as

$$\frac{v}{u} + 1 = \frac{1}{f}$$

instead of

$$\frac{v}{u} + 1 = \frac{v}{f}$$

The mistake is obvious if dimensions are considered because v/u and 1 are dimensionless but $1/f$ has the dimension L^{-1}.

As regards checking units, an example of a unit which is difficult to remember is that for thermal conductivity, k; see Chapter 17. We need an equation containing k. Now k is given by

$$\text{Heat flow } (F) = \frac{\Delta Q}{\Delta t} = kA(\theta_2 - \theta_1)/l$$

whence $k = Fl/A\,(\theta_2 - \theta_1)$ and the units are

$$\text{W} \times \text{m}/(\text{m}^2 \times \text{K}) \text{ or } \text{W}\,\text{m}^{-1}\,\text{K}^{-1}.$$

Exercise 3.1

1 What are the dimensions of:
   (a) density, (b) area, (c) cubic feet per minute,
   (d) power?

2 What are the dimensions of:
   (a) distance/velocity, (b) force × time, (c) angle moved through per second?

3 What are the dimensions of magnetic flux density?
   (Chapter 24 gives $\Phi = BA$, PD $= Blv$, PD $= d\Phi/dt$ and Chapter 20 gives PD $= W/Q$.)

4 The equation relating current I through a semiconductor diode to the applied potential difference V at temperature T is

   $$I = I_0 e^{-eV/kT}$$

   where e in the ^{eV} is the electron charge and k is the Boltzmann constant. What are the dimensions of k?

5 The surface tension of a liquid is measured in $\text{N}\,\text{m}^{-1}$. What are the dimensions of surface tension?

Exercise 3.2

1 Evaluate α and β in the equation $E = Cm^{\alpha}v^{\beta}$, where E is kinetic energy, m is mass, v is velocity and C is a dimensionless constant.

2 The force of attraction F between two particles of masses m_1 and m_2 situated a distance d apart is given by $F = Gm_1m_2/d^2$. Show that the dimensions of G are $M^{-1}L^3T^{-2}$.

3 The minimum velocity needed for a body to escape from the earth is given by $v = \sqrt{(2GM/R)}$ where M is the mass of the earth and R is its radius. Show that the equation is dimensionally correct. The dimensions of G are $M^{-1}L^3T^{-2}$.

Conversion of units

Students usually remember conversion factors, e.g. 1000 for changing metres to millimetres; but it is not always obvious whether to divide or multiply by a factor. Common sense should be used. 'Am I changing to smaller units? Will I therefore get more of them?' 1 metre changed to smaller millimetre units will become 1000 units. A density of $1\,\text{g}\,\text{cm}^{-3}$ ($1\,\text{g}$ per cm^3 of substance) will give many more (100^3 times more) when volume is $1\,\text{m}^3$, i.e. $1\,\text{g}\,\text{cm}^{-3}$ is equivalent to $10^6\,\text{g}\,\text{m}^{-3}$. Changing to kg the answer will become smaller by 1000, i.e. $10^3\,\text{kg}\,\text{m}^{-3}$.

Exercise 3.3

1 Convert
   (a) $30\,\text{km}\,\text{h}^{-1}$ to $\text{m}\,\text{s}^{-1}$ (b) $0.01\,\text{m}^2$ to mm^2
   (c) $400\,\text{nm}$ to μm (d) $120\,000\,\text{min}^{-1}$ to s^{-1}

2 The conductance σ of a conductor is $0.01\,\Omega^{-1}$. Convert this to $\text{m}\Omega^{-1}$.

Equations where conversion factors cancel

Consider the Boyle's law equation

$$p_1V_1 = p_2V_2$$

where p_1 and V_1 are initial pressurre and volume of a gas and p_2 and V_2 are new values. Perhaps $V_1 = 3.0\,\text{m}^3$, with $p_1 = 1.0$ atmosphere (1 bar) and then the pressure is changed to 2.0 atmosphere (i.e. 2 bar). We are asked to calculate V_2.

All our equations work with SI units. Now 1 bar $= 10^5$ SI units of pressure ($\text{N}\,\text{m}^{-2}$ or Pa).

$$1.0 \times 10^5 \times 3.0 = 2.0 \times 10^5 \times V_2$$

But the 10^5 on each side cancels, so that we get $V_2 = 1.5\,\text{m}^3$, whether p_1 and p_2 are in Pa or bar. All that is necessary here is that p_1 and p_2 have the same units.

A useful example of conversion factors cancelling is in Chapter 28, Example 1.

An unusual unit – the mole

Avogadro's number (N_A) is the number of normal carbon atoms (^{12}C atoms) that together have a mass of 12 grams. (This is very close to the number of normal hydrogen atoms (^{1}H) having a mass of 1 gram.)

The mole is one of the base units of the SI system. It is an amount of substance defined not by any property of the substance but by the number of particles it contains. This number equals Avogadro's number.

1 mole contains Avogadro's number of particles.

The particles must be named, e.g. atoms of oxygen or molecules of oxygen.

The unified atomic mass unit

This unit of mass is denoted by the symbol 'u' and is used for the masses of very small particles such as atoms. $1\,\text{u}$ is one twelfth of the mass of a normal carbon atom and is very close to the mass of a normal hydrogen atom.

So 1 mole of normal carbon atoms (N_A atoms) has a mass of $N_A \times 12\,\text{u}$, but N_A of these atoms have a mass of 12 grams, so $N_A \times 12\,\text{u} = 12\,\text{gram}$, meaning that

$$1\,\text{u} = \frac{1}{N_A}\,\textbf{gram}$$

For 1 mole of substance whose particles each have a mass of A atomic mass units the mass equals $N_A \times A$ atomic mass units which is $N_A \times A \times \dfrac{1}{N_A}$ grams or simply A grams.

For most purposes A may be taken as equal to the mass number (see Chapter 28) of the particles

concerned, e.g. 1 mole of ^{235}U atoms has a mass close to 235 grams. (Note the unfortunate emphasis on grams not kilograms!)

1 mole has a mass of A grams

or the mass per mole (the molar mass) $= A$ gram per mole (g mol^{-1}).

Weight

'Weight' is a force and the term should be used to describe the force on a body caused by gravity. A body's weight is related to its mass (m) by the formula $W = mg$ where g is the acceleration due to gravity (gravitational field strength) of the Earth. The units for W, m and g will normally be newton, kilogram and m s^{-2} (see Equation 5.5) and g can be taken as $10\,\text{m s}^{-2}$ (see 'Motion under gravity' in Chapter 5).

The term 'kilogram force' can be used for the weight of a 1 kg mass but 'kg force' is not an SI unit.

Exercise 3.4: Examination questions

1 (a) State what is meant by 'an equation is homogeneous with respect to its units'.

(b) Show that the equation $x = ut + \frac{1}{2}at^2$ is homogeneous with respect to its units.

(c) Explain why an equation may be homogeneous with respect to its units but still be incorrect. [Edexcel 2000]

2 When a body is moving through a resisting medium such as air it experiences a drag force D which opposes the motion. D is given by the expression

$$D = \tfrac{1}{2}C\rho A v^2$$

Where ρ is the density of the resisting medium, A is the effective cross-sectional area of the body, i.e. that area perpendicular to its velocity v. C is called the drag coefficient.

Show that C has no dimensions.

[WJEC 2000, part]

3 Coulomb's law for the force F between two charges q_1 and q_2, separated by a distance r, may written as

$$F = \frac{kq_1, q_2}{r^2},$$

where k is a constant.

(a) For the case when force, charge and distance are expressed in the SI units newton N, coulomb C and metre m respectively, deduce a unit for k in terms of N, C and m.

(b) (i) Write down an equation expressing the relationship between the constant k and the permittivity of free space ε_0.

(ii) The value of ε_0 is $8.85 \times 10^{-12}\,\mathrm{F\,m^{-1}}$. Hence obtain the numerical value and unit (in terms of farad F and metre m) of k. [CCEA 2000]

Section B
Mechanics

4
Statics

Representation of a force

A force is a vector quantity – that is, has magnitude and direction. We can thus represent a force by a line in the appropriate direction and of length proportional to the magnitude of the force (see Fig. 4.1).

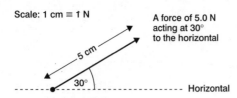

Fig. 4.1   Representation of a force

Addition of forces

Vector quantities such as forces are added using the parallelogram rule (see Fig. 4.2) – the resultant is the appropriate diagonal of the parallelogram.

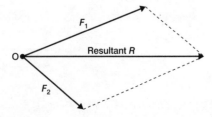

Fig. 4.2   Addition of vectors (e.g. forces)

Example 1

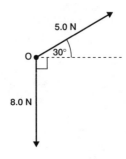

Fig. 4.3   Information for Example 1

Fig. 4.3 shows two forces acting at a point O. Find the magnitude and direction of the resultant force.

Method

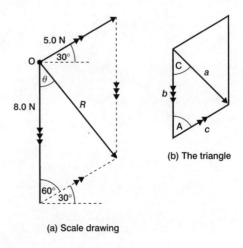

(b) The triangle

(a) Scale drawing

Fig. 4.4   Solution to Example 1

Referring to Fig. 4.4a, we could find the resultant R and angle θ by scale drawing and this is often sufficient.

There are also two ways of accurately calculating the values required:

(i) by use of the sine and cosine rule (see Chapter 2) as outlined below
(ii) by calculating the components at right angles of forces A and B and combining these components using Pythagoras (see Resolution of forces section, see page 22).

Referring to Fig. 4.4b (see Chapter 2) we see that

$$a^2 = b^2 + c^2 - 2bc \cos A$$

We have $a = R, b = 8.0, c = 5.0$ and $A = 60°$. So

$$R^2 = 8^2 + 5^2 - 2 \times 8 \times 5 \times \cos 60° = 49$$
$$\therefore \quad R = 7.0 \, \text{N}$$

To find θ, we know (see Chapter 2)

$$\frac{a}{\sin A} = \frac{c}{\sin C}$$

We have $a = 7.0, A = 60°, c = 5.0$ and $C = \theta$. So

$$\frac{7}{\sin 60°} = \frac{5}{\sin \theta}$$
$$\therefore \qquad \theta = 38.2°$$

Answer

The resultant is of magnitude 7.0 N at an angle of 38° to the 8.0 N force, as shown in Fig. 4.4a.

Example 2

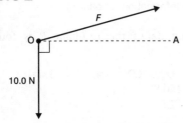

Fig. 4.5   Information for Example 2

Refer to Fig. 4.5, Two forces of magnitude 10.0 N and F newtons produce a resultant of magnitude 30.0 N in the direction OA. Find the magnitude and direction of F.

Method

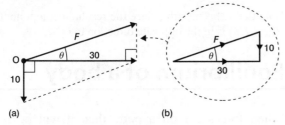

(a)   (b)

Fig. 4.6   Solution to Example 2

Referring to Fig. 4.6 then, in diagram (b):

$$F^2 = 10^2 + 30^2 = 1000$$
$$\therefore \quad F = 31.6 \, \text{N}$$

Also

$$\tan \theta = \frac{10}{30} = 0.3333$$

or $\qquad \theta = 18.4°$

Answer

F is of magnitude 31.6 N at an angle of 18.4° to the resultant force as shown.

Exercise 4.1

1

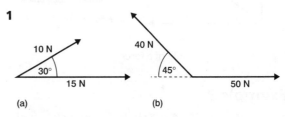

(a)   (b)

Fig. 4.7   Information for Question 1

Find the resultant of the forces in (a) Fig. 4.7a, (b) Fig. 4.7b.
Note that for $\theta > 90°$, $\cos \theta = -\cos(180 - \theta)$.

2

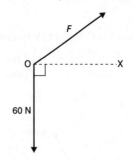

Fig. 4.8   Information for Question 2

Refer to Fig. 4.8. Forces of 60.0 N and F newtons act at a point O. Find the magnitude and direction of F if the resultant force is of magnitude 30.0 N along OX.

3

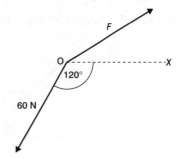

Fig. 4.9   Information for Question 3

Refer to Fig. 4.9 and repeat Question 2.

Resolution of forces

A single force can be formed by combining two (or more) forces so it follows that a single force can be replaced by, or *resolved* into, two components. This is usually done at right angles (see Fig. 4.10) because the separate components V and H have no effect on each other – i.e. V has no effect in the direction of H.

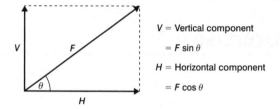

V = Vertical component

$= F \sin \theta$

H = Horizontal component

$= F \cos \theta$

Fig. 4.10   Components of a force

Example 3

Two coplanar forces A and B act at a point O, as shown in Fig. 4.11. Calculate the component of the resultant force

(a) along OX

(b) along OY

Use your answers for OX and OY to calculate

(c) the magnitude and direction of the resultant force due to the addition of forces A and B.

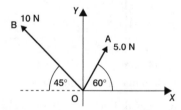

Fig. 4.11   Information for Example 3

Method

Refer to Figs 4.12 and compare with Fig. 4.10.

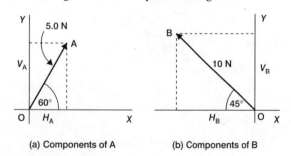

(a) Components of A

(b) Components of B

Fig. 4.12   Solution to Example 3

In diagram (a) we can see:

Vertical component of A along OY,
$V_A = 5 \sin 60° = 4.33\,\text{N}$
Horizontal component of A along OX,
$H_A = 5 \cos 60° = 2.50\,\text{N}$

In diagram (b) we can see:

Vertical component of B along OY,
$V_B = 10 \sin 45° = 7.07\,\text{N}$
Horizontal component of B along OX,
$H_B = -10 \cos 45° = -7.07\,\text{N}$

Note the minus sign, since H_B is in the opposite direction to OX.

The total component forces along OX and OY can be found by adding the separate components along OX and OY. Therefore

Resultant component along OX $= H_A + H_B$
$= -4.57\,\text{N}$
Resultant component along OY $= V_A + V_B$
$= 11.4\,\text{N}$

For part (c) we combine OX and OY as shown in Fig 4.13.

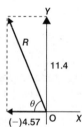

Fig. 4.13   Information for part (c)

The magnitude of the resultant R is found using Pythagoras:

$$R^2 = OX^2 + OY^2 = 4.57^2 + 11.4^2$$
$$= 150.8$$
$$\therefore \quad R = 12.3$$

Also $\tan \theta = 11.4/4.57 = 2.49$
$$\therefore \qquad \theta = 68.2°$$

Answer

(a) $-4.6\,\text{N}$
(b) $11\,\text{N}$
(c) $12\,\text{N}$ at an angle of $68°$ as shown.

Note that (a) is negative since the resultant component $(H_A + H_B)$ is in the opposite direction to OX.

Equilibrium of a body

When forces act on a body then it will be in equilibrium provided that:

(i) no net forces act on the body and

(ii) no net turning effect exists (that is the sum of clockwise moments and anticlockwise moments cancel out – see Principle of Moments, p. 24).

Example 4

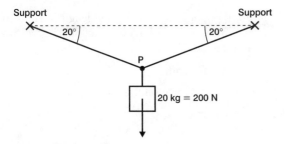

Fig. 4.14   Information for Example 4

A mass of 20.0 kg is hung from the midpoint P of a wire, as shown in Fig. 4.14. Calculate the tension in the wire. Assume $g = 10\,\mathrm{m\,s^{-2}}$.

Method

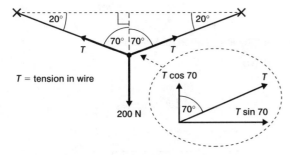

Fig. 4.15   Solution to Example 4

Fig. 4.15 shows the forces acting at the point P. The vertical component of tension T is $T\cos 70°$ in each case, so for equilibrium in a vertical direction

$$2T\cos 70° = 200$$

$$\therefore \qquad T = 292\,\mathrm{N}$$

Note that the horizontal component of tension is $T\sin 70°$ in each case, but these forces are in opposite directions and so cancel each other. This ensures equilibrium in the horizontal direction.

Answer

292 N.

Example 5

A body of mass 1.5 kg is placed on a plane surface inclined at 30° to the horizontal. Calculate the friction and normal reaction forces which the plane must exert if the body is to remain at rest. Assume $g = 10\,\mathrm{m\,s^{-2}}$.

Method

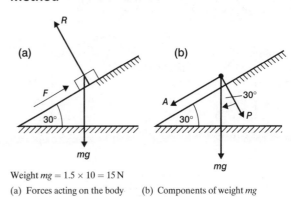

Weight $mg = 1.5 \times 10 = 15\,\mathrm{N}$

(a) Forces acting on the body   (b) Components of weight mg

Fig. 4.16   Solution to Example 5

The body exerts a downward force mg on the plane, as shown in Fig. 4.16, so the plane must exert an equal and opposite (upwards) force if the body is to remain at rest. It is convenient to resolve mg into a component P, perpendicular to the plane, and a component A, along the plane, as shown in Fig. 4.16b. Now

$$P = mg\cos 30 = 15 \times 0.866 = 13\,\mathrm{N}$$

$$A = mg\sin 30 = 15 \times 0.500 = 7.5\,\mathrm{N}$$

So, as shown in Fig. 4.16a, the plane must provide a normal reaction R equal to 13 N and a force F, due to friction, equal to 7.5 N. When R and F are added vectorially, they provide a vertically upwards force equal to mg.

Answer

7.5 N, 13 N.

Exercise 4.2

(Assume $g = 10\,\mathrm{m\,s^{-2}}$.)

1

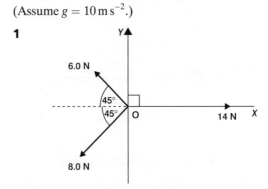

Fig. 4.17   Information for Question 1

Three forces are applied to the point O as shown in Fig. 4.17. Calculate

(a) the component in directions OX and OY respectively

(b) the resultant force acting at O.

23

2 Refer to Fig. 4.18 and calculate (a) the tension in the string, (b) the value of m.

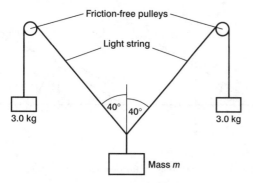

Fig. 4.18   Information for Question 2

3 A body of mass 3.0 kg is placed on a smooth (i.e. frictionless) plane inclined at 20° to the horizontal. A force of (a) 5.0 N, (b) 20 N is applied to the body parallel to the line of greatest slope of the plane and in a direction up the plane. Calculate the net force acting on the body in each case.

Turning effect of forces

A force can produce a turning effect, or moment, about a pivot. This can be a clockwise or anticlockwise turning effect.

Moment of a force (Nm) =
    Force (N) × perpendicular distance (m)
            of line of action of the force
            from the pivot.

Referring to Fig. 4.19, force F_1 produces a clockwise moment $F_1 d_1$ about pivot P and forces F_2 and F_3 produce anticlockwise moments $F_2 d_2$ and $F_3 d_3$ respectively about P.

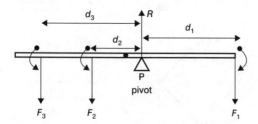

Fig. 4.19   Turning effect of forces about pivot P

The Principle of Moments states that for a body to be in equilibrium then:

Sum of clockwise moments = sum of anticlock-
                         wise moments.

So, referring to Fig. 4.19, in equilibrium (i.e. no turning):

$$F_1 d_1 = F_2 d_2 + F_3 d_3$$

Note also that, in equilibrium, the net force on the body must be zero. Thus, upwards reaction force R at pivot point P is given by:

$$R = F_1 + F_2 + F_3$$

Example 6

A hinged trapdoor of mass 15 kg and length 1.0 m is to be opened by applying a force F at an angle of 45° as shown in Fig. 4.20. Calculate:

(a) the value of F and

(b) the horizontal force on the hinge.

Assume $g = 10 \, \mathrm{m \, s^{-2}}$

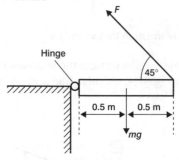

Fig. 4.20   Information for Example 6

Method

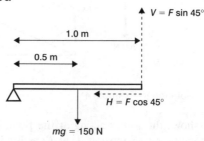

Fig. 4.21   Solution to Example 6

Fig. 4.21 is a simplified diagram showing the forces acting in which F has been resolved into its horizontal (H) and vertical (V) components. Weight of trapdoor $mg = 150 \, \mathrm{N}$.

(a) At equilibrium, taking moments about hinge (pivot) P:

    clockwise moments = anticlockwise moments
    $mg \times 0.5 = V \times 1.0$
    $mg \times 0.5 = F \times \sin 45 \times 1.0$
    $150 \times 0.5 = F \times 0.707 \times 1.0$
    $\therefore \quad F = 106 \, \mathrm{N}$

(b) Horizontal component
    $H = F \cos 45° = 106 \times 0.707 = 75.0 \, \mathrm{N}$.

Answer

(a) 0.11 kN, (b) 75 N.

Example 7

This example is about body mechanics.

Fig. 4.22 shows the forearm extended horizontally and holding an object of mass $M = 2.0\,\text{kg}$. The forearm pivots about the elbow joint J and the mass of the forearm $m = 1.4\,\text{kg}$ which acts effectively at a distance $0.18\,\text{m}$ from the elbow joint. The forearm and object are supported by an upwards force T provided by the biceps muscle and which acts 60 mm from the joint. Calculate:

(a) the magnitude of T

(b) the force acting at the elbow (pivot) joint J.

Assume $g = 10\,\text{m s}^{-2}$.

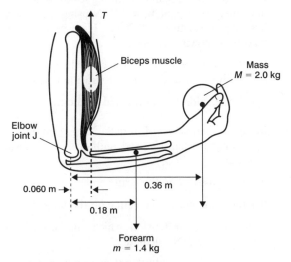

Fig. 4.22 Information for Example 7

Method

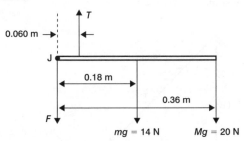

Fig. 4.23 Solution to Example 7

Fig. 4.23 is a schematic diagram showing the forces acting on the forearm. Force F acts at the elbow joint J (pivot). For the forearm, $mg = 14\,\text{N}$ and for the object $Mg = 20\,\text{N}$.

(a) At equilibrium, taking moments about the elbow joint (pivot)

clockwise moments = anticlockwise moments

$$(14 \times 0.18) + (20 \times 0.36) = T \times 0.06$$

$$\therefore \quad T = 162\,\text{N}$$

Note that the force F has no moment about the joint J since its line of action passes through J.

(b) In equilibrium:

total upwards force = total downwards force
$$162 = F + 14 + 20$$
$$\therefore \quad F = 128\,\text{N}$$

This (downwards) force F is effectively provided through the long bone connecting the elbow joint and the shoulder.

Answer

(a) 0.16 kN   (b) 0.13 kN.

Stability and toppling

When a body is in contact with a surface it will be in stable equilibrium provided that the vertical line passing through its centre of gravity lies within the base of contact with the surface.

Example 8

A uniform block of height 50 cm and of square cross section $40\,\text{cm} \times 40\,\text{cm}$ is placed on a rough plane surface as shown in Fig. 4.24 and the inclination of the plane is gradually increased. Calculate the angle of inclination of the plane at which the block topples over. You may assume that friction forces are sufficient to prevent the block from sliding down the plane.

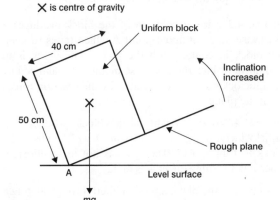

Fig. 4.24 Information for Example 8

Method

We assume that one edge of the block is perpendicular to the line of inclination of the plane. Suppose the plane is gradually tilted (anticlockwise) so that eventually the block will topple about a line through the point A.

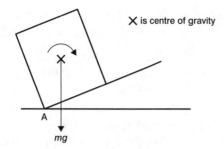

X is centre of gravity

A

mg

(a) Clockwise moment about A due to weight mg.

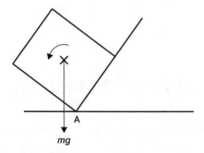

A

mg

(b) Anticlockwise moment about A – block topples.

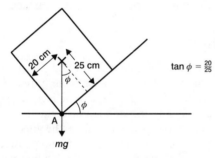

20 cm   25 cm $\tan \phi = \frac{20}{25}$

ϕ

A

ϕ

mg

(c) Block about to topple.

Fig. 4.25   (a), (b) and (c) Solution to Example 8

In Fig. 4.25a the weight *mg* of the block produces a clockwise moment about point A which tends to keep the block in contact with the plane. The block is in equilibrium since the clockwise moment is balanced by an anticlockwise moment caused by the reaction forces from the plane.

In Fig. 4.25b the weight of the block produces an anticlockwise moment about point A which causes the block to topple over. The block is not in equilibrium since a net turning effect acts on it.

In Fig. 4.25c the plane has been tilted through an angle ϕ such that the vertical line passing through its centre of gravity passes through point A. The block is (just) in equilibrium but for angles of tilt greater than ϕ the block will topple.

$$\tan \phi = 20/25 = 0.80$$
$$\phi = 38.7°$$

Answer

39°

Exercise 4.3

1

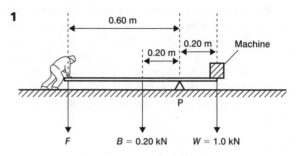

0.60 m

0.20 m   Machine

0.20 m

P

F   B = 0.20 kN   W = 1.0 kN

Fig. 4.26   Information for Question 1

Fig. 4.26 shows a man attempting to lift a piece of machinery of weight $W = 1.0\,\text{kN}$ using a uniform iron bar of weight $B = 0.20\,\text{kN}$. He uses a pivot P placed as shown. Calculate:

(a) the magnitude of the force F which he must apply downwards if he is to lift the machinery

(b) the reaction force provided by the pivot.

2 In Fig. 4.27 an object M of mass 20 kg is supported by a hinged weightless rod and string as shown. Calculate

(a) the tension T in the string and

(b) the horizontal force acting on the hinge.

If the maximum tension which the string can withstand is 500 N calculate

(c) the maximum **additional** mass which may be added to mass M prior to the string breaking.

Assume $g = 10\,\text{m}\,\text{s}^{-2}$.

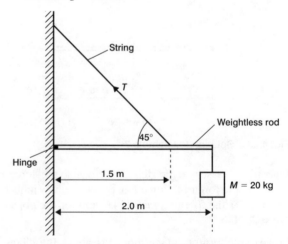

String

T

Weightless rod

45°

Hinge

1.5 m   M = 20 kg

2.0 m

Fig. 4.27   Information for Question 2

3 Fig. 4.28a shows the arm extended horizontally and supporting an object of mass $M = 6.0\,\text{kg}$ at a distance 0.80 m from the shoulder joint J. The deltoid muscle, which acts in tension, provides the necessary force T at an angle of $20°$ to the horizontal and at $0.15\,m$ from J, as shown in Fig 4.28b.

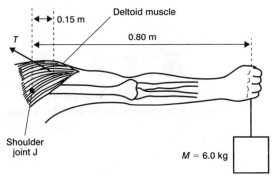

(a) Arm supporting mass

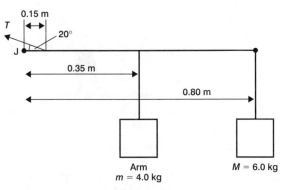

(b) Schematic diagram of forces acting

Fig. 4.28   Information for Question 3

The arm is of mass $m = 4.0\,kg$ acting at a distance of $0.35\,m$ from the shoulder joint. Calculate:

(a) the magnitude of the force T provided by the deltoid muscle

(b) the magnitude of
   (i) the horizontal force and
   (ii) the vertical force

acting at the shoulder joint.

Assume $g = 10\,m\,s^{-2}$.

4

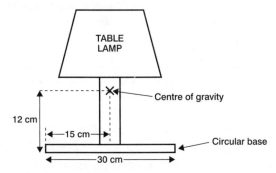

Fig. 4.29   Information for Question 4

A reading lamp has a round base of diameter $30.0\,cm$ and its centre of gravity is $12.0\,cm$ above

its base as shown in Fig. 4.29. Calculate the angle through which its base may be tilted before it topples. (Hint – see Example 8.)

Exercise 4.4:
Examination questions

(Assume $g = 10\,m\,s^{-2}$.)

1

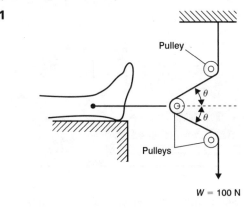

Fig. 4.30   Diagram for Question 1

A weight of $100\,N$ and a system of pulleys is used to apply a force in leg traction as shown in Fig 4.30. The magnitude of the force can be changed by changing the angle θ. Determine the value of the traction force applied to the leg if

  (i) $\theta = 60°$
  (ii) $\theta = 30°$

What is, theoretically, the maximum value of the traction force using $W = 100\,N$?

2   Fig. 4.31 illustrates a crane.

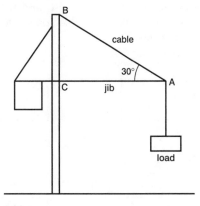

Fig. 4.31

For the purposes of this question, assume that the jib AC has negligible weight. AB is a cable which makes an angle of $30°$ with the jib, which is horizontal. The jib carries a load of $2000\,N$. The load is in equilibrium.

(i) Calculate the tension in the cable AB.

(ii) Calculate the compression force in the jib AC.

[CCEA 2001, part]

3 A skier unfortunately breaks a bone in the lower part of the leg whilst attempting a jump. While the bone is healing, a steady force is applied to the leg. This is called traction. Unless this is done the muscles would pull the fractured parts together so tightly that the leg, when healed, would be shorter than it was before the injury.

Figure 4.32 shows one arrangement for providing the traction. The pulley system is in equilibrium in the position shown.

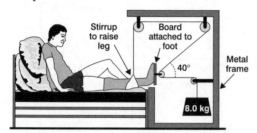

Fig. 4.32

(a) State fully the conditions that must be satisfied for a system to be in translational and rotational equilibrium.

(b) In Fig. 4.32 all the pulleys are frictionless so that the tension in the rope is the same everywhere.

(i) Determine the magnitude of the total horizontal force exerted on the leg by the system

(ii) Determine the magnitude of the total upward force exerted on the leg by the system.

(iii) Explain briefly why the force calculated in (i) does not move the patient towards the bottom of the bed. [AEB 1999]

4 The rectangular objects, **A**, **B**, **C** and **D** are each 2 cm long and 1 cm high. Which one of the bodies is in equilibrium?

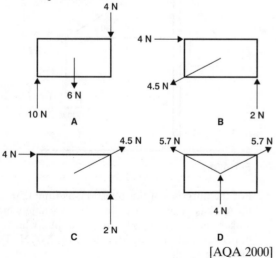

[AQA 2000]

5 Two campers have to carry a heavy container of water between them. One way to make this easier is to pass a pole through the handle as shown.

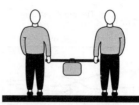

(a) The container weighs 400 N and the weight of the pole may be neglected. What force must each person apply?

An alternative method is for each person to hold a rope tied to the handle as shown below.

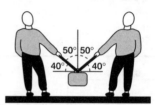

(b) Draw a free-body force diagram for the container when held by the ropes.

(c) The weight of the container is 400 N and the two ropes are at 40° to the horizontal. Show that the force each rope applies to the container is about 300 N.

(d) Suggest **two** reasons why the first method of carrying the container is easier.

(e) Two campers using the rope method find that the container keeps bumping on the ground. A bystander suggests that they move further apart so that the ropes are more nearly horizontal. Explain why this would not be a sensible solution to the problem.

[Edexcel 2001]

6 A uniform plank of weight 60 N is 2000 mm long and rests on a support that is 600 mm from end E.

At what distance from E must a 160 N weight be placed in order to balance the plank?

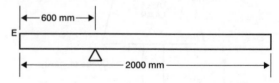

A 150 mm **B** 225 mm **C** 375 mm **D** 450 mm

[OCR 2001]

7 Fig. 4.33a shows a side view of a kitchen wall cupboard. Its lower edge rests against the wall at A. It is fastened by screws at a height h vertically above A. The mass of the cupboard is 10 kg and its centre of gravity is 0.15 m from the wall.

Fig. 4.33b is a free-body force diagram for the cupboard.

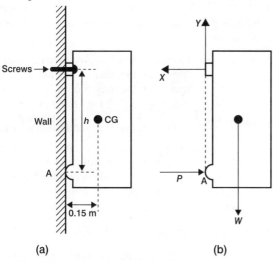

(a)   **(b)**

Fig. 4.33

(a) State the magnitude of force Y.

(b) Explain why forces X and P must have equal magnitude.

(c) Calculate the moment of the weight of the cupboard about point A.

(d) Calculate the value of force X when $h = 0.60\,\text{m}$.

(e) In principle the fixing screws could be positioned anywhere between point A and the top of the cupboard. Sketch a graph to show how the size of force X would depend on h for values of h from zero up to 0.60 m.

(f) Explain why in practice the screws are usually situated as high in the cupboard as possible.
[Edexcel 2000]

8 (a) Define the moment of a force about a point.

(b) Figure 4.34 shows a model bridge consisting of a uniform plank of wood. The plank is 1.0 m long and weighs 10 N. A toy car of weight 5 N is placed on it. The bridge is suspended from a rigid support by two strings and is in equilibrium. The plank does not touch the shaded blocks.

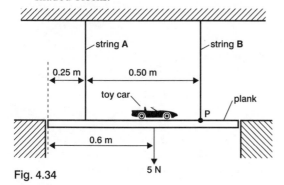

Fig. 4.34

(i) Show and label the forces acting on the bridge.

(ii) By taking moments about point **P**, calculate the tension in string **A**.

(iii) Calculate the tension in string **B**.
[AQA 2001]

9 Fig. 4.35 is a drawing of a mobile crane which is supported on wheels at A and B. The weight of the base of the crane is $28 \times 10^4\,\text{N}$.

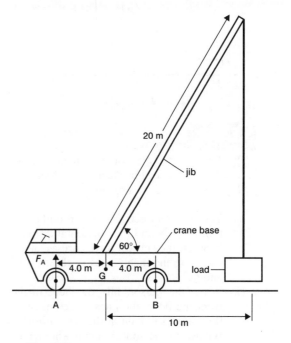

Fig. 4.35

G is the centre of mass of the base. The jib has a weight of $3 \times 10^4\,\text{N}$ uniformly distributed along its length. F_A is the total upwards force from the ground on the wheels at A when the crane is lifting a load of $15 \times 10^4\,\text{N}$.

(a) (i) Write down expressions for:
      1. the total of all clockwise moments about B;
      2. the total of all anti-clockwise moments about B.

    (ii) Calculate the value of F_A.

(b) State how you would calculate the maximum load which the crane could support in this configuration without toppling.

(c) The jib in Fig. 4.35 is inclined at 60° to the horizontal. Suggest how the angle of the jib could be changed in order to support the greatest possible load from near ground level without causing the crane to topple.
[OCR 2001]

10 (a) State the **two** conditions required for an object to be in equilibrium under the action of a system of forces.

(b) A person stands upright on one foot with the ball of the foot in contact with the floor and the heel raised just clear of the floor. The foot is in equilibrium under the action of three vertical forces, P, Q and R, as shown in Fig. 4.36.

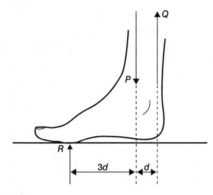

Fig. 4.36

P = force exerted on foot by bone in lower leg
Q = force produced by Achilles tendon
R = reaction force of ground

(i) The reaction force R of the ground on the foot is 625 N and the horizontal distances between the vertical forces are as shown. Calculate the magnitudes of the forces P and Q.

(ii) When the person lifts the heel further from the floor, the lines of action of the three forces remain in the same positions relative to the foot. Explain whether the magnitudes of P, Q and R will change when the heel is raised. [OCR 2001]

11

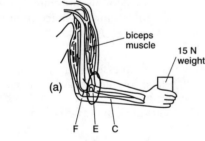

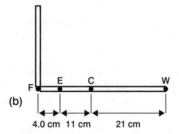

Fig. 4.37 Information for Question 11

Fig. 4.37a shows some muscles and bones in the arm. Fig. 4.37b shows the appropriate distances, where C is the centre of gravity of the lower arm including the hand, and F is the fulcrum at the elbow joint.

(i) On Fig. 4.37b draw labelled arrows to represent the directions of the forces exerted by the biceps muscle (E), the weight of the lower arm (C) and the 15 N weight in the hand (W).

(ii) If the weight of the lower arm including the hand is 20 N, show that the force exerted by the biceps muscle, to maintain the arm in this position, is approximately 0.2 kN.

(iii) Use your answer to calculate the reaction force at the fulcrum and draw its direction on Fig. 4.37b.

12 (a) State the Principle of Moments.

(b) Some tests are carried out on the stability of a table-lamp.

(i) A string is attached to the lamp, as shown, and pulled with a steadily increasing force, F. When F reaches 7.2 N the lamp is about to **tilt**, pivoting about the point P.

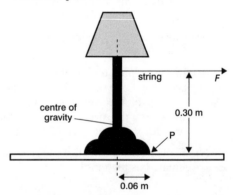

(I) Calculate the moment (torque) of F about P when $F = 7.2$ N.

(II) By considering when the lamp is about to **tilt**, calculate its weight. Its centre of gravity is shown on the diagram.

(ii)

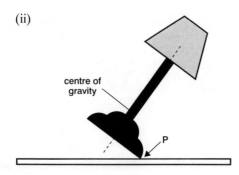

The lamp is now tilted, as shown, and released. Explain, in terms of moments, whether it will fall over or return to the upright. Feel free to add to the diagram.

(iii) State **two** ways in which the lamp could be redesigned to make it more stable.

[WJEC 2001]

13 (a) If an object is to be in equilibrium under the action of a number of coplanar forces, **two** conditions must apply. State these conditions.

(b) Define the term **couple** as used in mechanics.

(c) (i) A wheel of radius 0.50 m rests on a level road at point **C** and makes contact with the edge **E** of a kerb of height 0.20 m, as shown in Fig. 4.38

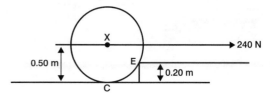

Fig. 4.38

A horizontal force of 240 N, applied through the axle of the wheel at **X**, is required just to move the wheel over the kerb.

Show that the weight of the wheel is 180 N. [CCEA 2001]

5
Velocity, acceleration and force

Velocity and speed

Velocity is a vector and speed is a scalar. Sometimes this difference is not properly recognised, so we must remember it.

Example 1

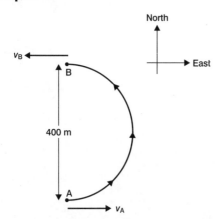

Fig. 5.1   Information for Example 1

A car takes 80 s to travel at constant speed in a semicircle from A to B as shown in Fig 5.1. Calculate (a) its speed, (b) its average velocity, (c) the change in velocity from A to B.

Method

(a) Speed $= \dfrac{\text{Distance}}{\text{Time}}$

$= \dfrac{\text{Length of semicircular arc}}{\text{Time}}$

$= \dfrac{\pi \times 200}{80}$

$= 2.5\pi \, \text{m s}^{-1}$

(b) Average velocity $= \dfrac{\text{Total displacement}}{\text{Time}}$

$= \dfrac{400}{80}$

$= 5.0 \, \text{m s}^{-1} \text{ north}$

(c) The speed at A and B is $2.5\pi \, \text{m s}^{-1}$, but velocities are different. Taking velocity to the 'right' (east) as positive, then

Velocity at A, $v_A = +2.5\pi$

Velocity at B, $v_B = -2.5\pi$

Change in velocity $= v_B - v_A = -5.0\pi \, \text{m s}^{-1}$

Note: the negative sign indicates the change is to the left.

Answer

(a) $2.5\pi \, \text{m s}^{-1}$, (b) $5.0 \, \text{m s}^{-1}$ north, (c) $5.0\pi \, \text{m s}^{-1}$ to the left.

Example 2

A ship travels due east at $3.0 \, \text{m s}^{-1}$. If it now heads due north at the same speed, calculate the change in velocity.

Method*

Initial velocity $\vec{u} = 3 \, \text{m s}^{-1}$ east

Final velocity $\vec{v} = 3 \, \text{m s}^{-1}$ north

The change in velocity is

$$\vec{v} - \vec{u} = \vec{v} + (-\vec{u})$$

We have $\vec{v} = 3 \, \text{m s}^{-1}$ north and $-\vec{u} = 3 \, \text{m s}^{-1}$ west. Fig 5.2 shows that vector addition of $\vec{v}$ and $-\vec{u}$ is a vector of magnitude $\sqrt{18} = 4.2$ in direction north-west.

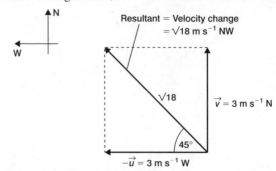

Fig. 5.2   Solution to Example 2

Answer

Velocity change $= 4.2 \, \text{m s}^{-1}$ north-west.

To denote the vector nature of velocity we sometimes put an arrow above it.

Components of velocity

Since velocity is a vector quantity it can be *resolved* into two components at right angles, in the same way as force (see Chapter 4). Fig 5.3 shows the relationship between total velocity R and its horizontal component H and vertical component V.

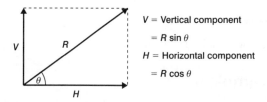

V = Vertical component
= $R \sin \theta$
H = Horizontal component
= $R \cos \theta$

Fig. 5.3   Components of velocity

Example 3

A shell is fired at $400\,\mathrm{m\,s^{-1}}$ at an angle of $30°$ to the horizontal. If the shell stays in the air for 40 s, calculate how far it lands from its original position. Assume that the ground is horizontal and that air resistance may be neglected.

Method

Refer to Fig 5.4. We require the range S. The horizontal component of the initial velocity is

$$H = 400 \cos 30° = 347\,\mathrm{m\,s^{-1}}$$

The horizontal component H remains unchanged if air resistance is negligible. So range S is given by

$$S = H \times \text{Time of flight}$$
$$= 347 \times 40 = 13\,880\,\mathrm{m}$$

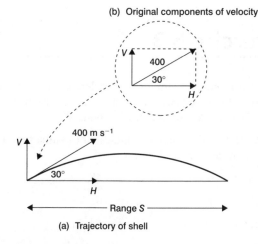

(b) Original components of velocity

(a) Trajectory of shell

Fig. 5.4   Solution to Example 3

Answer

The shell lands 14 km from its original position.

Exercise 5.1

1

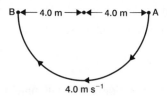

Fig. 5.5   Diagram for Question 1

An object moves along a semicircular path AB of radius 4.0 m as shown in Fig 5.5, at a constant speed of $4.0\,\mathrm{m\,s^{-1}}$. Calculate (a) the time taken, (b) the average velocity, (c) the change in velocity.

2

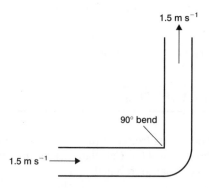

Fig. 5.6   Diagram for Question 2

Water enters and leaves a pipe, as shown in Fig 5.6, at a steady speed of $1.5\,\mathrm{m\,s^{-1}}$. Find the change in velocity.

3   A shell is fired at $500\,\mathrm{m\,s^{-1}}$ at an angle of θ degrees to the horizontal. The shell stays in the air for 80 s and has a range of 24 km. Assuming that the ground is horizontal and that air resistance may be neglected, calculate (a) the horizontal component of the velocity, (b) the value of θ.

Acceleration

Uniform acceleration means a constant rate of change of velocity – for example $4\,\mathrm{m\,s^{-1}}$ per second ($4\,\mathrm{m\,s^{-2}}$).

Example 4

A car moving with velocity $5.0\,\mathrm{m\,s^{-1}}$, in some direction, accelerates uniformly at $2.0\,\mathrm{m\,s^{-2}}$ for 10 s. Calculate (a) the final velocity, (b) the distance travelled during the acceleration.

Method

(a) Increase in velocity $= 2 \times 10 = 20 \, \text{m s}^{-1}$

Original velocity $u = 5.0 \, \text{m s}^{-1}$

$\therefore$ Final velocity $v = 5 + 20 = 25 \, \text{m s}^{-1}$

Alternatively, to find v, use

$$v = u + at \qquad (5.1)$$

where acceleration $a = +2.0$, time $t = 10 \, \text{s}$ and initial velocity $u = 5.0$. So

$$v = u + at = 5 + 2 \times 10$$
$$= 25 \, \text{m s}^{-1}$$

(b) Distance travelled s = Average velocity $\times$ Time, or

$$s = \tfrac{1}{2}(u + v) \times t \qquad (5.2)$$

$$= \tfrac{1}{2}(5 + 25) \times 10 = 1.5 \times 10^2 \, \text{m}$$

Answer

(a) $25 \, \text{m s}^{-1}$, (b) $1.5 \times 10^2 \, \text{m}$.

Exercise 5.2

1 A body starts from rest ($u = 0$) and accelerates at $3.0 \, \text{m s}^{-2}$ for $4.0 \, \text{s}$. Calculate (a) its final velocity, (b) the distance travelled.

2 Calculate the quantities indicated:

(a) $u = 0, v = 20, t = 8.0, a = $ ____

(b) $u = 10, v = 22, a = 1.5, \quad t = $ ____, $s = $ ____

(c) $u = 15, v = 10, a = -0.5, t = $ ____, $s = $ ____

Equations of motion

These are:

$$v = u + at \qquad (5.1)$$
$$v^2 = u^2 + 2as \qquad (5.3)$$
$$s = ut + \tfrac{1}{2}at^2 \qquad (5.4)$$

The meaning of the symbols was given earlier. These equations are obtained by combining Equations 5.1 and 5.2, and are recommended because they are more convenient to use.

Example 5

A car moving with velocity $10 \, \text{m s}^{-1}$ accelerates uniformly at $2.0 \, \text{m s}^{-2}$. Calculate its velocity after travelling $200 \, \text{m}$.

Method

We have $u = 10$, $a = 2.0$ and $s = 200$. We require v, so Equation 5.3 is used

$$v^2 = u^2 + 2as = 10^2 + 2 \times 2 \times 200$$
$$= 900$$
$$\therefore \quad v = 30 \, \text{m s}^{-1}$$

Note that since t is unknown it would be more difficult to use Equations 5.1 and 5.2.

Answer

Velocity acquired $= 30 \, \text{m s}^{-1}$.

Example 6

How far does a body travel in the fourth second if it starts from rest with a uniform acceleration of $2.0 \, \text{m s}^{-2}$?

Method

We have $u = 0$, $a = 2.0$ and require distance travelled between $t_1 = 3.0 \, \text{s}$ and $t_2 = 4.0 \, \text{s}$. Let s_1 and s_2 be distances travelled in $3 \, \text{s}$ and $4 \, \text{s}$ respectively. From equation 5.4

$$s_1 = ut_1 + \tfrac{1}{2}at_1{}^2 = 0 \times 3 + \tfrac{1}{2} \times 2 \times 3^2$$
$$= 9.0 \, \text{m}$$
$$s_2 = ut_2 + \tfrac{1}{2}at_2^2 = 0 \times 4 + \tfrac{1}{2} \times 2 \times 4^2$$
$$= 16.0 \, \text{m}$$
$$\therefore \quad \text{Distance travelled } s_2 - s_1 = 7.0 \, \text{m}.$$

Answer

The body travels $7.0 \, \text{m}$ in the fourth second.

Exercise 5.3

1 It is required to uniformly accelerate a body from rest to a velocity of $12 \, \text{m s}^{-1}$ in a distance of $0.20 \, \text{m}$. Calculate the acceleration.

2 Calculate the quantities indicated (assume that all quantities are in SI units):

(a) $u = 0, \quad a = 10, \quad s = 45, \quad t = $ ____

(b) $u = 15, \quad a = -1.5, \quad v = 6, \quad s = $ ____

(c) $u = 20, \quad a = -2.0, \quad s = 84, \quad t = $ ____

3 In an electron gun, an electron is accelerated uniformly from rest to a velocity of $4.0 \times 10^7 \, \text{m s}^{-1}$ in a distance of $0.10 \, \text{m}$. Calculate the acceleration.

Motion under gravity – vertical motion

Gravitational attraction produces a force which, on earth, causes a free-fall acceleration g of approximately $9.8\,\mathrm{m\,s^{-2}}$. For simplicity we take $g = 10\,\mathrm{m\,s^{-2}}$ here. The force is called the 'weight' of the object concerned.

'Free' vertical motion is simply uniformly accelerated motion, assuming negligible opposing forces, in which $a = g = \pm 10\,\mathrm{m\,s^{-2}}$ depending on the direction chosen as positive.

Example 7

An object is dropped from a height of 45 m. Calculate (a) the time taken to reach the ground, (b) its maximum velocity. Neglect air resistance. (Assume $g = 10\,\mathrm{m\,s^{-2}}$.)

Method

We have $u = 0$, $s = +45$ and $a = g = +10\,\mathrm{m\,s^{-2}}$ if we take downwards as the positive direction. We require t and v.

(a) To find t, rearrange Equation 5.4
$$t^2 = \frac{2s}{a} = \frac{2 \times 45}{10}$$
$$\therefore \quad t = 3.0\,\mathrm{s}$$

(b) To find v, use Equation 5.1
$$v = u + at = 0 + 10 \times 3$$
$$\therefore \quad v = 30\,\mathrm{m\,s^{-1}}$$

Answer

(a) $3.0\,\mathrm{s}$, (b) $30\,\mathrm{m\,s^{-1}}$.

Example 8

A cricket ball is thrown vertically upwards with a velocity of $20\,\mathrm{m\,s^{-1}}$. Calculate (a) the maximum height reached, (b) the time taken to return to earth. Neglect air resistance.

Method

We now take upwards as positive. So
$$u = +20, \quad a = g = -10.$$

(a) At the maximum height, distance s from ground level, the velocity v is zero.
From Equation 5.3
$$v^2 = u^2 + 2as$$
$$0^2 = 20^2 + 2 \times (-10) \times s$$
$$\therefore \quad s = 20\,\mathrm{m}$$

(b) On its return to earth, after time t, we have $s = 0$. So, using Equation 5.4
$$s = ut + \tfrac{1}{2}at^2$$
$$0 = 20 \times t + \tfrac{1}{2} \times (-10) \times t^2$$
$$\therefore \quad t = 4.0\,\mathrm{s}$$

(Note that $t = 0$ is also, obviously, a solution when $s = 0$.)
Alternatively find the time to reach its maximum height, (when its velocity is zero) which is $2.0\,\mathrm{s}$, and double it.

Answer

(a) $20\,\mathrm{m}$, (b) $4.0\,\mathrm{s}$.

Exercise 5.4

(Assume $g = 10\,\mathrm{m\,s^{-2}}$.)

1 A ball is dropped from a cliff top and takes $3.0\,\mathrm{s}$ to reach the beach below. Calculate (a) the height of the cliff, (b) the velocity acquired by the ball.

2 With what velocity must a ball be thrown upwards to reach a height of $15\,\mathrm{m}$?

3 A man stands on the edge of a cliff and throws a stone vertically upwards at $15\,\mathrm{m\,s^{-1}}$. After what time will the stone hit the ground $20\,\mathrm{m}$ below?

Motion under gravity – projectile motion

This includes objects which have horizontal as well as vertical motion, e.g. shells and bullets. We resolve any initial velocity into its horizontal and vertical components, which are then *treated separately. The vertical component determines the time of flight* (and any vertical distances) *and the horizontal component determines the range*.

Example 9

A stone is projected horizontally with velocity $3.0\,\mathrm{m\,s^{-1}}$ from the top of a vertical cliff $200\,\mathrm{m}$ high. Calculate (a) how long it takes to reach the ground, (b) its distance from the foot of the cliff, (c) its vertical and horizontal components of velocity when it hits the ground. Neglect air resistance.

Method

As in Fig 5.3 resolve initial velocity into its components (Fig. 5.7b):

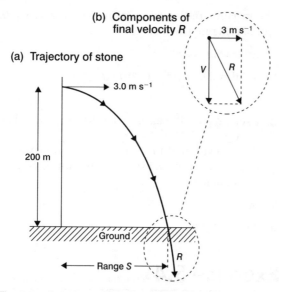

(b) Components of final velocity R

(a) Trajectory of stone

3.0 m s⁻¹

200 m

Ground

Range S

R

Fig. 5.7   Solution to Example 9 (not to scale)

initial vertical component $= 0$
initial horizontal component $= 3.0\,\text{m s}^{-1}$

(a) The vertical motion decides the time of flight. Taking downwards as positive we have $u = 0$, $s = 200$, $a = g = +10\,\text{m s}^{-2}$. To find t use

$$s = ut + \tfrac{1}{2}at^2$$
$$\therefore \quad 200 = 0 \times t + \tfrac{1}{2} \times 10 \times t^2$$
$$\therefore \qquad t = \sqrt{40} = 6.3\,\text{s}$$

(b) The horizontal component of velocity is unchanged (see Example 3). So

$$\text{Range } S = \text{Horizontal velocity} \times \text{Time}$$
$$= 3.0 \times \sqrt{40} = 19\,\text{m}$$

(c) The vertical component of velocity when the stone hits the ground is required. From part (a)

$$v^2 = u^2 + 2as$$
$$= 0^2 + 2 \times 10 \times 200$$
$$\therefore \quad v = \sqrt{4000} = 63\,\text{m s}^{-1}$$

The horizontal component remains at $3.0\,\text{m s}^{-1}$.

Note that to find the resultant velocity R of the stone on hitting the ground we must add the components vectorially, as shown in Fig 5.7b.

Answer

(a) 6.3 s, (b) 19 m, (c) $63\,\text{m s}^{-1}$, $3.0\,\text{m s}^{-1}$.

Exercise 5.5

(Neglect air resistance.)

1 Repeat Example 9 for a stone having a horizontal velocity of $4.0\,\text{m s}^{-1}$ and a cliff which is 100 m high.

2

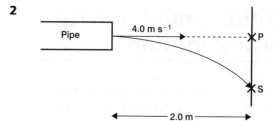

Pipe   4.0 m s⁻¹   P

S

2.0 m

Fig. 5.8   Diagram for Question 2

Water emerges horizontally from a hose pipe with velocity of $4.0\,\text{m s}^{-1}$ as shown in Fig 5.8. The pipe is pointed at P on a vertical surface 2.0 m from the pipe. If the water strikes at S, calculate PS.

3 A shell is fired from a gun with a velocity of $600\,\text{m s}^{-1}$ at an angle of $40°$ to the ground which is horizontal. Calculate (a) the time of flight, (b) the range, (c) the maximum height reached ($g = 10\,\text{m s}^{-2}$).

Force, mass and acceleration

A net force F (N) applied to a mass m (kg) produces an acceleration a (m s^{-2}) given by

$$F = ma \qquad\qquad (5.5)$$

By net force we mean the resultant force arising from applied forces, friction, gravitational forces and so on.

Example 10

A car of mass 900 kg is on a horizontal and slippery road. The wheels slip when the total push of the wheels on the road exceeds 500 N. Calculate the maximum acceleration of the car.

Method

It is the push of the road on the car wheels which is responsible for acceleration. This is equal in magnitude, but opposite in direction, to the push of the wheels on the road. We have $m = 900$ and $F = 500$ so, from Equation 5.5

$$a = \frac{F}{m} = \frac{500}{900} = 0.556\,\text{m s}^{-2}$$

Answer

The maximum acceleration is $0.556\,\text{m s}^{-2}$.

Example 11

A car of mass 1000 kg tows a caravan of mass 800 kg and the two have an acceleration of $2.0\,\text{m s}^{-2}$. If the only external force acting is that between the driving wheels and the road, calculate (a) the value of this force and (b) the tension in the coupling between the car and the caravan.

Method

(a) For car and caravan combined we have

$$m = 1000 + 800 = 1800$$

and $a = 2.0$. From Equation 5.5 the force F required is

$$F = ma = 1800 \times 2 = 3600 \, \text{N}$$

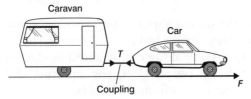

Fig. 5.9   Solution to Example 11

(b) Refer to Fig 5.9. T is the tension in the coupling and is the force accelerating the caravan. So for the caravan alone we have $m = 800 \, \text{kg}$ and $a = 2.0$.
So

$$T = ma = 800 \times 2 = 1600 \, \text{N}$$

Note that the net force on the car alone is $F - T = 3600 - 1600 = 2000 \, \text{N}$. This gives the car an acceleration of $2.0 \, \text{m s}^{-2}$.

Answer

(a) 3.6 kN, (b) 1.6 kN.

Example 12

An aircraft of mass $20 \times 10^3 \, \text{kg}$ lands on an aircraft-carrier deck with a horizontal velocity of $90 \, \text{m s}^{-1}$. If it is brought to rest in a distance of 100 m, calculate the (average) retarding force acting on the plane.

Method

We must first find the (negative) acceleration a of the plane. We have $u = 90$, $v = 0$, $s = 100$ and from Equation 5.3

$$v^2 = u^2 + 2as$$

$$\therefore \quad 0^2 = 90^2 + 2 \times a \times 100$$

$$\therefore \quad a = -40.5 \, \text{m s}^{-2}$$

The negative sign indicates the plane is slowing down. Force F required, since $m = 20 \times 10^3$, is given by

$$F = ma = 20 \times 10^3 \times (-40.5)$$

$$= -81 \times 10^4 \, \text{N}$$

The negative sign indicates that the force is in the opposite direction to the original velocity.

Answer

Retarding force is $81 \times 10^4 \, \text{N}$ on average.

Exercise 5.6

1   Calculate the quantities indicated (assume that all quantities are in SI units):

(a) $a = 2.5$, $m = 3.0$, $F =$ ____

(b) $F = 15$, $m = 30$, $a =$ ____

(c) $a = 2.5$, $F = 7.5$, $m =$ ____

2   A force of 24 N acts on a mass of 6.0 kg initially at rest. Calculate (a) the acceleration, (b) the distance travelled prior to achieving a velocity of $20.0 \, \text{m s}^{-1}$.

3   A lorry of mass $3.0 \times 10^3 \, \text{kg}$ pulls two trailers each of mass $2.0 \times 10^3 \, \text{kg}$ along a horizontal road. If the lorry is accelerating at $0.80 \, \text{m s}^{-2}$, calculate (a) the net force acting on the whole combination, (b) the tension in the coupling between lorry and first trailer, (c) the tension in the coupling between first and second trailers.

4   A metal ball of mass 0.50 kg is dropped from the top of a vertical cliff of height 90 m. When it hits the beach below it penetrates to a depth of 6.0 cm. Calculate (a) the velocity acquired by the ball just as it hits the sand, (b) the (average) retarding force of the sand. Neglect air resistance; $g = 10 \, \text{m s}^{-2}$.

5   What net force must be applied to an object of mass 5.0 kg, initially at rest, for it to acquire a velocity of $12 \, \text{m s}^{-1}$ over a distance of 0.10 m?

Non-uniform acceleration

So far we have assumed constant acceleration, that is, the change in velocity with time is constant. For non-uniform acceleration the change of velocity with time is not constant throughout the motion. The slope of the velocity versus time graph at a given time is the instantaneous acceleration.

Often a movement can be considered to be made up of two or more stages, in each of which acceleration is constant (see below).

In each case the area enclosed by the velocity–time graph is equal to the distance travelled. This is explained in more detail in Chapter 30.

Example 13

A car accelerates from rest for 30 s, then travels at constant velocity for 20 s before decelerating for 20 s and coming back to rest. The velocity–time graph for its motion is shown in Fig 5.10.

(a) Use the graph to calculate the car's acceleration at the following times:
   (i) 10 s   (ii) 25 s   (iii) 40 s   (iv) 60 s

(b) Estimate the total displacement after
   (i) 30 s   (ii) 50 s   (iii) 70 s

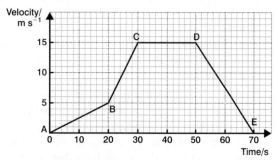

Fig. 5.10   Information for Example 13

Method

(a) The acceleration a at any time is the gradient of the velocity–time graph at that moment.
   (i) At time 10 s, gradient of AB = 5/20 = 0.25 m s^{-2}
   (ii) At time 25 s, gradient of BC = 10/10 = 1.0 m s^{-2}
   (iii) At time 40 s, gradient of CD = 0 m s^{-2}
   (iv) At time 60 s, gradient of DE = −15/20 = −0.75 m s^{-2}

Note that the slope of DE is negative, thus indicating a negative acceleration, or deceleration.

(b) The total distance travelled at any time is the area under the velocity–time graph.
   (i) Area up to time of 30 s is:
      area under AB + area under BC = 50 + 100 = 150 m
   (ii) Area up to time of 50 s is:
      area under AB + area under BC + area under CD = 50 + 100 + 300 = 450 m
   (iii) Area up to time of 70 s is:
      area under AB + area under BC + area under CD + area under DE = 50 + 100 + 300 + 150 = 600 m

Answer

(a) (i) 0.25 m s^{-2} (ii) 1.0 m s^{-2} (iii) 0
   (iv) −0.75 m s^{-2}

(b) (i) 150 m   (ii) 450 m   (iii) 600 m

Terminal velocity

Liquids and gases can exert a viscous drag force which opposes the motion of objects which pass through them. As shown in Fig 5.11, for a sphere of radius r (m) moving with velocity v (m s^{-1}) through a medium of viscosity η (Pa s) the resistive force F (N) opposing its motion is, assuming laminar flow conditions, given by Stokes' law:

$$F = 6\pi\eta rv \qquad (5.6)$$

This means that a sphere falling under gravity will eventually reach a terminal velocity at which time the gravitational force is balanced by the viscous drag force (we neglect any upthrust due to buoyancy effects from the liquid).

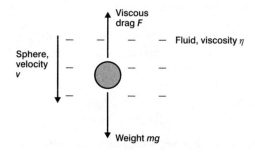

Fig. 5.11   Viscous drag on a falling sphere

Example 14

A spherical dust particle of diameter 20 μm falls, from rest, under gravity and in air until it attains a steady velocity.

(a) Calculate the value of this terminal velocity.

(b) Sketch a graph of the particle's velocity versus time, indicating the regions of maximum and minimum acceleration.

Assume the following values:

   viscosity of air $\eta = 1.8 \times 10^{-5}$ kg m^{-1} s^{-1}
   density of dust $\rho = 2.0 \times 10^{3}$ kg m^{-3}
   acceleration due to gravity $g = 10$ m s^{-2}

Neglect the effect of the upthrust due to buoyancy effects of the air on the particle.

Method

(a) If the dust particle has mass m then, when it has reached the terminal velocity, its weight mg is balanced by the viscous drag force due to the air. Thus:

   Weight mg = Viscous drag force $6\pi\eta rv$

Since the dust particle is spherical, then $m = 4/3 \times \pi r^3 \rho$, where r is the radius of the particle $= 10 \times 10^{-6}$ m and ρ its density.

Thus, substituting for m:

$$\tfrac{4}{3} \times \pi \times r^3 \rho g = 6\pi \eta r v$$

or

$$v = 2r^2 \rho g / 9\eta \qquad (5.7)$$

Inserting the values for r, ρ, g and η gives

terminal velocity $v = 2.47 \times 10^{-2}\,\mathrm{m\,s^{-1}}$

It is worth noting that the terminal velocity as stated in equation (5.7) is proportional to the square of the radius of the particle, so that larger particles attain a higher terminal velocity. In this case streamline flow may break down and the expression $F = 6\pi \eta r v$ no longer applies.

(b) The velocity–time graph is drawn in Fig 5.12. Note that the acceleration will be a maximum at commencement of the motion, where it will have a value equal to g (if the upthrust of the surrounding air can be neglected). Once terminal velocity is attained, the (minimum) acceleration will be zero. These two facts can be seen from the gradient of the velocity–time graph as shown in Fig 5.12.

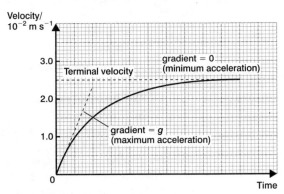

Fig. 5.12   Solution to Example 14

Answer

(a) $2.5 \times 10^{-2}\,\mathrm{m\,s^{-1}}$

Exercise 5.7

1   Fig 5.13 is a velocity–time graph for a moving body.

   (a) Calculate the value of the acceleration at each of the stages AB, BC and CD of its motion.

   (b) Calculate the distance travelled in each stage and the total distance covered.

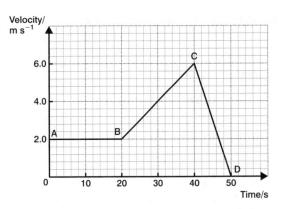

Fig 5.13   Information for Question 1

2   When a spherical drop of oil of density $9.0 \times 10^2\,\mathrm{kg\,m^{-3}}$ is allowed to fall through a gas of viscosity $1.2 \times 10^{-5}\,\mathrm{kg\,m^{-1}\,s^{-1}}$ it reaches a terminal velocity of $0.25\,\mathrm{m\,s^{-1}}$. Calculate the radius of the drop. Assume $g = 10\,\mathrm{m\,s^{-2}}$ and that the upthrust from the gas due to buoyancy effects is negligible

Thinking, braking and stopping distances

When a motorist has to brake his stopping distance is determined by the initial speed, his reaction time (the interval between receiving a stimulus and acting on it) and the deceleration due to the brakes.

Example 15

A car is travelling at a speed of $20\,\mathrm{m\,s^{-1}}$ and the driver has a personal reaction time of $0.80\,\mathrm{s}$. If the maximum deceleration which the brakes can apply to the car is $5.0\,\mathrm{m\,s^{-2}}$ calculate:

(a) the distance travelled prior to the driver applying the brakes (the thinking distance)

(b) the distance travelled during the braking and prior to stopping (the braking distance)

(c) the total stopping distance

Method

(a) Prior to braking the car travels at a constant speed of $20\,\mathrm{m\,s^{-1}}$ for $0.80\,\mathrm{s}$. Thus thinking distance $=$ speed $\times$ time $= 20 \times 0.80 = 16\,\mathrm{m}$

(b) We have $u = 20$, $a = -5.0$ (note the negative acceleration) and $v = 0$. We require the braking distance s. Rearranging Equation (5.3).

$$s = \frac{(v^2 - u^2)}{2a} = \frac{(0^2 - 20^2)}{2 \times -5.0} = 40\,\mathrm{m}.$$

(c) Stopping distance = thinking distance
$$+ \text{braking distance}$$
$$= 16 + 40 = 56 \,\text{m}$$

Answer

(a) 16 m, (b) 40 m, (c) 56 m.

Exercise 5.8

1 A motorist with a personal reaction time of 0.60 s is driving along a straight road at a speed of $12 \,\text{m s}^{-1}$ when he sees a pedestrian walk out in front of his car at a distance of 20 m away. If the car and driver have a total mass of 900 kg and the average braking force is 5.4 kN, determine

(a) the thinking distance,

(b) the braking distance,

(c) his stopping distance.

2 A motorist has a personal reaction time of 1.0 s. If he is travelling at $30 \,\text{m s}^{-1}$, at what rate must he be able to decelerate if he is to stop in a distance of 120 m?

Exercise 5.9:
Examination Questions

(Assume $g = 10 \,\text{m s}^{-2}$ except where stated.)

1 A shot putter throws a shot forward with a velocity of $10 \,\text{m s}^{-1}$ with respect to himself, in a direction $50°$ to the horizontal. At the same time the shot-putter is moving forward horizontally with a velocity of $3.0 \,\text{m s}^{-1}$. Calculate the magnitude and direction of the resultant velocity of the shot.

2 (a) Physical quantities can be classified as scalar quantities or vector quantities. Explain the difference, giving an example of each

(b) A light aircraft flies at a constant airspeed of $45 \,\text{m s}^{-1}$ on a journey towards a destination due north of its starting point. A wind is blowing at a constant speed of $20 \,\text{m s}^{-1}$ from the west. Find, by drawing or by calculation:
  (i) the direction in which the aircraft should point;
  (ii) the speed of the aircraft over the ground.
  [OCR 2001]

3 A car, originally travelling at a speed of $30 \,\text{m s}^{-1}$, decelerates uniformly to rest in a time of 20 s. Calculate the distance travelled by the car in the first 10 s.

4 A car accelerates uniformly from rest for a period of 8.0 s in which time it travels a distance of 48 m. Calculate the acceleration.

5 (a) (i) *'Dividing distance travelled by time taken gives a body's speed.'* Comment on this statement, and write an improved version. Use a bus journey as an example.
  (ii) A body, starting from rest, travels in a straight line with a constant acceleration, a, for a time t.
   (I) Sketch a velocity–time graph for the body.
   (II) Deduce, from the graph, that the distance, s, travelled by the body in time t is given by
   $$s = \tfrac{1}{2}at^2.$$

(b) On a building site, bags of cement, each of mass 50 kg, are placed on a sloping ramp and allowed to slide down it. The diagram shows the three main forces acting on a bag.

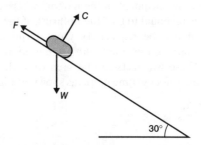

  (i) Give the name of each of the three forces shown.
  (ii) (I) Calculate W, given that the mass of a bag is 50 kg.
   (II) Calculate the component of W which acts down the slope.
   (III) The acceleration of a bag down the slope is $2.0 \,\text{m s}^{-2}$. Calculate the value of F. Explain your reasoning.
   (IV) By considering the direction at right angles to the slope, calculate the value of C. Explain your reasoning.
  (iii) Calculate the time it takes for a bag to travel 36 m, as measured along the sloping surface. The bags start from rest.
   [WJEC 2001]

6 A stone is dropped from the top of a tower of height 40 m. The stone falls from rest and air resistance is negligible.

How long does it take for the stone to fall the last 10 m to the ground?
(Use $g = 10 \,\text{m s}^{-2}$.)

A 0.38 s   B 1.4 s   C 2.5 s   D 2.8 s [OCR 2001]

7 An aircraft is travelling horizontally at $250\,\mathrm{m\,s^{-1}}$ when a part of the fuselage falls off. If the aircraft is travelling at a height of $4.5\,\mathrm{km}$, calculate

  (a) the time it takes for the fuselage portion to fall to the ground (neglect air resistance)

  (b) the horizontal distance it will have travelled in this time

  (c) its velocity just prior to impact with the ground.

8 In this question, all effects of air resistance can be neglected.

An athlete in the javelin event runs along a horizontal track and launches the javelin at an angle of $40.0°$ to the horizontal. The javelin rises to a maximum height and then falls to ground level. It hits the ground $4.00\,\mathrm{s}$ after launching, at a point a horizontal distance of $75.2\,\mathrm{m}$ from the launch point.

  (a) (i) Show that the horizontal component of the launch velocity is $18.8\,\mathrm{m\,s^{-1}}$.
    (ii) Calculate the magnitude of the launch velocity.

  (b) The length of the track from the start of the athlete's run to the launch point is $33.5\,\mathrm{m}$. For this run, the athlete starts from rest and accelerates uniformly at $1.50\,\mathrm{m\,s^{-2}}$ over the complete length of the track.
    (i) Calculate the speed of the athlete when she reaches the launch point.
    (ii) Comment on any difference between your answer to (b)(i) and the value quoted in (a)(i).

In the calculations in parts (c) and (d), treat the javelin as a point mass.

  (c) The javelin leaves the athlete's hand at a height of $1.80\,\mathrm{m}$ above the ground. Calculate the maximum height above the ground reached by the javelin.

  (d) For the javelin striking the ground at the end of its flight, calculate
    (i) the vertical component of the velocity
    (ii) the magnitude of the velocity. (Make use of information in (a)(i).)

[CCEA 2001]

9 A ball is to be kicked so that, at the highest point of its path, it just clears a fence a few metres away. The trajectory of the ball is shown in Fig 5.14

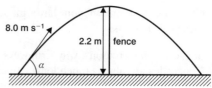

Fig. 5.14

The ground is level and the fence is $2.2\,\mathrm{m}$ high. The ball is kicked from ground level with an initial velocity of $8.0\,\mathrm{m\,s^{-1}}$ at an angle α to the horizontal. Air resistance is to be neglected.

  (i) Show that, if the ball just clears the fence, the angle of projection α must be $55°$.
  (ii) Find the horizontal velocity of the ball as it passes over the fence.
  (iii) Calculate the total time for which the ball is in the air from the instant it is kicked until it reaches the ground.

Assume $g = 9.8\,\mathrm{m\,s^{-2}}$. [CCEA 2000]

10 A railway locomotive of mass of $80\,000\,\mathrm{kg}$ exerts a tractive force of $400\,\mathrm{kN}$ at the rails. The locomotive hauls 8 coaches, each of mass $80\,000\,\mathrm{kg}$, as well as itself.

The total force of friction on the train is $150\,\mathrm{kN}$.

Which one of A to D is the acceleration of the train measured in $\mathrm{m\,s^{-2}}$?

A 0.35   B 0.78   C 1.80   D 3.40

[OCR Nuff 2001]

11

The diagram above shows a Concorde preparing for take-off. Its engines are turning but its brakes are still on, and it is not yet moving.

  (a) Mark on the diagram, using arrows and letters, the directions in which the following forces act *on* the aircraft:
    (i) its weight, labelled 'W',
    (ii) the force caused by the engines, labelled 'T',
    (iii) the *total* force exerted by the runway, labelled 'F'.

Describe the sum of these three forces.

Here is some data about the Concorde during take-off.

Total mass	$185\,000\,\mathrm{kg}$
Average thrust per engine	$170\,\mathrm{kN}$
Number of engines	4
Take-off speed	$112\,\mathrm{m\,s^{-1}}$.

  (b) Calculate its acceleration.

  (c) Calculate the time it takes to achieve take-off speed from rest.

  (d) Show that the distance it travels from rest to the take-off point is about $1700\,\mathrm{m}$.

Give *one* assumption you have made in doing these calculations.

Suggest *one* reason why runways used by Concordes are always much longer than 1700 m.
[Edexcel S-H 2000]

12 (a) Force and acceleration are both vector quantities.
  (i) State what is meant by a vector quantity.
  (ii) Name *one* other vector quantity.

(b) Two men are trying to drag a concrete block by pulling on ropes tied to a metal ring which is attached to the block. One man exerts a pull of 260 N to the East and the other exerts a pull of 150 N to the South, as shown.

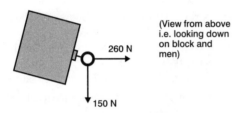

(View from above i.e. looking down on block and men)

260 N

150 N

  (i) Calculate the magnitude (size) and direction of the resultant of the forces applied by the men. Explain how you arrive at your answer.
  (ii) Instead of the block moving, the ring suddenly breaks away from the block. Calculate the initial acceleration of the ring if its mass is 6.0 kg.
[WJEC 2001]

13 The diagram shows three trucks which are part of a train. The mass of each truck is 84 000 kg.

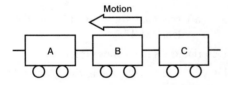

Motion

A   B   C

The train accelerates uniformly in the direction shown from rest to 16 m s^{-1} in a time of 4.0 minutes. Calculate the resultant force on each truck.

The force exerted by truck B on truck C is 11 200 N. Draw a free-body force diagram for truck B, showing the magnitudes of all the forces. Neglect any frictional forces on the trucks.
[Edexcel 2001, part]

14 A car is taken for a short test-drive along a straight road. A velocity vs. time graph for the first 40 seconds of the drive is given below.

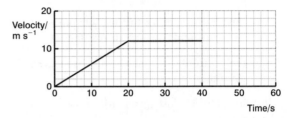

  (i) Calculate the acceleration of the car during the first 20 seconds.
  (ii) Calculate the car's displacement after
    (I) 20 seconds
    (II) 40 seconds.
  (iii) At 40 seconds from the start of the drive the car starts to slow down at a uniform rate. During the deceleration it travels a further 90 m before coming to rest. Complete the velocity vs. time graph above to show this final stage of the drive.   [WJEC 2001, part]

15 (a) A vehicle on a straight road starts from rest and accelerates at 1.5 m s^{-2} for 20 s. It then travels for 200 s at constant velocity, and finally decelerates uniformly, coming to rest after a further 30 s.
    (i) Sketch a velocity–time graph for the whole 250 s period. Label the velocity and time axes with appropriate values.
    (ii) Find the total distance travelled in the 250 s period. Hence calculate the average speed for the whole journey.
    (iii) Sketch a displacement–time graph for the whole 250 s period. Label the time axis with appropriate values.

In the following parts ((b) and (c)) of this question take the acceleration of free fall *g* as 10 m s^{-2} and ignore air resistance.

(b) A stone is projected with a vertical component of velocity of 30 m s^{-1} from the edge of the top of a tower 200 m high. It follows the trajectory shown in Fig 5.15

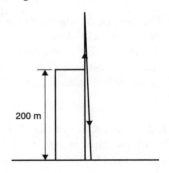

200 m

Fig. 5.15

Calculate
  (i) the time after projection at which the stone reaches its maximum height,
  (ii) the maximum height reached above the ground,
  (iii) the total flight time until the stone reaches the ground.

(c) Another stone is projected into the air from ground level at a velocity of $25\,\mathrm{m\,s^{-1}}$ at an angle of 35° to the horizontal (Fig 5.16).

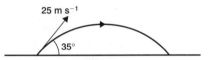

Fig. 5.16

Calculate
  (i) the horizontal range,
  (ii) the magnitude and direction of the velocity 0.60 s after projection.

(d) It is possible to project the stone in (c) with the same speed of $25\,\mathrm{m\,s^{-1}}$, but at a different angle to the horizontal, so that it has exactly the same horizontal range as in (c)(i). Copy Fig. 5.16 into your answer booklet, and on the same diagram draw a labelled sketch of the trajectory obtained with this alternative angle of projection. (No calculation is required.) State, with a reason, whether the time of flight with this second trajectory is less than, the same as, or greater than the time of flight when the stone is projected with a velocity of $25\,\mathrm{m\,s^{-1}}$ at 35° to the horizontal, as in (c). (No calculation is required.)   [CCEA 2001]

16 Some people think that all raindrops fall at the same speed; others think that their speed depends on their size.

(a) Calculate the speed of a raindrop after it has fallen freely from rest for 0.2 s.

(b) The raindrop falls for longer than 0.2 s. Explain why its acceleration does not remain uniform for the whole of its fall.

(c) Show that the mass of a 0.5 mm diameter spherical raindrop is less than 1×10^{-7} kg. $1.0\,\mathrm{m^3}$ of water has a mass of 1.0×10^3 kg.

(d) Calculate the raindrop's terminal velocity. Assume that the upthrust from the air is negligible. Explain your working clearly.

Viscosity of air = $1.8 \times 10^{-5}\,\mathrm{kg\,m^{-1}\,s^{-1}}$.

(e) Sketch a graph to show how the raindrop's velocity increases from rest to terminal velocity. Add a scale to the velocity axis.

(f) Explain how the terminal velocity would be different for a larger raindrop.
          [Edexcel S-H 2001]

17 A student uses a deodorant spray which spreads many small droplets into the air. The diagram below shows one of these droplets falling with terminal velocity.

(a) On a copy of the diagram, draw labelled arrows to represent the forces acting. Assume that the upthrust of the air is negligible.

●

What is the relationship between the forces when the droplet is falling with terminal velocity?

(b) After reaching terminal velocity the droplet falls 25 mm in 6.0 s. Calculate the terminal velocity.

Hence *estimate* the time for this droplet to reach the floor.

(c) Write down an expression for the weight of the droplet in terms of radius r and density ρ.

(d) The viscous drag F acting on a droplet of radius r falling with terminal velocity v through a medium of viscosity η, is given by the expression

$$F = 6\pi\eta r v$$

Show that the radius of the droplet is given by

$$r = \sqrt{\frac{9\eta v}{2\rho g}}$$

Hence calculate the radius of the falling droplet.

$\eta = 1.8 \times 10^{-5}\,\mathrm{N\,s\,m^{-2}}$
$\rho = 920\,\mathrm{kg\,m^{-3}}$

In the calculations above, the upthrust of the air is assumed to be negligible. Explain why this is a reasonable assumption.

Density of air = $1.2\,\mathrm{kg\,m^{-3}}$.
          [Edexcel S-H 2000]

18 A van driver, making an emergency stop from a speed of $18\,\mathrm{m\,s}^{-1}$, requires a thinking distance of $12\,\mathrm{m}$ and a braking distance of $22\,\mathrm{m}$.

(a) Showing your calculations, determine:
 (i) the reaction time of the driver;
 (ii) the average deceleration of the van during braking.

(b) The same van, with the same driver, is following a car on a motorway. Both vehicles are travelling at $30\,\mathrm{m\,s}^{-1}$, and the distance between the front of the van and the rear of the car is $15\,\mathrm{m}$. Determine, using a suitable calculation, whether or not the van will collide with the car if the car driver makes an emergency stop. Assume that the decelerations of the car and the van under braking are equal. [OCR 2001]

19 The following table gives data taken from the Highway Code for 'Typical Stopping Distances' of a car when braking.

Speed/ miles per hour	Speed/ $\mathrm{m\,s}^{-1}$	Thinking distance/ m	Braking distance/ m	Stopping distance/ m	Deceleration/ $\mathrm{m\,s}^{-2}$
20	8.9	6	6	12	6.6
30	13.4	9	14	23	6.4
50	22.4	15	38	53	
70	31.3	21	75	96	6.5

The 'thinking distance' is the distance the car moves while the driver is reacting before the brakes are applied.

(a) Calculate the thinking time for a speed of *20 miles per hour*.

Explain why the thinking distance varies with speed.

(b) The 'braking distance' is the distance the car travels while decelerating once the brakes have been applied.
 (i) Show that the deceleration is about $7\,\mathrm{m\,s}^{-2}$ while braking from a speed of *50 miles per hour*.
 (ii) Calculate the braking force which produces this deceleration for a car of mass $900\,\mathrm{kg}$.

(c) Brakes depend for their operation on the friction between brake pads and a steel disc connected to the wheel. A text book states that the magnitude of this friction does not depend on how fast the car is going, provided the wheels do not lock.

Use the data in the table to discuss whether the results are consistent with this statement.

(d) With extra passengers the mass of the car is much greater. If the braking force remains the same, explain how this would affect braking distances. [Edexcel S-H 2000]

6
Energy, work and power

Energy

Mechanical energy exists in two basic forms:

(1) Kinetic energy (KE) is energy due to motion. $KE = \frac{1}{2}mv^2$ (m is the mass, v the velocity of the body.)

(2) Potential energy (PE) is energy stored – e.g. in a compressed spring or due to the position of a body in a force field. Gravitational $PE = mgh$ (m is mass, g is acceleration due to gravity, h is height above a datum), and is energy stored in a gravitational field (see Chapter 9) due to the elevated position of the body.

Work and energy

Work is done when energy is transferred from one system to another. It involves a force acting over a distance. We define

> **Work done (J) = Force (N) × Distance (m)**
> **(6.1)**

Also   Work done = Energy transferred (J)

Example 1

A body of mass 5.0 kg is initially at rest on a horizontal frictionless surface. A force of 15 N acts on it and accelerates it to a final velocity of $12\,\text{m s}^{-1}$. Calculate (a) the distance travelled, (b) the work done by the force, (c) the final KE of the body.
Compare (b) and (c) and comment.

Method

(a) We have $m = 5.0$, $u = 0$, $F = 15$ and $v = 12$. To find distance s we must find acceleration a. From Equation 5.5

$$a = \frac{F}{m} = \frac{15}{5} = 3.0\,\text{m s}^{-2}$$

Using Equation 5.3
$$v^2 = u^2 + 2as$$
$$\therefore \quad 12^2 = 0^2 + 2 \times 3 \times s$$
$$\therefore \quad s = 24\,\text{m}$$

(b) From Equation 6.1
$$\text{Work done} = F \times s = 15 \times 24 = 360\,\text{J}$$

(c) $KE = \frac{1}{2}mv^2 = \frac{1}{2} \times 5 \times 12^2$
$$= 360\,\text{J}$$

The answers to (b) and (c) are the same. This is because, in the absence of friction forces and on a horizontal surface all the work done by the force becomes KE of the moving body.

Answer

(a) 24 m, (b) 3.6×10^2 J, (c) 3.6×10^2 J.

Example 2

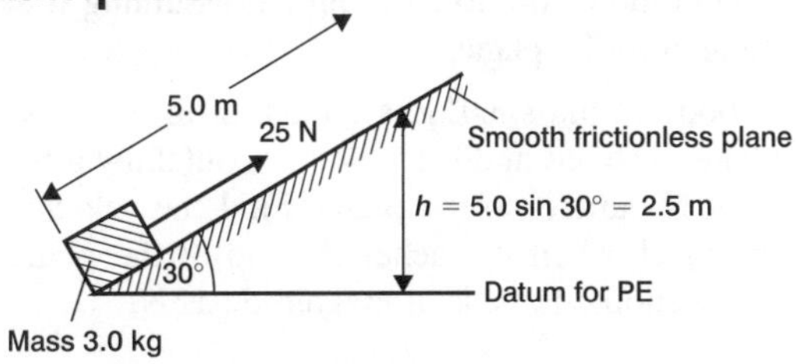

Fig. 6.1   Information for Example 2

Refer to Fig. 6.1. A block of mass 3.0 kg is pulled 5.0 m up a smooth plane, inclined at 30° to the horizontal, by a force of 25 N parallel to the plane. Find the velocity of the block when it reaches the top of the plane.

Method

Work done on the body becomes KE and PE. So, if final velocity of block is v, then

$$\text{Work done} = (\text{KE} + \text{PE}) \text{ gained by block, or}$$
$$F \times s = \frac{1}{2}mv^2 + mgh$$

We have $F = 25$, $s = 5.0$, $m = 3.0$, $g = 10$ and $h = 5.0\sin 30° = 2.5$. So

$$25 \times 5 = \frac{1}{2} \times 3 \times v^2 + 3 \times 10 \times 2.5$$
$$\therefore \quad v = \sqrt{\frac{100}{3}} = 5.8\,\text{m s}^{-1}$$

Note that in the above we do *not* subtract from the 25 N force the component of weight acting 'down' the plane ($mg \sin 30°$, from Chapter 4) since gravitational effects have been accounted for in the PE term.

Answer

$5.8 \,\mathrm{m\,s^{-1}}$.

Exercise 6.1

(Assume $g = 10\,\mathrm{m\,s^{-2}}$.)

1 Calculate the kinetic energy of the following:

(a) a car: mass $900\,\mathrm{kg}$, velocity $20\,\mathrm{m\,s^{-1}}$;

(b) an aeroplane mass $20 \times 10^3\,\mathrm{kg}$, velocity $200\,\mathrm{m\,s^{-1}}$;

(c) an electron; mass $9.0 \times 10^{-31}\,\mathrm{kg}$, velocity $20 \times 10^6\,\mathrm{m\,s^{-1}}$.

2 A force of 15 N is applied to a body of mass 3.0 kg, initially at rest on a smooth horizontal surface, for a time of 3.0 s. Calculate (a) the final velocity, (b) the distance travelled, (c) the work done, (d) the final KE of the body.

3 A block of mass 10 kg is pulled 20 m up a smooth plane inclined at 45° to the horizontal. The block is initially at rest and reaches a velocity of $2.0\,\mathrm{m\,s^{-1}}$ at the top of the plane. Calculate the magnitude of the force required, assuming it acts parallel to the plane.

4 A body of mass 5.0 kg is pulled 4.0 m up a *rough* plane, inclined at 30° to the horizontal, by a force of 50 N parallel to the plane. Find the velocity of the block when it reaches the top of the plane if the frictional force is of magnitude 12 N.

Energy interchange

The principle of conservation of energy states that the total amount of energy in an isolated system remains constant.

If dissipative effects, e.g. friction, are ignored then we have simply KE and PE interchange.

Example 3

A ball of mass 0.50 kg falls from a height of 45 m. Calculate (a) its initial PE, (b) its final KE, (c) its final velocity. Neglect air resistance. (Assume $g = 10\,\mathrm{m\,s^{-2}}$.)

Method

(a) We have $m = 0.50, h = 45, g = 10$. So
$$\mathrm{PE} = mgh = 0.5 \times 10 \times 45 = 225\,\mathrm{J}$$

(b) All the PE has been converted to KE just prior to striking the ground. So
$$\text{Final KE} = 225\,\mathrm{J}$$

(c) Let final velocity be v. Since $\mathrm{KE} = \frac{1}{2}mv^2$ then
$$225 = \frac{1}{2} \times 0.5 \times v^2$$
$$\therefore \quad v = 30\,\mathrm{m\,s^{-1}}$$

Note that the final velocity does not depend on the mass because it cancels out (since $\frac{1}{2}mv^2 = mgh$, $v^2 = 2gh$).

Note that, as in Example 7, Chapter 5, we could have solved this using the equations of motion with acceleration $a = g = 10\,\mathrm{m\,s^{-2}}$. This is because air resistance is negligible.

Answer

(a) $2.3 \times 10^2\,\mathrm{J}$, (b) $2.3 \times 10^2\,\mathrm{J}$, (c) $30\,\mathrm{m\,s^{-1}}$.

Example 4

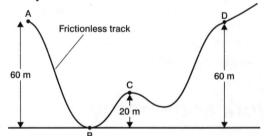

Fig. 6.2   Information for Example 4

Refer to Fig. 6.2. A truck of mass 150 kg is released from rest at A and moves along the frictionless track. Calculate (a) its maximum KE, (b) its maximum velocity, (c) its velocity at C. (Assume $g = 10\,\mathrm{m\,s^{-2}}$.) Explain what happens when it reaches D.

Method

(a) Maximum KE is when PE is a minimum. This occurs at B. We have
$$\text{Gain in KE} = \text{Loss in PE } (mgh)$$
Since $m = 150, g = 10, h = 60$,
$$\text{Loss in PE} = 150 \times 10 \times 60 = 90\,000\,\mathrm{J}$$
$$\therefore \quad \text{KE gain} = 90\,\mathrm{kJ}$$

(b) Maximum velocity v occurs for maximum KE.
$$\therefore \quad \tfrac{1}{2}mv^2 = \tfrac{1}{2} \times 150 \times v^2 = 90\,000$$
$$\therefore \quad v = \sqrt{1200} = 34.6\,\mathrm{m\,s^{-1}}$$

(c) At C, the drop in height h_1 is 40 m below A. So if velocity at C is v_1 then

$$\text{Gain in KE } \left(\tfrac{1}{2}mv_1{}^2\right) = \text{Loss in PE } (mgh_1)$$
$$\tfrac{1}{2} \times 150 \times v_1{}^2 = 150 \times 10 \times 40$$
$$\therefore \qquad v_1 = \sqrt{800} = 28.3\,\text{m s}^{-1}$$

Note, once again, that v_1 does not depend on the mass of the truck.

The truck arrives at D with zero KE, hence zero velocity, since its PE at D equals its PE at A. It then starts to roll back to C, B and A.

Answer

(a) 90 kJ, (b) 35 m s^{-1}, (c) 28 m s^{-1}.

Exercise 6.2

(Assume $g = 10\,\text{m s}^{-2}$.)

1 An object of mass 0.30 kg is thrown vertically upwards and reaches a height of 8.0 m. Calculate (a) its final PE, (b) the velocity with which it must be thrown, neglecting air resistance.

2 A cricket of mass 2.5 g has a vertical velocity of 2.0 m s^{-1} when it jumps. Calculate (a) its maximum KE and (b) the maximum vertical height it could reach.

3 A ball of mass 0.20 kg drops from a height of 10 m and rebounds to a height of 7.0 m. Calculate the energy lost on impact with the floor. Neglect air resistance.

4 A simple pendulum oscillates with an amplitude of 30°. If the length of the string is 1.0 m, calculate the velocity of the pendulum bob at its lowest point.

5

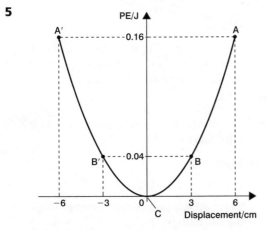

Fig. 6.3   Information for Question 5

Fig. 6.3 shows the PE versus displacement graph for a body, of mass 0.10 kg, oscillating about the point C. If the body has total energy 0.16 J, calculate its velocity at (a) A and A′, (b) B and B′, (c) C. Neglect friction, air resistance, etc.

'Lost' energy

If dissipative forces, e.g. friction and air resistance, cannot be neglected, then some energy will be 'lost' in the sense that it is converted to other forms (e.g. heat).

Example 5

A ball of mass 0.20 kg is thrown vertically upwards with a velocity of 15 m s^{-1}. If it reaches a height of 10 m, calculate the percentage loss in energy caused by air resistance. (Assume $g = 10\,\text{m s}^{-2}$.)

Method

The final PE is less than the initial KE due to transfer of energy to the surrounding air. We have $m = 0.20$, $u = 15, h = 10$ and $g = 10$, so

$$\text{Initial KE} = \tfrac{1}{2}mu^2 = \tfrac{1}{2} \times 0.2 \times 15^2 = 22.5\,\text{J}$$
$$\text{Final PE} = mgh = 0.2 \times 10 \times 10$$
$$= 20.0\,\text{J}$$
$$\therefore \quad \text{Energy transfer} = 22.5 - 20.0 = 2.5\,\text{J}$$
$$\text{Percentage loss} = \frac{\text{Energy transfer}}{\text{Initial KE}} \times 100$$
$$= \frac{2.5}{22.5} \times 100 = 11.1$$

Answer

11% of the initial KE is transferred to the surrounding air.

Example 6

A block of mass 6.0 kg is projected with a velocity of 12 m s^{-1} up a rough plane inclined at 45° to the horizontal. If it travels 5.0 m up the plane, calculate (a) the energy dissipated via frictional forces, (b) the magnitude of the (average) friction force. (Assume $g = 10\,\text{m s}^{-2}$.)

Method

(a)

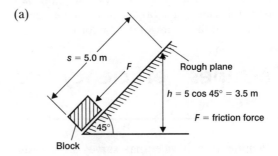

Fig. 6.4   Solution to Example 6

Refer to Fig. 6.4. Initial KE becomes PE and energy dissipated via work done ($F \times s$) against friction force F. So

$$\tfrac{1}{2}mu^2 = mgh + F \times s$$

We have $m = 6.0$, $u = 12$, $g = 10$ and $h = 3.5$, so

$$\tfrac{1}{2} \times 6 \times 12^2 = 6 \times 10 \times 3.5 + F \times s$$

$$\therefore \qquad F \times s = 222\,\text{J}$$

(b) $\qquad F \times s = 222\text{J}$ and $s = 5.0\,\text{m}$

$$\therefore \qquad F = 44\,\text{N}$$

Answer

(a) 0.22 kJ,   (b) 44 N.

Exercise 6.3

(Assume $g = 10\,\text{m s}^{-2}$.)

1 An object of mass 1.5 kg is thrown vertically upwards with a velocity of $25\,\text{m s}^{-1}$. If 10% of its initial energy is dissipated against air resistance on its upward flight, calculate (a) its maximum PE, (b) the height to which it will rise.

2 A cricket ball of mass 0.20 kg is thrown vertically upwards with a velocity of $20\,\text{m s}^{-1}$ and returns to earth at $15\,\text{m s}^{-1}$. Find the work done against air resistance during the flight.

3

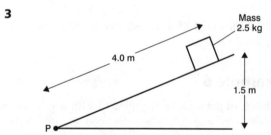

Fig. 6.5   Diagram for Question 3

Fig. 6.5 shows a mass of 2.5 kg initially at rest on a rough inclined plane. The mass is now released and acquires a velocity of $4.0\,\text{m s}^{-1}$ at P, the base of the incline. Find (a) the work done against friction, (b) the (average) friction force.

Machines – efficiency and power

A machine is a device that serves to transfer energy from one system to another. The useful energy output will be less than the energy input due to energy 'lost', e.g. in work done against friction. We define

> **Efficiency (%)**
> $$= \frac{\textbf{Useful energy (power) output}}{\textbf{Energy (power) input}} \;(\times\,\textbf{100})\;\;(6.2)$$

Power is the rate of transfer of energy – i.e. the work done in unit time. The SI unit for power is the watt ($1\text{W} = 1\,\text{J s}^{-1}$).

Example 7

A 1.0 kW motor drives a pump which raises water through a height of 15 m. Calculate the mass of water lifted per second, assuming the system is (a) 100% efficient, (b) 75% efficient. (Assume $g = 10\,\text{m s}^{-2}$.)

Method

(a) Each second the motor supplies 1000 J of energy, which we assume is all converted to gravitational PE of the water. We have $h = 15$ and require mass m. So

$$1000 = mgh = m \times 10 \times 15$$

$$\therefore \quad m = 6.7\,\text{kg}$$

(b) Only 75% of 1000 J, that is 750 J, becomes available to lift water. So, if the new mass is m_1,

$$750 = m_1 gh = m_1 \times 10 \times 15$$

$$\therefore \quad m_1 = 5.0\,\text{kg}$$

Answer

(a) $6.7\,\text{kg s}^{-1}$, (b) $5.0\,\text{kg s}^{-1}$.

Example 8

20×10^3 kg of water moving at $2.2\,\text{m s}^{-1}$ is incident on a water wheel each second. Calculate the maximum power output from the mill, assuming 40% efficiency.

Method

Energy input is KE of the water. In *one second* we have $m = 20 \times 10^3$ and $v = 2.2\,\text{m s}^{-1}$. So

$$\text{KE input} = \tfrac{1}{2}mv^2 = \tfrac{1}{2} \times 20 \times 10^3 \times 2.2^2$$

$$= 48.4 \times 10^3\,\text{J}$$

Of this energy 40% becomes useful energy output. So rearranging equation (6.2):

$$\text{Useful energy output} = \frac{40}{100} \times 48.4 \times 10^3$$

$$= 19.4 \times 10^3\,\text{J}$$

Answer

Maximum power output=19 kW.

Example 9

A boat travels at a constant velocity of $8.0 \, \text{m s}^{-1}$. If the engine develops a useful power output of $20 \, \text{kW}$, calculate the push exerted by the propeller on the water. Why is the boat not accelerating?

Method

When a motive force drives an object, such as the boat, then the useful power output, or motive power, is given by:

> **Motive Power P = Rate of doing work**
>
> $\qquad$ **= Driving force × distance/time**
>
> $\qquad$ **= Driving force F × velocity v**
>
> $\hfill$ **(6.3)**

We have $P = 20 \times 10^3$, $v = 8.0$, and require F. So

$$20 \times 10^3 = F \times 8$$

$$\therefore \qquad F = 2.5 \times 10^3 \, \text{N}$$

The boat is not accelerating because the resistance to motion of the boat as it passes through water is equal to $2.5 \, \text{kN}$. The net force on the boat is thus zero, so it does not accelerate.

Answer

$2.5 \, \text{kN}$.

Example 10

A train of mass $10 \times 10^3 \, \text{kg}$, initially at rest, accelerates uniformly at $0.50 \, \text{m s}^{-2}$. Calculate the power required at time $5.0 \, \text{s}$ and $8.0 \, \text{s}$, assuming (a) no resistive forces, (b) resistive forces of $1.0 \, \text{kN}$ act.

Method

Equation 6.3 tells us we require force F and velocity v at a given time in order to calculate the instantaneous power P.

(a) To find F use Equation 5.5, with $a = 0.50$ and $m = 10 \times 10^3$. So

$$F = ma = 0.5 \times 10 \times 10^3 = 5 \times 10^3 \, \text{N}$$

To find v after $t = 5.0 \, \text{s}$ and $t = 8.0 \, \text{s}$ use Equation 5.1 with $u = 0$ and $a = 0.50$. Thus

after $5 \, \text{s}$, $\quad v = 0 + 0.5 \times 5 = 2.5 \, \text{m s}^{-1}$

after $8 \, \text{s}$, $\quad v = 0 + 0.5 \times 8 = 4.0 \, \text{m s}^{-1}$

From Equation 6.3 we have

after $5 \, \text{s}$, $\quad P = F \times v = 5 \times 10^3 \times 2.5$

$\qquad\qquad\qquad = 12.5 \times 10^3$

after $8 \, \text{s}$, $\quad P = F \times v = 5 \times 10^3 \times 4$

$\qquad\qquad\qquad = 20 \times 10^3$

(b) The engine must apply an additional force of $1.0 \times 10^3 \, \text{N}$ in excess of that in part (a). So the engine must apply a constant force of $6.0 \times 10^3 \, \text{N}$. From equation 6.3

after $5 \, \text{s}$, $\quad P = F \times v = 6 \times 10^3 \times 2.5 = 15 \times 10^3$

after $8 \, \text{s}$, $\quad P = F \times v = 6 \times 10^3 \times 4.0 = 24 \times 10^3$

Answer

(a) $12.5 \, \text{kW}$, $20 \, \text{kW}$ (b) $15 \, \text{kW}$, $24 \, \text{kW}$.

Example 11

A car of mass $1.2 \times 10^3 \, \text{kg}$ moves up an incline at a steady velocity of $15 \, \text{m s}^{-1}$ against a frictional force of $0.6 \, \text{kN}$. The incline is such that it rises $1.0 \, \text{m}$ for every $10 \, \text{m}$ along the incline. Calculate the output power of the car engine. (Assume $g = 10 \, \text{m s}^{-2}$.)

Method

The car engine does work against friction forces and in raising the PE of the car as it moves up the incline. So, referring to energy transfer per second gives

$$\text{Power } P = \left(\begin{array}{c} \text{Rate of doing} \\ \text{work against friction} \end{array} \right) + \left(\begin{array}{c} \text{Rate of} \\ \text{gain of PE} \end{array} \right)$$

$$= F \times v + mgh$$

where $\quad F$ (frictional force) $= 0.60 \times 10^3$

$\qquad\quad v$ (velocity) $\qquad\qquad = 15$

$\qquad\quad m$ (mass of car) $\qquad = 1.2 \times 10^3$

$\qquad\quad g \qquad\qquad\qquad\qquad = 10$

$\qquad\quad h$ (gain in height per second) $= 15 \times \dfrac{1}{10} = 1.5$

$$\therefore \quad P = (0.6 \times 10^3 \times 15) + (1.2 \times 10^3 \times 10 \times 1.5)$$

$$= 27 \times 10^3 \, \text{W}$$

Answer

Output power $= 27 \, \text{kW}$.

Exercise 6.4

(Assume $g = 10 \, \text{m s}^{-2}$.)

1 Calculate the power rating of a pump if it is to lift $180 \, \text{kg}$ of water per minute through a height of $5.0 \, \text{m}$, assuming
(a) 100%, (b) 50%, (c) 70% efficiency.

2 $200 \, \text{kg}$ of air moving at $15 \, \text{m s}^{-1}$ is incident each second on the vanes of a windmill. Estimate the maximum output power of the mill. Why is this not achieved in practice?

3 A hydroelectric power station is driven by water falling on to a system of wheels from a height of $100 \, \text{m}$. If the output power of the station is

10 MW (10×10^6 W), calculate the rate at which water must impinge on the wheels, assuming (a) 100%, (b) 50% efficiency.

4 A car of mass 900 kg, initially at rest, accelerates uniformly and reaches $20 \, \text{m s}^{-1}$ after 10 seconds. Calculate the power developed by the engine after (a) 5.0 s, (b) 10 s, (c) a distance of 50 m from the start position. Assume that resistive forces are negligible.

5 Repeat Question 4 with a constant resistive force of 0.50 kN acting.

6 A car pulls a caravan of mass 800 kg up an incline of 8% (8 up for 100 along the incline) at a steady velocity of $10 \, \text{m s}^{-1}$. Calculate the tension in the tow bar, assuming (a) resistive forces are negligible, (b) a resistive force of 0.40 kN acts on the caravan.

7 A car has a maximum output power of 20 kW and a mass of 1500 kg. At what maximum velocity can it ascend an incline of 10%, assuming (a) no dissipative forces, (b) a constant resistance of 1.0 kN opposing its motion.

Exercise 6.5:
Examination Questions

(Assume $g = 10 \, \text{m s}^{-2}$ except where stated.)

1 A force of 0.35 kN is needed to move a vehicle of mass 1.5×10^3 kg at constant speed along a horizontal road. Calculate the work done, against frictional forces, in travelling a distance of 0.30 km along the road.

2 A heavy sledge is pulled across snowfields. The diagram shows the direction of the force F exerted on the sledge. Once the sledge is moving, the average horizontal force needed to keep it moving at a steady speed over level ground is 300 N.

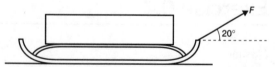

(a) Calculate the force F needed to produce a horizontal component of 300 N on the sledge.

(b) (i) Explain why the work done in pulling the sledge **cannot** be calculated by multiplying F by the distance the sledge is pulled.

(ii) Calculate the work done in pulling the sledge a distance of 8.0 km over level ground.

(iii) Calculate the average power used to pull the sledge 8.0 km in 5.0 hours.

(c) The same average power is maintained when pulling the sledge uphill. Explain **in terms of energy transformations** why it would take longer than 5.0 hours to cover 8.0 km uphill.
[AQA, 2001]

3 Part of a bobsled run is shown in Fig 6.6

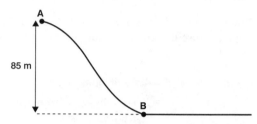

Fig. 6.6

The point **A** on the track is at a height of 85 m above point **B**. From **B** onwards, the run is horizontal. The bobsled, of mass 250 kg, starts from rest at **A**. It then slides down the slope to **B** and beyond.

(a) Assuming that no energy is lost as the bobsled slides down the slope, calculate the speed with which the sled is travelling as it passes point **B**.

(b) At **B** the brakes on the bobsled are applied to give a constant retarding force. The sled comes to rest having travelled 120 m along the horizontal part of the run. Calculate the magnitude of the deceleration produced by the brakes.

(c) Calculate the work done in stopping the bobsled.

(d) As the bobsled passes point **B**, it possesses kinetic energy. When it stops further along the run, its kinetic energy is zero. Account for this loss of kinetic energy in terms of the principle of conservation of energy.
[CCEA 2000]

4 The diagram shows part of a roller coaster ride. In practice, friction and air resistance will have a significant effect on the motion of the vehicle, but you should ignore them throughout this question.

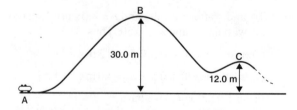

The vehicle starts from rest at A and is hauled up to B by a motor. It takes 15.0 s to reach B, at which point its speed is negligible. Complete the box in the diagram below, which expresses the conservation of energy for the journey from A to B.

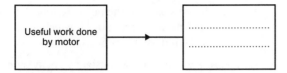

(a) The mass of the vehicle and the passengers is 3400 kg. Calculate

 (i) the useful work done by the motor.

 (ii) the power output of the motor.

At point B the motor is switched off and the vehicle moves under gravity for the rest of the ride.

Describe the overall energy conversion which occurs as it travels from B to C.

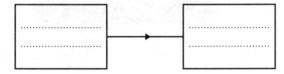

(b) Calculate the speed of the vehicle at point C.

(c) On another occasion there are fewer passengers in the vehicle; hence its total mass is less than before. Its speed is again negligible at B. State with a reason how, if at all, you would expect the speed at C to differ from your previous answer.   [Edexcel 2001]

5

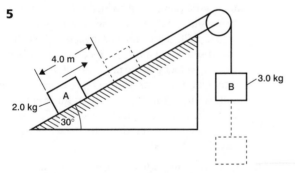

Fig. 6.7   Diagram for Question 5

Fig. 6.7 shows two blocks A and B connected by a light inextensible cord passing over a frictionless pulley. Block A starts from rest and moves up the smooth plane which is inclined at 30° to the horizontal. Calculate, at the moment that A has moved 4.0 m along the plane:

(a) the total kinetic energy of the system;

(b) the speed of the blocks A and B.

6 A ball of mass 0.2 kg is projected horizontally from the top of a wall 5.0 m above ground level at a speed of 6.0 m s^{-1}. Just before it hits the ground it has a speed of 10 m s^{-1}. Calculate how much energy has been dissipated as it fell through the air.

7 A sledge of mass 0.20×10^3 kg starts from rest at the top of a hill and slides down the hill without any driving force being applied. By the time it has fallen through a vertical height of 50 m it has acquired a speed of 20 m s^{-1}. Calculate the energy dissipated by frictional forces in this time.

8 Twin engine aircraft use less fuel than those with four engines. Recent improvements in engine reliability mean that they are now considered safe for long commercial flights over water. An aircraft powered by two Rolls-Royce Trent engines demonstrated its endurance by flying non-stop round the world. During this flight it used 1.7×10^5 litres of aviation fuel.

Each litre of fuel releases 38 MJ when combined with oxygen in the air.

(a) Calculate the total amount of energy released during the flight.

(b) The flight lasted 47 hours. Calculate the average input power to the engines.

(c) The distance covered by the aircraft was 41 000 km. Calculate the aircraft's average speed.

(d) The *maximum* thrust of each engine is 700 kN. Multiply the total maximum thrust by the average speed and comment on your answer.
     [Edexcel 2000]

9 (a) The movement of sea-water through turbines in a narrow harbour entrance is to be used to generate electricity. The harbour has vertical sides and encloses a surface area of 6.0×10^4 m^2. During a 3-hour period around low tide, sluice gates are opened and the water level in the harbour falls by 5.0 m.

 (i) Calculate, for the 3-hour period, the gravitational potential energy lost by the water leaving the harbour. The density of sea-water is 1050 kg m^{-3}.

 (ii) The generating system converts this potential energy to electrical energy with an efficiency of 40%. Calculate the mean electrical power generated.

(b) A small tidal power scheme and a large complex of coastal wind turbines can be built for approximately the same cost. The maximum electrical power output of each would be similar. Explain **two** advantages of choosing the tidal scheme option.
     [OCR 2001]

10 (a) A wind turbine has blades of total effective area 55 m^2 and is used in a head-on wind of speed 10 m s^{-1}. The density of air is 1.2 kg m^{-3}.

(i) Calculate the volume of air striking the blades every second, and hence show that about 650 kg of air strikes the blades every second.

(ii) Calculate the total kinetic energy of the air arriving at the blades every second.

(iii) The wind turbine can convert only 40% of this kinetic energy into electrical energy. Calculate the electrical power output of the wind turbine.

(b) A town with an electrical power requirement of 100 MW needs a new electricity generating station. The choice is between one oil-fired station or a collection of wind turbines. The power output from a small oil-fired generating station is about 100 MW; the useful power output of a single turbine is 20 kW.

Discuss the advantages and disadvantages of these types of power supply.   [AQA 2001]

11 A hiker of mass 80 kg completes a 2.5 hour uphill trek involving a change of height of 900 m.

(a) Calculate the average rate at which the hiker's body must produce the energy required for the change in height. Assume that the body has an overall efficiency of 20%.

(b) In addition, walking causes the hiker to use energy at an average rate of 230 W.
Estimate the total energy requirement for the whole expedition.   [OCR 2001]

12 (a) (i) (1) Define the term **work** as used in Physics.
(2) Give the unit of work in terms of the SI base units kg, m and s.
(ii) (1) Define **power**.
(2) Give the unit of power in terms of the SI base units kg, m and s.

(b) An object of mass 1.5 kg is pulled up a frictionless slope at a steady speed by an electric motor. The slope makes an angle of 25° with the horizontal, as shown in Fig 6.8.

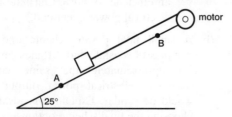

Fig. 6.8

The distance along the slope between the points **A** and **B** is 5.0 m.
(i) Calculate the work done in moving the object from **A** to **B**.

(ii) It takes 6.0 s for the object to go from **A** to **B**. Calculate the power which must be delivered by the motor.   [CCEA, 2001]

13 A cyclist is cycling with a constant velocity along a horizontal road as shown in the diagram below. The cyclist and bicycle should be regarded as a single object of total mass 70 kg throughout this question.

The arrow, labelled R, represents the direction of the resistive forces acting on the cyclist and bicycle. Draw labelled arrows, on the diagram above, to indicate the magnitude and directions of the other three forces acting on this object.

(a) The cyclist is cycling along this horizontal road with a constant velocity of 3.6 m s^{-1} producing a forward force of 4.0 N. Calculate the work done against the resistive forces each second.

(b) An advertisement in a newspaper suggests that an electric power unit (which is easy to fit) will make sure that cycling can always be fun – even uphill. The unit is labelled 120 W. The advertisement claims that a cyclist can achieve a steady velocity of 3.6 m s^{-1} up a hill of 1 in 12 with half the energy being provided by the unit. A hill of 1 in 12 means that for every 12 m along the road, the hill rises vertically 1.0 m.
Use suitable calculations to decide whether this claim is valid.

(c) It is preferable for the airflow past the cyclist to be laminar. Explain the meaning of the word laminar and state why this is preferable.   [Edexcel S-H 2001]

14

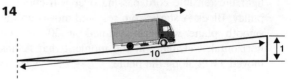

Fig. 6.9   Diagram for Question 14

A lorry of mass of mass 2.4×10^4 kg climbs a hill of incline 1 in 10, as shown in Fig 6.9, at a constant speed of 8.0 m s^{-1}. If the power of the lorry's engine is 24×10^4 W, calculate

(a) the driving force exerted by the engine

(b) the frictional resistance to the lorry's motion.

15 The diagram shows a car travelling at a constant velocity along a horizontal road.

(a) (i) Draw and label arrows on the diagram representing the forces acting on the car.
(ii) Referring to Newton's Laws of motion, explain why the car is travelling at constant velocity.

(b) The car has an effective power output of 18 kW and is travelling at a constant velocity of 10 m s^{-1}. Show that the total resistive force acting is 1800 N.

(c) The total resistive force consists of two components. One of these is a constant frictional force of 250 N and the other is the force of air resistance, which is proportional to the square of the car's speed.

Calculate
(i) the force of air resistance when the car is travelling at 10 m s^{-1},
(ii) the force of air resistance when the car is travelling at 20 m s^{-1},
(iii) the effective output power of the car required to maintain a constant speed of 20 m s^{-1} on a horizontal road.

[AQA 2001]

16 (Assume $g = 9.8$ m s^{-2} for this question)

(a) State the *Principle of Conservation of Energy*.

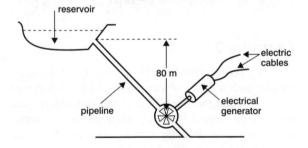

(b) In a hydroelectric scheme, water is conveyed through a long pipeline from the reservoir to the generator. In passing through the pipeline, the water descends a vertical height of 80 m. The generator produces 12 MW of power. The *overall efficiency* of the scheme is 60%.
(i) Explain what is meant by *overall efficiency*.
(ii) Show that the mass of water reaching the generator in one second is 2.55×10^4 kg
(iii) If the efficiency of the generator **alone** is 84%, calculate
(I) the power of the water reaching the generator,
(II) the speed with which the water reaches the generator.
(iv) Assuming the water is initially at rest, and there is no change in the level of the reservoir,
(I) calculate the power loss in the pipe.
(II) Hence estimate a value for the average force with which the pipe resists the flow of water. Explain your reasoning.

(c) (i) What is the efficiency of the pipe in conveying the energy from the reservoir to the generator?
(ii) Show how the efficiency of the pipeline and that of the generator are consistent with an overall efficiency of 60%.

[WJEC 2000]

7
Linear momentum

Momentum

The (linear) momentum of a body is defined by

$$\text{Momentum} = \text{Mass} \times \text{Velocity} \qquad (7.1)$$

Example 1

A body A of mass 5 kg moves to the right with a velocity of $4\,\text{m s}^{-1}$. A body of mass 3 kg moves to the left with a velocity of $8\,\text{m s}^{-1}$. Calculate (a) the momentum of A, (b) the momentum of B, (c) the total momentum of A and B.

Method

We use Equation 7.1
(a) Momentum of A = Mass × Velocity
$$= 5 \times 4 = +20\,\text{kg m s}^{-1}$$

(b) Velocity, and hence momentum, are *vector* quantities. We assumed in part (a) that motion to the right is positive in sign. So motion to the left is negative.

$$\text{Momentum of B} = \text{Mass} \times \text{Velocity} = 3 \times -8$$
$$= -24\,\text{kg m s}^{-1}$$

(c) $\begin{pmatrix}\text{Total momentum} \\ \text{of A and B}\end{pmatrix} = \begin{pmatrix}\text{Momentum} \\ \text{of A}\end{pmatrix}$
$$+ \begin{pmatrix}\text{Momentum} \\ \text{of B}\end{pmatrix}$$
$$= 20 - 24 = -4\,\text{kg m s}^{-1}$$

Answer

(a) $20\,\text{kg m s}^{-1}$,　(b) $-24\,\text{kg m s}^{-1}$,　(c) $-4\,\text{kg m s}^{-1}$

Exercise 7.1

1　A body has a mass of 2.5 kg. Calculate (a) its momentum when it has a velocity of $3.0\,\text{m s}^{-1}$, (b) its velocity when it has a momentum of $10.0\,\text{kg m s}^{-1}$.

2　An object A has mass 2 kg and moves to the left at $5\,\text{m s}^{-1}$. An object B has mass 4 kg and moves to the right at $2.5\,\text{m s}^{-1}$. Calculate (a) the momentum

of A, (b) the momentum of B, (c) the total momentum of A and B.

Conservation of momentum

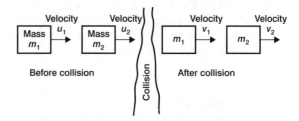

Fig. 7.1　Conservation of linear momentum

Provided that no *external* forces (such as friction) are acting, then, when bodies collide, the total momentum before collision is the same as that after collision. With reference to Fig. 7.1, this means

$$m_1u_1 + m_2u_2 = m_1v_1 + m_2v_2 \qquad (7.2)$$

Example 2

A 2.0 kg object moving with a velocity of $8.0\,\text{m s}^{-1}$ collides with a 4.0 kg object moving with a velocity of $5.0\,\text{m s}^{-1}$ along the same line. If the two objects join together on impact, calculate their common velocity when they are initially moving (a) in the same direction, (b) in opposite directions.

Method

(a) Fig. 7.2a shows the situation before and after impact. Since $v_1 = v_2 = v$, Equation 7.2 gives
$$m_1u_1 + m_2u_2 = m_1v + m_2v$$
$$= (m_1 + m_2)v$$
So
$$2 \times 8 + 4 \times 5 = 6v$$
$$\therefore \quad v = \frac{36}{6} = 6.0\,\text{m s}^{-1}$$

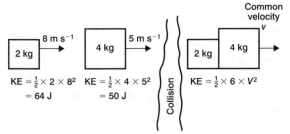

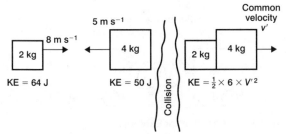

(b) Objects moving in opposite directions

Fig. 7.2   Diagram for Examples 2 and 3

(b) Fig. 7.2b illustrates the situation. As in Example 1, the 4 kg mass now has a negative velocity, so $u_2 = -5$. Hence, if v' is the common velocity,

$$2 \times 8 - 4 \times 5 = 6v'$$

$$\therefore \quad v' = -\frac{4}{6} = -0.67 \,\text{m s}^{-1}$$

Answer

(a) $6.0 \,\text{m s}^{-1}$, (b) $-0.67 \,\text{m s}^{-1}$.

Note: The negative value of v' means that the combined masses move to the left after collision. This is because the momentum of the 4 kg mass is larger than that of the 2 kg mass. Since the objects *join together* and their *two momenta are about the same*, there is a *small common velocity* after impact. During this collision a large fraction of the initial kinetic energy is converted to other forms of energy.

Exercise 7.2

1 A truck of mass 1.0 tonne moving at $4.0 \,\text{m s}^{-1}$ catches up and collides with a truck of mass 2.0 tonne moving at $3.0 \,\text{m s}^{-1}$ in the same direction. The trucks become coupled together. Calculate their common velocity.
(1 tonne = 1000 kg)

2 Repeat Question 1 but assume the trucks are moving in the same line and in opposite directions.

3 A pile-driver of mass 380 kg moving at $20 \,\text{m s}^{-1}$ hits a stationary stake of mass 20 kg. If the two move off together, calculate their common velocity.

Collisions and energy

Momentum is conserved in a collision. Total energy is also conserved but kinetic energy might not be. In general some kinetic energy will be converted to other forms (e.g. sound, work done during plastic deformation).

> An *inelastic collision* is one in which kinetic energy is not conserved.
> An *elastic collision* is one in which kinetic energy is conserved.
> A *completely inelastic collision* is one in which the objects stick together on impact.

Example 3

Calculate the KE converted to other forms during the collisions in (a) and (b) of Example 2.

Method

Refer to Figs 7.2a and b which show the kinetic energy of the various objects before and after collision.

(a) Before collision, total KE $= 64 + 50 = 114 \,\text{J}$
After collision, since $v = 6$,

$$\text{Total KE} = \tfrac{1}{2} \times 6 \times 6^2 = 108 \,\text{J}$$

$$\therefore \quad \text{KE converted} = 114 - 108 = 6 \,\text{J}$$

(b) Before collision, total KE $= 114 \,\text{J}$
After collision, since $v' = -0.67$,

$$\text{Total KE} = \tfrac{1}{2} \times 6 \times (-0.67)^2 = 1.3 \,\text{J}$$

$$\therefore \quad \text{KE converted} = 114 - 1.3 = 112.7 \,\text{J}$$

Answer

(a) 6 J, (b) 113 J.

Example 4

A 2.0 kg object moving with velocity $6.0 \,\text{m s}^{-1}$ collides with a stationary object of mass 1.0 kg. Assuming that the collision is perfectly elastic, calculate the velocity of each object after the collision.

Method

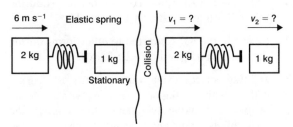

Fig. 7.3   An elastic collision (for Example 4)

Fig. 7.3 shows the situation before and after collision. We must find v_1 and v_2, the final velocities of the 2 kg and 1 kg objects respectively. This means we need two equations.

Since momentum is conserved, Equation 7.2 gives

$$m_1u_1 + m_2u_2 = m_1v_1 + m_2v_2$$

i.e. $2 \times 6 + 1 \times 0 = 2v_1 + v_2$

or $v_2 = 12 - 2v_1$ (7.3)

Kinetic energy is also conserved. So

$$\tfrac{1}{2}m_1u_1{}^2 + \tfrac{1}{2}m_2u_2{}^2 = \tfrac{1}{2}m_1v_1{}^2 + \tfrac{1}{2}m_2v_2{}^2$$

or

$$\tfrac{1}{2} \times 2 \times 6^2 + \tfrac{1}{2} \times 1 \times 0^2 = \tfrac{1}{2} \times 2 \times v_1{}^2 + \tfrac{1}{2} \times 1 \times v_1{}^2$$

$$\therefore \quad 6^2 + 0 = v_1{}^2 + \tfrac{1}{2}v_2{}^2 \qquad (7.4)$$

Note that we have two simultaneous equations (see Chapter 2), so we can substitute for v_2 from Equation 7.3 into Equation 7.4. This gives $v_1 = 2.0 \,\mathrm{m\,s}^{-1}$ and hence $v_2 = 8.0 \,\mathrm{m\,s}^{-1}$ (see Chapter 2).

Answer

The velocities are $2.0 \,\mathrm{m\,s}^{-1}$ and $8.0 \,\mathrm{m\,s}^{-1}$ in the original direction. Note the following values of KE:

Before collision:	m_1 has 36 J,	m_2 has 0 J
After collision:	m_1 has 4 J,	m_2 has 32 J

So the total KE remains unchanged at 36 J before and after collision.

Energy interchange is via the elastic spring which stores energy on compression during impact. This potential energy is converted to KE when the objects separate.

Exercise 7.3

1. Calculate the KE converted to other forms during the collision in Question 1 Exercise 7.2.

2. Calculate the KE converted to other forms during the collision in Question 2 Exercise 7.2.

3. A 2.0 kg object moving with a velocity of $8.0 \,\mathrm{m\,s}^{-1}$ collides with a 3.0 kg object moving with a velocity of $6.0 \,\mathrm{m\,s}^{-1}$ along the same direction. If the collision is completely inelastic, calculate the decrease in KE during collision.

Explosions

When an object explodes it does so as a result of some *internal* force. Thus the total momentum of the separate parts will be the same as that of the original body. This is often zero.

Example 5

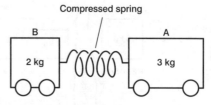

Fig. 7.4   Information for Example 5

Fig. 7.4 shows two trolleys A and B initially at rest, separated by a compressed spring. The spring is now released and the 3.0 kg trolley moves with a velocity of $1.0 \,\mathrm{m\,s}^{-1}$ to the right. Calculate (a) the velocity of the 2.0 kg trolley, (b) the total KE of the trolleys.

Neglect the mass of the spring and any friction forces.

Method

(a) Both trolleys are initially at rest so their momentum is zero. So

$$0 = m_A v_A + m_B v_B$$

where $m_A = 3.0$, $v_A = 1.0$, $m_B = 2.0$ and v_B is required. So

$$0 = 3v_A + 2v_B$$

or $v_B = -1.5 \,\mathrm{m\,s}^{-1}$

The negative sign indicates that trolley B moves to the left.

(b) Total KE is the sum of the separate KE of each trolley. So

$$\begin{aligned}
\text{Total KE} &= \tfrac{1}{2}m_A v_A{}^2 + \tfrac{1}{2}m_B v_B{}^2 \\
&= \tfrac{1}{2} \times 3 \times 1^2 + \tfrac{1}{2} \times 2 \times (-1.5)^2 \\
&= 3.75 \,\mathrm{J}
\end{aligned}$$

Note that the KE is of course positive in each case. The initial KE is zero, and the final KE comes from the potential energy stored in the compressed spring.

Answer

(a) $1.5 \,\mathrm{m\,s}^{-1}$ to the left, (b) 3.8 J.

Exercise 7.4

1 A shell of mass 1.6 kg is fired from a gun with a velocity of $250\,\text{m s}^{-1}$. Assuming that the gun is free to move, calculate its recoil velocity if it has a mass of 1000 kg.

2 A space satellite has total mass 500 kg. A portion of mass 20 kg is ejected at a velocity of $10\,\text{m s}^{-1}$. Calculate the recoil velocity of the remaining portion. Neglect the initial velocity of the satellite.

3 A radioactive nucleus of mass 235 units travelling at $400\,\text{km s}^{-1}$ disintegrates into a nucleus of mass 95 units and a nucleus of mass 140 units. If the nucleus of mass 95 units travels backwards at $200\,\text{km s}^{-1}$, what is the velocity of the nucleus of mass 140 units?

Impulse and force

If a force $F\,(\text{N})$ acts on a body of mass m (kg) for a time $t\,(\text{s})$ so that the velocity of the body changes from $u\,(\text{m s}^{-1})$ to $v\,(\text{m s}^{-1})$, then provided SI units are used:

$$F = \left(\begin{array}{c}\textbf{Rate of change}\\ \textbf{of momentum}\end{array}\right) = \frac{(mv - mu)}{t}$$

$$(7.5)$$

Rearranging Equation 7.5 gives

$$F \times t = mv - mu \qquad (7.6)$$

The product $F \times t$ is called the *impulse* of the force. It equals the change of momentum of the body.

Example 6

A stationary golf ball is hit with a club which exerts an average force of 80 N over a time of 0.025 s. Calculate (a) the change in momentum, (b) the velocity acquired by the ball if it has a mass of 0.020 kg.

Method

(a) Change of momentum = Impulse

$$= F \times t$$

$$= 80 \times 0.025$$

$$= 2.00\,\text{N s}$$

Note that the unit N s is the same as kg m s^{-1}.

(b) Change of momentum $= mv - mu$

We have $m = 0.020$, $u = 0$ and require v.

$$\therefore \quad 2.00 = 0.020 \times v - 0$$

$$\therefore \quad v = 100\,\text{m s}^{-1}$$

Answer

(a) $2.00\,\text{kg m s}^{-1}$, (b) $100\,\text{m s}^{-1}$.

Example 7

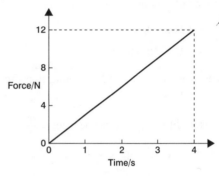

Fig. 7.5 Diagram for Example 7

Fig. 7.5 shows how the force acting on a body varies with time. The increase in momentum of the body, measured in N s, as a result of this force acting for four seconds is:

A 48   **B** 24   **C** 12   **D** 6.0   **E** 3.0

Method

From equation (7.6):

$$\text{Change of momentum} = \text{Impulse} = F \times t$$

In this case since **force F is changing with time then $F \times t$ corresponds to the area under the graph F versus t.** Since F versus t is a straight line passing through the origin then:

$$\text{area} = \tfrac{1}{2}\text{base} \times \text{height} = \tfrac{1}{2} \times 4 \times 12$$

$$= 24$$

Answer

B

Example 8

The outboard motor of a small boat has a propellor which sends back a column of water of cross-sectional area $0.030\,\text{m}^2$ at a speed of $8.0\,\text{m s}^{-1}$. Assuming the boat is held at rest calculate:

(a) the rate (in kg s^{-1}) at which water is propelled backwards

(b) the rate of change of momentum of the water (assuming it was originally at rest)

(c) the force exerted by the motor on the boat.

Assume the density of water $= 1.0 \times 10^3 \, \text{kg m}^{-3}$.

Method

(a) Volume of water sent back per second = area of cross-section × speed

$$= 0.03 \times 8.0$$
$$= 0.24 \, \text{m}^3 \, \text{s}^{-1}$$

Mass of water sent back per second = volume per second × density

$$= 0.24 \times 1.0 \times 10^3$$
$$= 0.24 \times 10^3 \, \text{kg s}^{-1}$$

(b) Using equation (7.5) and assuming that the water was originally at rest:

$$\text{Rate of change of momentum} = \frac{m}{t}(v - u)$$

$$= 0.24 \times 10^3 \times 8.0$$
$$= 1920 \, \text{kg m s}^{-2}$$

since $m/t = 0.24 \times 10^3$, $v = 8.0$ and $u = 0$.

(c) From equation (7.5):

$$\text{Force} = \text{rate of change of momentum}$$

$$= 1920 \, \text{N}$$

The force exerted by the motor, via the propellor, on the boat arises as a reaction to the force needed to change the momentum of the water.

Answer

(a) $0.24 \times 10^3 \, \text{kg s}^{-1}$, (b) $1.9 \times 10^3 \, \text{kg m s}^{-2}$,

(c) $1.9 \, \text{kN}$.

Note that the units kg m s^{-2} and N are effectively the same.

Exercise 7.5

1 A squash ball of mass $0.024 \, \text{kg}$ is hit with a racket and acquires a velocity of $10 \, \text{m s}^{-1}$. Its initial velocity is zero. If the time of contact with the racket head is $0.040 \, \text{s}$, calculate the average force exerted on the ball.

2 A machine gun fires bullets at a rate of 360 per minute. The bullets have a mass of $20 \, \text{g}$ and a speed of $500 \, \text{m s}^{-1}$. Calculate the average force exerted by the gun on the person holding it.

3

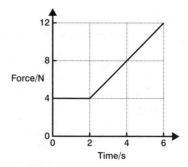

Fig. 7.6 Graph for Question 3

Fig. 7.6 shows how the force acting on a body changes with time. Calculate the change in momentum of the body.

Exercise 7.6: Examination questions

1 A bullet of mass $15 \, \text{g}$ is fired horizontally from a gun with a velocity of $250 \, \text{m s}^{-1}$. It hits, and becomes embedded in, a block of wood of mass $3000 \, \text{g}$, which is freely suspended by long strings, as shown in Fig. 7.7. Air resistance is to be neglected.

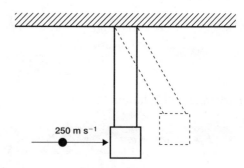

Fig. 7.7

(i) Calculate the magnitude of the momentum of the bullet as it leaves the gun.
(ii) Calculate the magnitude of the initial velocity of the wooden block and bullet after impact.
(iii) Use your answer to (ii) to calculate the kinetic energy of the wooden block and embedded bullet immediately after the impact.
(iv) Hence calculate the maximum height above the equilibrium position to which the wooden block, with the embedded bullet, rises after impact (assume $g = 10 \, \text{m s}^{-2}$).

[CCEA 2000, part]

2

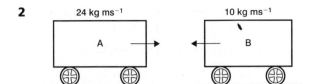

Fig. 7.8   Diagram for Question 2

Two trucks, A and B, are about to collide head on; their values of linear momentum are as shown in Fig. 7.8. After the collision the two trucks separate and move away from each other, at which time truck A has a linear momentum of $8.0\,\text{kg m s}^{-1}$.

Calculate:

(a) the original combined momentum of the trucks

(b) the momentum of truck B after collision and state its direction of travel.

3

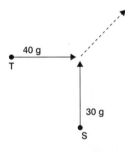

Fig. 7.9   Diagram for Question 3

Two particles, S of mass 30 g and T of mass 40 g, both travel at a speed of $35\,\text{m s}^{-1}$ in directions at right angles as shown in Fig. 7.9. The two particles collide and stick together. Calculate their speed after impact.

4   A train of mass 5.0×10^5 kg accelerates uniformly from rest on a straight horizontal track to a speed of $20\,\text{m s}^{-1}$ in 45 s.

(a) Calculate the force causing this acceleration.

(b) During a subsequent shunting operation, the train, travelling at $0.50\,\text{m s}^{-1}$, collides with a stationary train of mass 2.0×10^5 kg. Immediately after the collision, the two trains move together as a single unit. Forces, other than those generated by the impact, can be neglected. Calculate:
   (i) the speed of the combined trains after the impact;
   (ii) the kinetic energy lost in the collision.
   [OCR 2000]

5   A supermarket trolley of mass 10 kg travels at $2\,\text{m s}^{-1}$ towards a stationary trolley of mass 20 kg. The two trolleys collide, link and move off together.

(a) Which one of **A** to **D** below is the total momentum of the two trolleys, in kg m s^{-1}, after they have linked?

A 20   **B** 30   **C** 40   **D** 60

(b) Which one of the statements, **A** to **D**, about the total kinetic energy of the two trolleys immediately after the collision is correct.

A The total kinetic energy is zero.
B The total kinetic energy is greater than zero but less than 20 J.
C The total kinetic energy is exactly 20 J.
D The total kinetic energy is greater than 20 J.
   [OCR Nuff 2001]

6   (a) State the principle of *conservation of momentum* and the principle of *conservation of energy*. Give one example of the use of each principle.

(b) A moving ball of mass M and speed v collides head-on with a stationary ball of different mass.
   (i) After the collision, the first ball is stationary and 10% of the kinetic energy is lost. Show that the mass of the second ball is $10\,M/9$.
   (ii) In another collision between the two balls from the same starting conditions, no kinetic energy is lost. Determine the final velocities of the balls.

(c) A rubber ball is dropped on to flat ground from a height of 2.0 m. Calculate how long it takes for the ball to first hit the ground. The ball loses 10% of its kinetic energy at each bounce. Calculate the time taken for the ball to come to rest. Ignore air resistance (assume $g = 10\,\text{m s}^{-2}$).
   [Hint: $1 + x + x^2 + x^3 \ldots \ldots = 1/(1-x)$]
   [OCR Spec 2000]

7   (a)   (i) State the principle of conservation of momentum.
   (ii) Explain briefly how an elastic collision is different from an inelastic collision.

(b) Describe and explain what happens when a moving particle collides elastically with a stationary particle of equal mass.

(c) Figure 7.10 shows an astronaut undertaking a space-walk. The astronaut is tethered by a rope to a spacecraft of mass 4.0×10^4 kg. The spacecraft is moving at constant velocity.

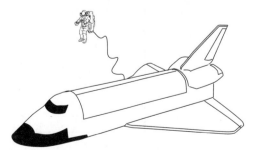

Fig. 7.10

The astronaut and spacesuit have a total mass of 130 kg. The change in velocity of the astronaut after pushing off is $1.80\,\mathrm{m\,s^{-1}}$.

(i) Determine the velocity change of the spacecraft.

(ii) The astronaut pushes for 0.60 s in achieving this speed. Calculate the average power developed by the astronaut. Neglect the change in motion of the spacecraft.

(iii) The rope eventually becomes taut. Suggest what would happen next.

[AQA 2000]

8 A stationary Uranium nucleus of mass 238 units decays into a Thorium nucleus of mass 234 units and an alpha particle of mass 4 units with speed $14 \times 10^6\,\mathrm{m\,s^{-1}}$. Calculate the recoil speed of the Thorium nucleus.

9 A stationary atomic nucleus disintegrates into an α-particle of mass 4 units and a daughter nucleus of mass 234 units. Calculate the ratio

$$\frac{\text{KE of } \alpha\text{-particle}}{\text{KE of daughter nucleus}}$$

10 (a) Collisions can be described as *elastic* or *inelastic*. State what is meant by an inelastic collision.

(b) A ball of mass 0.12 kg strikes a stationary cricket bat with a speed of $18\,\mathrm{m\,s^{-1}}$. The ball is in contact with the bat for 0.14 s and returns along its original path with a speed of $15\,\mathrm{m\,s^{-1}}$.

Calculate

(i) the momentum of the ball before the collision,

(ii) the momentum of the ball after the collision,

(iii) the total change of momentum of the ball,

(iv) the average force acting on the ball during contact with the bat,

(v) the kinetic energy lost by the ball as a result of the collision.   [AQA 2001]

11

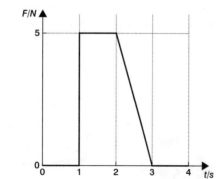

Fig. 7.11   Graph for Question 11

The graph of Fig. 7.11 shows the variation of force F acting on a body over a time t.

Calculate the change in momentum of the body

(a) after 2 s   (b) after 4 s

12 A tennis ball, moving horizontally at a high speed, strikes a vertical wall and rebounds from it.

(a) Describe the energy transfers which occur during the impact of the ball with the wall.

(b) The graph shows how the horizontal push of the wall on the tennis ball varies during the impact.

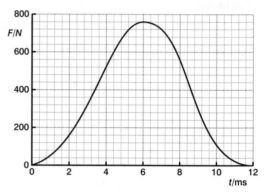

(i) What is represented by the area under the graph?

(ii) Estimate the value of this area and hence deduce the change of velocity of a tennis ball of mass 57.5 g which makes such an impact.

(iii) If the kinetic energy of the tennis ball is unchanged by this impact, with what speed did it strike the wall?

[Edexcel 2001]

13 (a) (i) What is the relationship between *force* and *momentum* as expressed by Newton's second law?

(ii) State Newton's third law.

(b) An astronaut uses a gas-gun to move around in space. The gun fires gas from a nozzle of area $150\,\mathrm{mm}^2$ at a speed of $210\,\mathrm{m\,s^{-1}}$. The density of the gas is $0.850\,\mathrm{kg\,m^{-3}}$ and the mass of the astronaut and associated equipment is 160 kg.

Calculate

(i) the mass of the gas leaving the gun in one second,

(ii) the initial acceleration of the astronaut, i.e. when starting from rest.

[WJEC 2000]

14 (a) A bullet of mass 5.0 g takes 2.0 ms to accelerate uniformly from rest along the 0.60 m length of a rifle barrel.

(i) Calculate the speed with which the bullet leaves the barrel.

(ii) The rifle recoils against the shoulder of the person firing it. Calculate the magnitude of the recoil force.

(b) A jet of water is directed at a vertical, rigid wall with a horizontal velocity of $15\,\mathrm{m\,s^{-1}}$. The cross-sectional area of the jet is $600\,\mathrm{mm^2}$. After the jet strikes the wall, the motion of the water is parallel to the wall. Calculate the magnitude of the force on the wall due to the jet.

Assume density of water $= 1000\,\mathrm{kg\,m^{-3}}$.

[CCEA 2001]

15 A ship is powered by a water jet propulsion unit, driven by a diesel engine. When the ship is stationary and the engine is running at full power the unit takes in water and expels it as a jet of cross-sectional area $0.30\,\mathrm{m^2}$ at a speed v.

Take the density of water to be $1050\,\mathrm{kg\,m^{-3}}$.

(a) Write down an expression for the mass of water flowing in the jet in one second.

(b) The kinetic energy given to the jet in one second is $1.5 \times 10^6\,\mathrm{J}$. Calculate:
(i) the magnitude of v;
(ii) the momentum gained by the water in the jet in one second.
(iii) State the magnitude of the thrust exerted by the jet on the ship.

(c) State **two** reasons why the output power of the diesel engine must be greater than $1.5 \times 10^6\,\mathrm{W}$. [OCR 2001]

16 (a) Express the SI unit of power in terms of the base units kg, m and s.

(b) The diameter of the rotor of a wind turbine is $36\,\mathrm{m}$. The rotor rotates about a horizontal axis, as shown in Fig. 7.12

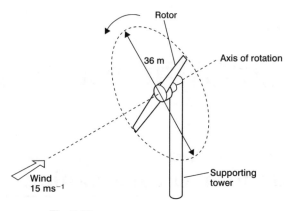

Fig. 7.12

The axis points directly into a wind which is blowing at $15\,\mathrm{m\,s^{-1}}$. Assume that the air emerges from the rotor at a mean axial speed of $13\,\mathrm{m\,s^{-1}}$. Take the density of air to be $1.2\,\mathrm{kg\,m^{-3}}$.

Show that:

(i) the mass of air incident in one second on the circle swept by the rotor is $1.83 \times 10^4\,\mathrm{kg}$;
(ii) the kinetic energy lost by this air is $5.1 \times 10^5\,\mathrm{J}$.

(c) Calculate the horizontal force exerted by the air on the rotor in a direction parallel to its axis of rotation.

(d) Suggest why the supporting tower for the wind turbine must be very rigid.

(e) The turbine converts the kinetic energy lost by the air into electrical energy with an efficiency of 40%. Calculate how many such turbines would be needed to provide the output of a conventional $500\,\mathrm{MW}$ power station.

[OCR 2001]

8
Circular motion

Uniform circular motion

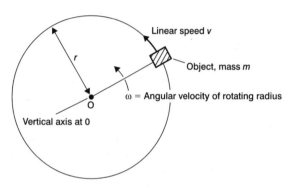

Fig. 8.1   Object moving in uniform circular motion

Fig. 8.1 shows a body moving with *uniform speed* at a *fixed distance* from a *fixed axis*. It is in uniform circular motion.

The body has a constant angular velocity ω defined by:

$$\omega = \frac{\text{angular displacement (radians)}}{\text{time taken (s)}}$$

Linear and angular motion

In Fig. 8.1 an object moves with uniform speed v (m s^{-1}) around the circumference of a circle, centre O. The rotating radius, of length r (m), has angular velocity ω (rad s^{-1}) such that

$$v = r\omega \qquad (8.1)$$

If T is the time for one revolution then, since time = distance ÷ speed:

$$T = \frac{2\pi r}{v} = \frac{2\pi}{\omega} \qquad (8.2)$$

since $v = r\omega$.

Example 1

A pulley wheel rotates at 300 rev min^{-1}. Calculate (a) its angular velocity in rad s^{-1}, (b) the linear speed of a point on the rim if the pulley has a radius of 150 mm, (c) the time for one revolution.

Method

(a) $300 \text{ rev min}^{-1} = \dfrac{300}{60} \text{ rev s}^{-1} = 5.00 \text{ rev s}^{-1} (\text{or Hz})$

The frequency f of rotation is thus 5.00 Hz. Now in one revolution the radius rotates through 2π rad. Thus the angular velocity ω of the rotating radius is given by

$$\omega = 2\pi f = 2\pi \times 5 = 10\pi \text{ rad s}^{-1}$$

(b) Use Equation 8.1, in which $\omega = 10\pi$ and $r = 150 \text{ mm} = 0.150 \text{ m}$. Thus

$$v = r\omega = 0.15 \times 10\pi$$

$$= 1.50\pi \text{ m s}^{-1}$$

(c) Since $f = 5.00 \text{ Hz}$ then each revolution takes $\dfrac{1}{5.00} = 0.200 \text{ s}$.

$$\therefore \quad T = 0.200 \text{ s}.$$

Note we could have used Equation 8.2.

Answer

(a) 31.4 rad s^{-1}, (b) 4.71 m s^{-1}, (c) 0.200 s.

Exercise 8.1

1   The turntable on a record player rotates at 45 rev min^{-1}. Calculate (a) its angular velocity in rad s^{-1}, (b) the linear speed of a point 14 cm from the centre, (c) the time for one revolution.

2   A car moves round a circular track of radius 1.0 km at a constant speed of 120 km h^{-1}. Calculate its angular velocity in rad s^{-1}.

Centripetal acceleration and force

The object in Fig. 8.1 has uniform speed, but its velocity is constantly changing, since its direction is changing. It is constantly accelerating *towards the centre* O, with magnitude a (m s^{-2}) given by

$$a = r\omega^2 = \frac{v^2}{r} \qquad (8.3)$$

A net inward force is needed to provide this acceleration. For a body of mass m the magnitude of the 'centripetal' force F is given by

$$F = mr\omega^2 = m\frac{v^2}{r} \qquad (8.4)$$

This force can be provided, for example, by the tension in a string, by gravitational or electrostatic attraction, or by friction.

Example 2

An object of mass 0.30 kg is attached to the end of a string and is supported on a smooth horizontal surface. The object moves in a horizontal circle of radius 0.50 m with a constant speed of 2.0 m s^{-1}. Calculate (a) the centripetal acceleration, (b) the tension in the string.

Method

(a) Use Equation 8.3 with $v = 2.0$ and $r = 0.50$. The centripetal acceleration a is given by

$$a = \frac{v^2}{r} = \frac{2^2}{0.5} = 8.0 \text{ m s}^{-2}$$

(b) Use Equation 8.4 with $m = 0.30$, $v = 2.0$ and $r = 0.50$. So

$$F = m\frac{v^2}{r} = \frac{0.3 \times 2^2}{0.5} = 2.4 \text{ N}$$

This force is provided by the tension in the string.

Answer

(a) 8.0 m s^{-2}, (b) 2.4 N.

Example 3

An object of mass 4.0 kg is whirled round in a vertical circle of radius 2.0 m with a speed of 5.0 m s^{-1}. Calculate the maximum and minimum tension in the string connecting the object to the centre of the circle. Assume acceleration due to gravity $g = 10$ m s^{-2}.

Method

Use Equation 8.4 with $m = 4.0$, $v = 5.0$ and $r = 2.0$. Thus the centripetal force F is given by

$$F = m\frac{v^2}{r} = \frac{4 \times 5^2}{2} = 50 \text{ N}$$

Thus a net inward force of 50 N must act on the body during its rotation. In Fig. 8.2a the body is at the bottom of the vertical circle. So

$$T_1 - mg = 50$$

$$\therefore \quad T_1 = 50 + mg = 50 + 40 = 90 \text{ N}$$

This is the maximum tension in the string.

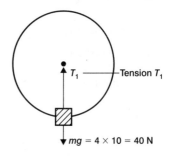

(a) Body at bottom of circle

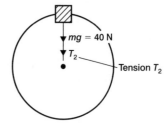

(b) Body at top of circle

Fig. 8.2   Forces acting on a body moving in a vertical circle

At the top of the vertical circle, in Fig. 8.2b,

$$T_2 + mg = 50$$

$$\therefore \quad T_2 = 50 - mg = 50 - 40$$

$$= 10 \text{ N}$$

This is the minimum tension in the string.

Answer

Maximum tension = 90 N,
Minimum tension = 10 N.

Example 4

A car travels over a humpback bridge of radius of curvature 45 m. Calculate the maximum speed of the car if its road wheels are to stay in contact with the bridge. Assume $g = 10$ m s^{-2}.

Method

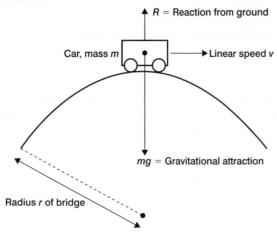

R = Reaction from ground

Car, mass m

Linear speed v

mg = Gravitational attraction

Radius r of bridge

Fig. 8.3   Forces acting on a car

Fig. 8.3 shows the forces acting on the car when its wheels are in contact with the bridge. *A net inward force equal to mv^2/r must always exist.* So

$$mg - R = m\frac{v^2}{r}$$

As v increases, so R must decrease, since mg is constant. In the limiting case, when the wheels are just about to leave the ground, $R = 0$, so

$$mg = m\frac{v^2}{r}$$

The mass m cancels out and is not required. So maximum speed v is given by

$$v^2 = rg$$

We have $r = 45$ and $g = 10$, so

$$v = \sqrt{rg} = \sqrt{450} = 21.2$$

Answer

The maximum speed is $21\,\text{m s}^{-1}$.

Exercise 8.2

1   A car of mass 1.0×10^3 kg is moving at $30\,\text{m s}^{-1}$ around a bend of radius 0.60 km on a horizontal track. What centripetal force is required to keep the car moving around the bend, and where does this force come from?

2   An object of mass 6.0 kg is whirled round in a vertical circle of radius 2.0 m with a speed of $8.0\,\text{m s}^{-1}$. Calculate the maximum and minimum tension in the string connecting the object to the centre of the circle.

    If the string breaks when the tension in it exceeds 360 N, calculate the maximum speed of rotation,

in m s^{-1}, and state where the object will be when the string breaks.
Assume $g = 10\,\text{m s}^{-2}$.

3   A car travels over a humpback bridge at a speed of $30\,\text{m s}^{-1}$. Calculate the minimum radius of the bridge if the car road wheels are to remain in contact with the bridge. What happens if the radius is less than the limiting value? Assume $g = 10\,\text{m s}^{-2}$.

The conical pendulum

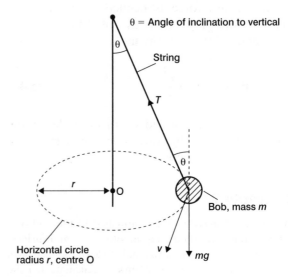

θ = Angle of inclination to vertical

String

T

r

O

Bob, mass m

v   mg

Horizontal circle radius r, centre O

Fig. 8.4   The conical pendulum

Fig. 8.4 shows the forces acting on a conical pendulum in which the bob sweeps out a horizontal circle, centre O and radius r, with linear speed v. Resolving forces on the mass m gives

(vertically) $T \cos\theta = mg$	**(8.5)**

(horizontally) $T \sin\theta = m\dfrac{v^2}{r}$	**(8.6)**

Example 5

A conical pendulum consists of a small bob of mass 0.20 kg attached to an inextensible string of length 0.80 m. The bob rotates in a horizontal circle of radius 0.40 m, of which the centre is vertically below the point of suspension. Calculate (a) the linear speed of the bob in m s^{-1}, (b) the period of rotation of the bob, (c) the tension in the string. Assume $g = 10\,\text{m s}^{-2}$.

Method

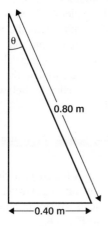

Fig. 8.5 Diagram for Example 5

We are given $m = 0.20$, $r = 0.40$, $g = 10$. Also we are given θ since, from Fig. 8.5,

$$\sin \theta = \frac{0.40}{0.80} = 0.50$$

$$\therefore \qquad \theta = 30°$$

(a) To find v divide Equation 8.6 by 8.5 to give

$$\tan \theta = \frac{v^2}{rg}$$

$$\therefore \qquad v^2 = rg \tan 30° = 0.4 \times 10 \times 0.577$$

$$\therefore \qquad v = 1.52 \, \text{m s}^{-1}$$

(b) $\qquad$ Periodic time $T = \dfrac{\text{Circumference of circle}}{\text{Linear speed}}$

$$\therefore \qquad T = \frac{2\pi r}{v} = \frac{2 \times 3.14 \times 0.4}{1.52}$$

$$= 1.65 \, \text{s}$$

(c) Rearranging Equation 8.5 gives

$$T = \frac{mg}{\cos \theta} = \frac{0.2 \times 10}{\cos 30°}$$

$$= 2.31 \, \text{N}$$

Answer

(a) $1.5 \, \text{m s}^{-1}$, (b) $1.7 \, \text{s}$, (c) $2.3 \, \text{N}$.

Exercise 8.3

1 A conical pendulum consists of a bob of mass 0.50 kg attached to a string of length 1.0 m. The bob rotates in a horizontal circle such that the angle the string makes with the vertical is 30°. Calculate (a) the period of the motion, (b) the tension in the string. Assume $g = 10 \, \text{m s}^{-2}$.

Exercise 8.4: Examination questions

(Assume $g = 10 \, \text{m s}^{-2}$ except where stated.)

1 The Earth rotates about a vertical axis every 8.6×10^4 s. For a body on the equator calculate:

(a) its angular velocity

(b) its linear speed

(c) its acceleration due to the rotation of the earth's axis.

Assume the Earth has radius 6.4×10^6 m.

2 (a) A body is attached to a piece of string and whirled in a horizontal circle of radius r at a constant angular velocity ω.
   (i) **1.** Define **angular velocity**.
      2. State the SI unit of angular velocity.
   (ii) Write down the equation relating the linear speed v of the body and its angular velocity.

(b) A fan turns at 900 revolutions per minute.
   (i) Find the angular velocity at any point on one of the fan blades. Give your answer in terms of the SI unit you quoted in (a) (i) **2**.
   (ii) The distance from the axis of rotation of the fan to the tip of one of the blades is 20 cm. Find the linear speed of the tip.
      [CCEA 2000]

3 An aircraft flies with its wings tilted as shown in Fig. 8.6 in order to fly in a horizontal circle of radius r. The aircraft has mass 4.00×10^4 kg and has a constant speed of $250 \, \text{m s}^{-1}$.

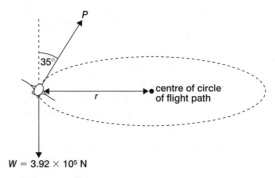

$W = 3.92 \times 10^5$ N

Fig. 8.6

With the aircraft flying in this way, two forces acting on the aircraft in the vertical plane are the force P acting at an angle of 35° to the vertical and the weight W.

(a) State the vertical component of P for horizontal flight.

(b) Calculate P.

(c) Calculate the horizontal component of P.

(d) Use Newton's second law to determine the acceleration of the aircraft towards the centre of the circle.

(e) Calculate the radius r of the path of the aircraft's flight. [OCR 2000]

4

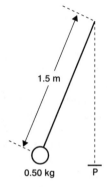

Fig. 8.7   Diagram for Question 4

The diagram shows a simple pendulum with a length of 1.5 m and a bob of mass 0.50 kg. When it passes through the lowest point P it has a speed of $2.0 \, \text{m s}^{-1}$. Calculate the tension in the string as the bob passes through point P.

5 A simple pendulum is of length 0.5 m and the bob has mass 0.25 kg. Find the greatest value for the tension in the string when the pendulum is set in oscillation by drawing the bob to one side through an angle of 5.0° and releasing from rest. Explain where in the cycle the tension is greatest. [WJEC spec 2000]

6 (a) A girl of mass 30 kg sits at the edge of a roundabout (merry-go-round) of radius 2.0 m. A boy turns the roundabout by gripping its edge and running round so that a point on the edge moves with a steady speed of $2.5 \, \text{m s}^{-1}$.
   (i) Calculate the angular velocity of the roundabout.
   (ii) Calculate the magnitude of the minimum force required to prevent the girl from sliding off the roundabout.
   (iii) The maximum centripetal force that the girl can provide is 180 N. Trying to make the girl slide off, the boy runs faster. At what speed must he make a point on the edge of the roundabout move in order to make the girl slide off?

(b) A mass of 2.0 kg, attached to a string, is whirled in a vertical circle of radius 0.40 m at

a constant angular velocity. The magnitude of the angular velocity is such that the string **just** remains taut when the mass is vertically above the centre of rotation.
   (i) Calculate the angular velocity of the mass.
   (ii) Find the tension in the string when the mass is vertically below the centre of rotation. [CCEA 2001]

7 A metal sphere of mass M is attached to one end of a light inextensible string.

(a) The sphere is whirled in a circle in a **vertical** plane at constant angular velocity. The radius of the circle is 400 mm. The arrangement is illustrated in Fig. 8.8.

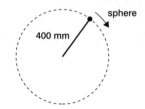

Fig. 8.8

During the rotation of the sphere, the tension T in the string varies with time t as shown in Fig. 8.9.

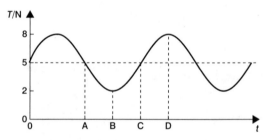

Fig. 8.9

On Fig. 8.9 **A**, **B C** and **D** are instants of time corresponding to certain points on the graph of T against t.
   (i) On Fig. 8.8, mark the positions of the sphere corresponding to each of the instants **A**, **B C** and **D**. Label these points **a**, **b**, **c** and **d** respectively.
   (ii) Use the data above and information from Fig. 8.9 to show that the mass of the sphere is 0.30 kg. Take $g = 10 \, \text{m s}^{-2}$.
   (iii) Calculate the linear speed of the sphere as it moves round the circular path.
   (iv) Calculate the angular velocity of the string.

(b) The sphere is now whirled in a circle in a **horizontal** plane. The length L of the string is gradually increased, but the linear speed of the sphere is kept constant. On a copy of

Fig. 8.10, sketch a graph to show the variation of the tension T in the string with its length L.

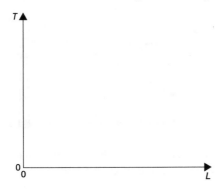

Fig. 8.10

[CCEA 2000]

8 (a) A rally car crosses a straight hump-backed bridge of vertical radius of curvature 60 m. Calculate the maximum speed of the car if the car is to remain in contact with the road while it is crossing the bridge.

(b) Later, the car travels along a banked curve on a horizontal road.
Explain, without calculation:
 (i) why banking the road helps the car to travel round the curve;
 (ii) why there is a certain speed at which the car experiences no sideways frictional force in a plane parallel to the road surface. [OCR 2000]

9 A car of mass 1000 kg travels over a humpback bridge of radius of curvature 50 m at a constant speed of 15 m s^{-1}. Calculate the magnitude and direction of the force exerted by the car on the road when it is at the top of the bridge. Assume $g = 10$ m s^{-2}.

10 (a) What is a **centripetal force**? Describe and explain one example where such a force exists.

(b) A motor car travels with uniform speed along a straight, level road. The diameter of each wheel of the car is 560 mm, and the angular velocity of the wheel about the axle is 59.6 rad s^{-1}.
 (i) What is the angular velocity of a point on the wheel midway between the axle and the outer edge of the tyre?
 (ii) Show that the speed of the car is about 60 kilometres per hour.

(c) As the car in (b) proceeds at its constant speed of 60 kilometres per hour, it passes over a hump-backed bridge. The bridge may be considered to be the arc of a circle in a vertical plane. The car travels over the bridge, just without losing contact with the road.
 (i) Calculate the radius of curvature of the bridge.
 (ii) If the car were travelling with a speed slightly greater than 60 kilometres per hour, describe and explain qualitatively what would happen to the car as it crosses the bridge.

(d) A three-bladed fan rotates at a constant angular speed. One of the blades of the fan has a distinguishing mark. The fan is illuminated using a stroboscope, which gives short pulses of bright light at regular intervals. The flashing frequency can be varied. The flashing frequency is reduced from a high value to a value at which the fan appears stationary for the first time, and the mark on the blade is visible. This occurs at a flashing frequency of 50 flashes per second. The radius of each blade of the fan is 150 mm. Calculate
 (i) the rate of rotation of the fan in revolutions per minute,
 (ii) the angular speed of a fan blade in radians per second,
 (iii) the instantaneous speed of the tip of a fan blade in metres per second.

(e) A metal sphere **M** of mass 1.35 kg is suspended from a rigid support by a light string of length 1.50 m. The sphere is made to move with uniform speed in a horizontal circle of radius 0.90 m, as shown in Fig. 8.11.

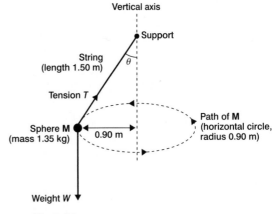

Fig. 8.11

The tension in the string is T and the weight of the sphere is W. The angle between the string and the vertical is θ.
 (i) Write down expressions for the vertical and horizontal components of the tension.

(ii) One of the components in (i) effectively supports the weight of the sphere, and the other provides the centripetal force to move it in a horizontal circle. Identify the component responsible for supporting the weight of the sphere. Hence find the magnitude of the tension in the string.

(iii) Calculate the linear speed of the sphere as it moves in the horizontal plane.

(iv) Calculate the time required for the sphere to make one complete revolution of its horizontal motion.   [CCEA 2001]

11 One of the rides at a theme park has a number of chairs each suspended from a pair of chains from the edge of a framework. The framework revolves so that the chairs swing outwards as they move round in circles (see diagram in next column).

The framework has radius 4.0 m. The chains are 5.0 m long.

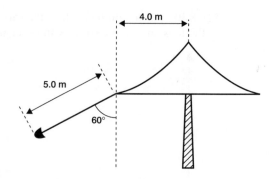

For safety, the angle θ of the chains with the vertical must not go above 60°.

The diagram above shows a chair swung outwards as the canopy revolves at the maximum safe rate. On the diagram draw the forces acting on the chair. Hence find a value for the maximum rate of rotation (angular velocity) of the framework. Show your reasoning clearly.   [Edexcel S-H 2000]

9
Gravitation

Gravitational force

Bodies attract each other solely as a result of the matter they contain. The gravitational force F (N) between two particles m_1 (kg) and m_2 (kg) placed distance r (m) apart is given by

$$F = \frac{Gm_1m_2}{r^2} \qquad (9.1)$$

where G is the universal gravitational constant and has value $6.7 \times 10^{-11}\,\mathrm{N\,m^2\,kg^{-2}}$.

Example 1

Calculate the gravitational attraction force between bodies of mass $3.0\,\mathrm{kg}$ and $2.0\,\mathrm{kg}$ placed with their centres $50\,\mathrm{cm}$ apart.

Method

We assume that the bodies are uniform spheres so they act, for this purpose, as if they are point masses (particles) located at their centres. We have $G = 6.7 \times 10^{-11}$, $m_1 = 3.0$, $m_2 = 2.0$ and $r = 0.50$. From equation 9.1

$$F = \frac{Gm_1m_2}{r^2} = \frac{6.7 \times 10^{-11} \times 3 \times 2}{(0.5)^2}$$

$$= 1.6 \times 10^{-9}\,\mathrm{N}$$

This is a very small force. To get an appreciable force one or both of the objects must be very large. Our weight* is the result of the gravitational attraction force from the Earth.

Answer

$1.6 \times 10^{-9}\,\mathrm{N}$.

Example 2

Two 'particles' of mass $0.20\,\mathrm{kg}$ and $0.30\,\mathrm{kg}$ are placed $0.15\,\mathrm{m}$ apart. A third particle of mass $0.050\,\mathrm{kg}$ is placed between them on the line joining the first two particles. Calculate (a) the gravitational force acting on the third

*We neglect effects due to the Earth's rotation which make a difference of about 0.3%.

particle if it is placed $0.050\,\mathrm{m}$ from the $0.30\,\mathrm{kg}$ mass and (b) where along the line it should be placed for no gravitational force to be exerted on it.

Method

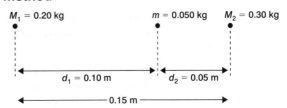

Fig. 9.1   Solution to Example 2

Refer to Fig. 9.1

(a) Both masses M_1 and M_2 attract m. Using Equation 9.1, we have for mass M_1 an attractive force F_1 (towards M_1) given by

$$F_1 = \frac{GM_1m}{d_1^{\,2}} = \frac{6.7 \times 10^{-11} \times 0.2 \times 0.05}{(0.1)^2}$$

$$= 6.7 \times 10^{-11}\,\mathrm{N}$$

For mass M_2 an attractive force F_2 (towards M_2) exists given by

$$F_2 = \frac{GM_2m}{d_2^{\,2}} = \frac{6.7 \times 10^{-11} \times 0.3 \times 0.05}{(0.05)^2}$$

$$= 40.2 \times 10^{-11}\,\mathrm{N}$$

Thus the net force F (towards M_2) is given by

$$F = F_2 - F_1 = 33.5 \times 10^{-11}\,\mathrm{N}$$

(b) Suppose mass m is x from M_1 and thus $(0.15 - x)$ from M_2. Then

$$F_1 = \frac{GM_1m}{x^2} \quad \text{and} \quad F_2 = \frac{GM_2m}{(0.15 - x)^2}$$

For no gravitational force to act on mass m, $F_1 = F_2$. Thus

$$\frac{GM_1m}{x^2} = \frac{GM_2m}{(0.15 - x)^2} \qquad (9.2)$$

Note that G and m cancel out, so that d is independent of m. Substituting $M_1 = 0.2$ and $M_2 = 0.3$ into Equation 9.2 gives

$$\frac{0.2}{x^2} = \frac{0.3}{(0.15 - x)^2}$$

Taking square roots and cross multiplying gives

$$\sqrt{2} \times (0.15 - x) = \sqrt{3} \times x$$

This gives $x = 0.067\,\text{m}$.

Answer

(a) $34 \times 10^{-11}\,\text{N}$, (b) $0.067\,\text{m}$ from M_1 (0.20 kg).

Exercise 9.1

(Assume $G = 6.7 \times 10^{-11}\,\text{N}\,\text{m}^2\,\text{kg}^{-2}$.)

1 Calculate the gravitational attraction force between two 'particles', each of mass 20 kg, placed 1.0 m apart.

2 Consider the Earth as a uniform sphere of radius 6.4×10^6 m and mass 6.0×10^{24} kg. Find the gravitational force on a mass of 5.0 kg placed on the surface of the Earth. (Assume the Earth can be replaced by a point mass acting at its centre.) Compare this with the weight of a 5.0 kg mass on Earth.

3 Two small spheres of mass 4.0 kg and M kg are placed 80 cm apart. If the gravitational force is zero at a point 20 cm from the 4 kg mass along the line between the two masses, calculate the value of M.

4 The mass of the Earth is 6.0×10^{24} kg and that of the moon is 7.4×10^{22} kg. If the distance between their centres is 3.8×10^8 m, calculate at what point on the line joining their centres is no gravitational force. Neglect the effect of other planets and the sun.

Gravitational field strength

The gravitational field strength g $(\text{N}\,\text{kg}^{-1})$ is defined as the gravitational force acting on unit mass placed at the point in question. It equals the acceleration due to gravity g $(\text{m}\,\text{s}^{-2})$ at this point.

Example 3

Assuming that the Earth is a uniform sphere of radius 6.4×10^6 m and mass 6.0×10^{24} kg, find the gravitational field strength g at a point (a) on the surface, (b) at height 0.50 times its radius above the Earth's surface.

Method

(a) We assume that the Earth can be replaced by a point mass acting at its centre. Then in Equation 9.1, $F = g$ if $m_1 = 1$. If M is the mass of the planet,

$$g = \frac{GM}{r^2} \qquad (9.3)$$

This is a general expression.
We have $G = 6.7 \times 10^{-11}$, $M = 6.0 \times 10^{24}$ and $r = 6.4 \times 10^6$.
Substituting in Equation 9.3 gives $g = 9.8\,\text{N}\,\text{kg}^{-1}$.
Note: this equals the acceleration due to gravity at the Earth's surface.

(b) We now have distance $r_1 = 1.5r$. Equation 9.3 tells us $g \propto 1/(\text{distance})^2$. If g_1 is the new value, then

$$\frac{g_1}{g} = \frac{r^2}{r_1^2} = \frac{r^2}{(1.5r)^2} = 0.444$$

$$g_1 = 0.444g = 4.36\,\text{N}\,\text{kg}^{-1}$$

Answer

(a) $9.8\,\text{N}\,\text{kg}^{-1}$, (b) $4.4\,\text{N}\,\text{kg}^{-1}$.

Example 4

The acceleration due to gravity at the Earth's surface is $9.8\,\text{m}\,\text{s}^{-2}$. Calculate the acceleration due to gravity on a planet which has (a) the same mass and twice the density, (b) the same density and twice the radius.

Method

Acceleration due to gravity equals the gravitational field strength g. Equation 9.3 tells us that g depends on mass M and radius r of the planet.

(a) In this case the radius r_1 of the planet differs from Earth radius r. Let the density of Earth be ρ and of the planet be 2ρ. Since both have the same mass M,

$$M = \underbrace{\tfrac{4}{3}\pi r^3 \rho}_{\text{for Earth}} = \underbrace{\tfrac{4}{3}\pi r_1^3 \times 2\rho}_{\text{for planet}}$$

or $\quad r^3 = 2r_1^3 \quad$ giving $\quad \dfrac{r}{r_1} = 2^{1/3}$

From Equation 9.3 we see that $g \propto 1/r^2$ for two planets of the same mass. So, if g_1 is the gravitational field strength on the planet,

$$\frac{g_1}{g} = \frac{r^2}{r_1^2} = (2)^{2/3}$$

since $r = (2)^{1/3} r_1$. As $g = 9.8$,

$$g_1 = (2)^{2/3} \times 9.8 = 15.6$$

(b) The new planet has radius $2r$. Let its mass be M_2. It has density ρ, therefore

$$M_2 = \tfrac{4}{3}\pi(2r)^2\rho = 8M$$

since $M = \tfrac{4}{3}\pi r^3 \rho$. From Equation 9.3 we see that $g \propto M/r^2$. If g_2 is the gravitational field strength on the planet,

$$\frac{g_2}{g} = \frac{M_2}{(2r)^2} \div \frac{M}{r^2} = 2$$

since $M_2 = 8M$. Thus $g_2 = 2g = 19.6$.

Answer

(a) $15.6\,\mathrm{m\,s^{-2}}$, (b) $19.6\,\mathrm{m\,s^{-2}}$.

Exercise 9.2

(Assume $G = 6.7 \times 10^{-11}\,\mathrm{N\,m^2\,kg^{-2}}$.)

1 The gravitational field strength on the surface of the moon is $1.7\,\mathrm{N\,kg^{-1}}$. Assuming that the moon is a uniform sphere of radius $1.7 \times 10^6\,\mathrm{m}$, calculate (a) the mass of the moon, (b) the gravitational field strength $1.0 \times 10^6\,\mathrm{m}$ above its surface.

2 The acceleration due to gravity at the Earth's surface is $9.8\,\mathrm{m\,s^{-2}}$. Calculate the acceleration due to gravity on a planet which has (a) the same mass and twice the radius, (b) the same radius and twice the density, (c) half the radius and twice the density.

3 If the Earth has radius r and the acceleration due to gravity at its surface is $9.8\,\mathrm{m\,s^{-2}}$, calculate the acceleration due to gravity at a point that is distance r above the surface of a planet with half the radius and the same density as the Earth.

Gravitational potential and escape speed

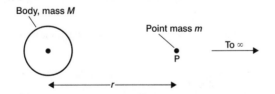

Fig. 9.2 Gravitational potential at P

Refer to Fig. 9.2. The gravitational potential U at point P due to the gravitational attractive force of mass M is given by

$$U = -\frac{GM}{r} \qquad (9.4)$$

The negative sign indicates that work must be done to take a mass from P to infinity (where the potential is zero). U is the work done per kg.

Example 5

Assuming that the Earth is a uniform sphere of radius $6.4 \times 10^6\,\mathrm{m}$ and mass $6.0 \times 10^{24}\,\mathrm{kg}$, calculate (a) the gravitational potential at (i) the Earth's surface and (ii) a point $6.0 \times 10^5\,\mathrm{m}$ above the Earth's surface, (b) the work done in taking a $5.0\,\mathrm{kg}$ mass from the Earth's surface to a point $6.0 \times 10^5\,\mathrm{m}$ above it, (c) the work done in taking a $5.0\,\mathrm{kg}$ mass from the Earth's surface to a point where the Earth's gravitational effect is negligible.

Method

(a) We use Equation 9.4 in which $G = 6.7 \times 10^{-11}$ and $M = 6.0 \times 10^{24}$.

(i) We have $r = r_1 = 6.4 \times 10^6$. So, if U_1 is the potential here,

$$U_1 = \frac{-GM}{r_1} = \frac{-6.7 \times 10^{-11} \times 6.0 \times 10^{24}}{6.4 \times 10^6}$$
$$= -6.28 \times 10^7\,\mathrm{J\,kg^{-1}}$$

(ii) We have $r = r_2 = (6.4 + 0.6) \times 10^6$ m. If U_2 is the potential at r_2, Equation 9.4 gives $U_2 = -5.74 \times 10^7\,\mathrm{J\,kg^{-1}}$.

(b) *The work required W, per kg,* is the difference in gravitational potential, so

$$W = U_2 - U_1 = 0.54 \times 10^7\,\mathrm{J}$$

Note: we subtract U_1 from U_2 since there is an *increase* in gravitational potential as we move away from the Earth. For a $5.0\,\mathrm{kg}$ mass we require $5.0 \times 0.54 \times 10^7 = 2.7 \times 10^7\,\mathrm{J}$. (We cannot use the simple form mgh to calculate work required, since g changes appreciably between the two points.)

(c) The work required W', per kg, is given by

$$W' = \text{Potential at } \infty - \text{Potential at Earth's surface}$$
$$= 0 - (-6.28 \times 10^7)$$
$$= 6.28 \times 10^7\,\mathrm{J}$$

For a $5.0\,\mathrm{kg}$ mass the work required is

$$5 \times 6.28 \times 10^7 = 31.4 \times 10^7\,\mathrm{J}$$

Answer

(a) (i) $-6.3 \times 10^7\,\mathrm{J\,kg^{-1}}$ and (ii) $-5.7 \times 10^7\,\mathrm{J\,kg^{-1}}$, (b) $2.7 \times 10^7\,\mathrm{J}$, (c) $31 \times 10^7\,\mathrm{J}$.

Example 6

Calculate the minimum speed which a body must have to escape from the moon's gravitational field, given that the moon has mass $7.7 \times 10^{22}\,\mathrm{kg}$ and radius $1.7 \times 10^6\,\mathrm{m}$.

Method

As the body moves away from the moon's surface, its kinetic energy decreases because its gravitational potential increases. Referring to Fig. 9.2 we see that the work required to take a body of mass m from P to infinity is GMm/r. Suppose the body has speed v at point P, then it will have just enough kinetic energy to escape, provided that

$$\tfrac{1}{2}mv^2 = \frac{GMm}{r}\quad,$$

or

$$v = \sqrt{\frac{2GM}{r}} \qquad (9.5)$$

We have $G = 6.7 \times 10^{-11}$, $M = 7.7 \times 10^{22}$ and $r = 1.7 \times 10^6$.

Substituting into Equation 9.5 gives

$$v = 2.46 \times 10^3\,\mathrm{m\,s^{-1}}.$$

Answer

Escape speed $= 2.5 \times 10^3\,\mathrm{m\,s^{-1}}$.

Exercise 9.3

(Assume $G = 6.7 \times 10^{-11}\,\mathrm{N\,m^2\,kg^{-2}}$.)

1 The gravitational potential difference between two points is $3.0 \times 10^3\,\mathrm{J\,kg^{-1}}$. Calculate the work done in moving a mass of 4.0 kg between the two points.

2 The moon has mass 7.7×10^{22} kg and radius 1.7×10^6 m. Calculate (a) the gravitational potential at its surface and (b) the work needed to completely remove a 1.5×10^3 kg space craft from its surface into outer space. Neglect the effect of the Earth, planets, sun, etc.

3 A planet has radius 5.0×10^5 m and mean density $3.0 \times 10^3\,\mathrm{kg\,m^{-3}}$. Calculate the escape speed of bodies on its surface.

4 A neutron star has radius 10 km and mass 2.5×10^{29} kg. A meteorite is drawn into its gravitational field. Calculate the speed with which it will strike the surface of the star. Neglect the initial speed of the meteorite.

Satellites and orbits

Satellites are objects which are in orbit around a larger mass as a direct result of gravitational attraction. Our planets are satellites of the sun and the moon is a satellite of the Earth.

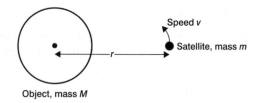

Fig. 9.3   A satellite in orbit

The centripetal acceleration and force (see Chapter 8) is provided by gravitational attraction. Fig. 9.3 shows a satellite of mass m in circular orbit of radius r around an object of mass M. Suppose v is the speed of rotation and T is the period of rotation. The centripetal force F required is:

$$F = \frac{mv^2}{r} \qquad (8.4)$$

This force is provided by gravitational attraction, and

$$F = \frac{GMm}{r^2} \qquad (9.1)$$

Also, in circular motion, we have

$$T = \frac{2\pi r}{v} \qquad (8.2)$$

These three equations are used to solve problems on satellites in orbit.

Equating (8.4) and (9.1) gives

$$\frac{mv^2}{r} = \frac{GMm}{r^2}$$

or

$$v^2 = GM/r \qquad (9.6)$$

Substituting $v^2 = 4\pi^2 r^2/T^2$ from Equation (8.2) gives

$$T^2 = \frac{4\pi^2 r^3}{GM} \qquad (9.7)$$

Since G and M are constant then $T^2 \propto r^3$ – this is Kepler's 3rd law and can be applied to any satellite in orbit around a massive body.

In a *Geostationary orbit* a satellite orbits a planet and stays directly above the same point on the planet (see Example 7).

Example 7

Satellites which orbit the Earth with a time period of 24 hours are used for communication purposes since they appear stationary above a given point on Earth. Calculate the height of such a satellite above the Earth's surface.

Assume mass of Earth $M = 6.0 \times 10^{24}$ kg, $G = 6.7 \times 10^{-11}$ N m^2 kg^{-2} and the radius of the Earth $R = 6.4 \times 10^6$ m.

Method

We use Equation 9.7 in which $T = 24$ hrs $= 24 \times 3600$ $= 8.64 \times 10^4$ s, $G = 6.7 \times 10^{-11}$ and $M = 6.0 \times 10^{24}$. Let the radius of the 'synchronous' orbit be r. Rearranging Equation 9.7 gives

$$r^3 = \frac{GMT^2}{4\pi^2}$$

$$= \frac{6.7 \times 10^{-11} \times 6.0 \times 10^{24} \times (8.64)^2 \times 10^8}{4\pi^2}$$

$$\therefore \quad r^3 = \frac{3.00 \times 10^{24}}{39.48}$$

$$\therefore \quad r = 42.4 \times 10^6 \text{ m}$$

Since the Earth has a radius of 6.4×10^6 m then the height above the Earth's surface is $(42.4 - 6.4) \times 10^6$ m.

Answer

36×10^6 m.

Example 8

Use the following data to calculate the time, in Earth years, for Mars to orbit the sun.

(Average) radius of Earth's orbit $R = 15 \times 10^{10}$ m
(Average) radius of Mars' orbit $r = 23 \times 10^{10}$ m

Method

From Equation 9.7, since G and M are constant, then $T^2 \propto r^3$. Let $t = $ time, in Earth years, for Mars to orbit the sun. Since the orbit time of the Earth is 1, then

$$\frac{t^2}{1^2} = \frac{r^3}{R^3} = \frac{(23 \times 10^{10})^3}{(15 \times 10^{10})^3}$$

$$\therefore \quad t^2 = (23/15)^3 = 3.60$$

$$\therefore \quad t = 1.9$$

Answer

1.9 Earth years.

Exercise 9.4

Assume $G = 6.7 \times 10^{-11}$ N m^2 kg^{-2} and the Earth has mass $M = 6.0 \times 10^{24}$ kg and radius $r = 6.4 \times 10^6$ m.

1 Given G, M and r calculate:

   (a) the period of a satellite orbiting close to the Earth's surface

   (b) the height above the Earth's surface of a weather satellite which orbits the Earth every 2.0 hours.

2 Use Kepler's 3rd Law to calculate R (in m) and T (in Earth years) for the following planets as they orbit the Sun:

	Earth	Venus	Saturn
(Average) radius of orbit/10^{10} m	15	11	R
Time of orbit/ Earth years	1.0	T	29

Exercise 9.5: Examination questions

(Assume $G = 6.7 \times 10^{-11}$ N m^2 kg^{-2} unless stated.)

1 Write a word equation which states Newton's law of gravitation.
   Mars may be assumed to be a spherical planet with the following properties:

   Mass m_M of Mars $= 6.42 \times 10^{23}$ kg
   Radius r_M of Mars $= 3.40 \times 10^6$ m

   Calculate the force exerted on a body of mass 1.00 kg on the surface of Mars. Take $G = 6.67 \times 10^{-11}$ N m^2 kg^{-2} [Edexcel 2001]

2

Fig. 9.4   Diagram for Question 2

X and Y are the centres of two small spheres of masses m and $4m$ respectively. The gravitational field strengths due to the two spheres at a point Z, lying on the line between X and Y (see Fig. 9.4), are equal in magnitude. Show that

   $ZY = 2ZX$

3 On the ground, the gravitational force on a satellite is W.
   What is the gravitational force on the satellite when at a height $R/50$, where R is the radius of the Earth?

   A $1.04W$ **B** $1.02W$ **C** $0.98W$ **D** $0.96W$
                                              [OCR 2001]

4 Outside a uniform sphere of mass M, the gravitational field strength is the same as that of a point mass M at the centre of the sphere.

The Earth may be taken to be a uniform sphere of radius r. The gravitational field strength at its surface is g.

What is the gravitational field strength at a height h above the ground?

A $\dfrac{gr^2}{(r+h)^2}$ **B** $\dfrac{gr}{(r+h)}$

C $\dfrac{g(r-h)}{r}$ **D** $\dfrac{g(r-h)^2}{r^2}$ [OCR 2000]

Questions 5 and 6

These questions are about the gravitational field and potential near the planets Mars and Earth.

5 Mars has a radius of approximately 0.5 of that of the Earth and has a mass of approximately 0.1 of the Earth. The gravitational field strength at the surface of the Earth is approximately $10\,\text{N}\,\text{kg}^{-1}$.

Which one of **A** to **D** below is the best estimate, in $\text{N}\,\text{kg}^{-1}$, of the gravitational field strength at the surface of Mars?

A 2   **B** 4   **C** 8   **D** 20

[OCR Nuff 2001]

6 The gravitational potential at the surface of the Earth is $-6.3 \times 10^7\,\text{J}\,\text{kg}^{-1}$.

Which one of **A** to **D** below is the gravitational potential, in $\text{J}\,\text{kg}^{-1}$, at a point one Earth radius above the surface of the Earth?

A -1.6×10^7 **B** -3.1×10^7
C -1.3×10^8 **D** -2.5×10^8

[OCR Nuff 2001]

7 Fig. 9.5 shows the final equilibrium position of two of the spheres in an experiment to determine the universal gravitational constant, G.

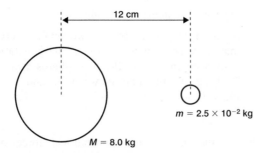

Fig. 9.5   Diagram for Question 7

(a) Calculate the magnitude of the gravitational force which is exerted by the larger sphere on the smaller sphere. Is this an attractive or a repulsive force?

(b) Originally, the smaller sphere was 6.0 cm further away from the larger sphere. Calculate by how much the potential energy of the smaller sphere has changed during its movement from its original position.

8 A binary star consists of two stars of masses $24 \times 10^{30}\,\text{kg}$ and $6.0 \times 10^{30}\,\text{kg}$, their centres being $3.0 \times 10^9\,\text{m}$ apart. The graph shows how the net gravitational potential varies with distance from the centre of the more massive star along the line joining their centres.

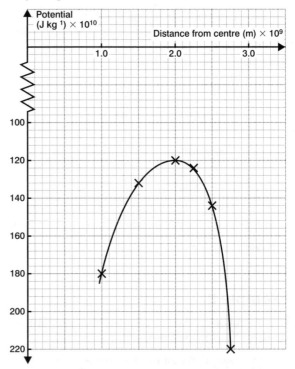

(a)* Use the graph to determine where, along the line of centres, the gravitational field strength (intensity) is zero. Explain your reasoning.

(b) Verify your answer to part (a) by an independent calculation.   [WJEC 2000]

9 Calculate the speed with which a body must be projected from the Earth's surface so as to completely escape from the Earth's gravitational effect (the escape speed).

Assume the Earth has mass $M = 6.0 \times 10^{24}\,\text{kg}$ and radius $r = 6.4 \times 10^6\,\text{m}$.

10 The escape speed v from the surface of a planet can be calculated from $v = \sqrt{2gr}$, where g is the acceleration of free fall at the planet's surface and r is the planet's radius.

For Earth the escape speed $v = 11\,\text{km}\,\text{s}^{-1}$.

(Author's hint: the gravitational field strength equals $(-)$ the slope of the gravitational potential versus distance graph.)

(a) Calculate the escape speed for a planet of the same mass as the Earth but twice its radius.

(b) The escape speed is independent of the mass of the object being launched. Explain why it is nevertheless desirable to keep the mass of a space probe as small as possible.

[Edexcel 2000]

11 (a) Define:
  (i) gravitational potential (at a point);
  (ii) velocity of escape.

(b) Use the data below to show that the radius of the orbit of a geostationary satellite is about 4.2×10^7 m.

mass of Earth $= 6.0 \times 10^{24}$ kg
gravitational constant $= 6.7 \times 10^{-11}$ N m² kg⁻²

(c) Fig. 9.6 shows how the gravitational potential V_G in the Earth's field varies with distance r from the Earth's centre for regions close to the orbit of a geostationary satellite.
With the aid of Fig. 9.6, determine:
  (i) the work required to lift a rocket of mass 200 kg from $r = 4.0 \times 10^7$ m to $r = 4.4 \times 10^7$ m;
  (ii) the velocity of escape from a satellite orbit at $r = 4.2 \times 10^7$ m.

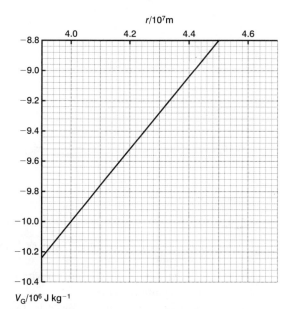

Fig. 9.6 [OCR 2001]

12 (a) Show that the speed v of a particle in a circular orbit of radius r around a planet of mass M is given by the expression

$$v = \sqrt{\frac{GM}{r}}$$

where G is the gravitational constant.

(b) In SI units the value of G is 6.7×10^{-11}. State an SI unit for G.

(c) Fig. 9.7 shows two of the moons, P and Q, of Jupiter. The moons move in circular orbits around the planet. The inner moon P is 1.3×10^8 m from the centre of the planet and the outer moon Q is 2.4×10^{10} m from the centre. The speed of Q is 2.3×10^3 m s⁻¹.

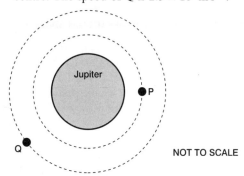

NOT TO SCALE

Fig. 9.7

  (i) Determine the mass M_J of Jupiter.
  (ii) Calculate the orbital speed v of P.
  (iii) Calculate the ratio
  $$\frac{\text{gravitational field strength of Jupiter at P}}{\text{gravitational field strength of Jupiter at Q}}.$$

[OCR 2001]

13 This question is about the potential dangers of 'space junk', such as disused satellites and rocket parts left orbiting the Earth.

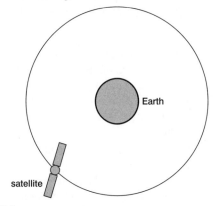

Fig. 9.8

(a) (i) Draw an arrow on Fig. 9.8 to represent the resultant force acting on the satellite in circular orbit around the Earth.
  (ii) Show that for a circular orbit of radius r, around a planet of mass M, a satellite must have an orbital speed v, given by

$$v = \sqrt{\frac{GM}{r}}$$

where G is the universal gravitational constant.

(b) The lowest Earth-orbiting satellites have an orbital period of about 90 minutes.
   (i) Show that the radius at which they orbit the Earth is about 6.7×10^6 m;
   $$G = 6.7 \times 10^{-11} \, \mathrm{N \, m^2 \, kg^{-2}}$$
   mass of Earth $= 6.0 \times 10^{24}$ kg
   (ii) Show that the orbital speed is about $7.8 \times 10^3 \, \mathrm{m \, s^{-1}}$;
   (iii) Show that the kinetic energy of a 1000 kg satellite in this orbit is about 3.0×10^{10} J.

(c) 1 tonne of the explosive TNT yields 4.1×10^9 J. By comparing this value to the kinetic energy of a satellite in Earth orbit suggest why 'space junk' presents a significant risk to future space missions.
[OCR Nuff 2001]

14 A space station is in a stable circular orbit at a distance of 20 000 km from the Earth's centre. The radius of the orbit of geostationary satellites is 42 000 km.

(a) (i) Use this information and Kepler's third law to show that the orbital period of the space station is approximately 8 hours.
   (ii) Use the value 8 hours from (i) to estimate the gravitational field strength at the space station. State your result with an appropriate SI unit.

(b) In its stable circular orbit, the space station is subject to a gravitational force. State and explain whether work is done by this force.
[OCR 2000]

15 Landsat is a satellite which orbits at a height of 9.18×10^5 m above the Earth's surface.
Calculate the period of Landsat using the following data. Hence determine the number of orbits it makes per day.
Useful data:
   (Radius of orbit)$^3 \propto$ (period of orbit)^2
   Radius of the Earth $= 6.37 \times 10^6$ m
   At 3.59×10^7 m above the Earth's surface, a satellite would be in a geostationary orbit.
[Edexcel 2001]

16 (a) Satellites used for telecommunications are frequently placed in a geostationary orbit. State **three** features of the motion of a satellite in a geostationary orbit.

(b) The planet Mars has radius 3.39×10^6 m and mass 6.50×10^{23} kg. The length of a day on Mars is 8.86×10^4 s (24.6 hours).
   (i) A satellite is to be placed in geostationary orbit about Mars. At what height above the surface of Mars should the satellite be placed? Show clearly how you obtain your answer.
   (ii) Calculate the acceleration of free fall on the surface of Mars.

(c) Mars has two moons, Phobos and Deimos, which move in circular orbits about the planet. The radii of these orbits are 9.38×10^3 km and 23.5×10^3 km respectively. The orbital period of Phobos is 0.319 days. Calculate the orbital period of Deimos. Take $G = 6.67 \times 10^{-11} \, \mathrm{N \, m^2 \, kg^{-2}}$ [CCEA 2001]

17 (a) (i) Define
      1. electric field strength,
      2. electric potential.
   (ii) State how electric field strength at a point may be determined from a graph of the variation of electric potential with distance from the point.

(b) The moon Charon (discovered in 1978) orbits the planet Pluto. Fig. 9.9 shows the variation of the gravitational potential ϕ with distance d above the surface of Pluto along a line joining the centres of Pluto and Charon.

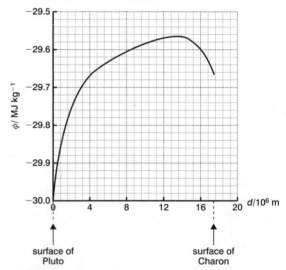

Fig. 9.9

The gravitational potential is taken as being zero at infinity.
   (i) Suggest why all values of gravitational potential are negative.
   (ii) By reference to your answer to (a)(ii), suggest why the gradient at a point on the graph of Fig. 9.9 gives the magnitude of the acceleration of free fall at that point.
   (iii) Use Fig. 9.9 to determine, giving an explanation of your working,
      1. the distance from the surface of Pluto at which the acceleration of free fall is zero,
      2. the acceleration of free fall on the surface of Charon.

(c) A lump of rock of mass 2.5 kg is ejected from the surface of Charon such that it travels towards Pluto.

(i) Using data from Fig. 9.9, determine the minimum speed with which the rock hits the surface of Pluto.

(ii) Suggest why, if the rock travels from Pluto to Charon, the minimum speed on reaching Charon is different from that calculated in (i). [OCR 2001]

Section C
Matter

10
Elasticity

Hooke's law

Specimens in which extension e (m) is proportional to the applied force F (N) are said to obey Hooke's law. In this case

$$F = ke \qquad (10.1)$$

where k is the force constant or stiffness constant $(N\,m^{-1})$ of the specimen and depends upon the dimensions of the specimen.

Hooke's law is often obeyed by springs and specimens of metals in tension (and compression). In this proportional region we also define Young's modulus E $(N\,m^{-2})$ – see below – which is the same for all specimens of the same material, irrespective of their dimensions. Specimens may be stretched beyond their proportional limit, in which case Hooke's law is no longer obeyed.

Work done in stretching a specimen

Fig. 10.1 shows typical force extension graphs for (a) a spring and (b) a metal specimen. Work is done on the specimen when it is extended (or compressed). *The work done is equal to the area under the force-extension (or compression) graph.* Within the proportional limit:

$$\text{Work done} = \tfrac{1}{2} F \times e \qquad (10.2)$$

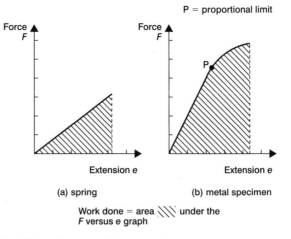

P = proportional limit

(a) spring

(b) metal specimen

Work done = area \\\\ under the F versus e graph

Fig. 10.1   Force extension graphs

where F (N) is the force required to produce an extension e (m). The work done becomes potential energy, termed strain energy, stored within the specimen. Up to the elastic limit this energy is recoverable.

Example 1

A spring is stretched by applying a force to it. Table 10.1 is a table of values of extension e against stretching force F for the spring.

Table 10.1

Force F (N)	0	0.20	0.40	0.60	0.80
Extension (mm)	0	5.0	10	15	20

(a) Draw a graph of extension (x axis) versus stretching force (y axis) and calculate the force constant of the spring.

(b) Calculate the work required to stretch the spring (i) initially by 5 mm and (ii) from an initial extension of 10 mm to a final extension of 15 mm.

If the spring is now replaced by two identical springs placed side by side and next to each other, calculate:

(c) the extension of the double spring if a stretching force of 1.2 N is applied to the combination.

Method

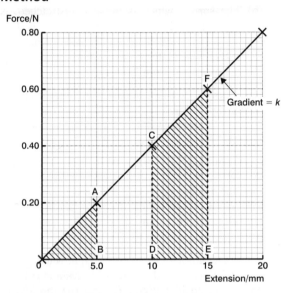

Fig. 10.2   Solution to Example 1

(a) The graph is shown in Fig. 10.2. Since this is a straight line passing through the origin, the spring obeys Hooke's Law. The force constant k of the spring can be found by rearranging Equation (10.1) and is the gradient of the straight line:

$$k = F/e = 0.80/(20 \times 10^{-3}) = 40 \, \text{N m}^{-1}$$

(b) (i) The work required is the area OAB under the graph up to the 5.0 mm point. Since the graph is a straight line and $F = 0.20 \, \text{N}$ when $e = 5.0 \times 10^{-3}$ then, using Equation (10.2):

$$\text{Work done} = \text{area under graph}$$
$$= 1/2 \times 0.20 \times 5.0 \times 10^{-3}$$
$$= 0.50 \, \text{mJ}$$

(ii) The work required is the area CDEF under the graph. This is given by:

$$\text{Work done} = \text{area CDEF}$$
$$= 1/2 \times (\text{CD} + \text{EF}) \times \text{DE}$$
$$= 1/2 \times (0.40 + 0.60) \times 5.0 \times 10^{-3}$$
$$= 2.5 \, \text{mJ}$$

(c) In this case the double spring has a force constant of *twice* the single spring, since each of the springs (in parallel) effectively takes half of the stretching force of 1.2 N (in this case effectively 0.60 N each). Thus, since the force constant of the double spring is now $80 \, \text{N m}^{-1}$, the extension e for an applied force of 1.2 N is given by rearranging Equation (10.1):

$$e = F/k = 1.20/80 = 0.015 \, \text{m, or } 15 \, \text{mm}.$$

Note that we could have obtained this answer by assuming each spring takes half of the stretching force.

Note that *if the springs had been in series*, instead of in parallel, the springs would have each taken the total force and *the extension would have been the sum of the separate extensions*.

Example 2

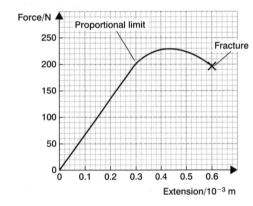

Fig. 10.3   Information for Example 2

Fig. 10.3 shows a force–extension graph for a metal specimen.

Calculate:

(a) the force constant of the specimen
(b) the work done in stretching the specimen up to:
   (i) the proportional limit   (ii) fracture.

Method

(a) The force constant k is found by rearranging Equation 10.1 and is the gradient of the straight line portion:

$$k = F/e = 200/0.3 \times 10^{-3}$$
$$= 667 \times 10^{3} \, \text{N m}^{-1}$$

(b) Work done = Area under force–extension graph, where force is in newtons and extension in metres.
   (i) Area under linear portion of graph

$$= \tfrac{1}{2} \, \text{Height} \times \text{Base}$$
$$= \tfrac{1}{2} \times 200 \times 0.3 \times 10^{-3}$$
$$= 3.0 \times 10^{-2} \, \text{J}$$

Note: we could have used Equation 10.2, with $F = 200 \, \text{N}$ and $e = 0.3 \times 10^{-3}$ m.

(ii) We must add to (a) the area under the graph beyond the proportional limit and up to fracture. This is found by 'counting squares' on the graph paper and is approximately equal to

$$66 \times 10^{-3} \, \text{J} = 6.6 \times 10^{-2} \, \text{J}.$$

So, total work done up to fracture equals

$$6.6 \times 10^{-2} + 3.0 \times 10^{-2} = 9.6 \times 10^{-2} \, \text{J}$$

Answer

(a) $6.7 \times 10^5 \, \text{N m}^{-1}$

(b) (i) $3.0 \times 10^{-2} \, \text{J}$, (ii) $9.6 \times 10^{-2} \, \text{J}$.

Example 3

A mass of 3.5 kg is gradually applied to the lower end of a vertical wire and produces an extension of 0.80 mm.

Calculate (a) the energy stored in the wire and (b) the loss in gravitational potential energy of the mass during loading. Account for the difference between the two answers. Assume that the proportional limit is not exceeded and $g = 10 \, \text{m s}^{-2}$.

Method

(a) We have

$$F = 3.5 \times g = 35 \, \text{N}$$

and

$$e = 0.80 \times 10^{-3} \, \text{m}$$

Equation 10.2 gives

$$\text{Work done} = \tfrac{1}{2} Fe = \tfrac{1}{2} \times 35 \times 0.8 \times 10^{-3}$$
$$= 14 \times 10^{-3} \, \text{J}$$

This is stored as elastic 'strain' energy.

(b) Loss in PE $= mgh$. We have $m = 3.5$, $g = 10$ and $h = 0.80 \times 10^{-3}$.

$$\therefore \quad \text{loss in PE} = 3.5 \times 10 \times 0.8 \times 10^{-3}$$
$$= 28 \times 10^{-3} \, \text{J}$$

The energy stored is only *half* the loss in gravitational PE because the wire needs a *gradually* increasing load, from zero to 35 N, to extend it. The remaining gravitational PE is given to the loading system (e.g. the hand as it gradually attaches the load to the wire). Note that if the load is *suddenly* applied the initial extension would be 1.6 mm; that is, *twice* the equilibrium extension.

Answer

(a) $14 \times 10^{-3} \, \text{J}$, (b) $28 \times 10^{-3} \, \text{J}$.

Exercise 10.1

1   A spring, which obeys Hooke's Law, is stretched by applying a gradually increasing force.

(a) A force of 4.0 N is needed to increase its length by 16 cm. Calculate the force constant of this spring.

(b) The spring, which is initially unstretched, is stretched by 2.0 cm. The applied force is then increased until the spring is stretched by 5.0 cm. Calculate the work done in increasing the extension from 2.0 cm to 5.0 cm.

2   The following tensile test data were obtained using a metal specimen:

Load/10^2 N	0	2.0	4.0	4.5	5.0	5.5
Extension/mm	0	0.10	0.20	0.24	0.30	0.40

Plot the load–extension graph and calculate the work done in stretching the specimen up to (a) the proportional limit (load $= 4.0 \times 10^2$ N), (b) fracture (load $= 5.5 \times 10^2$ N).

3   A metal column shortens by 0.25 mm when a load of 120 kN is placed upon it. Calculate (a) the energy stored in the column and (b) the loss in gravitational PE of the load. Explain why the values in (a) and (b) differ. Assume that the proportional limit is not exceeded

Stress and strain

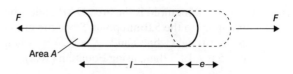

Fig. 10.4   A solid specimen under tension

Refer to Fig. 10.4 in which a specimen of original length l (m) and cross-sectional area A (m^2) is subjected to a tensile force F (N), so that its extension is e (m). We define

$$\text{Tensile stress } \sigma = \frac{F}{A} \quad (\text{N m}^{-2} \text{ or Pa})^* \tag{10.3}$$

$$\text{Tensile strain } \epsilon = \frac{e}{l} \quad (\text{no units}) \tag{10.4}$$

*$1 N m^{-2} = 1 Pa$ (pascal).

Example 4

A metal bar is of length 2.0 m and has a square cross-section of side 40 mm. When a tensile force of 80 kN is applied, it extends by 0.046 mm. Calculate (a) the stress, (b) the strain in the specimen.

Method

We have $l = 2.0$, $A = (40 \times 10^{-3})^2 = 16 \times 10^{-4}$. $F = 80 \times 10^3$ and $e = 0.046 \times 10^{-3}$. So Equations 10.3 and 10.4 give

(a) $\sigma = \dfrac{F}{A} = \dfrac{80 \times 10^3}{16 \times 10^{-4}} = 5.0 \times 10^7 \, \mathrm{N\,m^{-2}}$

(b) $\epsilon = \dfrac{e}{l} = \dfrac{0.046 \times 10^{-3}}{2.0} = 2.3 \times 10^{-5}$

Answer

(a) $5.0 \times 10^7 \, \mathrm{N\,m^{-2}}$, (b) 2.3×10^{-5}.

Exercise 10.2

1 A metal bar has circular cross-section of diameter 20 mm. If the maximum permissible tensile stress is 80 $\mathrm{MN\,m^{-2}}$ $(80 \times 10^6 \, \mathrm{N\,m^{-2}})$, calculate the maximum force which the bar can withstand.

2 A metal specimen has length 0.50 m. If the maximum permissible strain is not to exceed 0.10% (1.0×10^{-3}), calculate its maximum extension.

3 A metal bar of length 50 mm and square cross-section of side 20 mm is extended by 0.015 mm under a tensile load of 30 kN. Calculate (a) the stress, (b) the strain in the specimen.

Young's modulus

Up to a certain load, called the limit of proportionality,* extension is proportional to applied force, so that strain is proportional to stress. The slope of the stress–strain graph in the linear region is called Young's modulus E. So we define

$$E = \frac{\sigma}{\epsilon} \ (\mathrm{N\,m^{-2}}) \qquad (10.5)$$

Work done per unit volume

The work done *per unit volume* (sometimes termed the energy density) is equal to the area

Sometimes no distinction is made between this and the elastic limit.

under the *stress–strain* graph. Within the proportional limit:

> **work done per unit volume**
> **= area under stress–strain graph**
> $= \frac{1}{2}\sigma \times \epsilon$ $\qquad$ (10.6)

where σ is the stress $(\mathrm{N\,m^{-2}})$ required to produce strain ϵ.

Example 5

A steel bar is of length 0.50 m and has a rectangular cross-section 15 mm by 30 mm. If a tensile force of 36 kN produces an extension of 0.20 mm, calculate Young's modulus for steel. Assume that the limit of proportionality is not exceeded.

Method

From Equations 10.5, 10.3 and 10.4

> $E = \dfrac{\textbf{Stress}}{\textbf{Strain}} = \dfrac{(\textbf{Force} \div \textbf{Area})}{(\textbf{Extension} \div \textbf{Original length})}$
> $\qquad\qquad\qquad\qquad\qquad\qquad (10.7)$

We have

$$\text{Force} = 36 \times 10^3 \, \mathrm{N}$$
$$\text{Area} = 15 \times 30 = 450 \, \mathrm{mm^2} = 450 \times 10^{-6} \, \mathrm{m^2}$$
$$\text{Extension} = 0.20 \, \mathrm{mm} = 0.20 \times 10^{-3} \, \mathrm{m}$$

Original length $= 0.50$ m

So Equation 10.7 gives

$$E = \frac{(36 \times 10^3) \div (450 \times 10^{-6})}{(0.2 \times 10^{-3}) \div 0.5}$$
$$= 2.0 \times 10^{11} \, \mathrm{N\,m^{-2}}$$

Answer

Young's modulus for steel $= 2.0 \times 10^{11} \, \mathrm{N\,m^{-2}}$.

Example 6

An aluminium alloy strut in the landing gear of an aircraft has a cross-sectional area of 60 $\mathrm{mm^2}$ and a length of 0.45 m. During landing the strut is subjected to a compressive force of 3.6 kN. Calculate by how much the strut will shorten under this force. Assume that Young's modulus for the alloy is 90 $\mathrm{GN\,m^{-2}}$ and the proportional limit is not exceeded.

Method

Equation 10.7 gives

$$E = \frac{F \div A}{e \div l}$$

In this case the strut is compressed. Since materials in general have the same value for the elastic modulus in tension as in compression, it is necessary only to replace extension e in the above equation by compression c.

We have $A = 60\,\text{mm}^2 = 60 \times 10^{-6}\,\text{m}^2$, $l = 0.45$, $F = 3.6 \times 10^3$, $E = 90 \times 10^9$, and require the compression c. So

$$90 \times 10^9 = \frac{(3.6 \times 10^3) \div (60 \times 10^{-6})}{(c \div 0.45)}$$

Rearranging gives $c = 0.30 \times 10^{-3}\,\text{m}$.

Answer

The strut shortens by 0.30 mm.

Example 7

A vertical steel wire of length 0.80 m and radius 1.0 mm has a mass of 20 kg applied to its lower end. Assuming that the proportional limit is not exceeded, calculate (a) the extension, (b) the energy stored per unit volume in the wire. Take the Young modulus for steel as $2.0 \times 10^{11}\,\text{N m}^{-2}$ and g as $10\,\text{m s}^{-2}$.

Method

(a) Rearranging Equation 10.7 gives

$$e = \frac{F \div A}{E \div l} = \frac{Fl}{EA}$$

We have $F = 20 \times g = 200\,\text{N}$,

$$A = \pi \times (\text{radius})^2 = \pi \times (1.0 \times 10^{-3})^2$$
$$= \pi \times 10^{-6}\,\text{m}^2$$
$$l = 0.80$$

and $E = 2.0 \times 10^{11}$

So $e = \dfrac{Fl}{EA} = \dfrac{200 \times 0.80}{2.0 \times 10^{11} \times \pi \times 10^{-6}}$

$$= 0.255 \times 10^{-3}\,\text{m}$$

(b) We have

$$\sigma = \frac{F}{A} = 200/(\pi \times 10^{-6}) = \frac{2.0}{\pi} \times 10^8\,\text{N m}^{-2}$$

and $\varepsilon = \sigma/E = 200/(\pi \times 10^{-6} \times 2.0 \times 10^{11})$

$$= \frac{1.0}{\pi} \times 10^{-3}$$

From Equation 10.6

work done per unit volume $= \frac{1}{2}\sigma \times \varepsilon$

$$= \frac{1}{2} \times \frac{2.0}{\pi^2} \times 10^5 = 1.01 \times 10^4\,\text{J m}^{-3}$$

Answer

(a) 0.25 mm, (b) $1.0 \times 10^4\,\text{J m}^{-3}$.

Exercise 10.3

(Assume that the proportional limit is not exceeded.)

1   A vertical copper wire is 1.0 m long and has radius 1.0 mm. A load of 180 N is attached to the bottom end and produces an extension of 0.45 mm. Calculate (a) the tensile stress, (b) the tensile strain, (c) the value of Young's modulus for copper.

2   A steel strut has a cross-sectional area of $25 \times 10^3\,\text{mm}^2$ and is 2.0 m long. Calculate the magnitude of the compressive force which will cause it to shorten by 0.30 mm. Assume that E for steel is $200\,\text{GN m}^{-2}$.

3   A bronze wire of length 1.5 m and radius 1.0 mm is joined end-to-end to a steel wire of identical size to form a wire 3.0 m long. Calculate (a) the resultant extension if a force of 200 N is applied, (b) the force required to produce an extension of 0.30 mm. Assume that E for bronze is $1.0 \times 10^{11}\,\text{N m}^{-2}$; for steel $2.0 \times 10^{11}\,\text{N m}^{-2}$.
Hint: (a) total force acts on each wire, extension equals the sum of extensions, (b) $e \propto 1/E$ for each wire, or use $e \propto F$.

4   A load of 0.12 kN is gradually applied to a copper wire of length 1.5 m and area of cross-section $8.0\,\text{mm}^2$. Calculate (a) the extension, (b) the energy stored per unit volume in the wire. Take the Young modulus for copper as $1.2 \times 10^{11}\,\text{N m}^{-2}$.

5   A steel bar has a rectangular cross-section 50 mm by 40 mm and is 2.0 m long. Calculate the work done in extending it by 6.0 mm. Take E for steel as $2.0 \times 10^{11}\,\text{N m}^{-2}$.

Temperature effects

When the temperature of a rod changes then its length will, if unrestrained, change such that:

$$\Delta l = \alpha l \Delta T \qquad \qquad (10.8)$$

where Δl is the change in length, in metres, α the linear expansivity (unit $= {}^\circ\text{C}^{-1}$ or K^{-1}), l the original length in metres and ΔT the rise in temperature, in ${}^\circ\text{C}$ or K.

If, during a temperature change, the rod is to be prevented from changing in length, large forces are often required.

Example 8

A solid copper rod is of cross-sectional area $15\,\text{mm}^2$ and length $2.0\,\text{m}$. Calculate (a) its change in length when its temperature rises by $30\,^\circ\text{C}$, (b) the force needed to prevent it from expanding by the amount in (a). Take the linear expansivity α for copper as $20 \times 10^{-6}\,\text{K}^{-1}$ and the Young modulus E for copper as $1.2 \times 10^{11}\,\text{N}\,\text{m}^{-2}$. Assume that the proportional limit is not exceeded.

Method

(a) We have $l = 2.0$, $\Delta T = +30\,^\circ\text{C}$ (+ sign for temperature rise) and $\alpha = 20 \times 10^{-6}$. Equation 10.7 gives

$$\Delta l = \alpha l \Delta T = 20 \times 10^{-6} \times 2 \times 30$$
$$= 12 \times 10^{-4}\,\text{m}$$

(b) A compressive force F (N) must be supplied which is sufficient to decrease the length by

$$\Delta l = 12 \times 10^{-4}\,\text{m}$$

Rearranging Equation 10.7 gives

$$F = \frac{EeA}{l}$$

We have $E = 1.2 \times 10^{11}$, $e = \Delta l = 12 \times 10^{-4}$, $A = 15\,\text{mm}^2 = 15 \times 10^{-6}\,\text{m}^2$ and $l = 2.0$.

$$\therefore \quad F = \frac{EeA}{l}$$
$$= \frac{1.2 \times 10^{11} \times 12 \times 10^{-4} \times 15 \times 10^{-6}}{2.0}$$
$$= 1080\,\text{N}$$

Answer

(a) $1.2\,\text{mm}$, (b) $1.1\,\text{kN}$.

Exercise 10.4

(For steel, take $\alpha = 12 \times 10^{-6}\,\text{K}^{-1}$ and $E = 2.0 \times 10^{11}$ $\text{N}\,\text{m}^{-2}$. Assume that the proportional limit is not exceeded.)

1 Calculate the force required to extend a steel rod of cross-sectional area $4.0\,\text{mm}^2$ by the same amount as would occur due to a temperature rise of $60\,\text{K}$. Hint: let length $= l$; this cancels out.

2 A section of railway track consists of a steel bar of length $15\,\text{m}$ and cross-sectional area $80\,\text{cm}^2$. It is rigidly clamped at its ends on a day when the temperature is $20\,^\circ\text{C}$. If the temperature falls to $0\,^\circ\text{C}$, calculate (a) the force the clamps must exert to stop the bar contracting and (b) the strain energy stored in the bar.

Exercise 10.5: Examination questions

(Assume $g = 10\,\text{m}\,\text{s}^{-2}$.)

1 A certain spring, which obeys Hooke's law, has a force constant k of $60\,\text{N}\,\text{m}^{-1}$.

(a) You are to draw a graph of stretching force F against extension x for this spring, for a range of x from 0 to $25\,\text{mm}$.
  (i) Use the space below to make any calculations to help you draw this graph.
  (ii) On a copy of Fig. 10.5, label the axes appropriately, and draw the graph.

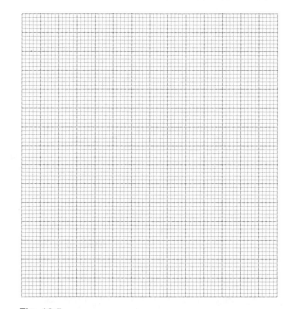

Fig. 10.5

(b) (i) Use your graph in (a)(ii) to determine the work required to stretch the spring from an initial extension of $5\,\text{mm}$ to a final extension of $25\,\text{mm}$. Show clearly how you obtain your result.
  (ii) State the principle of the method you used for your calculation in (i), and explain how you used it in obtaining your answer. [CCEA 2001]

2 A load of 4.0 N is suspended from a parallel two-spring system as shown in the diagram.

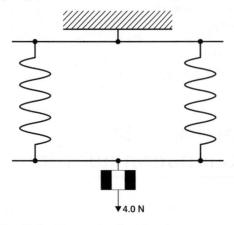

Fig. 10.6   Diagram for Question 2

The spring constant of each spring is $20\,\mathrm{N\,m^{-1}}$. The elastic energy, in J, stored in the system is

A 0.1   **B** 0.2   **C** 0.4   **D** 0.8   [AQA 2000]

3 Many specialist words are used to describe the properties of materials. Some of these words are listed below:

Brittle, ductile, elastic, hard, malleable, plastic, stiff

It is important for engineers to know how different materials behave. One common test which could be performed is to measure the extension of a sample when an increasing force is applied. A force–extension graph for copper wire is shown below.

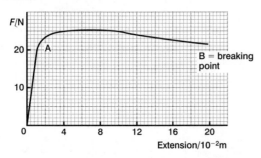

From the list above, choose *two* words which can explain the behaviour of copper. Explain the meaning of each word with reference to the copper wire graph.

(a) Calculate the stiffness of the copper wire.

(b) Estimate the energy required to break this sample of copper wire.   [Edexcel S-H 2000]

4 Two steel wires A and B of the same length are each put under the same tension. Wire A has twice the radius of wire B. The ratio of the stored energy of A to the stored energy of B is

A 4:1   **B** 2:1   **C** 1:1   **D** 1:2   **E** 1:4

5 (a) Fig. 10.7 shows a vertical nylon filament with a weight suspended from its lower end.

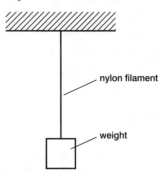

Fig. 10.7

The cross-sectional area of the filament is $8.0 \times 10^{-7}\,\mathrm{m^2}$.
The Young modulus of nylon is $2.0 \times 10^9\,\mathrm{Pa}$.
The ultimate tensile stress of nylon is $9.0 \times 10^7\,\mathrm{Pa}$.

Calculate:
(i) the maximum weight W the filament can support without breaking;
(ii) the weight W' which will extend the filament by 0.50% of its original length.

(b) The information in (a) gives the Young modulus of nylon for small stresses. By reference to the molecular structure and tensile properties of nylon, suggest why this value is inappropriate for large stresses.
[OCR 2001]

6 An object of mass 0.5 kg is suspended by a length of copper wire from a rigid support. The object is raised to a point adjacent to the support, and at the same level, and released from rest. Find the minimum cross-sectional area of the wire if it is not to break. Assume that Hooke's law applies throughout.

The Young modulus for copper is $1.1 \times 10^{11}\,\mathrm{Pa}$ and its tensile strength (i.e. the maximum stress that can be applied without breaking) is $3.0 \times 10^8\,\mathrm{Pa}$.   [WJEC spec 2000]

7 (a) (i) State Hooke's law.
(ii) Explain why wires used as guitar strings must have elastic properties.

(b) The data below are for a thin steel wire suitable for use as a guitar string.

ultimate tensile stress:	$1.8 \times 10^9\,\mathrm{Pa}$
Young modulus:	$2.2 \times 10^{11}\,\mathrm{Pa}$
cross-sectional area:	$2.0 \times 10^{-7}\,\mathrm{m^2}$

In a tensile test, a specimen of the wire, of original length 1.5 m, is stretched until it breaks.

Assuming the wire obeys Hooke's law throughout, calculate:

  (i) the extension of the specimen immediately before breaking;

  (ii) the elastic strain energy released as the wire breaks.   [OCR 2001]

8 A wire of length 3.0 m is hung vertically from a rigid support, and a mass of 0.15 kg is attached to its lower end. Fig. 10.8 shows the arrangement. The wire obeys Hooke's law for all extensions in this question.

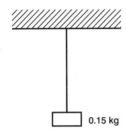

Fig. 10.8

(a) The Young modulus of the material of the wire is 2.0×10^{11} Pa. The diameter of the wire is 0.30 mm. Calculate the extension produced in the wire.

(b) Calculate the elastic strain energy stored in the wire.   [CCEA 2001, part]

9 A specimen fibre of glass has the same dimensions as a specimen of copper wire.

The length of each specimen is 1.60 m and the radius of each is 0.18 mm. Force–extension graphs for both specimens are shown in Fig. 10.9.

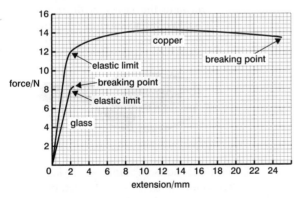

Fig. 10.9

(a) (i) State which of the two materials is brittle.

  (ii) Explain which feature of Fig. 10.9 leads you to your answer in (i).

(b) Using the graphs and the data given, determine

  (i) the area of cross-section of each specimen,

  (ii) the Young modulus of the glass,

  (iii) the ultimate tensile stress for copper,

  (iv) an approximate value for the work done to stretch the copper wire to its breaking point.   [OCR 2000]

10 (a) A metal wire of original length L and cross-sectional area A is stretched by a force F, causing an extension e.

  (i) Write down expressions for the **strain** of the wire and the **stress** in it.

  (ii) Assuming that the extension is such that Hooke's law is obeyed, obtain an expression for the Young modulus E of the metal of the wire in terms of A, e, F and L.

  (iii) Find the relation between the force constant k of the wire (the constant of proportionality in the Hooke's law equation) and the Young modulus E of the metal of the wire.

  (iv) Explain why one refers to the **Young modulus** of the **metal** of the wire, but to the **force constant** of the **wire** itself.

(b) Describe, in detail, an experiment to determine the Young modulus of copper. Your answer should include a clearly labelled diagram, an outline of the method, headings for a table of results that would be taken, and the method of analysis of the results to obtain the value of the Young modulus. Mention two safety precautions which should be taken.

(c) A uniform rod of length 0.80 m and weight 150 N is suspended from a horizontal beam by two vertical wires, as sketched in Fig. 10.10.

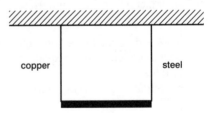

Fig. 10.10

The wire at the left-hand end of the rod is copper, of original length 2.00 m and area of cross-section 0.25 mm². That at the right-hand end is steel, of the same original length but of area of cross-section 0.090 mm². The Young modulus of copper is 1.3×10^{11} Pa and that of steel is 2.1×10^{11} Pa.

  (i) Find the extension in each wire, assuming that the wires remain vertical and that Hooke's law is obeyed.

  (ii) Because the wires extend by different amounts, the suspended rod is not exactly horizontal. It is required to return the rod to the horizontal position by attaching an additional load to it. Find the **minimum** additional load required to do this, and state the point on the rod where this additional load should be attached.

(iii) Strain energy is stored in each of the supporting wires. For the situation where the suspended rod has been made horizontal by attaching the additional load in (ii), decide whether this energy is the same for each of the wires, or whether the greater amount of energy is stored in the copper wire or the steel wire. Explain your reasoning. [CCEA 2000, part]

11 The graph shows part of the stress–strain relationship for steel. No values are given on the stress axis.

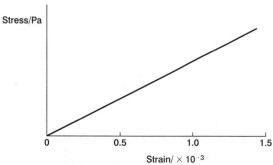

Calculate the energy density for steel when subject to a strain of 1.3×10^{-3}. The Young modulus for steel is 2.2×10^{11} Pa.

Nylon and steel have similar values for their ultimate tensile stress.

Why are steel cables preferred to Nylon ones in the manufacture of the supporting cables for a suspension bridge? [Edexcel 2001, part]

12 (a) Glass is described as a *brittle* material with an *amorphous* structure. Explain the two terms in italics.
   (i) *brittle*   (ii) *amorphous*.

(b)

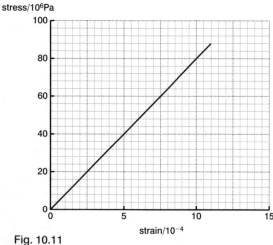

Fig. 10.11

Fig. 10.11 shows a graph of tensile stress against tensile strain for a glass fibre

Use Fig. 10.11 to calculate
 (i) the Young modulus for glass;
 (ii) the strain energy per unit volume just before the fibre breaks, i.e. where the graph line ends. State your answer with a suitable SI unit.
 (iii) the extension just before a fibre of unstretched length 0.50 m breaks.
[OCR 2000]

13 This question is about the plastic deformation of aluminium.

Aluminium expands when its temperature rises. To a good approximation, the increase Δl in the length l of an aluminium rod is given by

$$\Delta l = \alpha l \Delta T$$

where ΔT is the rise in temperature. The constant α is called the *linear expansivity* of aluminium. Its numerical value is given at the end of the question.

When aluminium cools, it contracts by the same amount.

(a) To confirm that you understand the process, verify that an aluminium rod which is 5.0 m long at 40 °C increases its length by about 7 mm when heated to 100 °C. Numerical data are given at the end of the question.

(b) (i) An aluminium rod 0.50 m long is heated so that its length increases by 0.20%. How big was the rise in temperature?
   (ii) The rod is now clamped at its ends, so that it cannot contract as it cools. Calculate the stress in the rod when it has cooled to its original temperature.
   (iii) The rod has a cross-sectional area of 2.0 mm². Calculate the tension in the rod.

(c)

Fig. 10.12

Fig. 10.12 shows an aluminium frying pan. The aluminium undergoes plastic deformation for strains in excess of 0.2%

Explain why pouring cold water into the hot frying pan causes its base to become permanently curved.

Numerical data
linear expansivity of aluminium = 23×10^{-6} K^{-1}
Young modulus of aluminium = 71×10^{9} Pa
[OCR Nuff spec 2000]

Section D
Oscillations and waves

11
Simple harmonic motion

Definition of SHM

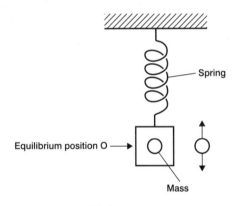

Fig. 11.1   Vertical oscillations

Fig. 11.1 shows a mass on the end of a spring. When displaced vertically it will perform simple harmonic motion because a restoring force acts which is proportional to the displacement of the mass from its equilibrium position O. Thus *its acceleration is always directed towards the point O and is proportional to the displacement from that point*.

Fig. 11.2 illustrates some characteristics of the motion. Fig. 11.2c is the displacement–time graph of the vertical SHM shown in Fig. 11.2a. Fig. 11.2b shows the rotating radius or 'phasor' representation of SHM – the point R moves in uniform circular motion with angular velocity ω. It can be shown that the motion of R projected on to the vertical diameter XY is the same as the SHM shown in Fig. 11.2a and c. Note that the amplitude of SHM is the radius OR, and the 'phase angle' $\theta = \omega t$.

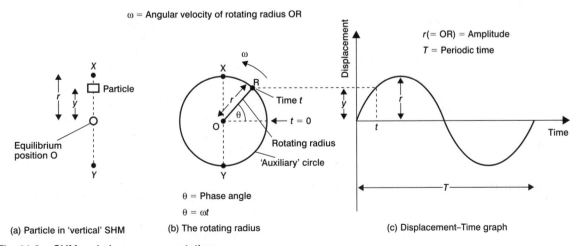

Fig. 11.2   SHM and phasor representation

From the definition of SHM, the acceleration a of the particle is related to its displacement y from the equilibrium position by

$$a = -\text{Constant} \times y$$

where the negative sign indicates that the acceleration is in the opposite direction to the displacement. Also, it can be shown that

$$a = -\omega^2 y \qquad (11.1)$$

where ω is the angular velocity of the rotating radius in Fig. 11.2b and may be called the angular frequency of the simple harmonic motion.

Example 1

A body oscillates vertically in SHM with an amplitude of 30 mm and a frequency of 5.0 Hz. Calculate the acceleration of the particle (a) at the extremities of the motion, (b) at the centre of the motion, (c) at a position midway between the centre and the extremity.

Method

We have frequency $f = 5.0$ Hz. Thus the angular velocity ω of the rotating radius in Fig. 11.2b is, from Chapter 8, given by

$$\omega = 2\pi f = 10\pi \, \text{rad s}^{-1}$$

We use Equation 11.1 to find the acceleration a.

(a) At the top of the motion we ascribe y a positive value, so $y = +0.030$ m. Thus

$$a = -\omega^2 y = -100\pi^2 \times 0.03$$
$$= -3\pi^2 \, \text{m s}^{-2}$$

Note that acceleration a is negative (downwards) when displacement y is positive (upwards).

Similarly at the bottom of the oscillation $y = -0.030$ m, so $a = +3\pi^2 \, \text{m s}^{-2}$.

Note: a is positive (upwards) when y is negative (downwards).

(b) We have $y = 0$, so $a = -\omega^2 y = 0$.

(c) At a position halfway upwards $y = +0.015$ m. So

$$a = -\omega^2 y = -100\pi^2 \times 0.015$$
$$= -1.5\pi^2 \, \text{m s}^{-2}$$

At a position halfway downwards $y = -0.015$ m and $a = +1.5\pi^2 \, \text{m s}^{-2}$.

Answer

(a) $\mp 3.0\pi^2 \, \text{m s}^{-2}$, (b) 0, (c) $\mp 1.5\pi^2 \, \text{m s}^{-2}$.

Example 2

A horizontal platform vibrates vertically in SHM with a period of 0.20 s and with slowly increasing amplitude. What is the maximum value of the amplitude which will allow a mass, resting on the platform, to remain in contact with the platform? Assume acceleration due to gravity $g = 10 \, \text{m s}^{-2}$.

Method

When the platform moves downwards the mass will remain in contact with it only so long as the platform accelerates downwards with value less than or equal to g. The maximum downwards acceleration of the platform is at the top of its motion. If the amplitude is r, then, using Equation 11.1,

$$a = -\omega^2 y = -\omega^2 r$$

at the top of the motion. Now $\omega = 2\pi/T$ where T, the period of the motion, equals 0.20 s.

When the mass is on the point of leaving the platform $a = -g$ (negative indicates downwards), so

$$-g = -\omega^2 r = -\left(\frac{2\pi}{T}\right)^2 r \qquad (11.2)$$

We have $g = 10$, $T = 0.2$ and require r.

Rearranging Equation 11.2 gives

$$r = g\left(\frac{T}{2\pi}\right)^2 = 10 \times \left(\frac{0.2}{2\pi}\right)^2$$
$$= 0.010 \, \text{m}$$

Answer

Maximum amplitude = 10 mm.

Exercise 11.1

1   A body oscillates in SHM with an amplitude of 2.0 cm and a periodic time of 0.25 s. Calculate (a) its frequency, (b) the acceleration at the extremities and at the centre of the oscillation, (c) the acceleration when it is displaced 0.5 cm above the centre of the oscillation. Note: $f = 1/T$.

2   The piston in a particular car engine moves in approximately SHM with an amplitude of 8.0 cm. The mass of the piston is 0.80 kg and the piston makes 100 oscillations per second. Calculate (a) the maximum value of the acceleration of the piston, (b) the force needed to produce this acceleration.

3   A body of mass 0.40 kg has a maximum force of 1.2 N acting on it when it moves in SHM with an

amplitude of 30 mm. Calculate (a) the frequency, (b) the periodic time of the motion.

4 A small mass rests on a horizontal platform which vibrates vertically in SHM with a constant amplitude of 30 mm and with a slowly increasing frequency. Find the maximum value of the frequency which will allow the mass to remain in contact with the platform. Assume $g = 10\,\text{m}\,\text{s}^{-2}$.

Mass on a spring

When a mass m (kg) is attached to the end of a spring of force constant k $(\text{N}\,\text{m}^{-1})$, the periodic time T (s) of oscillations is given by

$$T = 2\pi\sqrt{\frac{m}{k}} \qquad (11.3)$$

Since $T = 2\pi/\omega$ we have

$$\omega^2 = \frac{k}{m} \qquad (11.4)$$

Example 3

A mass of 0.2 kg is attached to the lower end of a light helical spring and produces an extension of 5.0 cm. Calculate (a) the force constant of the spring.

The mass is now pulled down a further distance of 2.0 cm and released. Calculate (b) the time period of subsequent oscillations, (c) the maximum value of the acceleration during the motion. Assume $g = 10\,\text{m}\,\text{s}^{-2}$.

Method

(a) We assume that the spring obeys Hooke's law. From Chapter 10, Equation 10.1, an applied force F(N) produces a change in length e (m) given by

$$F = ke \qquad (11.5)$$

where k $(\text{N}\,\text{m}^{-1})$ is the force constant of the spring. We have $F = mg$ where $m = 0.2$ kg and $g = 10$. Since $e = 5.0 \times 10^{-2}$ m, Equation 11.5 gives

$$0.2 \times 10 = k \times 5.0 \times 10^{-2}$$
$$\therefore \quad k = 40\,\text{N}\,\text{m}^{-1}$$

(b) We use Equation 11.3 with $m = 0.2$ and $k = 40$.

$$T = 2\pi\sqrt{\frac{m}{k}} = 2\pi\sqrt{\frac{0.2}{40}} = 0.44\,\text{s}$$

Note that T is independent of the initial displacement (2.0 cm in this case).

(c) From Equation 11.4

$$\omega^2 = \frac{k}{m} = \frac{40}{0.2} = 200\,\text{rad}^2\,\text{s}^{-2}$$

For the maximum acceleration we use Equation 11.1, with displacement y at its maximum value of 2.0×10^{-2} m.

$$a = -\omega^2 y = -200 \times 2.0 \times 10^{-2} = -4.0\,\text{m}\,\text{s}^{-2}$$

The negative sign indicates direction.

Note an alternative way to find a. At maximum displacement, the net force acting on the mass is

$$F = k \times \text{Displacement} = 40 \times 2.0 \times 10^{-2}$$
$$= 0.80\,\text{N}$$

Thus the maximum acceleration a is given by

$$a = \frac{\text{Force}}{\text{Mass}} = \frac{0.80}{0.20} = 4.0\,\text{m}\,\text{s}^{-2}$$

Answer

(a) $40\,\text{N}\,\text{m}^{-1}$, (b) 0.44 s, (c) $4.0\,\text{m}\,\text{s}^{-2}$.

The simple pendulum

The periodic time T (s) of 'small angle' oscillations of a simple pendulum of length l (m) is given by

$$T = 2\pi\sqrt{\frac{l}{g}} \qquad (11.6)$$

where g is the acceleration due to gravity.

Example 4

Calculate the frequency of oscillation of a simple pendulum of length 80 cm. Assume $g = 10\,\text{m}\,\text{s}^{-2}$.

Method

We use Equation 11.6 with $l = 0.80$ and $g = 10$.

$$T = 2\pi\sqrt{\frac{l}{g}} = 2\pi\sqrt{\frac{0.80}{10}}$$
$$= 1.777$$

Now frequency $f = \frac{1}{T} = \frac{1}{1.777} = 0.56\,\text{Hz}$

Answer

0.56 Hz.

Example 5

Two simple pendulums of length 0.40 m and 0.60 m are set off oscillating in step. Calculate (a) after what further time the two pendulums will once again be in step, (b) the number of oscillations made by each pendulum during this time. (Assume $g = 10\,\mathrm{m\,s^{-2}}$.)

Method

(a) The two pendulums become out of step since they have different periodic times. Let T_1 be the periodic time of the pendulum of length $l_1 = 0.40\,\mathrm{m}$ and T_2 that of the pendulum of length $l_2 = 0.60\,\mathrm{m}$. Using Equation 11.6

$$T_1 = 2\pi\sqrt{\frac{l_1}{g}} = 2\pi\sqrt{\frac{0.4}{10}} = 1.257\,\mathrm{s}$$

$$T_2 = 2\pi\sqrt{\frac{l_2}{g}} = 2\pi\sqrt{\frac{0.6}{10}} = 1.539\,\mathrm{s}$$

During the required time interval the shorter pendulum will complete one more oscillation than the longer pendulum. Let t be the time interval between the pendulums falling in step. If n equals the number of oscillations of the shorter pendulum, $(n-1)$ equals the number of oscillations of the longer pendulum. Thus

$$t = nT_1 = (n-1)T_2$$

So, since $T_1 = 1.257\,\mathrm{s}$ and $T_2 = 1.539\,\mathrm{s}$,

$$n \times 1.257 = (n-1) \times 1.539$$

$$\therefore \quad n = \frac{1.539}{0.282} = 5.46$$

But $t = nT_1$, so

$$t = 5.46 \times 1.257 = 6.86\,\mathrm{s}$$

(b) The shorter pendulum makes $n = 5.5$ oscillations and the longer pendulum $(n-1) = 4.5$ oscillations.

Answer

(a) 6.9 s, (b) 5.5 and 4.5 oscillations.

Exercise 11.2

(Assume $g = 10\,\mathrm{m\,s^{-2}}$.)

1 A mass of 0.60 kg is hung on the end of a vertical light spring of force constant $30\,\mathrm{N\,m^{-1}}$. Calculate (a) the extension produced, (b) the time period of any subsequent oscillations, (c) the number of oscillations in 1 minute.

2
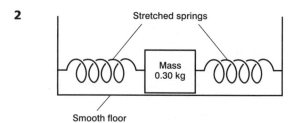
Fig. 11.3 Diagram for Question 2

Refer to Fig. 11.3, in which the 0.30 kg mass is tethered by two identical springs of force constant $2.5\,\mathrm{N\,m^{-1}}$. If the mass is now displaced by 20 mm to the left of its equilibrium position and released, calculate (a) the time period and frequency of subsequent oscillations, (b) the acceleration at the centre and extremities of the oscillation.

Note: effective force constant is twice that for one spring.

3 Calculate the length of a simple pendulum of periodic time (a) 1.0 s, (b) 0.5 s. If the two are set off oscillating in step, calculate (c) the number of times they will be in step over a 60 s period.

4 Two simple pendulums, of slightly different length, are set off oscillating in step. The next time they are in step is after a time of 20 s has elapsed, during which time the longer pendulum has completed exactly 10 oscillations. Find the length of each pendulum.

Displacement, velocity and acceleration variation with time

Refer to Fig. 11.2. The following relationships apply in SHM:

(1) The displacement y is related to time t by

$$y = r\sin\theta = r\sin\omega t \qquad (11.7)$$

This assures $y = 0$ when $t = 0$. The maximum displacement equals the amplitude r.

(2) The instantaneous velocity v is given by $v = r\omega\cos\omega t$ so that v is related to the displacement y by

$$v = \omega\sqrt{(r^2 - y^2)} \qquad (11.8)$$

Note that $v = 0$ at the extremities of the oscillation, when $y = r$. Also v has maximum value $\pm \omega r$ when $y = 0$, at the centre of the oscillation.

(3) The instantaneous acceleration a is given by $a = -\omega^2 r \sin \omega t$. Since $y = r \sin \omega t$ then:

$$a = -\omega^2 y \qquad \qquad \textbf{(11.1)}$$

This agrees with the original definition.

Example 6

A body vibrates in SHM in a vertical direction with an amplitude of 50 mm and a periodic time of 4.0 s.

(a) Calculate the displacement after (i) 2.5 s, (ii) 5.0 s, assuming that the displacement is zero at time zero.

(b) Calculate the time it takes the body to move to its maximum upwards displacement from a position 30 mm below it.

Method

(a) The angular velocity ω of the motion is given by $\omega = 2\pi/T$ where $T = 4.0$ s. So $\omega = 0.5\pi \, \text{rad s}^{-1}$. We use Equation 11.7 with $r = 50 \times 10^{-3}$ m to find displacement.

  (i) We have $t = 2.5$, so

$$y = r \sin \omega t = 50 \times 10^{-3} \sin (0.5\pi \times 2.5)$$
$$= 50 \times 10^{-3} \sin 1.25\pi$$

Now $\pi \, \text{rad} = 180°$, so $1.25\pi = 225°$, and

$$y = 50 \times 10^{-3} \sin 225° = -35 \times 10^{-3} \, \text{m}$$

Note: we assumed that the body was initially moving in a positive direction. The negative sign indicates a displacement in the opposite direction to this.

  (ii) We have $t = 5.0$ s, so

$$y = r \sin \omega t = 50 \times 10^{-3} \sin (0.5\pi \times 5)$$
$$= 50 \times 10^{-3} \sin 450°$$

We subtract multiples of 360°, which means that previous whole oscillations are ignored. Subtracting 360° from 450°, we have

$$y = 50 \times 10^{-3} \sin 90°$$
$$= 50 \times 10^{-3} \, \text{m}$$

(b) The body moves from an upwards displacement of 20 mm to 50 mm. Referring to Fig. 11.2 we have $y_1 = 20 \times 10^{-3}$ m and $y_2 = 50 \times 10^{-3}$ m. We use Equation 11.7 to find θ_1 and θ_2 and the corresponding times t_1 and t_2 for the rotating radius of Fig. 11.2(b). Thus:

$$y_1 = r \sin \theta_1$$
$$20 \times 10^{-3} = 50 \times 10^{-3} \sin \theta_1$$

or $\qquad \theta_1 = 0.412 \, \text{rad}$

Since $\theta = \omega t$, then

$$t_1 = \theta_1/\omega = 0.412/0.5\pi = 0.262 \, \text{s}.$$

Similarly:

$$y_2 = r \sin \theta_2$$
$$50 \times 10^{-3} = 50 \times 10^{-3} \sin \theta_2$$

This gives $\theta_2 = 0.5\pi \, \text{rad}$ and $t_2 = 1.0$ s.

(Note this time of 1.0 s corresponds to the time it takes to travel 1/4 of a period, from zero displacement to its first maximum.)

Hence, time taken:

$$t_2 - t_1 = 1.0 - 0.262 = 0.738 \, \text{s}.$$

Answer

(a) (i) -35 mm, (ii) 50 mm. (b) 0.74 s.

Example 7

A body moves in SHM with an amplitude of 30 mm and a frequency of 2.0 Hz. Calculate the values of (a) acceleration at the centre and extremities of the oscillation, (b) velocity at these positions, (c) velocity and acceleration at a point midway between the centre and extremity of the oscillation.

Method

We have $\omega = 2\pi f$ and $f = 2.0$. So $\omega = 4.0\pi \, \text{rad s}^{-1}$.

(a) We use Equation 11.1. At the centre $y = 0$ so $a = 0$.

At the extremities the displacement equals 30×10^{-3} m. So

$$a = -\omega^2 y = -(4\pi)^2 \times 30 \times 10^{-3}$$
$$= -0.48\pi^2 \, \text{m s}^{-2}$$

When y is positive a is negative and vice versa.

(b) We use Equation 11.8. At the centre $y = 0$ and $v = \pm \omega r$, depending on whether the body is moving upwards (+) or downwards (−) at that instant.
Since $r = 30 \times 10^{-3}$ and $\omega = 4.0\pi$,

$$v = \pm \omega r = \pm 0.12\pi \, \text{m s}^{-1}$$

At the extremities $v = 0$.

(c) At the midway point $y = 15 \times 10^{-3}$ m. Since $r = 30 \times 10^{-3}$, Equation 11.8 gives

$$v = \omega \sqrt{(r^2 - y^2)} = 4\pi \sqrt{(30^2 - 15^2)} \times 10^{-3}$$
$$= 0.33 \, \text{m s}^{-1}$$

This can be positive or negative depending on which way the body is moving.

Equation 11.1 gives

$$a = -\omega^2 y = -(16\pi^2) \times 15 \times 10^{-3}$$
$$= -0.24\pi^2 \,\mathrm{m\,s^{-2}}$$

When y is positive a is negative and vice versa.

Answer

(a) $0, \pm 0.48\pi^2 \,\mathrm{m\,s^{-2}}$, (b) $\pm 0.12\pi \,\mathrm{m\,s^{-1}}, 0$
(c) $\pm 0.33 \,\mathrm{m\,s^{-1}}, \pm 0.24\pi^2 \,\mathrm{m\,s^{-2}}$.

Exercise 11.3

1 A body is vibrating in SHM in a vertical direction with an amplitude of 40 mm and a frequency of 0.50 Hz. Assume at $t = 0$ the displacement is zero and it is moving upwards.

    (a) Calculate the values of displacement, velocity and acceleration at each of the following times (in seconds): 0.00, 0.25, 0.50, 0.75, 1.00, 1.25, 1.50, 1.75, 2.00. Sketch the graphs of displacement, velocity and acceleration against time.

    (b) Calculate the time it takes to travel from an upwards displacement of 20 mm to one of 30 mm in the same cycle. Compare this value of time with that taken from readings on the displacement-time graph.

2 A body vibrates in SHM with an amplitude of 30 mm and frequency of 0.50 Hz. Calculate (a) the maximum acceleration, (b) the maximum velocity, (c) the magnitude of acceleration and velocity when the body is displaced 10 mm from its equilibrium position.

State the value of the constants r (in metres) and ω (in $\mathrm{rad\,s^{-1}}$) in the equation $y = r\sin\omega t$ which describes the motion of the body.

Energy in SHM

There is a continuous interchange between kinetic energy (KE) and potential energy (PE) during vibration. Assuming no energy losses, the total energy is constant. At the centre of the oscillation we take PE as zero, so all the energy here is KE. Thus at the centre of the oscillation

$$\text{Total energy} = \text{KE} = \tfrac{1}{2}mv^2$$

Now $v = (\pm)\omega r$ at the centre, so

$$\boxed{\text{Total energy} = \text{KE} = \tfrac{1}{2}m\omega^2 r^2 \qquad (11.9)}$$

Example 8

A body of mass 0.10 kg oscillates in SHM with an amplitude of 5.0 cm and with a frequency of 0.50 Hz. Calculate (a) the maximum value and (b) the minimum value of its kinetic energy. State where these occur.

Method

(a) The maximum KE is at the centre of the motion. We use Equation 11.9 in which $m = 0.10$ kg, $\omega = 2\pi f = 2\pi \times 0.50 = \pi \,\mathrm{rad\,s^{-1}}$ and amplitude $r = 5.0 \times 10^{-2}$ m.

$$\begin{aligned}\text{KE} &= \tfrac{1}{2}m\omega^2 r^2 \\ &= \tfrac{1}{2} \times 0.1 \times \pi^2 \times (5.0 \times 10^{-2})^2 \\ &= 12 \times 10^{-4} \,\mathrm{J}\end{aligned}$$

(b) The minimum value of KE is at the extremities of the motion. Since velocity v is zero here KE is zero.

Answer

(a) 12×10^{-4} J, at centre, (b) zero, at extremities.

Exercise 11.4

1 A mass of 0.50 kg vibrates in SHM with a maximum KE of 3.0 mJ. If its amplitude is 20 mm, calculate the frequency of the motion.

2 A mass oscillates in SHM on the end of a spring of force constant $40 \,\mathrm{N\,m^{-1}}$. If the amplitude of the motion is 30 mm, calculate the maximum KE of the mass. (Hint: $\omega^2 = k/m$.)

3 A body oscillates in SHM with a total energy of 2.0 mJ. Calculate the total energy if (separately)

    (a) the amplitude is doubled (frequency being constant);

    (b) the frequency is halved (amplitude being constant);

    (c) the amplitude and frequency are both doubled.

Exercise 11.5: Examination questions

Assume $g = 10 \,\mathrm{m\,s^{-2}} \,(10 \,\mathrm{N\,kg^{-1}})$.

1 A motorist notices that when driving along a level road at $95 \,\mathrm{km\,h^{-1}}$ the steering wheel vibrates with an amplitude of 6.0 mm. If she speeds up or slows down, the amplitude of the vibrations becomes smaller.
Explain why this is an example of resonance.

Calculate the maximum acceleration of the steering wheel given that its frequency of vibration is 2.4 Hz. [Edexcel 2001]

2 A mass of 1.6 kg is suspended from a light vertical spring and oscillates with a period of 1.5 s. Calculate the force constant of the spring.

3 A 0.60 kg mass is suspended from a light helical spring which is attached to a peg, vibrating in simple harmonic motion, whose frequency of vibration can be varied as shown in Fig. 11.4(a). The variation in amplitude of vertical vibrations of the mass, as the frequency of vibration of the peg is varied, is shown in Fig 11.4(b).

Estimate the resonant frequency of the spring-mass system using Fig 11.4(b). Use this value to calculate the spring constant of the spring.

(a)

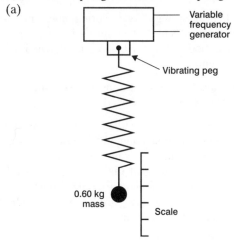

(b)

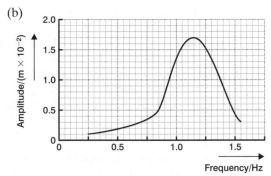

Fig 11.4 Information for Question 3

4 A fly of mass 0.25 g is trapped in a spider's web of negligible mass. When the fly struggles, it is noted that the web vibrates with a frequency of 16 Hz. The system of fly and web may be assumed to behave in the same way as a loaded helical spring.

(a) Calculate the effective force constant k of the web.

(b) Find the frequency of vibration if a bluebottle of mass 1.0 g were trapped at the same point in the same web, instead of the fly.

[CCEA 2000, part]

5 The mass of an empty car is 800 kg. It is supported on four identical springs. An evenly distributed load of mass 400 kg causes the car to compress each spring by a distance of 0.070 m. Each spring provides an upwards force F, given by $F = kx$, where x is the compression of the spring and k is the spring constant.

(a) Calculate the value of the spring constant k for **one** spring.

(b) The loaded car is pushed downwards and then released. Calculate the period of oscillation of the car on its springs. Neglect the effects of damping.

(c) Predict one disadvantage of a car designed with:
   (i) a very long period of oscillation;
   (ii) a very short period of oscillation.

[OCR 2001]

6 (a) A light helical spring is suspended vertically. The unstretched length of the spring is 200 mm. When a mass of 500 g is attached to the lower end, the total length becomes 240 mm.

Calculate the period of small vertical oscillations of the mass.

(b) With a mass M attached to the spring, the frequency of vertical oscillations is f. Calculate the new frequency of vibrations, as a multiple of the original frequency, if the mass were increased to $4M$.

7 This question is about oscillations of a tethered trolley.

A trolley is tethered by two elastic cords on a horizontal runway. Fig. 11.5 shows the arrangement.

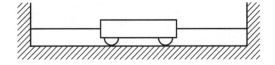

Fig. 11.5

The two identical elastic cords obey Hooke's law for trolley displacements up to and including 0.10 m. When released from an initial displacement of 0.10 m, the trolley executes simple harmonic motion. Fig. 11.6 shows the variation of overall restoring force, F, with trolley displacement, x.

(a) (i) How can you tell from the graph that Hooke's law is obeyed?
   (ii) Give a physical reason to explain why the gradient of the graph is negative.

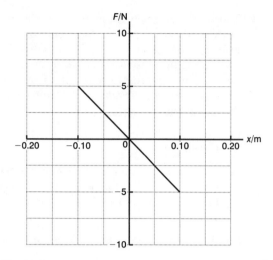

Fig. 11.6

(b) (i) Use the graph to find a value for the force constant of the system. Make your reasoning clear.

(ii) The trolley has a mass of 0.80 kg. Calculate the period of oscillation when the initial amplitude is 0.10 m.

(c) (i) What would be the period of oscillation of a trolley of mass 0.40 kg when tethered in the same way? Explain your answer.

(ii) What will be the period of oscillation of the 0.80 kg trolley when its amplitude is reduced to 0.05 m? Explain your answer.

(d) When the trolley displacement exceeds 0.10 m, one cord becomes slack. The other cord continues to obey Hooke's law as before.

(i) On Fig. 11.6, continue the force/displacement graph at both ends to show this behaviour. Draw these extensions as accurately as you can.

(ii) Does the trolley execute simple harmonic motion when displaced by 0.15 m? Explain your answer.   [OCR Nuff 2001]

8 A mass of 8.0 kg is suspended from a light vertical spring of force constant 2.0×10^3 N m^{-1}. The mass is displaced downwards by 9.0 mm and then released. Calculate

(a) the period of the resulting oscillations

(b) the maximum acceleration of the mass.

9 If the period of oscillation of a simple pendulum is doubled when the length of the pendulum is increased by 1.8 m, calculate the original length of the pendulum in metres.

10 A simple pendulum has a time period T at the surface of the Earth. If taken to another planet where the acceleration due to gravity is one half

that on Earth, what would the new time period be? Give your answer in terms of T.

11 Two simple pendulums of slightly different lengths are set off oscillating in phase. The time periods are 1.00 s and 0.98 s. Calculate the number of oscillations made by the *shorter* pendulum during the time interval it takes for the two pendulums to be once again moving in phase.

12 A body is oscillating in simple harmonic motion as described by the following expression:

$$y = 3 \sin (20\pi t)$$

Calculate the (minimum) time it takes the body to move from its mean position to its position of maximum displacement.

13 A helical spring has a spring constant (force constant) 50 N m^{-1}. The spring is hung vertically and a body of mass 0.40 kg is attached to the lower end.

(a) Calculate the extension of the spring. (Hooke's law is obeyed.)

(b) The body is then pulled down 20 mm from the equilibrium position and released. It oscillates in simple harmonic motion.
The magnitude of the acceleration of a body moving in simple harmonic motion is $\omega^2 x$, where x is the displacement from the equilibrium position.

Calculate

(i) the period and frequency of the subsequent oscillations,

(ii) the magnitude of the initial acceleration of the body when it is released,

(iii) the speed of the body when it is 5.0 mm below the equilibrium position,

(iv) the time taken for the body to move to the equilibrium position from a point 5.0 mm below it.   [CCEA 2000]

14 The movement of the tides may be assumed to be simple harmonic with a period approximately equal to 12 hours. The diagram overleaf shows a vertical wooden pole fixed firmly to the sea bed. A ring is attached to the pole at point R.

(a) What is the amplitude of this tide?

(b) High tide on a particular day is at 9 a.m. State the times of the next mid-tide and the next low tide.

(i) Next mid-tide:

(ii) Next low-tide:

(c) Calculate the time at which the falling water level reaches the ring R.

[Edexcel 2000, part]

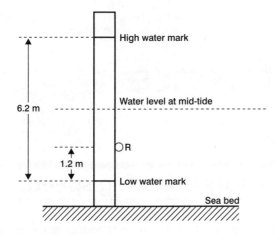

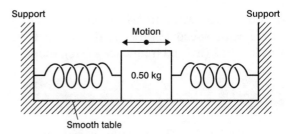

(c) Plot the points representing maxima and minima of kinetic energy on the graph grid below and sketch the graph of kinetic energy vs. time.

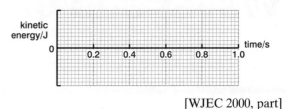

[WJEC 2000, part]

15 A metal sphere of mass 0.25 kg hangs from a spring. The top end of the spring is clamped. The sphere is raised 0.080 m above its equilibrium position and released.

A displacement vs. time graph for the motion is given below.

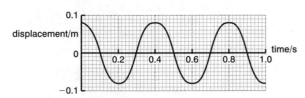

(a) Write down the periodic time of the motion.

(b) Calculate the maximum kinetic energy of the sphere.

16

Fig. 11.7   Diagram for Question 16

Fig. 11.7 shows a mass of 0.50 kg which is in contact with a smooth horizontal table. It is attached by two light springs to two fixed supports as shown. If the mass moves in linear simple harmonic motion with a period of 2.0 s and an amplitude of 4.0 cm, calculate the energy associated with this motion.

12
Waves and interference

Wave relationships

A progressive wave transfers energy from its source with speed c $(\mathrm{m\,s}^{-1})$. If the wave has wavelength λ (m) and frequency f (Hz), then

$$c = f\lambda \qquad\qquad (12.1)$$

The periodic time T (s) of the wave motion is related to frequency f by

$$f = \frac{1}{T} \qquad\qquad (12.2)$$

Equations 12.1 and 12.2 apply to longitudinal and transverse waves.

Example 1

A progressive wave travels a distance of 18 cm in 1.5 s. If the distance between successive crests is 60 mm, calculate (a) the frequency, (b) the periodic time of the wave motion.

Method

The speed c is given by

$$c = \frac{\text{Distance travelled (m)}}{\text{Time taken (s)}} = \frac{18 \times 10^{-2}}{1.5}$$

$$= 0.12\,\mathrm{m\,s}^{-1}$$

Now wavelength $\lambda = 60\,\mathrm{mm} = 0.060\,\mathrm{m}$. Rearranging Equation 12.1 gives

$$f = \frac{c}{\lambda} = \frac{0.12}{0.06} = 2.0\,\mathrm{Hz}$$

Rearranging Equation 12.2 gives

$$T = \frac{1}{f} = \frac{1}{2} = 0.50\,\mathrm{s}$$

Answer

(a) 2.0 Hz, (b) 0.50 s.

Factors affecting speed

The speed c of (longitudinal) sound waves in a solid is given by

$$c = \sqrt{E/\rho} \qquad\qquad (12.3)$$

where E is the Young's modulus $(\mathrm{N\,m}^{-2})$ of the material and ρ is the density $(\mathrm{kg\,m}^{-3})$.

The speed of propagation of transverse waves along a string or wire is given by

$$c = \sqrt{\frac{T}{m}} \qquad\qquad (12.4)$$

where T is the tension in the string, in newtons, m the mass per unit length of the string, in $\mathrm{kg\,m}^{-1}$.

Example 2

(a) Calculate the speed of propagation of longitudinal waves in a solid of Young's modulus $2.0 \times 10^{11}\,\mathrm{N\,m}^{-2}$ and density $7.8 \times 10^{3}\,\mathrm{kg\,m}^{-3}$.

(b) Calculate the time it takes the wave in the solid to travel 1.0 km and compare this with the time it takes sound to travel 1.0 km in air. Assume the speed of sound in air is $3.3 \times 10^{2}\,\mathrm{m\,s}^{-1}$.

Method

(a) We use Equation 12.3, in which $E = 2.0 \times 10^{11}$ and $\rho = 7.8 \times 10^{3}$. Thus

$$c = \sqrt{E/\rho} = \sqrt{(2.0 \times 10^{11}/7.8 \times 10^{3})}$$
$$= 5.06 \times 10^{3}\,\mathrm{m\,s}^{-1}$$

(b) The time taken t (s) is given by

$$t = \text{distance travelled (m)/speed } c$$

where distance travelled $= 1.0 \times 10^{3}\,\mathrm{m}$.

For the wave in the solid

$$t_{\text{solid}} = 1.0 \times 10^{3}/5.06 \times 10^{3} = 0.198\,\mathrm{s}$$

For the wave in air

$$t_{air} = 1.0 \times 10^3 / 3.3 \times 10^2 = 3.03 \, s$$

The wave in the solid takes much less time to travel 1.0 km since its speed is much greater.

Answer

(a) $5.1 \times 10^3 \, \text{m s}^{-1}$ (b) 0.20 s; 3.0 s

Example 3

A horizontal stretched elastic string has length 3.0 m and mass 12 g. It is subject to a tension of 1.6 N. Transverse waves of frequency 40 Hz are propagated down the string. Calculate the distance between successive crests of this wave motion.

Method

We use Equation 12.4, with mass per unit length of string $m = (12 \times 10^{-3}) \div 3.0 = 4.0 \times 10^{-3} \, \text{kg m}^{-1}$. Since $T = 1.6 \, \text{N}$,

$$c = \sqrt{\frac{T}{m}} = \sqrt{\frac{1.6}{4.0 \times 10^{-3}}}$$

$$= 20 \, \text{m s}^{-1}$$

The distance between successive crests is the wavelength λ. We have frequency $f = 40$ Hz. Rearranging Equation 12.1 gives

$$\lambda = \frac{c}{f} = \frac{20}{40} = 0.50 \, \text{m}$$

Answer

0.50 m.

Phase angle

Exercise 12.1

1 The speed of electromagnetic waves in air is $3.0 \times 10^8 \, \text{m s}^{-1}$. Calculate (a) the frequency of yellow light of wavelength 0.60×10^{-6} m, (b) the wavelength of radio waves of frequency 2.0×10^5 Hz.

2 Calculate the Young's modulus of aluminium, given that the speed of propagation of longitudinal waves is $5.0 \times 10^3 \, \text{m s}^{-1}$ and its density is $2.7 \times 10^3 \, \text{kg m}^{-3}$.

3 The speed of propagation of sound waves in steel is $5.1 \times 10^3 \, \text{m s}^{-1}$. Calculate the speed of sound in a solid with the same density but with half the Young's modulus.

4 The speed of transverse waves along a stretched wire is $50 \, \text{m s}^{-1}$. What is the speed when the tension in the wire is doubled?

5 A horizontal stretched elastic string is subject to a tension of 2.5 N. Transverse waves of frequency 50 Hz and wavelength 2.0 m are propagated down the string. Calculate (a) the speed of the waves, (b) the mass per unit length of the string.

Fig. 12.1 shows the displacement y at all points on a sine wave, at a fixed time, over a single wavelength. It shows how displacement* y varies with distance x. The particle at P lags behind the particle at O by phase angle ϕ (in radians) given by

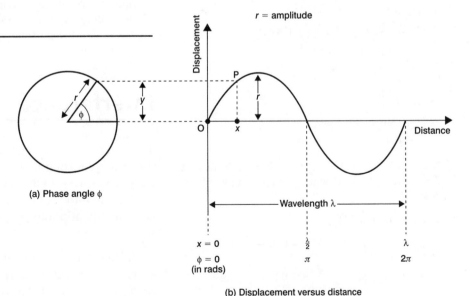

(a) Phase angle ϕ

(b) Displacement versus distance

*Displacement variation with time, at a given point on a sine wave, is dealt with in Chapter 11 on simple harmonic motion – see Fig 11.2 and Equation 11.7.

Fig. 12.1 Displacement at a fixed time

$$\phi = \frac{2\pi}{\lambda} \times x \qquad (12.5)$$

The relation between y and x in Fig 12.1b is:

$$y = r \sin \phi = r \sin \frac{2\pi}{\lambda} x = r \sin kx \qquad (12.6)$$

where $k = 2\pi/\lambda$ is called the wave number.

Example 4

A progressive wave has amplitude 0.40 m and wavelength 2.0 m. At a given time the displacement $y = 0$ at $x = 0$. Calculate

(a) the displacement at $x = 0.50$ m and 1.4 m;

(b) the phase angles at $x = 0.50$ m and 0.80 m;

(c) the phase difference between any two points which are 0.30 m apart on the wave.

Method

We have amplitude $r = 0.40$ and wavelength $\lambda = 2.0$.

(a) Using Equation 12.6, with

$$k = 2\pi/\lambda = 2\pi/2 = \pi \, \text{m}^{-1},$$

we have:

for $x = 0.5$,
$$y = r \sin kx = 0.4 \sin (\pi \times 0.5)$$
$$= 0.4 \sin 90° = 0.40 \, \text{m}$$

(Note here that $y = r$, since $x = \frac{1}{4}\lambda$.)

for $x = 1.4$,
$$y = r \sin kx = 0.4 \sin (\pi \times 1.4)$$
$$= 0.4 \sin 252° = -0.38 \, \text{m}$$

Note the negative sign which indicates a *downwards* displacement, assuming upwards is positive.

(b) Using Equation 12.5:

for $x = 0.5$, $\quad \phi = \frac{2\pi x}{\lambda} = 0.5\pi$

for $x = 0.8$, $\quad \phi = \frac{2\pi x}{\lambda} = 0.8\pi$

(c) We can replace Equation 12.5 by

$$\Delta\phi = \frac{2\pi}{\lambda} \times \Delta x$$

where $\Delta\phi$ is the phase difference in radians between two points spaced Δx (m) apart on the wave. We have $\Delta x = 0.30$ and $\lambda = 2.0$, so

$$\Delta\phi = \frac{2\pi}{\lambda} \times \Delta x = \frac{2\pi}{2} \times 0.3 = 0.3\pi$$

Note that this agrees with part (b), since two points at 0.5 m and 0.8 m have phase angles 0.5π and 0.8π.

Answer

(a) 0.40 m, −0.38 m; (b) 0.5π rad, 0.8π rad; (c) 0.3π rad.

Exercise 12.2

1 A wave on a stretched string has amplitude 5.0 cm and wavelength 30 cm. At a given time the displacement $y = 0$ at $x = 0$. Calculate (a) the wave displacements at $x = 10$ cm and $x = 50$ cm, (b) the phase angles at $x = 10$ cm and $x = 50$ cm.

2 A progressive wave has wavelength 20 cm. Calculate the minimum distance between two points which differ in phase by $60°$ ($\pi/3$ rad).

3 A transverse wave travels along a horizontal stretched string. In front of the string is a screen with two slots in it so that all an observer can see is the motion of two points on the string placed 3.0 m apart. The observer notes that the two points perform SHM with a period of 2.0 s, and that one point lags in phase by $90°$ compared with the other. Calculate (a) the frequency of the wave, (b) *two* possible values for the wavelength of the wave.

Interference

This phenomenon occurs for all types of waves – for example sound, water waves and electro-magnetic waves (light, microwaves and so on). To simplify the situation our initial treatment considers continuous waves, like sound or water waves.

Interference occurs due to superposition of waves – the resultant displacement being the sum of the separate displacements of the individual wave motions. Fig. 12.2 shows two sources S_1 and S_2 which emit waves of the same frequency and wavelength λ and of approximately the same

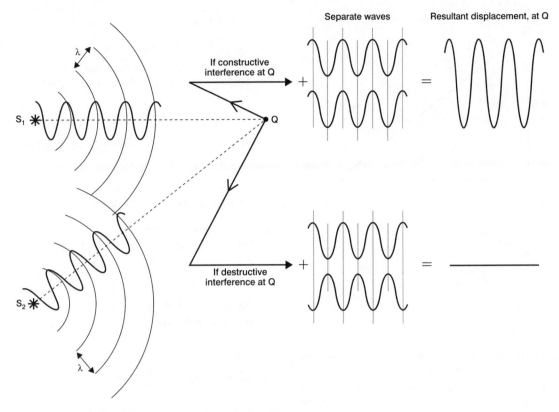

Fig. 12.2 Interference at Q between waves from two sources

amplitude. Regions of constructive and destructive interference exist. At a given point Q in the interference pattern

$$S_2Q - S_1Q = n\lambda \qquad \text{for constructive interference}$$

$$S_2Q - S_1Q = (n + \tfrac{1}{2})\lambda \qquad \text{for destructive interference}$$

where $n = 0, 1, 2, 3 \ldots$ This assumes the waves from S_1 and S_2 set off in phase.

When waves from two sources arrive at a point in phase there is constructive interference. If the waves arrive out of phase there is destructive interference.

Example 5

Fig. 12.3 shows two sources X and Y which emit sound of wavelength 2.0 m. The two sources emit in phase, and emit waves of equal amplitude. What does an observer hear (a) at Q, (b) at R.

Method

(a) Q is equidistant from X and Y, so XQ = YQ. Thus

$$XQ - YQ = 0$$

There is constructive interference at Q, since the two sets of waves arrive in phase. The resultant

amplitude of the sound at Q is twice that due to each source acting individually.

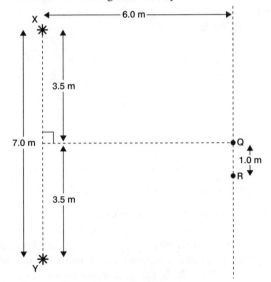

Fig. 12.3 Information for Example 5

(b) We must find the path difference XR − YR.

Refer to Fig. 12.4. Using Pythagoras' theorem, we see that

$$XR^2 = 4.5^2 + 6.0^2 = 56.25$$

$$\therefore \quad XR = 7.5 \text{ m}$$

Also $YR^2 = 2.5^2 + 6.0^2 = 42.25$

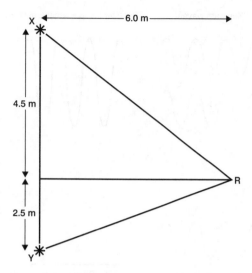

Fig. 12.4   Solution to Example 5

$\therefore$ YR $= 6.5$ m

So   XR $-$ YR $= 1.0$ m $= \frac{1}{2}\lambda$

since wavelength $\lambda = 2.0$ m. There is destructive interference at R because the two sets of waves arrive with a path difference of $\frac{1}{2}\lambda$, i.e. 180° out of phase. The resultant amplitude at R will be zero,* so that an observer will hear nothing at R.

Answer

(a) A sound of double amplitude, (b) nothing.

Example 6

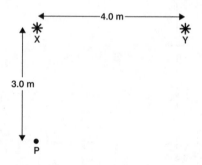

Fig. 12.5   Information for Example 6

Fig. 12.5 shows two sources X and Y which are identical and emit in phase. Calculate *two* possible values of wavelength for which (a) constructive interference, (b) destructive interference would occur at point P.

Method

We must calculate the path difference YP $-$ XP. Using Pythagoras' theorem,

$$YP^2 = 3^2 + 4^2 = 25$$

*This ignores any difference in amplitude of the waves which may occur because R is further from X than Y.

$\therefore$ YP $= 5.0$ m

$\therefore$ YP $-$ XP $= 2.0$ m

(a) For constructive interference   YP $-$ XP $= n\lambda$. Thus wavelength λ is given by $n\lambda = 2.0$ m

$\therefore$ $\lambda = \dfrac{2.0}{n}$ where $n = 0, 1, 2, 3, \ldots$

For $n = 0$, $\lambda = \infty$ which is not practical. For $n = 1$, $\lambda = 2.0$ m.

For $n = 2$, $\lambda = 1.0$ m. Clearly other (smaller) values of λ are also suitable.

(b) For destructive interference YP $-$ XP $= \left(n + \frac{1}{2}\right)\lambda$. Thus wavelength λ is given by $\left(n + \frac{1}{2}\right)\lambda = 2.0$.

$\therefore$ $\lambda = \dfrac{2.0}{\left(n + \frac{1}{2}\right)}$ where $n = 0, 1, 2, 3, \ldots$

For $n = 0$, $\lambda = 4.0$ m. For $n = 1$, $\lambda = 4/3$ m. Other (smaller) values of λ are also suitable.

Answer

(a) 2.0 m, 1.0 m, (b) 4.0 m, $\frac{4}{3}$ m.

Exercise 12.3

1   Referring to Fig. 12.3, suppose that source X is 180° out of phase with source Y. What does an observer hear (a) at Q, (b) at R?

2

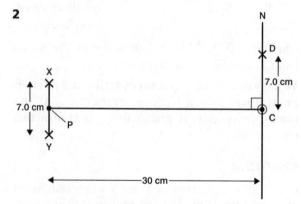

Fig. 12.6   Information for Question 2

Fig. 12.6 shows two identical microwave sources X and Y which emit in phase. There is constructive interference at C, which is on the perpendicular bisector of the line XY and 30 cm from P, the midpoint of XY. A detector moved from C towards N locates the first minimum at D. If CD $= 7.0$ cm calculate the wavelength of the microwaves emitted by X and Y.

3

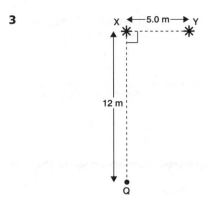

Fig. 12.7   Diagram for Question 3

X and Y in Fig. 12.7 are two identical sources of sound which emit in phase. Calculate the largest two values of wavelength (excluding $\lambda = \infty$) for which (a) constructive, (b) destructive interference will occur at Q. If the velocity of sound in air is $340\,\text{m}\,\text{s}^{-1}$, calculate the frequencies to which these wavelengths correspond.

4

Fig. 12.8   Diagram for Question 4

X and Y in Fig. 12.8 are two identical sources of sound which emit in phase. Calculate the lowest possible value of frequency of the sources for there to be (a) constructive, (b) destructive interference at Q. (Velocity of sound $= 340\,\text{m}\,\text{s}^{-1}$.)

Young's double-slit arrangement

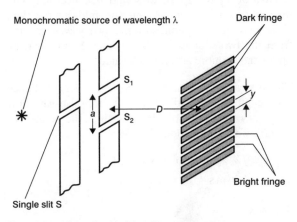

Fig. 12.9   Young's double-slit arrangement

Fig. 12.9 shows the set-up. The dark and bright fringes arise due to the interference of light emerging from two slits S_1 and S_2. In order that

the sources S_1 and S_2 are coherent (i.e. phase-linked and of the same frequency) they must receive light from the same point on the source – this is ensured by diffraction of light at the single slit S.

The fringe separation y, in metres, is given by

$$y = \frac{\lambda D}{a} \qquad (12.7)$$

where λ is the wavelength of source, in metres, D the distance, in metres, from slits to fringes and a the slit separation, in metres.

Example 7

In a Young's double-slit experiment, mercury green light of wavelength $0.54\,\mu\text{m}$ $(0.54 \times 10^{-6}\,\text{m})$ was used with a pair of parallel slits of separation $0.60\,\text{mm}$. The fringes were observed at a distance of $40\,\text{cm}$ from the slits. Calculate the fringe separation.

Method

We have $\lambda = 0.54 \times 10^{-6}$, $a = 0.60 \times 10^{-3}$ and $D = 0.40$. Using Equation 12.7

$$y = \frac{\lambda D}{a} = \frac{0.54 \times 10^{-6} \times 0.40}{0.60 \times 10^{-3}}$$

$$= 0.36 \times 10^{-3}\,\text{m}$$

Answer

Fringe separation $= 0.36\,\text{mm}$.

Example 8

In a Young's double-slit arrangement green monochromatic light of wavelength $0.50\,\mu\text{m}$ was used. Five fringes were found to occupy a distance of $4.0\,\text{mm}$ on the screen. Calculate the fringe separation if (independently) (a) red light of wavelength $0.65\,\mu\text{m}$ was used, (b) the slit separation was doubled, (c) the slits–screen distance was doubled.

Method

Five fringes occupy $4.0\,\text{mm}$. So the fringe separation is $4.0/5 = 0.80\,\text{mm}$.

(a) We see from Equation 12.7 that, for fixed D and a value, $y \propto \lambda$. If λ increases by a factor of $(0.65 \times 10^{-6}) \div (0.50 \times 10^{-6}) = 1.3$, then y will increase by a factor of 1.3. Thus y becomes

$$1.3 \times 0.80 = 1.04\,\text{mm}$$

(b) For given λ and D values, $y \propto 1/a$. So if a is doubled, y becomes halved. Thus y becomes

$$0.50 \times 0.80 = 0.40\,\text{mm}$$

(c) For given λ and a values $y \propto D$. So if D is doubled, y is doubled. Thus y becomes 1.6 mm.

Answer

(a) 1.0 mm, (b) 0.40 mm, (c) 1.6 mm.

Exercise 12.4

1 In a Young's double-slit experiment, sodium light of wavelength 0.59×10^{-6} m was used to illuminate a double slit with separation 0.36 mm. If the fringes are observed at a distance of 30 cm from the double slits, calculate the fringe separation.

2 In an experiment using Young's slits, six fringes* were found to occupy 3.0 mm when viewed at a distance of 36 cm from the double slits. If the wavelength of the light used is 0.59 μm, calculate the separation of the double slits.

3 When red monochromatic light of wavelength 0.70 μm is used in a Young's double-slit arrangement, fringes with separation 0.60 mm are observed. The slit separation is 0.40 mm. Find the fringe spacing if (independently)

(a) yellow light of wavelength 0.60 μm is used;

(b) the slit separation becomes 0.30 mm;

(c) the slit separation is 0.30 mm and the slits–fringe distance is doubled.

Formation of stationary (standing) waves

Stationary (standing) waves occur as a result of interference between progressive waves of the same frequency and wavelength travelling along the same line. They may be formed due to interference between waves from two separate sources, as shown in Fig. 12.10, or alternatively, due to interference between incident and reflected waves (see Example 10).

If the two progressive waves which form the stationary wave have equal amplitude r, then the nodes, which are positions of permanent destructive interference, have zero amplitude. The antinodes, which are positions of maximum constructive interference, have amplitude $2r$. As shown in Fig. 12.10, the separation of adjacent

*One fringe means one fringe separation.

nodes and of adjacent antinodes is $\lambda/2$, where λ is the wavelength of the progressive waves from which the stationary wave is formed.

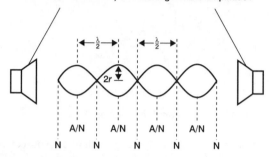

Two sources of progressive waves (e.g. loudspeaker or microwave transmitter) of wavelength λ and amplitude r

N = node
A/N = antinode

Fig. 12.10 Formation of a stationary (standing) wave in the region between two sources of progressive waves

Example 9

Two loudspeakers which are connected to the same oscillator face each other and are separated by a distance of about 3 m. A small microphone, placed approximately midway along the line between the two loudspeakers, records positions of minimum intensity, which are separated by 4.2 cm. If the oscillator is set at a frequency 4.0 kHz, calculate the speed of sound in air.

Method

The loudspeaker separation of about 3 m happens to be a convenient distance, but is irrelevant in so far as calculation of the speed of sound is concerned. The microphone is moved around the midway position, since here the amplitude of the two waves arriving from the two sources will be about the same, so the nodes can be more accurately located.

The nodes are 4.2 cm apart. So $\lambda/2 = 4.2$ cm, hence wavelength $\lambda = 8.4$ cm $= 8.4 \times 10^{-2}$ m. Also we know frequency $f = 4.0 \times 10^3$ Hz. To find the speed of sound c we use Equation 12.1, i.e.

$$c = f\lambda = 4.0 \times 10^3 \times 8.4 \times 10^{-2}$$
$$= 336 \text{ m s}^{-1}$$

Answer

Speed of sound = 0.34 km s^{-1}.

Note that the wavelength λ and speed c relate to the progressive waves which make up the stationary wave.

Example 10

A microwave transmitter is aimed at a metal plate, as shown in Fig. 12.11.

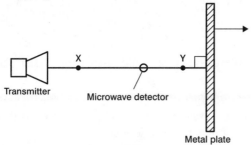

Fig. 12.11   Information for Example 10

(a) A small detector, moved along the line XY, travels 14 cm in moving from the first to the eleventh consecutive nodal position. Calculate the frequency of the microwaves emitted.

(b) The detector is now fixed in position and the metal plate is moved to the right, along the direction XY at a speed of $28 \, \text{cm s}^{-1}$. Explain what the detector observes.
Assume that the speed of electromagnetic waves is $3.0 \times 10^8 \, \text{m s}^{-1}$.

Method

(a) Between the first and eleventh nodes there are ten half-wavelengths. Thus $10 \times \lambda/2 = 14 \, \text{cm}$, so wavelength $\lambda = 2.8 \, \text{cm} = 2.8 \times 10^{-2} \, \text{m}$. We are given speed $c = 3.0 \times 10^8$; to find the frequency f we rearrange Equation 12.1:

$$f = \frac{c}{\lambda} = \frac{3 \times 10^8}{2.8 \times 10^{-2}}$$

$$= 1.07 \times 10^{10} \, \text{Hz}$$

(b) As the metal plate moves to the right, the stationary wave pattern moves also – and at the same speed, since there must always be a node* at the metal plate (it is a 'perfect' reflector). In 1 second a 28 cm 'length' of stationary wave will pass the detector, which will thus observe $(28 \div 1.4) = 20$ nodes and 20 antinodes.

Answer

(a) $1.1 \times 10^{10} \, \text{Hz}$, (b) the detector observes 20 successive maxima, followed by minima, each second.

Exercise 12.5

1 Two loudspeakers face each other and are separated by a distance of about 20 m. They are connected to the same oscillator, which gives a signal frequency of 800 Hz.

Electric field node.

(a) Calculate the separation of adjacent nodes along the line joining the two loudspeakers.

(b) A small microphone, moved at constant speed along this line, records a signal which varies periodically at 5.0 Hz. Calculate the speed at which the microphone moves.
Assume that the speed of sound is $340 \, \text{m s}^{-1}$.

2 A source S of microwaves faces a detector D. A metal reflecting screen is now placed beyond D with its plane perpendicular to the line from S to D. As the screen is moved slowly away from D, the detector registers a series of maximum and minimum readings, the screen being displaced a distance of 5.6 cm between the first and fifth minimum. Calculate the wavelength and frequency of the microwaves. Assume $c = 3.0 \times 10^8 \, \text{m s}^{-1}$.

Stationary waves in strings and wires

When a string or wire which is fixed at both ends is plucked, progressive transverse waves travel along the string or wire and are reflected at its ends. This results in the formation of stationary waves with certain allowed wavelengths and frequencies. Fig. 12.12 shows the fundamental, which has the largest wavelength and hence the smallest frequency, and the first two overtones.

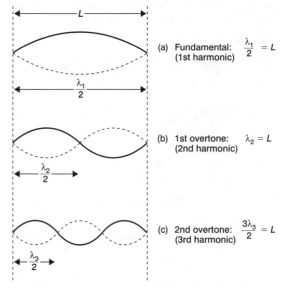

Fig. 12.12   Stationary waves in a string or wire fixed at both ends

Now the speed of transverse waves along a stretched string or wire is given by Equation 12.4

$$c = \sqrt{\frac{T}{M}} \qquad\qquad (12.4)$$

where T is the tension and m the mass per unit length. Thus the wavelengths and frequencies of the stationary waves in Fig. 12.12 are as follows:

Table 12.1

Mode	Wavelength	Frequency
Fundamental	$\lambda_1 = 2L$	$f_1 = \dfrac{c}{\lambda_1} = \dfrac{1}{2L}\sqrt{\dfrac{T}{m}}$
1st overtone	$\lambda_2 = L$	$f_2 = \dfrac{c}{\lambda_2} = \dfrac{1}{L}\sqrt{\dfrac{T}{m}}$
2nd overtone	$\lambda_3 = \frac{2}{3}L$	$f_3 = \dfrac{c}{\lambda_3} = \dfrac{3}{2L}\sqrt{\dfrac{T}{m}}$

Note that $f_2 = 2f_1$, so the first overtone is the second harmonic, and $f_3 = 3f_1$, so the second overtone is the third harmonic. If the string or wire is held at the centre, only even harmonics (2nd, 4th and so on) can occur.

Example 11

A horizontal string is stretched between two points a distance 0.80 m apart. The tension in the string is 90 N and its mass is 4.5 g. Calculate (a) the speed of transverse waves along the string and (b) the wavelengths and frequencies of the three lowest frequency modes of vibration of the string. (c) Explain how your answer to (b) would differ if the string is held lightly at its centre position.

Method

(a) To find the speed c we use Equation 12.4, with $T = 90$ and $m = (4.5 \times 10^{-3}) \div 0.80$:

$$c = \sqrt{\frac{T}{m}} = \sqrt{\frac{90}{(4.5 \times 10^{-3}) \div 0.80}}$$
$$= 126 \,\mathrm{m\,s^{-1}}$$

(b) The fundamental has wavelength $\lambda_1 = 2L = 1.6$ m. Its frequency f_1 is given by

$$f_1 = \frac{c}{\lambda_1} = \frac{126}{1.6} = 78.8\,\mathrm{Hz}$$

The first overtone has wavelength

$$\lambda_2 = L = 0.80\,\mathrm{m}$$

Its frequency f_2 is given by

$$f_2 = \frac{c}{\lambda_2} = \frac{126}{0.8} = 158\,\mathrm{Hz}$$

Alternatively, we could use $f_2 = 2f_1$ (see Table 12.1).

The second overtone has wavelength

$$\lambda_3 = \tfrac{2}{3}L = 0.533\,\mathrm{m}$$

Its frequency f_3 is given by

$$f_3 = \frac{c}{\lambda_3} = \frac{126}{0.533} = 236\,\mathrm{Hz}$$

Alternatively we could use $f_3 = 3f_1$ (see Table 12.1).

(c) If the string is held lightly at the centre, then only even harmonics are possible, i.e. those with the following wavelengths and frequencies:

$$\text{2nd harmonic} \quad \lambda_2 = L = 0.80\,\mathrm{m}$$
$$f_2 = 2f_1 = 158\,\mathrm{Hz}$$

$$\text{4th harmonic} \quad \lambda_4 = \frac{L}{2} = 0.40\,\mathrm{m}$$
$$f_4 = 4f_1 = 316\,\mathrm{Hz}$$

$$\text{6th harmonic} \quad \lambda_6 = \frac{L}{3} = 0.27\,\mathrm{m}$$
$$f_6 = 6f_1 = 474\,\mathrm{Hz}$$

and so on.

Answer

(a) $126\,\mathrm{m\,s^{-1}}$.
(b) Wavelengths: 1.6 m, 0.80 m, 0.53 m. Frequencies: 79 Hz, 0.16 kHz, 0.24 kHz.
(c) Even harmonics only, as detailed above.

Example 12

The fundamental frequency of vibration of a stretched wire is 120 Hz. Calculate the new fundamental frequency if (a) the tension in the wire is doubled, the length remaining constant, (b) the length of the wire is doubled, the tension remaining constant, (c) the tension is doubled and the length of the wire is doubled.

Method

In Table 12.1 we see that the fundamental frequency f_1 is given by

$$f_1 = \frac{1}{2L}\sqrt{\frac{T}{m}} \qquad\qquad (12.8)$$

For a particular wire the mass per unit length m is constant.

(a) For a constant length L and for a constant m we see from Equation 12.8 that $f_1 \propto \sqrt{T}$. Since the tension doubles, the new fundamental frequency f'_1 is $\sqrt{2}$ times the original. Thus

$$f'_1 = \sqrt{2} \times 120 = 170\,\mathrm{Hz}$$

(b) For a constant tension T and for a constant m we see from Equation 12.8 that $f_1 \propto 1/L$. Since the length doubles, the new fundamental frequency f''_1 is half the original, i.e. 60 Hz.

(c) For a constant m Equation 12.8 tells us that $f_1 \propto \sqrt{T}/L$. If the tension doubles and the length doubles, then the new fundamental frequency f'''_1 is $\sqrt{(2)}/2$ times the original. Thus

$$f'''_1 = \frac{\sqrt{2}}{2} \times 120 = 84.9 \, \text{Hz}$$

Answer

(a) 170 Hz, (b) 60.0 Hz, (c) 84.9 Hz.

Exercise 12.6

1   A horizontal wire of fixed length 0.90 m and mass per metre 4.5×10^{-3} kg m^{-1} is subject to a fixed tension of 50 N. Find the wavelengths and frequencies of the three lowest frequency modes of vibration when the wire is (a) free to vibrate at its midpoint, (b) lightly held at its midpoint.

2   A wire of cross-sectional area 0.20 mm^2 and made of steel of density 8.0×10^3 kg m^{-3} is subject to a tension of 60 N. Calculate (a) the mass per unit length of the wire, (b) the speed of transverse waves propagated down the wire, (c) the wavelength of waves with frequency 120 Hz, (d) the length of wire which, when fixed at its ends, gives a fundamental frequency of 120 Hz. Note: Mass = Length × Area × Density.

3   The fundamental frequency of vibration of a stretched wire is 150 Hz. Calculate the new fundamental frequency if (a) the tension in the wire is tripled, the length remaining constant, (b) the length of wire is halved, the tension remaining constant, (c) the tension is tripled and the length of wire is halved.

Stationary waves in pipes

When an air column is made to vibrate at one end, a progressive longitudinal (sound) wave travels along the air column and is reflected at its end so that a stationary longitudinal (sound) wave is formed.

Fig. 12.13 shows a 'closed' or 'stopped' pipe, which means it is closed at one end. The fundamental and the first two overtones are shown. Let c be the speed of progressive sound

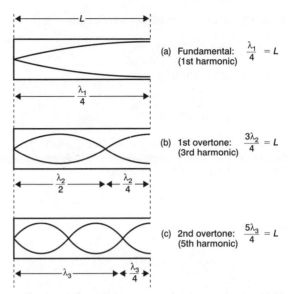

Note that a node exists at the closed end and an antinode at the open end

Fig. 12.13   Stationary waves in a 'closed' pipe

waves in air at the particular temperature. The wavelengths and frequencies of the stationary waves in Fig. 12.13 are as follows (Table 12.2):

Table 12.2 Closed pipe

Mode	Wavelength	Frequency
Fundamental	$\lambda_1 = 4L$	$f_1 = \dfrac{c}{\lambda_1} = \dfrac{c}{4L}$
1st overtone	$\lambda_2 = \frac{4}{3}L$	$f_2 = \dfrac{c}{\lambda_2} = \dfrac{3c}{4L}$
2nd overtone	$\lambda_3 = \frac{4}{5}L$	$f_3 = \dfrac{c}{\lambda_3} = \dfrac{5c}{4L}$

Note that $f_2 = 3f_1$, so that the first overtone is the third harmonic, and $f_3 = 5f_1$, so that the second overtone is the fifth harmonic.

Example 13

A closed organ pipe is of length 0.680 m. Calculate the wavelengths and frequencies of the three lowest frequency modes of vibration. Take the speed of sound to be 340 m s^{-1}.

Method

The pipe has length $L = 0.680$ m, and the speed of sound $c = 340$ m s^{-1}.

According to Table 12.2 the fundamental has wavelength $\lambda_1 = 4L = 2.72$ m. Its frequency f_1 is given by

$$f_1 = \frac{c}{\lambda_1} = \frac{340}{2.72} = 125 \, \text{Hz}$$

105

Similarly the first overtone has wavelength

$$\lambda_2 = 4L/3 = 0.907\,\text{m}$$

and frequency f_2 given by

$$f_2 = \frac{c}{\lambda_2} = \frac{340}{0.907} = 375\,\text{Hz}.$$

Alternatively we could use $f_2 = 3f_1$ (see Table 12.2).

The second overtone has wavelength

$$\lambda_3 = 4L/5 = 0.544\,\text{m}$$

and frequency f_3 given by

$$f_3 = \frac{c}{\lambda_3} = \frac{340}{0.544} = 625\,\text{Hz}$$

Alternatively we could use $f_3 = 5f_1$ (see Table 12.2).

Answer

The wavelengths are 2.72 m, 0.907 m and 0.544 m with frequencies 125 Hz, 375 Hz and 625 Hz respectively.

Exercise 12.7

(Assume that the speed of sound is $340\,\text{m s}^{-1}$)

1 Calculate the length of a closed pipe with a fundamental frequency of 250 Hz.

2 A tall vertical cylinder is filled with water and a tuning fork of frequency 512 Hz is held over its open end. The water is slowly run out. Calculate the position of the water level below the open end when (a) first resonance and (b) second resonance are heard.

3 An organ pipe, of length 0.500 m, is closed at one end. Calculate the values of the two lowest resonant frequencies of the pipe.

4   Loudspeaker

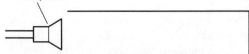

Fig 12.14   Diagram for Question 4

A small loudspeaker is mounted at one end of a tube as shown in Fig. 12.14, the other end of which is closed. The loudspeaker is connected to a signal generator of variable frequency and the frequency is gradually increased. The lowest frequency which will cause the air in the tube to resonate is 200 Hz. Calculate the values of the next two resonant frequencies.

Exercise 12.8:
Examination questions

1 A wave has a wavelength of 6.0 m and a frequency of 2.5 Hz. Calculate the wave-velocity.

2 Water waves moving across the surface of a pond travel a distance of 14 cm in 0.70 s. The horizontal distance between a crest and a neighbouring trough is 2.0 mm. Calculate the frequency of the waves.

3 (a) Describe the behaviour of the particles in a stretched cord during the passage of a transverse wave.

(b) A large explosion at the Earth's surface creates two waves, a compressional wave (P) with a speed of $6.0\,\text{km s}^{-1}$ and a shear wave (S) with a speed of $3.5\,\text{km s}^{-1}$. Both waves travel along the surface of the Earth to a seismological station; where the waves arrive with a 30 s interval between them.
Calculate the distance, measured along the Earth's surface, between the seismological station and the site of the explosion.
[OCR 2001]

4 The speed of sound in steel is $5.1 \times 10^3\,\text{m s}^{-1}$. If steel has a density of $7.8 \times 10^3\,\text{kg m}^{-3}$, calculate its Young's modulus.

5 In old Hollywood Western films the outlaws would sometimes be shown with their ears on the railway track listening for an approaching train.

(a) Calculate the speed of sound in the metal railway track.
(Young modulus of steel $= 2.0 \times 10^{11}\,\text{N m}^{-2}$; density of steel $= 8000\,\text{kg m}^{-3}$.)

(b) If the train produced a sudden noise on the railway track, the listeners would hear two noises. Explain why they would hear two noises. The noise was produced 2.0 km from the outlaws. Calculate the time interval between the two noises heard. The speed of sound in air is $330\,\text{m s}^{-1}$. [Edexcel S-H 2000]

6 (a) (i) State how the variation of amplitude with distance from the source differs for a progressive wave and a stationary wave.
(ii) State how the energy flow differs for a progressive wave and a stationary wave.

(b)

A transverse progressive wave is travelling in the x-direction. Graphs of displacement, y, against time are given below for **two** points in the path of the wave.

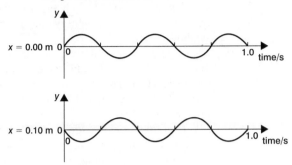

(i) Deduce from the graphs
   (I) the frequency of the waves,
   (II) **two** possible values for the wavelength, explaining your reasoning.
(ii) Use **one** of your wavelength values from part (i) (II) above to calculate a possible wave speed. [WJEC 2001]

7 (a) A transverse wave passes through a medium. The speed of propagation of the wave is $5.0\,\mathrm{m\,s^{-1}}$. Fig. 12.15 is a graph of the displacement s of a particle of the medium as a function of time t.

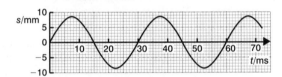

Fig. 12.15

(i) Using information from Fig. 12.15 deduce the amplitude, the period and the frequency of the wave.
(ii) Calculate the wavelength of the wave.

(b) In a simplified description of an earthquake, shock waves travel radially outwards as though from a point source. The waves are of two types, P-waves and S-waves. P-waves travel with a constant speed of $8.4\,\mathrm{km\,s^{-1}}$, and S-waves with a constant speed of $5.6\,\mathrm{km\,s^{-1}}$.
Following a particular earthquake, a monitoring station receives P-wave and S-wave signals separated by a time interval of 65 s.
(i) Calculate the distance of the source of the earthquake from the monitoring station.

(ii) The information obtained from the monitoring station is limited to the distance of the source from the station. However, similar information from a number of stations may be combined to locate the source of the earthquake accurately. What is the minimum number of stations required? Explain your answer with the aid of a diagram. [CCEA 2000]

8 (a) A transverse wave is passing through a medium. Fig. 12.16 is a graph showing the variation of displacement x with time t for a particle of the medium.

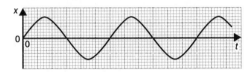

Fig. 12.16

Fig. 12.17

(i) On Fig. 12.16, indicate
   1. the amplitude A,
   2. the period T
   of the wave.
(ii) On Fig. 12.17, sketch a graph to show the variation of the displacement x with time t for a wave of equal amplitude and the same period as that in Fig. 12.16, but with a phase difference of $180°$. The x- and t-scales in Fig. 12.17 are the same as in Fig. 12.16.

(b) Two different sinusoidal waves of the same type are propagated in a medium under different conditions, so that their velocities are not the same. A point P in the medium is disturbed in turn by each wave.

Table 12.3, which is incomplete, gives some details about the waves.
Make appropriate calculations and deductions to complete the blanks in Table 12.3.
  [CCEA 2000]

Table 12.3   Information for Question 8(b)

wave	velocity $v/\mathrm{m\,s^{-1}}$	wavelength λ/m	frequency f/Hz	period t/s	phase at P at time t ϕ_t/degrees	phase at P 0.001 s after t $\phi_{(t+0.001)}$/degrees
1	330	1.32			0	
2		3.40			0	36.0

9 A progressive transverse wave has a frequency of 0.50 kHz. If the least distance between two points which have a phase difference of $\pi/3$ is 0.050 m, calculate the speed of the wave.

10 In Fig. 12.18 X and Y are two generators of water waves of wavelength 0.50 m. Each of the generators, when operating on its own, produces waves of amplitude 60 mm at P, which is 2.00 m from X.

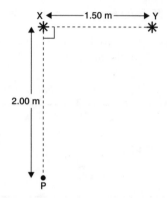

Fig. 12.18   Diagram for Question 10

Find the amplitude of the resulting disturbance at P when the generators X and Y are operating (a) in phase, (b) 180° out of phase.

11 (a) State the principle of superposition of waves.

(b) Monochromatic light from a source passes through a single slit S, and then through two narrow, parallel slits S_1 and S_2, separated by a distance a. The light falls on a screen a distance d from the slits. A fringe pattern is formed on the screen. Fig. 12.19 shows the arrangement, which is perfectly symmetrical about the line SO.

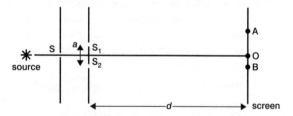

Fig. 12.19

(i) The light from slits S_1 and S_2 is said to be coherent. What is meant by **coherent** in this context?

(ii) In Fig. 12.19, A is a point on the screen where **constructive** interference occurs between waves coming from S_1 and S_2. B is a point where **destructive** interference occurs. State what would be observed on the screen at these points.

(iii) For the fringe pattern observed, write down the equation relating the wavelength λ of the light to the quantities d and a. Identify any other symbol(s) used.

(iv) The separation a of the slits is 0.80 mm and the distance d between slits and screen is 3.6 m. The slits are illuminated with light of wavelength 4.4×10^{-7} m.
  (1) Calculate the fringe separation.
  (2) A point C on the screen is 9.9 mm away from the central bright fringe at O. Show that a **bright** fringe is formed at C. Explain your working.
  (3) How far beyond C would the next **dark** fringe be?   [CCEA 2001]

12 Light of wavelength 600 nm falls on a pair of slits, forming fringes 3.0 mm apart on a screen.
What is the fringe spacing when light of wavelength 300 nm is used and the slit separation is halved?

A 0.75 mm **B** 1.5 mm   **C** 3.0 mm   **D** 6.0 mm
[OCR 2001]

13 Fig. 12.20 shows an arrangement for observing interference fringes from two narrow slits.

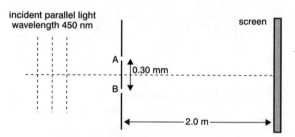

Fig. 12.20   (not to scale)

The incident parallel light is a monochromatic beam of wavelength 450 nm. The two slits **A** and **B** have their centres a distance 0.30 mm apart. The screen is situated a distance 2.0 m from the slits.

(a) Make a sketch of the interference pattern which you would expect to observe on the screen. Explain why the pattern has bright and dark regions.

(b) Calculate the spacing between fringes observed on the screen.

(c) How would you expect the pattern to change when, separately:
  (i) the light source is changed to one of wavelength 600 nm,
  (ii) the slit spacing is increased to 0.50 mm,
  (iii) the slits **A** and **B** are each made wider?

(d) (i)* Calculate the wavelength in glass of refractive index 1.50 of light which has a wavelength 450 nm in air.

(ii) A thin wedge of the glass is now introduced so that it gradually covers slit **A**, but not slit **B**. The arrangement is shown in Fig. 12.21. Suggest how you expect the pattern to change as the wedge is introduced. How many fringes will have passed the centre line, and in which direction, when a thickness of 0.050 mm of glass has been inserted over slit **A**?

Fig. 12.21   [OCR spec 2001]

14 Figure 12.22 shows a standing wave set up on a wire of length 0.87 m. The wire is vibrated at a frequency of 120 Hz.

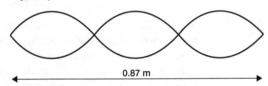

0.87 m

Fig. 12.22

(a) Calculate the speed of transverse waves along the wire.

(b) Show that the fundamental frequency of the wire is 40 Hz.   [AQA 2001]

15 The frequency of the fundamental note emitted by a plucked wire of length 1.00 m is 256 Hz. If the wire is shortened by 0.60 m, whilst kept at the same tension, calculate the new fundamental frequency.

16 The diagram shows an electron-microscope image of the world's smallest guitar.

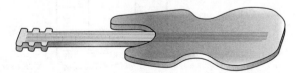

Its strings have a length of 10 millionths (10×10^{-6}) of a metre. They have a width of about 50 billionths (50×10^{-9}) of a metre – the

Authors' hint: $_1n_2 = \dfrac{c_1}{c_2} = \dfrac{\lambda_1}{\lambda_2}$; see Chapter 14.

size of approximately 100 atoms. Plucking the tiny strings would produce a high-pitched sound at the inaudible frequency of approximately 10 MHz. The guitar was made by researchers at Cornell University with a single silicon crystal; this tiny guitar is a playful example of nanotechnology.

(a) (i) Explain briefly why a vibrating string creates a sound wave.

(ii) Comment on the phrase "the inaudible frequency of approximately 10 MHz".

(b) (i) When the string of this guitar vibrates at its fundamental frequency (10 MHz), what is the wavelength of the waves on the string? State *one* assumption your are making.

(ii) What is the speed of the waves along the string?

(iii) The string has a mass per unit length of $4 \times 10^{-12}\,\text{kg}\,\text{m}^{-1}$. Calculate the tension in the string.   [Edexcel S-H 2000]

17 Stationary waves may be formed with light. A narrow beam of monochromatic light is incident normally on a mirror, and is reflected back along the same path. Superposition of the waves in the incident and reflected beams may set up a stationary wave, with a node at the surface of the mirror, as shown in Fig. 12.23.

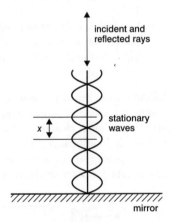

incident and reflected rays

stationary waves

x

mirror

Fig. 12.23   (not to scale)

(a) For light of wavelength 450 nm, what is the distance x between adjacent antinodes of the stationary wave pattern (Fig. 12.23)?

(b) It is possible to demonstrate the formation of the antinodes by placing a thin, transparent, photographic film at a very small angle θ the surface of the mirror, as shown

Fig. 12.24. When an antinode occurs at the film, there is blackening when the film is processed. There is no photographic action at the nodes. Thus, when the film is processed, a pattern of parallel dark lines is obtained on the film, as in Fig. 12.25.

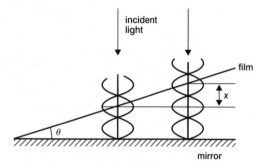

Fig. 12.24 (not to scale)

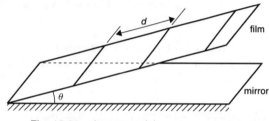

Fig. 12.25 (not to scale)

In such an experiment, the wavelength of the light used is 450 nm. The film is set an angle θ of 4.3×10^{-3} degrees to the mirror.

(i) Calculate the distance d between adjacent dark lines on the processed film (Fig. 12.25).

(ii) Describe what would happen to the pattern of lines if the angle between the film and the mirror were increased.

[CCEA 2001, part]

18 A student carries out the following experiment to determine the speed of sound in air.

A tube 0.46 m long, closed at one end, is set up with a small loudspeaker facing the open end. The loudspeaker is connected to a signal generator. The arrangement is shown in Fig. 12.26.

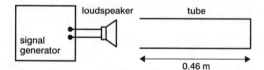

Fig. 12.26

The student gradually increases the frequency of the signal from the generator, from a very low value, until the column of air in the tube first resonates. This occurs at a reading on the signal generator of 180 Hz.

(a) (i) How does the student detect when the air in the tube is resonating?

(ii) Using the above data, obtain a value for the speed of sound in the air in the tube.

(b) The student then turns the dial of the signal generator to a higher frequency range, and detects another resonant frequency at a reading of 900 Hz.

(i) Show that the wavelength in air of a sound wave of this frequency is 0.37 m.

(ii) Fig. 12.27 is a sketch of the tube used in this experiment.

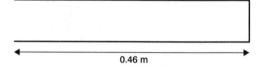

Fig. 12.27

On Fig. 12.27, mark the positions of the nodes and antinodes of the vibrations of the air particles in the tube when the air column is resonating at the frequency of 900 Hz. Indicate nodes with the letter N, and antinodes with the letter A.

[CCEA 2001]

19 (a) (i) State the difference between a **progressive wave** and a **stationary wave**.

(ii) State **two** of the conditions which must apply if a stationary wave is to be formed from two progressive waves.

(b) A vibrating tuning fork is held over the open end of a pipe, as shown in Fig. 12.28.

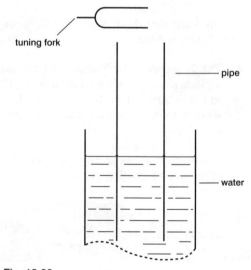

Fig. 12.28

The lower end of the pipe is immersed in water in a vertical cylinder. The pipe and tuning fork are slowly raised until a stationary wave is obtained at the first position of resonance.

(i) On Fig. 12.28, sketch the wave pattern for the first position of resonance. Indicate the positions of any nodes and antinodes by the letters **N** and **A** respectively.

(ii) The frequency of the tuning fork is 160 Hz. Taking the speed of sound in air as $340\,\text{m s}^{-1}$, calculate the length of the air column which will give the first position of resonance.

(iii) The tuning fork is replaced with one of frequency 480 Hz. How far, and in what direction, will the pipe need to be moved to obtain the first position of resonance for this pipe?　　　　[CCEA 2000]

13
Diffraction and the diffraction grating

Diffraction

When waves pass through an aperture or meet an obstacle, the waves spread to some extent into a region of geometrical shadow. This effect is called diffraction. Calculations are usually restricted to:

(i) the transmission grating
(ii) the limit of resolution for optical instruments

The optical diffraction grating

A transmission grating consists of many parallel equidistant slits of width and spacing of the order of the wavelength of light. If plane waves (parallel light) are incident on it, then, by superposition of the secondary wavelets from each slit, it can be shown that a transmitted wavefront is formed only along a few specified directions.

If the incident parallel beam is at normal incidence (see Figs. 13.1 and 13.2), then emergent parallel beams are seen only in directions such that

$$d \sin\theta = n\lambda \qquad (13.1)$$

where d is the spacing of the slits, $n(= 0, 1, 2, \ldots)$ the order of diffracted beam, λ the wavelength of incident light and θ the angle of diffracted beam to the normal.

The following examples involve use of Equation 13.1

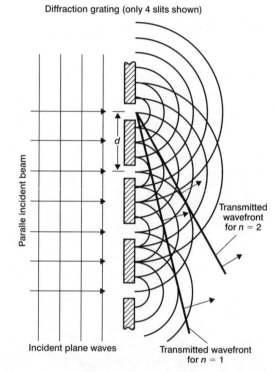

Fig. 13.1 Action of the diffraction grating: formation of transmitted wavefronts

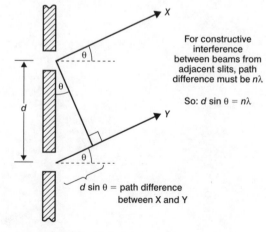

Fig. 13.2 Action of the diffraction grating showing $d \sin\theta = n\lambda$

Example 1

Monochromatic light of wavelength 600 nm is incident normally on an optical transmission grating of spacing 2.00 μm. Calculate (a) the angular positions of the maxima; (b) the number of diffracted beams which can be observed; (c) the maximum order possible.

Method

We are given

$$\lambda = 600 \times 10^{-9} \, \text{m}$$

$$d = 2.00 \times 10^{-6} \, \text{m}$$

(a) We substitute into Equation 13.1 as follows:

   (i) for $n = 1$

$$2.00 \times 10^{-6} \times \sin \theta_1 = 1 \times 600 \times 10^{-9}$$

   This gives $\sin \theta_1 = 0.3$ or $\theta_1 = 17.5°$

   (ii) for $n = 2$

$$2.00 \times 10^{-6} \times \sin \theta_2 = 2 \times 600 \times 10^{-9}$$

   Thus gives $\sin \theta_2 = 0.6$, or $\theta_2 = 36.9°$

   (iii) $n = 3$ gives $\sin \theta_3 = 0.9$, or $\theta_3 = 64.2°$

   (iv) $n = 4$ gives $\sin \theta_4 = 1.2$, which is impossible (see Chapter 2). Thus the fourth order is not observed.

(b) Fig. 13.3 is a schematic diagram showing the positions of the various maxima. Note that for $n = 0$ (the zeroth order) $\theta = 0$. Thus seven diffracted maxima are observed.

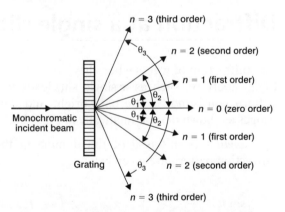

Fig. 13.3   Angular distribution of diffracted beams

(c) This has been covered in part (a), which shows that since $n = 4$ is impossible, the maximum order is 3. A quicker way to do this is as follows:

$$\sin \theta \leqslant 1$$

From Equation 13.1, $\sin \theta = n\lambda/d$. So

$$\frac{n\lambda}{d} \leqslant 1$$

or $\quad n \leqslant \dfrac{d}{\lambda} = \dfrac{2.00 \times 10^{-6}}{600 \times 10^{-9}} = 3.33$

$$\therefore \quad n \leqslant 3.33$$

Since n must be an integer its maximum value is 3.

Answer

(a) The angular positions are

$$n = 1 \qquad \theta_1 = 17.5°$$
$$n = 2 \qquad \theta_2 = 36.9°$$
$$n = 3 \qquad \theta_3 = 64.2°$$

   Note the trivial case of $\theta = 0$ for $n = 0$.

(b) There are seven diffracted beams.

(c) The maximum order is $n = 3$.

Example 2

Light consisting of wavelengths 420 nm and 650 nm is incident normally on a transmission grating of 6.00×10^5 lines m^{-1}. Calculate the angular separation of the wavelengths in the second-order spectrum.

Method

There are 6.00×10^5 lines per metre of grating, So the grating spacing d is given by

$$d = \frac{1}{6.00 \times 10^5} = 1.666 \times 10^{-6} \, \text{m}$$

Using Equation 13.1, for the second-order spectrum ($n = 2$) we have

(a) for $\lambda = 420 \times 10^{-9} \, \text{m}$

$$1.666 \times 10^{-6} \times \sin \theta_2 = 2 \times 420 \times 10^{-9}$$

   This gives $\sin \theta_2 = 0.504$ and $\theta_2 = 30.3°$.

(b) for $\lambda' = 650 \times 10^{-9}$

$$1.666 \times 10^{-6} \times \sin \theta'_2 = 2 \times 650 \times 10^{-9}$$

   This gives $\sin \theta'_2 = 0.780$ and $\theta'_2 = 51.3°$.

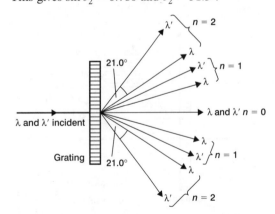

Fig. 13.4   Angular separation as in Example 2

A schematic diagram of the situation is given in Fig. 13.4. The angular separation

$$\theta'_2 - \theta_2 = 51.3 - 30.3° = 21.0°$$

Answer

The angular separation in the second-order spectrum is 21.0°.

Example 3

White light which has been passed through a certain filter has a range of wavelengths from 450 nm to 700 nm. It is incident normally on a diffraction grating. Show that if there are second- and third-order spectra, they will overlap.

Method

For any particular grating the angle of diffraction, for a given order, is greater for the longer wavelengths. This is seen by rearranging Equation 13.1:

$$\sin \theta = \frac{n\lambda}{d} \qquad (13.2)$$

Thus, for a given d and n value $\sin \theta \propto \lambda$.

We must therefore show that the second-order red (700 nm) has a higher θ value than the third-order blue (450 nm). For the given grating the d value is constant, so Equation 13.2 becomes

$$\sin \theta = \text{Constant} \times n\lambda$$

For $\lambda_1 = 700$ nm in the second order ($n = 2$)

$$\sin \theta_1 = \text{Constant} \times 2 \times 700 \times 10^{-9}$$
$$= \text{Constant} \times 1.40 \times 10^{-6}$$

For $\lambda_2 = 450$ nm in the third order ($n = 3$)

$$\sin \theta_2 = \text{Constant} \times 3 \times 450 \times 10^{-9}$$
$$= \text{Constant} \times 1.35 \times 10^{-6}$$

Since $\sin \theta_1 > \sin \theta_2$, then $\theta_1 > \theta_2$. So the second order at the red end overlaps with the third order at the blue end.

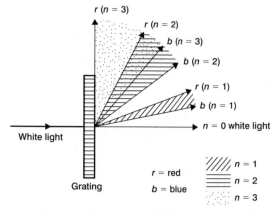

Fig. 13.5 Appearance of diffraction spectra using white light (upper half only is shown)

Fig. 13.5 is a schematic diagram of the white light diffraction spectra using a typical grating. The angular spread in a given order, and the maximum order, depend upon the grating spacing. However, the second- and third- and higher-order spectra (if present) will always overlap with each other as discussed above.

Exercise 13.1

1 What is the wavelength of light which gives a first-order maximum at an angle of $22° 30'$ when incident normally on a grating with 600 lines mm^{-1}?

2 Light of wavelength 600 nm is incident normally on a diffraction grating of width 20.0 mm, on which 10.0×10^3 lines have been ruled. Calculate the angular positions of the various orders.

3 A source emits spectral lines of wavelength 589 nm and 615 nm. This light is incident normally on a diffraction grating having 600 lines per mm. Calculate the angular separation between the first-order diffracted waves. Find the maximum order for each of the wavelengths.

4 When a certain grating is illuminated normally by monochromatic light of wavelength 600 nm, the first-order maximum is observed at an angle of $21.1°$. If the same grating is now illuminated with light with wavelength from 500 nm to 700 nm, find the angular spread of the first-order spectrum.

Diffraction at a single slit

The diffraction of waves when they are restricted by an aperture, such as a single slit, leads to a pattern consisting of alternate bright and dark fringes as shown diagrammatically in Fig. 13.6.

The *angular* positions θ of the minima in this diffraction pattern are given by:

$$\sin \theta_m = \frac{m\lambda}{w} \qquad (13.3)$$

where λ = wavelength of light used, w = width of slit and $m = 1, 2, 3 \ldots \ldots$

Example 4

Laser light of wavelength 650 nm is incident on a single rectangular slit of width 0.130 mm. The resulting diffraction pattern is viewed on a screen placed 3.00 m from the slit. Calculate:

(a) the distance between the centre of the central maximum and the first minimum

(b) the width of the central maximum.

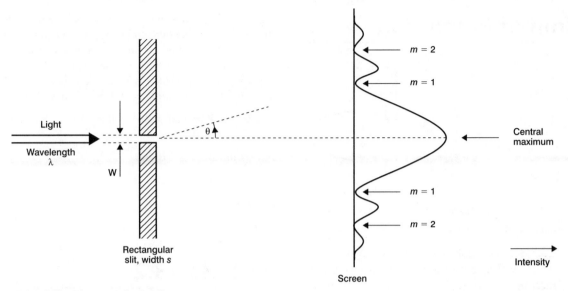

Fig. 13.6   Diffraction at a single slit

Method

(a) Referring to Fig. 13.7 we have:

wavelength $\lambda = 650 \times 10^{-9}\,\text{m}$

width of slit $w = 0.130 \times 10^{-3}\,\text{m}$

and $\qquad\qquad m = 1$

From Equation 13.3

$$\sin \theta_1 = 1 \times \frac{\lambda}{w} = \frac{650 \times 10^{-9}}{0.130 \times 10^{-3}} = 5.00 \times 10^{-3}$$

Since $\sin \theta_1$ is very small (see Chapter 2) then $\theta_1 = \sin \theta_1$. Thus

$$\theta_1 = 5.00 \times 10^{-3}\,\text{rad}\,(= 0.286°)$$

Thus, distance d between centre of central maximum and first minimum is given by:

$$r = L\theta_1 = 3.00 \times 5.00 \times 10^{-3}$$
$$= 15.0 \times 10^{-3}\,\text{m}$$

where L = distance from slit to screen ($= 3.00\,\text{m}$)

(b) The width R of the central maximum is the distance ($= 2r$) between the two first minima ($m = 1$) on either side of the central maximum. Thus

$$R = 2r = 2 \times 15.0 \times 10^{-3}$$
$$= 30.0 \times 10^{-3}\,\text{m}$$

Answer

(a) 15.0 mm, $\qquad\qquad$ (b) 30.0 mm.

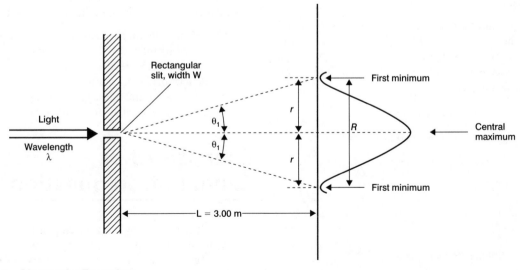

Fig. 13.7   Diagram for Example 4

Limit of Resolution

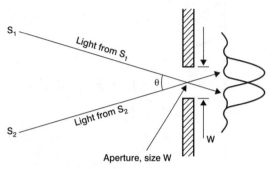

Fig. 13.8   Limit of resolution

Fig. 13.8 shows light from two separate point sources S_1 and S_2 entering an optical instrument in directions separated by an angle θ. Diffraction at the entrance to the instrument – which has an aperture of size (i.e. diameter) W – means that the images of the sources are broadened. According to the Rayleigh criterion the optical instrument cannot distinguish (i.e. resolve) between images of S_1 and S_2 which are less than an angular distance θ apart, given by:

$$\sin\theta \cong \lambda/W* \qquad\qquad (13.4)$$

where λ = wavelength of the light emitted by the sources.

Since θ is usually small we have (see Chapter 2):

$$\sin\theta = \theta \cong \lambda/W \qquad\qquad (13.5)$$

where θ is in radians (see Equation 2.15).

Example 5

Two point sources of light are placed 12 mm apart and emit light of wavelength $0.60\,\mu\text{m}$. Calculate the maximum distance at which the two sources can just be distinguished by an observer with an eye pupil diameter of 4.0 mm.

Method

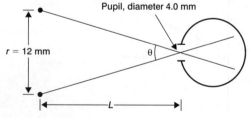

Fig. 13.9   Solution to Example 5

Strictly, for a circular aperture, $\sin\theta = 1.22\lambda/W$.

Fig. 13.9 shows the situation in which we require to find the minimum value of θ, and hence the maximum value of L, for which the two sources separated by a distance $r = 12 \times 10^{-3}$ m can just be distinguished. We use Equation 13.5, in which we assume θ is small. Since $\lambda = 0.60 \times 10^{-6}$ m and $W = 4.0 \times 10^{-3}$ m we have:

$$\theta \cong \lambda/W = 0.60 \times 10^{-6}/4.0 \times 10^{-3}$$
$$= 0.15 \times 10^{-3}\ \text{rad}$$

For small values of θ then

$$L = r/\theta = 12 \times 10^{-3}/0.15 \times 10^{-3} = 80\,\text{m}$$

Answer

80 m (approximately)

Exercise 13.2

1. Calculate the angular width of the central maximum if yellow light of wavelength $0.60\,\mu\text{m}$ is incident on a single slit of width
   (a) 0.10 mm   (b) 0.010 mm

2. A parallel beam of blue monochromatic light of wavelength $0.46\,\mu\text{m}$ is incident normally on a rectangular slit of width 0.20 mm and the resulting diffraction pattern is viewed on a screen 4.0 m beyond the slit and normal to the incident light. Calculate the distance from the centre of the diffraction pattern to the first minimum.

3. Calculate the value of θ for a human eye with a pupil diameter of 5.0 mm using light of wavelength $0.45\,\mu\text{m}$.

4. An observer can just distinguish between two point sources of light at a distance of 1.0 km. If the observer can just distinguish between rays of light with an angular separation of 2.0×10^{-4} radian, calculate the separation of the two point sources.

5. The Mount Palomar telescope has a resolving power such that $\theta = 0.10\,\mu\text{rad}$. Assuming this relates to light received from a source of wavelength $0.40\,\mu\text{m}$, estimate the diameter of the receiving dish of the telescope. (Hint: the receiving dish acts as the aperture for diffraction.)

Exercise 13.3: Examination questions

1. A diffraction grating has a spacing of 1.6×10^{-6} m. A beam of light is incident normally on the grating. The first order maximum makes an angle of $20°$ with the undeviated beam.

DIFFRACTION AND THE DIFFRACTION GRATING

What is the wavelength of the incident light?

A 210 nm   **B** 270 nm   **C** 420 nm   **D** 550 nm
[OCR 2000]

2 In the spectrum of the element strontium there is a red line, wavelength 600 nm.

When light from a strontium source is passed through a diffraction grating with 5.0×10^5 lines/metre which one of **A** to **D** below is the angle, in degrees, at which the **second** order red line is observed?

A 0.60   **B** 17   **C** 37   **D** 53
[OCR Nuff 2001]

3 Light from two different monochromatic sources is incident on a diffraction grating at normal incidence. One source has wavelength of 534 nm and gives a second order maximum at an angle of 32.3°. If light from the second source gives a second order maximum at an angle of 27.5°, calculate the wavelength of the second source.

4 A laser emits a narrow beam of red light towards a diffraction grating, beyond which is a curved white screen as shown in Fig. 13.10.

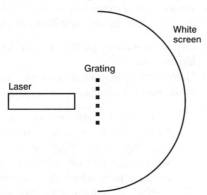

Fig. 13.10   (not to scale)

(a) A line of spots is observed on the screen. Explain why there is more than one spot.

(b) Use the data below to calculate the number of spots on the screen.

Wavelength of red light = 633 nm
Number of lines per mm on grating = 380 mm^{-1}
[OCR 2001]

5

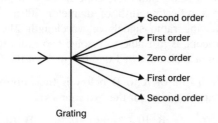

Fig. 13.11   Information for Question 5

Blue light of wavelength 480 nm is incident normally on a diffraction grating and is split into a number of beams as shown in Fig. 13.11.

If the angular separation of the second order beams is 44.6°, calculate the number of lines per millimetre of the grating.

6 A diffraction grating is used to analyse the visible light emitted by a discharge lamp containing atomic hydrogen. Fig. 13.12 illustrates the principle of the experiment.

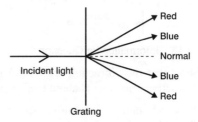

Fig. 13.12

A narrow beam of light is incident normally on the grating. The first-order spectrum of the diffracted light includes red and blue rays. These emerge symmetrically about the normal to the grating. The angle between the two red rays is 38.3°, and that between the two blue rays is 25.1°. The grating has 500 lines per millimetre.

(a) Show that the wavelength of the red light is 656 nm, and that of the blue light is 435 nm.

(b) For each colour of light, determine how many orders of diffraction are theoretically observable. [CCEA 2000, part]

7 A light source emits two distinct wavelengths, one of which is 450 nm. When light from the source is incident normally on a diffraction grating, it is observed that the fourth order image formed by the light of wavelength 450 nm lies at the same angle of diffraction as the third order image for the other wavelength. If the angle of diffraction for each image is 46°, calculate (a) the second wavelength emitted by the source, (b) the number of lines per metre of the grating.

8 The emission spectrum of a certain element contains just two wavelengths, a red and a violet. When the light is examined with a diffraction grating having 250 lines per millimetre, it is found that a line at 19.88° contains both red and violet light.

(a) At what other angles, if any, would lines containing both colours be found?

(b) Identify the line that occurs at the greatest diffraction angle, i.e. find its colour, order and the angle at which it occurs.

[It may be helpful to know that the **approximate** values of the two wavelengths are

$$\lambda_{red} \approx 7 \times 10^{-7}\,\text{m and } \lambda_{violet} \approx 4 \times 10^{-7}\,\text{m}$$

[WJEC spec 2000]

9 A beam of ultrasound of wavelength 0.14 mm is incident normally on a slit of width 3.0 mm as shown in Fig. 13.13

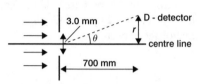

Fig. 13.13   Information for Question 9

(a) Show that the diffraction angle θ for the first diffraction minimum is about 0.05 radian.

(b) A small detector is used to study the diffraction pattern and detects the first diffraction minimum at point D as shown. Calculate the distance r of the detector from the centre line.

10 This question is about the diffraction of light by gratings.

A diffraction grating is made by securing extremely fine wire onto a frame (Fig. 13.14). The wire is $0.1\,\mu\text{m}$ thick and the gap between the wires is $2.0\,\mu\text{m}$.

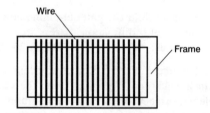

Fig. 13.14

(a) Calculate the angle for the *second* order maximum of the interference pattern when light of wavelength 600 nm is incident on the grating.

(b) A similar grating is made with wire of $1.05\,\mu\text{m}$ diameter and the width of the gap is also $1.05\,\mu\text{m}$. Explain why the second order maximum is missing with this grating.*

[OCR Nuff 2000]

11 This question is about a diffraction grating.

A diffraction grating has regularly-spaced slits, each having the same width as the opaque strips between the slits. It is illuminated by monochromatic light of wavelength $5.0 \times 10^{-7}\,\text{m}$.

(*Author's hint: calculate the angular position of the first minimum for the single slit diffraction pattern.)

(a) Explain why the second order interference maxima are missing. Support your answer mathematically.

(b) If the grating has 700 slits per millimetre $(7 \times 10^5\,\text{m}^{-1})$, there are no third order maxima either. Explain why this is so. Support your answer with appropriate calculations.

(c) Explain what has happened to the 'missing' energy; that is, the energy which we might have expected to be in the missing maxima.

(d) The wavelength is now reduced to $4.0 \times 10^{-7}\,\text{m}$. Explain what changes, if any, will take place to the interference patterns.

[OCR spec Nuff 2001]

12 A parallel beam of monochromatic light of wavelength 500 nm is incident normally on a long single slit of width 0.250 mm. At what distance from the slit should a screen be placed in order that the first dark fringes on either side of the central maximum be separated by 6.00 mm?

13 A person standing on the deck of an aircraft carrier can just distinguish between two lights on the wings of an aircraft at a distance of 10 km. If the person has a resolving power of 2.0×10^{-4} radians, calculate the distance between the lights.

14 (a) Define **resolving power** as applied to the human eye.

(b) Under certain lighting conditions, the diameter of the pupil of another student's eye is 6.0 mm.

   (i) Two small light sources are placed 4.0 mm apart at one end of a large assembly hall. They emit light of wavelength 640 nm. Find the maximum distance from which the student can just resolve the images of the two sources.

   (ii) The sources are then replaced by another pair of sources, separated by the same distance, but emitting light of wavelength appreciably less than 640 nm. In which direction should the student move so that she can again **just** resolve the images of the two sources?

[CCEA 2000, part]

15 The Arecibo radio telescope in Central America has a reflecting dish of diameter 300 m. When detecting radio signals of wavelength 21 cm the telescope is just able to resolve two radio sources both at a distance of $1.0 \times 10^{20}\,\text{m}$ from the Earth.

Which one of **A** to **D** below is the approximate separation, in m, of the two sources?

A 10^{23} **B** 10^{21} **C** 10^{19} **D** 10^{17}

[OCR Nuff 2000]

Section E
Geometrical optics

14
Refraction

Refractive index

Light, and other kinds of waves, can change direction when they pass from one medium to another. This is called *refraction* and occurs because of a change in the speed of propagation of wave energy.

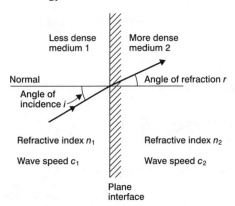

Fig. 14.1   Refraction

Referring to Fig. 14.1 then:

$$n_1 \sin i = n_2 \sin r \qquad (14.1)$$

$$\text{or} \quad \frac{\sin i}{\sin r} = \frac{n_2}{n_1} = {}_1n_2 \qquad (14.2)$$

where n_1 = absolute refractive index of medium 1 (wave passes from air to medium 1), n_2 = absolute refractive index of medium 2 (wave passes from air to medium 2) and ${}_1n_2$ = refractive index when wave passes from medium 1 to medium 2.

Example 1

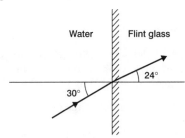

Fig. 14.2   Diagram for Example 1

As shown in Fig. 14.2, a beam of light travelling through water (absolute refractive index 1.3) is incident on a flint glass surface at an angle of $30°$ and is refracted at an angle of $24°$. Calculate:

(a) the absolute refractive index of flint glass

(b) the angle of incidence for an angle of refraction of $30°$

(c) the refractive index for light passing from water to flint glass

(d) the refractive index for light passing from flint glass to water.

Method

(a) We have $n_1 = 1.3$, $i = 30°$, $r = 24°$ and require n_2. Rearranging Equation 14.1:

$$n_2 = n_1 \frac{\sin i}{\sin r} = 1.3 \times \frac{\sin 30°}{\sin 24°} = \frac{0.65}{0.407}$$

$$= 1.60$$

(b) We have $n_1 = 1.3$, $n_2 = 1.6$, $r = 30°$ and require i. Rearranging Equation 14.1:

$$\sin i = \frac{n_2}{n_1} \sin r = \frac{1.6}{1.3} \times \sin 30° = \frac{0.80}{1.3}$$

which gives $i = 38°$.

(c) Water is medium 1 and flint glass is medium 2 and the refractive index required is $_1n_2$. From Equation 14.2

$$_1n_2 = \frac{n_2}{n_1} = \frac{1.6}{1.3} = 1.23$$

(d) Since light now passes from glass to water we require $_2n_1$ which is found from:

$$_2n_1 = \frac{n_1}{n_2} = \frac{1.3}{1.6}$$

$$= 0.81$$

Note that $\quad _1n_2 = \dfrac{1}{_2n_1}$

Answer

(a) 1.6, (b) 38°, (c) 1.2, (d) 0.81.

Example 2

A ray of light travelling through air $(n = 1.00)$ is incident at an angle of $40.0°$ on to the first face of a crown glass prism $(n = 1.52)$ of angle $60.0°$. Calculate

(a) the angle of emergence of the ray at the second face and

(b) the angle of deviation of the ray on passing through the prism.

Method

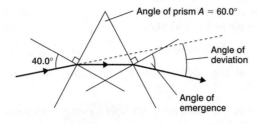

(a) Information for Example 2

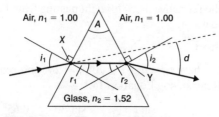

(b) Solution to Example 2

Fig. 14.3 Diagrams for Example 2

Fig. 14.3a illustrates the quantities required and Fig. 14.3b is useful when calculating these quantities. We must calculate r_1 and r_2 if we are to find i_2, the angle of emergence, and d, the angle of deviation.

(a) At X we have $i_1 = 40.0°$, $n_1 = 1.00$, $n_2 = 1.52$ and require r_1. From Equation 14.1:

$$n_1 \sin i_1 = n_2 \sin r_1$$

or $\quad \sin r_1 = \dfrac{n_1}{n_2} \sin i_1 = \dfrac{1.00}{1.52} \times \sin 40.0° = 0.423$

$\therefore \qquad r_1 = 25.0°$

We require r_2 which is found by noting

$$A = r_1 + r_2 \qquad\qquad (14.3)$$

Hence $\quad r_2 = A - r_1 = 60.0 - 25.0$

$$= 35.0°$$

To find i_2 on refraction at Y it is convenient to use Equation 14.1 in the form

$$n_1 \sin i_2 = n_2 \sin r_2$$

with $n_1 = 1.00$, $n_2 = 1.52$, $r_2 = 35.0°$. Hence

$$\sin i_2 = \frac{1.52}{1.00} \times \sin 35.0° = 0.872$$

$\therefore \quad$ angle of emergence $i_2 = 60.7°$

(b) Deviation angle d is found from:

$$d = (i_1 - r_1) + (i_2 - r_2)$$
$$= (40.0 - 25.0) + (60.7 - 35.0)$$
$$= 40.7°$$

Answer

(a) 60.7°, (b) 40.7°.

Exercise 14.1

(Assume refractive index of air = 1.00.)

1 A ray of light travelling through a liquid of absolute refractive index 1.4 is incident on the plane surface of a perspex block at an angle of 55°. Calculate the angle of refraction in the perspex if it has an absolute refractive index of 1.5.

2

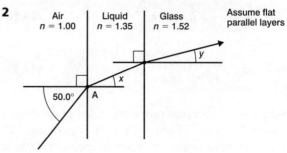

Fig. 14.4 Diagram for Question 2

In Fig. 14.4 a ray of light travelling through air is incident at A at an angle of 50.0° on to a glass surface which is coated with a layer of liquid. Use the information given below to find the angles x and y:

absolute refractive index of liquid $= 1.35$
absolute refractive index of glass $= 1.52$

3 A ray of light travelling through air is incident at an angle of 30.0° on to the first face of a perspex prism of angle 45.0°. If the perspex has refractive index 1.49, calculate the angle of emergence at the second face.

Refractive index, speed and wavelength

Refractive index is equal to the wave speed ratio. Referring to Fig. 14.1, when waves pass from medium 1 to medium 2 then:

$$_1n_2 = \frac{n_2}{n_1} = \frac{\sin i}{\sin r} = \frac{c_1}{c_2} \qquad (14.4)$$

where $c_1 =$ speed of waves in medium 1 and $c_2 =$ speed of waves in medium 2.

Note that the frequency f of the waves as they pass from medium 1 to medium 2 does not alter. Now:

$$c = f\lambda \qquad (12.1)$$

Thus the wavelength of the waves must change from λ_1 to a new value λ_2 as the wave passes from medium 1 to medium 2.

Example 3

During ultrasonic imaging, ultrasound is incident at an angle of 10.0°, in soft tissue, on to a plane soft tissue – bone boundary. If the angle of refraction in the bone is 27.4°, calculate:

(a) the speed of ultrasound in bone given that it is 1.54 km s^{-1} in soft tissue

(b) the refractive index when ultrasound travels from bone to soft tissue.

Method

(a) We have $i = 10.0°$, $r = 27.4°$, $c_1 = 1.54 \times 10^3$ and require c_2. Rearranging Equation 14.4:

$$c_2 = c_1 \frac{\sin r}{\sin i} = 1.54 \times 10^3 \times \frac{0.460}{0.174}$$

$$= 4.08 \times 10^3$$

(b) In this case bone is medium 1, with $i = 27.4°$ and soft tisue is medium 2 with $r = 10.0°$. Thus, from Equation 14.4:

$$_1n_2 = \frac{\sin i}{\sin r} = \frac{\sin 27.4°}{\sin 10.0°} = \frac{0.460}{0.174}$$

$$= 2.65$$

Answer

(a) 4.08 km s^{-1}, (b) 2.65.

Example 4

The speed of light in air is $3.00 \times 10^8 \text{ m s}^{-1}$ and the speed of light in a certain type of glass is $1.96 \times 10^8 \text{ m s}^{-1}$. Assuming that yellow light of wavelength 589 nm in air is used, calculate:

(a) the refractive index when yellow light passes from air into the glass

(b) the angle of refraction in glass when yellow light is incident at an angle of 50.0° in air

(c) the wavelength of yellow light in the glass.

Method

(a) We have $c_1 = 3.00 \times 10^8$, $c_2 = 1.96 \times 10^8$ and require $_1n_2$. From Equation 14.4:

$$_1n_2 = \frac{c_1}{c_2} = \frac{3.00 \times 10^8}{1.96 \times 10^8}$$

$$= 1.53$$

(b) We have $i = 50°$, $_1n_2 = 1.53$ and require r. Rearranging Equation 14.4:

$$\sin r = \frac{\sin i}{_1n_2} = \frac{\sin 50°}{1.53} = 0.500$$

$$\therefore \quad r = 30.0°$$

(c) Equation 12.1, or $c = f\lambda$, holds for both medium 1 and medium 2. Since f does not change, then

$$c_1 = f\lambda_1 \text{ and } c_2 = f\lambda_2.$$

Dividing the two equations gives:

$$\frac{c_1}{c_2} = \frac{\lambda_1}{\lambda_2} \qquad (14.5)$$

Rearranging Equation 14.5 gives:

$$\lambda_2 = \lambda_1 \times \frac{c_2}{c_1} = \frac{589 \times 1.96 \times 10^8}{3.00 \times 10^8}$$

$$= 385 \text{ nm}$$

Answer

(a) 1.53 (b) 30.0° (c) 385 nm

Exercise 14.2

1   Table 14.1 Velocity of ultrasound in various media

Medium	Velocity of sound (km s^{-1})
Air	0.330
Soft tissue	1.54
Bone	4.08

Table 14.1 gives the velocity of ultrasound for various media. Calculate the angle of refraction when ultrasound is incident at 12.0° on to the following boundaries:

(a) soft tissue to bone

(b) air to soft tissue.

2   When ultrasound passes from water to muscle it has an increase of 7.0% in its speed of propagation. If the angle of incidence at the interface between water and muscle is 15.0°, calculate the angle of refraction. (Hint: represent the speeds by c and $1.07c$).

3   A ray of monochromatic light is incident, in air, at an angle of 45.0° on to a plane air–water interface. The speed of light in air is 3.00×10^8 m s^{-1} and the speed of light in water is 2.25×10^8 m s^{-1}. Calculate:

(a) the refractive index of light when passing from air into water

(b) the angle of refraction in water

(c) the wavelength of the light in air, assuming it has wavelength 405 nm in water.

Critical angle

When light passes from a more (optically) dense medium to a less (optically) dense medium there is a maximum angle of incidence beyond which light will be totally internally reflected. This maximum angle is called the *critical angle* i_c, as shown in Fig. 14.5, in which n_1 is greater than n_2.

In Fig. 14.5b, $i = i_c$ and $r = 90°$. From Equation 14.2

$$\frac{\sin i_c}{\sin 90°} = \sin i_c = \frac{n_2}{n_1} \qquad (14.6)$$

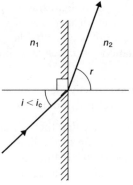

(a) Refraction with $n_1 > n_2$

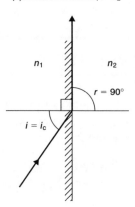

(b) Critical angle

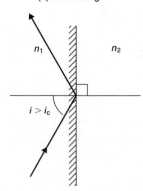

(c) Total internal reflection

Fig. 14.5   Critical angle and total internal reflection

Example 5

Calculate the critical angle for light passing from flint glass ($n_1 = 1.65$) to

(a) water ($n_2 = 1.33$) and
(b) air ($n_2 = 1.00$).

Method

(a) We use Equation 14.6 in which $n_1 = 1.65$, $n_2 = 1.33$ and we require i_c:

$$\sin i_c = \frac{n_2}{n_1} = \frac{1.33}{1.65} = 0.806$$

or $i_c = 53.7°$.

(b) We have $n_1 = 1.65$ and $n_2 = 1.00$. Equation 14.6 becomes:

$$\sin i_c = \frac{1.00}{n_1} = \frac{1.00}{1.65} = 0.606$$

or $i_c = 37.3°$.

Answer

(a) $53.7°$, (b) $37.3°$.

Fibre Optics

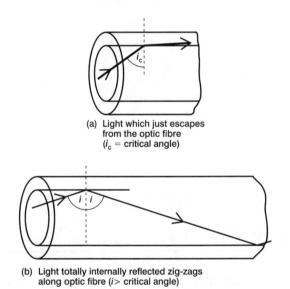

(a) Light which just escapes from the optic fibre (i_c = critical angle)

(b) Light totally internally reflected zig-zags along optic fibre ($i >$ critical angle)

Fig. 14.6   Fibre optic action

Fig. 14.6(a) and (b) show the action of an optic fibre in which we assume that the core has a higher refractive index (n_1) than the cladding (n_2). Light for which i is (just) greater than the critical angle i_c is totally internally reflected and can be retained within the fibre by repeated internal reflections as shown in Fig 14.6(b). Thus light is guided along the fibre.

A fibre with a sharp change of refractive index between the fibre (core) and the cladding, as described here, is known as a step index fibre.

Example 6

A step index fibre has a core of refractive index 1.50 and a cladding of refractive index 1.40. The fibre end face is perpendicular to the fibre axis. Rays of (monochromatic) light, initially in air of refractive index 1.00, are incident on the end face of the fibre at angles of (1) $0°$, (2) $20.0°$ and (3) $40.0°$, as shown in Fig 14.7.

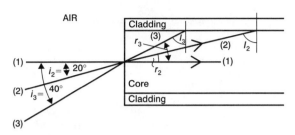

Fig. 14.7   Information for Example 6

(a) Calculate the critical angle at the core–cladding interface.

(b) Calculate the angle which these rays make at the core–cladding interface.

(c) What is the maximum angle at which light can enter the end face of the fibre and still be retained within it?

(d) If the fibre is of length 2.00 m, calculate the minimum and maximum distances that a light ray can travel in moving from one end of the fibre to the other.

Method

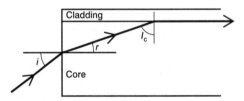

Fig. 14.8   Critical angle at the core–cladding interface

(a) Let the critical angle at the core–cladding interface be I_c, as shown in Fig. 14.8. We use Equation 14.6 in which $n_1 = 1.50$, $n_2 = 1.40$ and we require I_c:

$$\sin I_c = \frac{n_2}{n_1} = \frac{1.40}{1.50} = 0.933$$

or $I_c = 69.0°$

(b) (1) In the case of ray 1, it continues parallel to the axis of the fibre and never meets the cladding, assuming the fibre is straight.

(2) For ray 2, this is refracted at the air–core interface as shown in Fig. 14.7. We use Equation 14.1 in which $n_1 = n_{air} = 1.00$, $n_2 = n_{core} = 1.50$, angle of incidence for ray 2 is $i = i_2 = 20.0°$ and require angle of refraction $r = r_2$. Thus

$$n_{air} \sin 20.0° = n_{core} \sin r_2$$

Rearranging gives

$$\sin r_2 = \frac{\sin 20.0°}{1.50} = 0.228$$

Hence

$$r_2 = 13.1°$$

The angle I_2 at the interface is given by

$$I_2 + r_2 = 90°$$

Hence

$$I_2 = 76.9°.$$

Since I_2 is greater than the critical angle, this ray will be totally internally reflected and hence retained within the fibre – see also Fig. 14.6(b).

(3) For ray 3, this is refracted as shown in Fig. 14.7. We use Equation 14.1 with $n_1 = n_{air} = 1.00$, $n_2 = n_{core} = 1.50$, angle of incidence for ray 3 is $i = i_3 = 40.0°$ and require angle of refraction $r = r_3$. Thus:

$$n_{air} \sin 40.0° = n_{core} \sin r_3$$

Rearranging gives

$$\sin r_3 = \frac{\sin 40.0°}{1.50} = 0.429$$

Hence

$$r_3 = 25.4°$$

The angle I_3 at the interface is given by

$$I_3 + r_3 = 90°$$

Hence

$$I_3 = 64.6°.$$

Since I_3 is less than the critical angle, this ray will not be totally internally reflected and hence will escape from the core of the fibre, via the cladding, into the surrounding air.

(c) The maximum angle i occurs when light is incident on the core–cladding interface at (an angle just greater than) the critical angle $I_c = 69.0°$ which is shown in Fig. 14.8. Since $I_c + r = 90°$, then $r = 21.0°$. We use Equation 14.1 to find i:

$$n_{air} \sin i = n_{core} \sin r$$

Hence

$$1.00 \sin i = 1.50 \sin 21.0°$$

or $i = 32.5°$

(d)

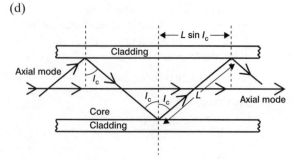

Fig. 14.9   Solution to Example 6

As shown in Fig. 14.9, the minimum distance that light can travel corresponds to the axial direction (axial mode) which is equal to the length of the fibre, that is, 2.0 m. The maximum distance that light can travel corresponds to repeated internal reflections at (just above) the critical angle. Fig. 14.9 shows that:

$$\frac{\text{maximum distance}}{\text{minimum distance}} = \frac{L}{L \sin I_c} = \frac{1}{\sin I_c}$$

or

$$\text{maximum distance} = \frac{\text{minimum distance}}{\sin I_c}$$

Since

$$\text{minimum distance} = 2.00 \text{ m and } I_c = 69.0°$$

then

$$\text{maximum distance} = \frac{2.00}{\sin 69.0°} = 2.14 \text{ m}$$

This difference in distance can lead to distortion in signals transmitted along step index fibres, since the signals arrive at different times at the output end of the fibre.

Answer

(a) 69.0°
(b) (1) –;   (2) 76.9°;   (3) 64.6°
(c) 32.5°
(d) 2.00 m (min); 2.14 m (max)

Exercise 14.3

1 The critical angle at an interface between crown glass and air is $i_c = 49°$. Calculate the refractive index of crown glass, assuming $n_{air} = 1.0$.

2 Calculate the angle of incidence of a ray of light on one face of a glass prism of angle 60.0° and made of a material of refractive index 1.50, if the ray is *just* totally internally reflected at the second face.

3 Calculate the critical angle for a boundary between a glass fibre, for which the refractive index is 1.60, and cladding, for which the refractive index is 1.50.

4

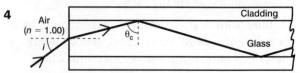

Fig. 14.10   Information for Question 4

Fig. 14.10 shows light incident on one end of an optical fibre and being refracted so that it is incident on the boundary with the cladding at (just greater than) the critical angle. The core is made of glass with refractive index 1.47 and the cladding is of refractive index 1.45. Calculate:

(a) the critical angle

(b) angle i.

5 Light travels through a glass optical fibre 30 m long. The refractive index of the glass is 1.50 and that of its cladding is 1.30. Calculate:

(a) the speed of light in the glass of the fibre

(b) the minimum and maximum distances light travels when trapped in the fibre

(c) the minimum and maximum times taken for light to traverse the fibre.

Assume speed of light in air $= 3.00 \times 10^8 \, \mathrm{m\,s^{-1}}$

6 Calculate the time taken to travel through a 40.0 m length of fibre by red light and by blue light, for which the fibre has refractive indices of 1.45 (red) and 1.47 (blue). Take the velocity of light in air to be $3.00 \times 10^8 \, \mathrm{m\,s^{-1}}$ and the refractive index of air to be 1.00. Consider the axial mode only. (Hint: see Equation 14.4.)

Note: the difference in travel times can lead to distortion in signals transmitted along fibres.

Exercise 14.4: Examination questions

Assume refractive index of air $= 1.00$
speed of light in air $= 3.00 \times 10^8 \, \mathrm{m\,s^{-1}}$ unless stated.

1 (a) Explain what is meant by *refraction*.

(b) A block of glass of refractive index 1.52 is surrounded by air. In an experiment, a beam of light is projected through the glass and strikes one of the faces (internally) at an angle of incidence of 30° (see diagram).

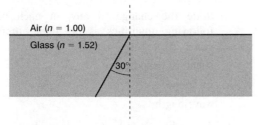

(i) Calculate the angle of refraction.
(ii) Show the refracted ray on the diagram, marking the angle of refraction.

(c) The experiment is repeated with a film of water on the face of the block (see diagram).

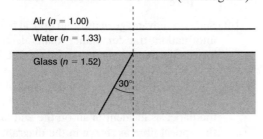

(i) Calculate the angle of refraction for the light passing into the water.

(ii) Calculate the angle of refraction for the light passing into the air from the water and comment on your answer.

(iii) Continue the ray in the diagram, showing its path through the water and into the air. [WJEC 2001]

2 The speed of light in air is slightly less than in a vacuum. This causes light entering the Earth's atmosphere from space to undergo refraction.

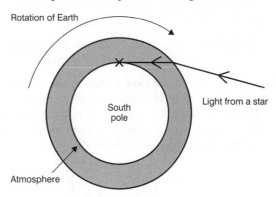

An observer at X is looking for the star to appear over his horizon as the Earth rotates. Because of this refraction he sees it appear slightly earlier than it would do if there was no atmosphere (see diagram above, which is NOT to scale). Assume that the atmosphere has a uniform density and a definite boundary.

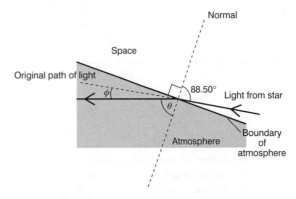

The second diagram (greatly exaggerated) shows the path of the light as it enters the atmosphere.

(a) Calculate θ, the angle of refraction shown in the diagram. (The refractive index of the atmosphere is 1.0003.)

(b) Calculate ϕ, the deviation of the light as it enters the atmosphere.

(c) Show that this angle of deviation causes the star to appear above the horizon about 2 minutes early. [Edexcel S-H 2000]

3 The speed of light in a vacuum is $3.0 \times 10^8 \, \text{m s}^{-1}$. The speed of light in a sample of glass is $2.0 \times 10^8 \, \text{m s}^{-1}$.

Which one of **A** to **D** below is the refractive index of the sample of glass?

A 0.67   **B** 0.44   **C** 1.5   **D** 2.3

[OCR Nuff 2001]

4 The speed of sound in water is $1.50 \times 10^3 \, \text{m s}^{-1}$ and the speed of sound in air is $330 \, \text{m s}^{-1}$. Calculate:

(a) the refractive index of sound passing from air into water

(b) the critical angle at an air–water interface.

In which direction must sound pass to be totally internally reflected at an air–water boundary?

5 The diagram shows a cross-section of one wall and part of the base of an empty fish tank, viewed from the side. It is made from glass of refractive index 1.5. A ray of light travelling in air is incident on the base at an angle of 35° as shown.

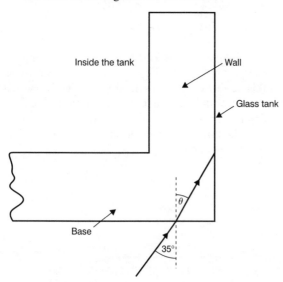

(a) Calculate the angle θ.

(b) (i) Calculate the critical angle for the glass–air interface.

    (ii) Hence, draw on the diagram the continuation of the path of the ray through the glass wall and out into the air. Mark in the values of all angles of incidence, refraction and reflection.

[AQA 2001]

6 Diamonds are highly valued as gems because of their brilliance. Most of the light incident on a well-cut diamond will be totally internally reflected due to their very high refractive index. Fake diamonds made of paste (flint glass) reflect a much smaller proportion of the incident light.

The diagrams below show the path of light through a diamond and through an identically shaped jewel made of paste.

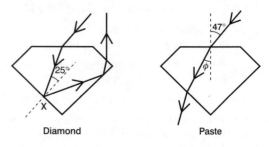

Diamond   Paste

(a) Calculate the angle ϕ for the ray of light passing through the paste jewel. (Refractive index of paste = 1.5.)

(b) The speed of light in the diamond is $1.24 \times 10^8 \, \text{m s}^{-1}$. Calculate the refractive index for diamond.

(c) Show that the ray of light in the diamond will be totally internally reflected at X.

[Edexcel S-H 2000]

7 A beam of light in air is incident on a short length of glass fibre as shown in Fig. 14.11.

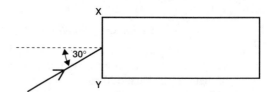

Fig. 14.11

(a) State the change, if any, in each of the following quantities as the light enters the glass:

speed of propagation ......

frequency ......

wavelength ......

(b) The refractive index from air to glass is 1.50.
    (i) Calculate the angle of refraction at the surface XY.
    (ii) Sketch on Fig. 14.11 the path of the beam as it passes through and emerges from the fibre.

(c) State one advantage of using an optical fibre for information transfer rather than electrically insulated wires (a cable).   [OCR 2000]

8 (a) The diagram shows a 'step index' optical fibre. A ray of monochromatic light, in the plane of the paper, is incident in air on the end face of the optical fibre as shown in the diagram.

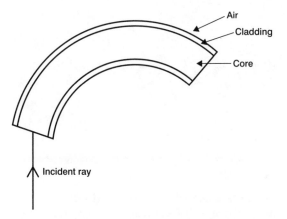

Incident ray

(i) Draw on the diagram the complete path followed by the ray until it emerges at the far end.

(ii) Name the process which occurs as the ray enters the end of the optical fibre.

(iii) The core has a refractive index of 1.50, clad in a material of refractive index 1.45. Calculate the critical angle of incidence at the core-cladding interface.

(b) (i) Give **one** reason why a cladding material is used in an optical fibre.

(ii) In part (a) (iii), the cladding material has a refractive index of 1.45. Explain why it would be advantageous to use cladding material of refractive index less than 1.45.

(c) State **one** use of optical fibres.   [AQA 2001]

9 A step-index optical fibre has a core made of glass of refractive index 1.52. The cladding is made of material of refractive index 1.47.

(a) Calculate the critical angle for the core-cladding boundary.

(b) Fig. 14.12 shows a section of the fibre containing the axis of the fibre.

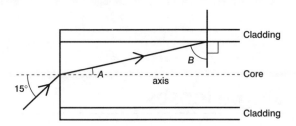

Fig. 14.12   (not to scale)

A beam of light enters the fibre at an angle of incidence of 15°.

(1) Calculate the angles A and B in Fig. 14.12.

(2) State whether this beam will be totally internally reflected at the core/cladding interface, or whether it will escape into the cladding. Indicate your answer by placing a tick in the appropriate box.

The beam is totally internally reflected ☐

The beam ecapes into the cladding   ☐

(c) The optical fibre is 15 km long. Assuming that the fibre is straight, calculate the shortest and the longest time for pulses of light, entering the fibre in different directions, to pass from one end of the fibre to the other.

[CCEA 2001, part]

15
Thin lenses and the eye

Single lenses

As shown in Fig. 15.1, a lens acts to produce an image I from an object O. The lens formula is

$$\frac{1}{v} + \frac{1}{u} = \frac{1}{f} \qquad (15.1)$$

where, as shown in Fig. 15.1, u is the object distance, v the image distance, and f the focal length of the lens.

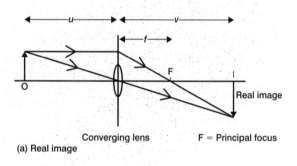

(a) Real image

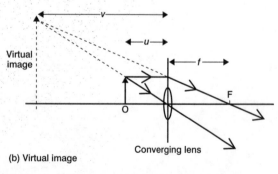

(b) Virtual image

Fig. 15.1 Formation of images by a converging lens

We shall use the real is positive, virtual is negative sign convention. This means that the focal length of a converging lens is positive and that of a diverging lens (see Fig. 15.2) is negative.

Correct signs *must* be used in the lens formula.

Example 1

An object is placed (a) 25.0 cm, (b) 10.0 cm from a converging lens of focal length 15.0 cm. Calculate the image distance and lateral magnification produced in each case, and state the type of image produced.

Method

We have a converging lens so the focal length $f = +15$.

(a) This is a real object, so $u = +25$. We arrange Equation 15.1 to find v:

$$\frac{1}{v} = \frac{1}{f} - \frac{1}{u} = \frac{1}{15} - \frac{1}{25} = \frac{2}{75}$$

$$\therefore \quad v = 37.5 \text{ cm}$$

Since v is positive the image is real. The situation is similar to that shown in Fig. 15.1a.

The lateral magnification m is defined by

$$m = \frac{\text{Height of image}}{\text{Height of object}} \qquad (15.2)$$

It can be shown that

$$m = \frac{\text{Image distance}}{\text{Object distance}} = \frac{v}{u} \qquad (15.3)$$

We have $v = +37.5$ and $u = +25$. Thus

$$m = \frac{v}{u} = \frac{37.5}{25}$$

$$= 1.50$$

The image is 1.50 times as long as the object.

(b) We have a real object so $u = +10$. Rearranging Equation 15.1 to find v:

$$\frac{1}{v} = \frac{1}{f} - \frac{1}{u} = \frac{1}{15} - \frac{1}{10} = -\frac{1}{30}$$

$$\therefore \quad v = -30.0 \text{ cm}$$

Note that v is negative, so that the image is virtual. The situation is similar to that shown in Fig. 15.1b. The lens acts as a simple magnifying glass. To find the lateral magnification m we use Equation 15.3 with $v = -30$ and $u = +10$:

$$m = \frac{v}{u} = \frac{-30}{10}$$

$$= -3.00$$

The image is 3.00 times as long as the object. The significance of the negative sign is that the image is virtual.

Answer

(a) 37.5 cm, 1.50 times, real, (b) 30.0 cm, 3.00 times, virtual

Example 2

When a real object is placed in front of a diverging lens of focal length 20.0 cm, an image is formed 12.0 cm from the lens. Calculate (a) the object distance, (b) the lateral magnification produced. Draw a sketch to show the arrangement.

Method

We have a diverging lens, so the focal length $f = -20$. A real object always produces a virtual image when using a diverging lens, so that $v = -12$.

(a) Rearrange Equation 15.1 to find u:

$$\frac{1}{u} = \frac{1}{f} - \frac{1}{v} = \frac{1}{-20} - \frac{1}{(-12)} = \frac{2}{60}$$

$$\therefore \quad u = 30.0 \, \text{cm}$$

(b) To find the lateral magnification m we use Equation 15.3, with $v = -12$ and $u = 30$:

$$m = \frac{v}{u} = \frac{-12}{30}$$

$$= -0.40$$

The image is 0.40 times as long as the object. The negative sign shows the virtual nature of the image. A sketch of the arrangement is given in Fig. 15.2. A diverging lens always produces a virtual, erect diminished image when viewing a real object.

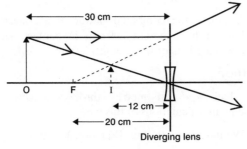

Fig. 15.2   Solution to Example 2

Answer

(a) 30.0 cm, (b) (–)0.40 times.

Example 3

A camera has a lens of focal length 50.0 mm. If it can form images of objects from infinity down to 1.50 m from the lens, calculate the distance through which it must be possible to move the lens.

Method

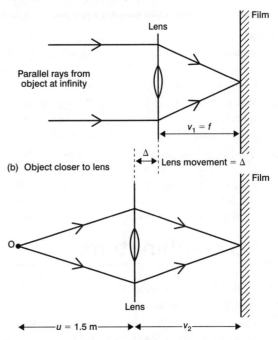

Fig. 15.3   Formation of images using a camera lens

The lens in the camera forms real images on the film, as shown in Fig. 15.3. The compound lens in the camera is thought of as a single thin, converging lens of focal length $f = 5.00$ cm.

As shown in Fig. 15.3a, when the object is at infinity, the lens must be at a distance

$$v_1 = f = 5.00 \, \text{cm}$$

from the lens. When the (real) object is at a distance

$$u = +1.50 \, \text{m} = +150 \, \text{cm}$$

from the lens, the image distance v_2, as shown in Fig. 15.3b, is given by rearranging Equation 15.1:

$$\frac{1}{v_2} = \frac{1}{f} - \frac{1}{u} = \frac{1}{5} - \frac{1}{150} = \frac{29}{150}$$

$$\therefore \quad v_2 = 5.17 \, \text{cm}$$

The required movement Δ of the lens is, as shown in Fig. 15.3, given by

$$\Delta = v_2 - v_1 = 5.17 - 5.00 = 0.17 \, \text{cm}$$

Answer

The lens must move by 0.17 cm

Exercise 15.1

1 An object placed 20 cm from a converging lens results in a real image formed 30 cm from the lens. Calculate the focal length of the lens.

2 When an object is placed 10 cm from a converging lens, an erect image which is three times as long as the object is obtained. Calculate (a) the image distance, (b) the focal length of the lens.

3 An erect image, twice as long as the object, is obtained when using a simple magnifying glass of focal length 10 cm. Calculate (a) the object distance, (b) the image distance. Hint: $v/u = -2$.

4 When a real object is placed 12 cm in front of a diverging lens, a virtual image is formed 8.0 cm from the lens. Find the focal length of the lens.

5 The focal length of a camera lens is 100 mm. Calculate how far from the film the lens must be set in order to photograph an object which is (a) 100 cm, (b) 500 cm from the lens. Hence calculate (c) the movement of the lens between these two positions.

Lens combinations

When two lenses are used, the image produced by the first lens acts as an object for the second lens. This means that, in certain circumstances, we can have a virtual object for the second lens.

Lens combinations – lenses in contact

In this particular case the combined lens system can be replaced by a single lens of focal length f given by

$$\frac{1}{f} = \frac{1}{f_1} + \frac{1}{f_2} \qquad (15.4)$$

where, as shown in Fig. 15.4, f_1 and f_2 are the individual focal lengths of the separate lenses. The thickness of the lenses is neglected.

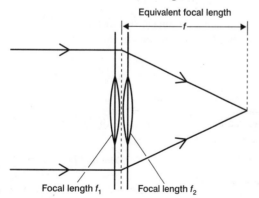

Equivalent focal length

Focal length f_1 Focal length f_2

Fig. 15.4 Lenses in contact

Example 4

A converging lens of focal length 30 cm is placed in contact with a diverging lens of focal length 20 cm. Calculate the focal length of the combination.

Method

We use Equation 15.4 in which for the converging lens $f_1 = +30$, and for the diverging lens $f_2 = -20$. Thus

$$\frac{1}{f} = \frac{1}{30} + \frac{1}{-20} = \frac{2-3}{60} = \frac{-1}{60}$$

$$\therefore \quad f = -60 \text{ cm}$$

The equivalent lens is diverging and has focal length 60 cm. Note that the combined lens must be diverging, since the single diverging lens is more powerful (i.e., has a shorter focal length) than the converging lens.

Answer

The combination is a diverging lens of focal length 60 cm.

Power of a lens

The power F of a lens, *in dioptres* (D), is defined by

$$F = \frac{1}{f} \qquad (15.5)$$

where f is the focal length of the lens, *in metres*.

For two lenses in contact as in Fig. 15.4, Equation 15.4 shows that we must *add* the powers of the lenses. Thus the combined lens system can be replaced by a single lens of power F given by

$$F = F_1 + F_2 \qquad (15.6)$$

where F_1 and F_2 are the individual powers of the separate lenses.

Example 5

(a) Calculate the power of a converging lens of focal length 250 mm.

(b) Calculate the combined power of a converging lens of focal length 200 mm in contact with a diverging lens of focal length 50 mm.

Method

We shall work in metres throughout this, and subsequent, calculations of this type.

(a) We have $f = +0.25$ m. Equation 15.5 gives

$$F = \frac{1}{0.25} = +4.0 \text{ D}$$

Note the positive sign, since we have a converging lens.

(b) We have $f_1 = 0.20$ m,

$$\text{so } F_1 = \frac{1}{f_1} = 5.0 \text{ D}$$

and $f_2 = -0.050$ m,

$$\text{so } F_2 = \frac{1}{f_2} = -20 \text{ D}.$$

Equation 15.6 gives the combined power

$$F = F_1 + F_2 = 5.0 - 20.0 = -15.0 \text{ D}.$$

Note that the combined lens has a negative power and focal length since the diverging lens has a higher power (has a shorter focal length) than the converging lens.

Answer

(a) $+4.0$ D, (b) -15.0 D.

Exercise 15.2

1 A converging achromatic doublet consists of a converging (crown glass) lens of focal length 20 cm and a diverging lens (made of flint glass). If the focal length of the doublet is 80 cm, calculate the focal length of the diverging lens.

2 (a) Calculate the focal length of a lens of power -2.0 D.
   (b) A lens of power -2.0 D is placed in contact with a converging lens of focal length 20 cm. Find the power of the combined lens system.

Correction of defective vision

The eye and accommodation

The eye has the ability to form clear images on the retina of objects at differing distances from the eye. In order to do this the focal length of the eye lens must be able to be changed – this is done by the action of the ciliary muscles. This effect is called accommodation. When the eye is focused on a distant object it is said to be 'unaccommodated'. In order to focus on objects close to the eye the focal length of the eye must be decreased – that is, its power must be increased.

Example 6

The cornea and lens of a normal, unaccommodated eye has a power of $+50$ D. Find (a) the lens to retina distance for this eye, (b) the power of the lens system required to clearly focus on objects at a point 25 cm from the eye.

Method

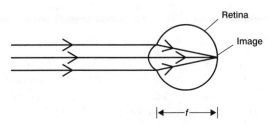

Fig. 15.5 Formation of an image of a distant object by the eye

(a) The combined cornea and eye lens system will form images of distant objects at the focus of the combined lens, as shown in Fig. 15.5. Thus the lens to retina distance is equal to the focal length f of the lens. Now

$$f = \frac{1}{F} = \frac{1}{50} = 0.020 \text{ m}$$

The lens to retina distance is 0.020 m.

(b) The eye lens must now accommodate (change its focal length) in order to clearly focus on objects close to the eye. This still results in images formed on the retina at a distance from the lens of 0.020 m.

Suppose that the new focal length of the combined lens is f'. For this lens, the (real) object distance u is 0.25 m and the (real) image distance v is 0.020 m (see Fig. 15.1a). Using Equation 15.1

$$\frac{1}{f'} = \frac{1}{0.020} + \frac{1}{0.25} = 54$$

From Equation 15.5 we see that $1/f'$ is the power of the lens.

The combined lens needs a power of 54 D.

Answer

(a) 0.020 m, (b) 54 D.

Eye defects

A 'normal' eye has a far point of infinity and a near point of 25 cm.

In *myopia* (near sightedness), the far point is closer than infinity and the near point may be closer than that for the normal eye. This may be due to the eyeball being too long, or the cornea too curved. A diverging lens is used to correct this defect as shown in Fig. 15.6.

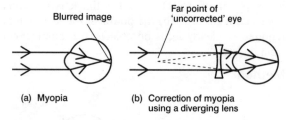

(a) Myopia

(b) Correction of myopia using a diverging lens

Fig. 15.6 Myopia and its correction

In *hypermetropia* (long sightedness), the far point is at infinity but the near point will be more than 25 cm from the eye. A converging lens is used to correct this defect as shown in Fig. 15.7.

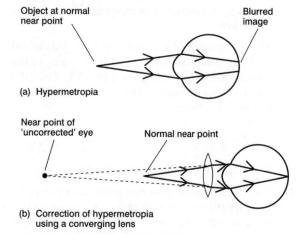

(a) Hypermetropia

(b) Correction of hypermetropia using a converging lens

Fig. 15.7 Hypermetropia and its correction

In *presbyopia* (old sight) the eye lens has become hard with age and is thus unable to change its shape and so accommodate over a sufficiently wide range. The situation may arise in which the near point is further away from the eye than 25 cm and the far point is closer to the eye than infinity. The elderly person may thus require two sets of spectacles to aid close and long distance work separately. The spectacles may take the form of bifocal lenses.

Example 7

A person with short sight has a far point of 250 cm and a near point of 15 cm.

(a) Calculate the power of the spectacle lens required to enable distant objects to be seen.

(b) Calculate the near point for the person when using this spectacle lens.

(c) State the range of distinct vision when wearing the spectacles.

Method

(a) A diverging lens is used. The power of the lens is such that it produces a virtual image at the far point of the eye (in this case 2.50 m), using an object at an infinite distance from the lens (see Fig. 15.6b). Thus we have $u = \infty$, $v = -2.50$ m. We require the power F of the correcting lens. Using the lens formula (Equation 15.1) and noting $F = 1/f$ then

$$F = \frac{1}{v} + \frac{1}{u} \qquad (15.7)$$

$$= \frac{1}{-2.50} + \frac{1}{\infty}$$

$$= -0.40 + 0$$

Hence power $F = -0.40$ D. (Focal length $f = -2.50$ m) Note that this is a diverging lens.

(b) When an object is placed at the person's 'corrected' near point it produces a virtual image at the original, 'uncorrected', near point. Thus we have, for the correction lens

u = 'corrected' near point distance

v = original, 'uncorrected', near point distance $= -0.15$ m

F = power of correction lens $= -0.40$ D

Using Equation 15.7 gives

$$\frac{1}{-0.15} + \frac{1}{u} = -0.40$$

$$\therefore \quad \frac{1}{u} = -0.40 + \frac{1}{0.15} = 6.267$$

Hence $u = 0.160$ m.

(c) The range of distinct (corrected) vision is from 0.16 m to infinity.

Answer

(a) -0.40 D, (b) 0.16 m, (c) 0.16 m to infinity.

Example 8

An elderly person with presbyopia has a near point of 0.400 m and a far point of 4.00 m. Calculate:

(a) the range of power which this person's eye lens has;

(b) the power of the spectacle lens required to enable objects at the normal near point to be seen;

(c) the range of distinct vision when using the spectacle lens in (b);

(d) the power of the spectacle lens required to enable objects at infinity to be seen.

Method

(a) The person can focus objects from 0.400 m up to 4.00 m from the eye. In each case the image is focused on the retina and the image distance is thus fixed (the length of the eyeball) – let this distance be v.

Referring to Fig. 15.1, suppose that the eye focuses on an object 0.400 m from it. We have

Object distance $= 0.400$ m
Image distance $= v$ (length of eyeball)

If we let the power of the cornea and eye lens be F_1, then using Equation 15.7

$$\frac{1}{v} + \frac{1}{0.400} = F_1$$

Similarly, when the eye focuses on an object 4.00 m from it we have

Object distance $= 4.00$ m
Image distance $= v$ (length of eyeball)

If the power of the cornea and eye lens is now F_2 then Equation 15.7 gives

$$\frac{1}{v} + \frac{1}{4.00} = F_2$$

By combining the two equations above we obtain

$$F_1 - F_2 = \left(\frac{1}{v} + \frac{1}{0.400}\right) - \left(\frac{1}{v} + \frac{1}{4.00}\right)$$

$$= \frac{1}{0.400} - \frac{1}{4.00} = \frac{9}{4.00}$$

$$= 2.25$$

The range of power of the person's eye lens is thus 2.25 D. This is called the *amplitude of accommodation*. For a young person the amplitude of accommodation is typically 11 D.

(b) A converging lens is used such that when the (real) object is placed at the normal near point we get a virtual image at the near point of the uncorrected eye (see Fig. 15.7b). Thus we have

$u = +0.25$ m (normal near point)
$v = -0.400$ m (virtual image at near point of uncorrected eye)

and require the power of the correction lens. Using Equation 15.7

$$F = \frac{1}{-0.400} + \frac{1}{0.25}$$

$$= -2.5 + 4.0 = +1.50$$

Hence power required $= +1.50$ D.

(c) The range of distinct (corrected) vision using the correction lens in (b) is obtained by noting that the far point of the uncorrected eye is 4.00 m. A real object placed as far as possible from the lens will produce a virtual image at 4.00 m away. Thus we have

$u =$ far point using correction lens
$v = -4.00$ m (virtual image at far point of the uncorrected eye)
$F = +1.50$

Equation 15.7 gives

$$\frac{1}{-4.00} + \frac{1}{u} = 1.50$$

$\therefore$ or $u = 0.571$ m.

The far point using the correction lens is 0.571 m from the eye. The range of distinct (corrected) vision is 0.25 m to 0.57 m.

(d) In this case a diverging lens must be used whose power is such that it produces a virtual image at the far point of the uncorrected eye – in this case 4.00 m – using an object at an infinite distance from the lens. Working in a similar manner to Example 7a we have $u = \infty$, $v = -4.00$ m and require the power F of the correcting lens. Equation 15.7 gives

$$F = \frac{1}{-4.00} + \frac{1}{\infty} = -0.25$$

The power of the correcting lens required is -0.25 D.

Answers

(a) 2.25 D, (b) +1.5 D, (c) 0.25 m to 0.57 m,
(d) −0.25 D.

Exercise 15.3

1 The combined lens of a normal, unaccommodated eye has a power of 56 D.

   (a) Calculate the lens to retina distance.

   (b) If the eye clearly focuses on an object 25 cm away, find the change in power required.

2 A short sighted person has a far point of 150 cm and a near point of 20 cm.

   (a) Calculate the power of the spectacle lens needed to clearly view an object at the normal far point.

   (b) Find the range of distinct vision when wearing the lens.

3 An elderly person, with presbyopia, has a near point without spectacles of 0.50 m and an amplitude of accommodation of 1.5 D.

Calculate:

(a) the far point without spectacles;

(b) the power of the spectacle lens needed to enable
   (i) distant objects to be seen,
   (ii) objects at the normal near point to be seen.

Exercise 15.4:
Examination questions

1 An illuminated object is placed 48.0 cm from a screen. A lens is to be placed between the object and screen in order to produce a real image on the screen.

(a) If the image is to be the same size as the object, what kind* of lens and what focal length would be needed?

(b) If the image is to be twice the size of the object, what kind* of lens and what focal length would be needed?

2 An object of height 6.0 mm is placed at a distance of 8.0 cm from a converging lens of focal length 12 cm. Calculate:

(a) the distance of the image from the lens (in cm) and

(b) the height of the image (in mm).

(c) Is the image real or virtual?

3 (a) Explain what is meant by the **principal focus** and the **focal length** of a **diverging** lens.

(b) Fig. 15.8 shows a **diverging** lens with principal foci **F** and **F′** and an object **OB** placed perpendicular to the principal axis.

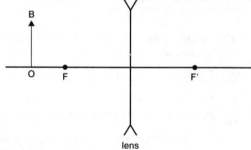

Fig. 15.8

On Fig. 15.8 draw suitable construction rays to locate the image of **OB**. Label this image **IM**. Show a suitable position of the eye for viewing the image.

(c) A **diverging** lens has focal length 200 mm. An object is placed 400 mm from the lens.
   (i) Find the position of the image and its linear magnification.

*Converging or diverging.

(ii) The object now commences to move towards the lens at a constant speed of $20\,\text{mm s}^{-1}$.
   (1) Find the position of the image 2.0 s after the object starts to move.
   (2) Calculate the average speed of the image during this time interval.
   (3) State the direction of movement of the image. [CCEA 2000]

4 (a) Explain what is meant by the terms
   (i) a diverging lens,
   (ii) the principal focus of a diverging lens,
   (iii) the focal length of a diverging lens.

(b) (i) The linear magnification of an image formed by a converging lens is given by the height of the image divided by the height of the object. State another expression for the linear magnification. Identify any symbols appearing in your expression.
   (ii) Describe an experiment to verify the expression you have given in your answer to (b) (i). Assume that an illuminated object is available. Your account should include a labelled sketch of the apparatus, an outline of the procedure, headings for a table of results which would be taken, and details of how the results would be processed to verify the expression.

(c) Draw a labelled ray diagram to show how an image is formed using a **diverging** lens. In your diagram, clearly identify the object, the image, the principal focus, and the position of the eye to view the image.

(d) A converging lens and a diverging lens each have a focal length of magnitude 150 mm. When a linear object of height 20.0 mm is placed 600 mm from the **diverging** lens, the image is of height D. The same object is used with the **converging** lens to produce an image, also of height D.
   Find
   (i) the height D of the image,
   (ii) the exact location of the image for each lens,
   (iii) the nature of the image for each lens.

(e) A slide of dimensions 35.0 mm × 24.0 mm is placed in a desk-top projector. The lens system in the projector may be assumed to be a single thin lens, which forms a real image of the slide. The image has dimensions 280 mm × 192 mm. The distance between the slide in the projector and its image is 567 mm. Calculate the focal length of the projector lens, and identify its type.

[CCEA 2001]

5 At the theatre, people in the audience sometimes use theatre glasses. These are special binoculars which help them to see the actors.

For each eye, the lenses in the theatre glasses are arranged as shown below.

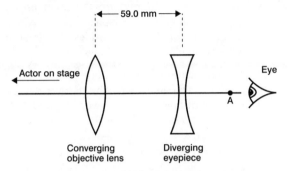

An actor is standing on a stage 30 m from a member of the audience who is using the glasses. The objective lens forms an image of the actor 80 mm behind this lens at A.

(a) (i) Explain why the focal length of the objective lens is 80 mm.
   (ii) Calculate the power of the objective lens.

(b) The eyepiece is a diverging lens of focal length 20.0 mm. The image at A formed by the objective lens acts as the object for the eyepiece.

The final position of the image formed by the eyepiece can be calculated using

$$\frac{1}{v} - \frac{1}{21} = \frac{1}{f}.$$

   (i) Suggest why the object distance is negative.
   (ii) Calculate the final position of the image formed by the eyepiece.
   (iii) Is this image real or virtual? Explain your answer.

(c) How would you adjust the glasses to move the final image further away from the eye? Give your reasoning. [Edexcel S-H 2000]

6 (a) The lens of a camera has a focal length of 35 mm. The lens can be moved with respect to the film to produce clear images of near or distant objects. The camera is first used to photograph an object at 400 mm from the lens. It is then used to photograph an object at infinity.
   (i) State the direction through which the lens is moved between taking these photographs.
   (ii) Calculate the distance through which the lens is moved.

(b) Many camera lenses can be fitted with a polarizing filter. The filter material is similar to that used in some sunglasses. Explain an advantage gained from the use of this filter. [OCR 2000]

7 (a) The lens in a camera has focal length 80 mm. The lens can be moved to provide accurate focusing.
   (i) Draw a labelled sketch to show the formation of a diminished, real image by the lens. Label the object and image, and the principal focus of the lens.
   (ii) A dog 0.40 m tall stands 1.6 m away from the camera. The lens is adjusted to give a clear image of the dog. Calculate the distance from the lens to the film, the linear magnification and the image height.

(b) (i) An object is placed at a principal focus of a diverging lens. Draw a ray diagram to illustrate the formation of the image. Label the object and image, and the principal focus of the lens. Show where the eye should be placed to view the image.
   (ii) A diverging lens has focal length 200 mm. An object 10.0 mm high is placed on the principal axis. The image produced is 4.0 mm high.
   Calculate
   (1) the linear magnification,
   (2) the distance of the image from the lens,
   (3) the distance of the object from the lens. [CCEA 2001, part]

8 A camera has a lens of focal length 50.0 mm and produces a sharp image on its film of an object 200 mm from the lens. Calculate:

(a) the power of the lens

(b) the distance from the lens to the film in the camera.

9 The focusing system of a person's eye has a total power of 57.0 D when viewing a distant object. Assuming that the focusing system of the eye may be treated as a single lens placed at the front of the eye, calculate:

(a) the distance from this lens to the retina

(b) the focal length of the focusing system when viewing an object 40.0 cm in front of the eye.

10 (a) (i) State what is meant by *accommodation* of the eye.
   (ii) Explain how the eye achieves this.

(b) The lens system of the eye of a young child is situated 1.7 cm from the retina. The power of the lens can be varied from 54 D to 60 D.
   (i) Determine the location of the child's near point.

(ii) Explain whether the child can focus on an object at infinity.

(iii) State from what defect of vision the child suffers. [OCR 2000]

11 A short-sighted person is prescribed a lens of power −1.25 D so that an object at infinity may be seen.

(a) Calculate the image distance for this lens alone when the object is at infinity.

(b) Explain why, when using this prescribed lens in spectacles, the image in (a) is formed at the person's far point.

12 (a) A student is unable to focus on objects that are more than 2.0 m away unless he is wearing his glasses. His glasses enable him to see a distant object clearly by forming a virtual image of this object at 2.0 m from his eyes.

(i) Explain whether the lenses are converging or diverging

(ii) State the focal length of the lenses in his glasses.

(iii) Hence, calculate the power of these lenses.

(iv) Draw a ray diagram of one of these lenses forming an image of an object that is 4.0 m away from the lens. Label the image.

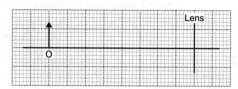

(b) During his next sight test, the optician finds that the student's sight has changed.

The student sees clearly when an additional lens of power +0.20 D is combined with his existing lenses.

(i) Calculate the power of this new lens combination.

(ii) Explain whether the student's sight when not wearing glasses has improved or worsened. [Edexcel S-H 2001]

13 (a) A bundle of light rays from a point on an object enters a person's eye. Which component of the eye provides the greatest converging effect on these rays?

(b) **Accommodation** is the ability of the eye to produce clear images of objects over a wide range of distance from the eye.

(i) Which components of the eye enable the process of accommodation?

(ii) Describe the mechanism of this process.

(c) When viewing objects, a person is said to have a **near point** and a **far point**. Explain what is meant by

(i) near point,

(ii) far point.

(d) A person has a near point distance of 12.0 cm and a far point distance of 320 cm.

(i) Calculate the power of spectacle lens needed to change the person's far point to the normal far point position.

(ii) Calculate the person's near point distance when wearing the spectacles in (i). [CCEA 2001]

14 A short-sighted person can only see objects clearly when they lie between his far point and a point 200 mm from his eye. In order to allow him to see distant objects clearly he is prescribed a diverging lens of focal length 300 mm.

(a) What is the person's far point without spectacles?

(b) Calculate the change in position of the person's near point when spectacles are used.

16
Optical instruments

Angular magnification

When an object is viewed, the *apparent* size of the object is determined by the length L of the image formed on the retina. As shown in Fig. 16.1, L is determined by the *visual angle* θ which the object subtends at the eye. Throughout this chapter we assume θ to be small, in which case L is directly proportional to θ.

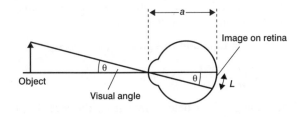

Note: $L = a\theta$

where L = length of image on retina (metres)

a = length of eyeball (metres)

θ = visual angle (radians); assumed small

Fig. 16.1   Visual angle

The purpose of an optical instrument is to increase the size of the visual angle. In doing so the final image, when viewed through the instrument, appears to be larger than when the object is viewed using the unaided or 'naked' eye. We define the angular magnification (or magnifying power) M of an optical instrument by

$$M = \frac{\beta}{\alpha} \qquad (16.1)$$

where β is the angle subtended at the eye by the *image* when using the instrument, and α is the angle subtended using the unaided eye by the *object* when at the appropriate distance.

The magnifying glass (simple microscope)

Using the unaided eye, the maximum apparent size of the object occurs when it is placed at the least distance of distinct vision D (typically 250 mm for adults) from the eye, as shown in Fig. 16.2.

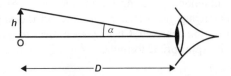

h = height of object O

Fig. 16.2   Visual angle α of an object at the least distance of distinct vision D

The angle subtended α, in radians (see Chapter 2), is given by

$$\alpha = \frac{h}{D} \qquad (16.2)$$

where h is the height of the object O.

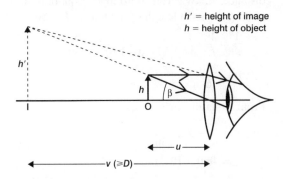

h' = height of image
h = height of object

Fig. 16.3   Visual angle β using a simple microscope

Fig. 16.3 shows the formation of an image I when the object O is placed distance u from the magnifying glass. *Since u is less than D, β is greater than α.* Thus

$$\beta = \frac{h'}{v} = \frac{h}{u} \qquad (16.3)$$

Combining Equations 16.1, 16.2 and 16.3 gives

$$M = \frac{\beta}{\alpha} = \frac{h/u}{h/D}$$

$$M = \frac{D}{u} \qquad (16.4)$$

This is a *general expression* and is true for whatever value of v (and u) we have. Note that *in normal adjustment the image distance v equals D*. It is convenient to use Equation 16.4, but you should also be able to work from first principles (see below).

Example 1

An object of height 2.00 mm is to be viewed using a simple magnifying glass of focal length 50.0 mm. If the final image is formed at the least distance of distinct vision (250 mm) from the eye, calculate the visual angle subtended (a) using the unaided eye, (b) using the magnifying glass. Hence calculate (c) the angular magnification achieved. Check your answer to part (c) using the appropriate formula.

Method

Using millimetres we have $h = 2.00$ and $D = 250$.

(a) From Equation 16.2

$$\alpha = \frac{h}{D} = \frac{2}{250}$$

$$= 8.00 \times 10^{-3} \, \text{rad}$$

(b) Referring to Fig. 16.3 we have image distance equal to 250 and since the image is virtual, $v = -250$. To find β we require the object distance u. We can rearrange Equation 15.1, putting the focal length of the lens as $f = 50$:

$$\frac{1}{u} = \frac{1}{f} - \frac{1}{v} = \frac{1}{50} - \frac{1}{(-250)} = \frac{6}{250}$$

$$\therefore \quad u = \frac{250}{6}$$

From Equation 16.3

$$\beta = \frac{h}{u} = 2 \times \frac{6}{250}$$

$$= 48.0 \times 10^{-3} \, \text{rad}$$

(c) From Equation 16.1

$$M = \frac{\beta}{\alpha} = \frac{48 \times 10^{-3}}{8 \times 10^{-3}} = 6.00$$

Note that the image is virtual, so we should, strictly speaking, write $M = -6.00$. It is common practice, however, to write only the numerical

value and to omit the sign. M does not depend on the height of the object, since h cancels in the derivation of Equation 16.4. To check our answer for M using Equation 16.4, we have $D = 250$ and $u = 250/6$, so

$$M = \frac{D}{u} = 250 \times \frac{6}{250} = 6.00$$

Answer

(a) $8.00 \times 10^{-3} \, \text{rad}$, (b) $48.0 \times 10^{-3} \, \text{rad}$, (c) 6.00 times.

Example 2

A man wishes to study a photograph in fine detail by using a lens as a simple magnifying glass in such a way that he sees an image magnified ten times and at a distance of 250 mm from the lens. What focal length lens should he use, and how far from the photograph should it be held?

Method

We have $M = 10.0$. Refer to Fig. 16.3. Using millimetres we have $v = -250$ and we have to *assume* that $D = -250$. We require u and f.

Rearranging Equation 16.4 gives us (without signs)

$$u = \frac{D}{M} = \frac{250}{10} = 25.0 \, \text{mm}$$

The photograph is held 25 mm from the lens. From Equation 15.1

$$\frac{1}{f} = \frac{1}{v} + \frac{1}{u} = \frac{1}{-250} + \frac{1}{25} = \frac{9}{250}$$

$$\therefore \quad f = \frac{250}{9} = 27.8 \, \text{mm}$$

Answer

A converging lens of focal length 27.8 mm is needed at 25.0 mm from the photograph.

Exercise 16.1

1 A man whose least distance of distinct vision is 250 mm views a stamp using a converging lens of focal length 30 mm. If the final image is located at the least distance of distinct vision, calculate (a) the distance of the stamp from the lens, (b) the angular magnification he achieves. Assume that the eye is close to the lens.

2 Repeat Question 1, but assume that the image is to be observed at infinity.

3 Repeat Question 1 for a man whose least distance of distinct vision is 180 mm.

The astronomical telescope

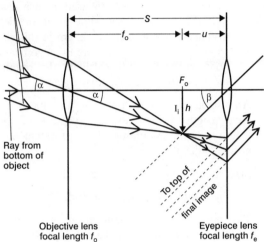

Rays from top of object

α

α

F_o

I_i h

β

Ray from bottom of object

To top of final image

Objective lens focal length f_o

Eyepiece lens focal length f_e

Fig. 16.4 Visual angle in the astronomical telescope

Fig. 16.4 shows how an astronomical telescope, used to observe a distant object such as a star, increases the visual angle. The object would subtend an angle α when viewed with the unaided eye. Use of the telescope leads to the formation of a final image which subtends an angle β at the eye (assuming that the eye is close to the eyepiece lens). From Fig. 16.4, in which the intermediate image I_i, is of height h and is formed at distance u from the eyepiece lens,

$$\alpha = \frac{h}{f_o} \tag{16.5}$$

and

$$\beta = \frac{h}{u} \tag{16.6}$$

where α and β, in radians, are small angles. Combining Equations 16.1, 16.5 and 16.6 gives the angular magnification (or magnifying power) M:

$$M = \frac{\beta}{\alpha} = \frac{h/u}{h/f_o}$$

or

$$M = \frac{f_o}{u} \tag{16.7}$$

Equation 16.7 is a *general expression* and can be used at whatever distance the final virtual image is formed from the eye. In *normal* adjustment, in which the final image is formed at infinity, we note two special characteristics:

(1) $u = f_e$, so that $M = f_o/f_e$.

(2) The objective and eyepiece lenses are separated by a distance $(f_o + f_e)$.

Example 3

An astronomical telescope has an objective lens of focal length 100 cm and an eyepiece lens of focal length 5.00 cm. Calculate the angular magnification and the separation of the lenses when the telescope is in normal adjustment,

Method

Referring to Fig. 16.4 we have $f_o = +100$ and $f_e = +5.00$.

In normal adjustment $u = f_e = +5.00$. Thus from Equation 16.7

$$M = \frac{f_o}{u} = \frac{f_o}{f_e} = \frac{100}{5}$$

$$= 20.0$$

The separation S of the lenses is given by

$$S = (f_o + u) = (f_o + f_e) = 100 + 5$$
$$= 105 \text{ cm}$$

Answer

$M = 20.0$, lens separation is 105 cm.

Example 4

An astronomical telescope consists of two thin converging lenses. When it is in normal adjustment the lenses are 650 mm apart and the angular magnification is 12.0. Calculate the focal length of the objective lens and the eyepiece lens.

Method

We have $M = 12$ and lens separation $S = 650$ mm. Now, in normal adjustment (see Example 3)

$$M = \frac{f_o}{f_e} \quad \therefore \quad f_o = 12f_e \tag{i}$$

and $\quad S = f_o + f_e \quad \therefore \quad f_o + f_e = 650 \tag{ii}$

We have two simultaneous equations (see Chapter 2) so we substitute for f_o from (i) to (ii):

$$12f_e + f_e = 650$$
$$\therefore \qquad f_e = 50 \text{ mm}$$
$$\therefore \qquad f_o = 12f_e = 600 \text{ mm}$$

Answer

The objective has focal length 600 mm, the eyepiece 50 mm.

Exercise 16.2

1 An astronomical telescope which is in normal adjustment consists of two thin converging lenses of focal length 60.0 cm and 3.00 cm. It is focused on a distant object which subtends an angle of 2.00×10^{-3} rad when viewed directly. Calculate (a) the angular magnification achieved, (b) the separation of the lenses, (c) the angle subtended by the final image.

2 An astronomical telescope has an object of focal length 90 cm and an eyepiece of focal length 5.0 cm. When, in normal adjustment, it is used to view a full moon, the final image subtends an angle of 0.10 radian at the eye lens. Calculate (a) the angular magnification, (b) the angle subtended by the moon when viewed directly.

Given that the distance between the moon and the Earth is 3.8×10^5 km, calculate (c) the diameter of the moon.

Exercise 16.3: Examination questions

1 A student with a least distance of distinct vision of 12.0 cm uses a 6.00 cm focal length converging lens as a magnifying glass in order to examine fine detail on a biological slide. Assuming that the lens is held close to the eye and that the image is viewed at the student's least distance of distinct vision, calculate:

(a) the distance x of the slide from the lens

(b) the angular magnification produced.

2 (a) A photograph is taken of a distant object using a camera with a lens of focal length 7.5 cm. If the negative is viewed at a distance of 25 cm using the naked eye, calculate the overall magnification achieved.

(b) In order to view the slide in more detail a simple magnifying glass is used in such a way that the image of the slide is at the least distance of distinct vision (25 cm in this case) from the lens/eye. If the overall magnification now achieved is 1.0, calculate the required focal length of the magnifying glass.

3 An astronomical telescope has an objective of focal length 900 mm and an eyepiece of focal length 18 mm. Assuming the instrument is in normal adjustment, calculate:

(a) the separation of the lenses and

(b) the magnifying power achieved.

4 The angular magnification of an astronomical telescope in normal adjustment is 4.00. If the distance between the lenses is 625 mm, calculate the focal length of:

(a) the eyepiece lens

(b) the objective lens.

5 An astronomical telescope in normal adjustment has an objective lens of focal length 20 cm and the separation of the lenses is 25 cm. The telescope views a distant object which subtends an angle of 5.0×10^{-2} rad at the objective lens. Calculate the angle, in rad, subtended by the final image at the eyepiece lens.

6 An astronomical telescope in normal adjustment has an objective of focal length 0.90 m and the lenses are separated by a distance of 1.00 m. When used to view the full moon the image subtends an angle of 51×10^{-3} rad at the eye lens. If the distance between the Earth and the moon is 3.8×10^5 km, calculate the diameter of the moon.

Section F
Heat

17
Thermal properties of matter

Heating matter

In this section we deal with the change in internal energy of a body due to the transfer of energy on a microscopic scale. Thermal energy is transferred from an area of higher temperature to an area of lower temperature. Heat supplied, for example by electrical means, results in a change of temperature and/or a change of state.

Energy transfer Q needed to raise the temperature of a mass m of a substance by $\Delta\theta$ (in Kelvin) is given by:

$$Q = mc\Delta\theta \qquad (17.1)$$

where c is the *specific heat capacity* of the substance.

Similarly, the total heat capacity of an object is the energy needed to raise the temperature of an object of mass m by 1 K. Total heat capacity $= mc$.

We shall apply the law of conservation of energy throughout our calculations on thermal energy transfer.

Electrical heating

As pointed out in Chapter 20, the heat supplied per second by an electrical heater is give by VI, where V is the potential difference (voltage) across the heater wire and I is the current through it. This results in the transfer of thermal energy into the surrounding material at a rate often described as the 'power dissipated'.

$$\text{Power} = V \times I \qquad (17.2)$$

If the resistance of the heater wire is R (see Chapter 20) we may use the relation $V = IR$ to obtain

$$\text{Power } P = I^2R \text{ or } \frac{V^2}{R} \qquad (17.3)$$

In time t seconds the energy Q transferred by the heater is

$$Q = VIt \text{ or } I^2Rt \text{ or } (V^2/R)t \qquad (17.4)$$

Example 1

A solid copper block of mass 5.0 kg is heated for 7 minutes exactly by an electric heater embedded in the block. A potential difference of 25 V is applied across the heater, and the current is recorded as 2.0 A. If the temperature of the block rises by 10 K, calculate the specific heat capacity of copper, assuming that no heat escapes from the apparatus and that the heat capacity of the heater itself is negligible (see Fig. 17.1).

Method

The energy Q supplied by the heater is

$$Q = VIt = 25 \times 2.0 \times 7 \times 60 = 21 \times 10^3 J$$

This is absorbed by the copper block, of mass $m = 5.0$ kg, and causes a temperature rise $\Delta\theta = 10$ K.

141

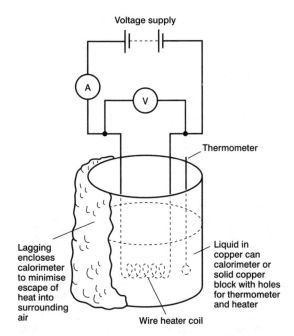

Voltage supply

Thermometer

Lagging encloses calorimeter to minimise escape of heat into surrounding air

Liquid in copper can calorimeter or solid copper block with holes for thermometer and heater

Wire heater coil

Fig. 17.1 Apparatus for simple electrical calorimetry (not to scale)

Now

$$Q = mc\Delta\theta = 5.0 \times c \times 10$$

where c is the specific heat capacity of the copper. We can equate the two terms since no significant amount of the heat supplied stays in the heater or escapes from the copper block. Thus:

$$21 \times 10^3 = 50 \times c$$

or $\quad c = 0.42 \times 10^3$

Answer

$0.42\,\mathrm{kJ\,kg^{-1}\,K^{-1}}$.

Example 2

50 g of water at 12 °C is placed in a copper calorimeter which weighs 0.10 kg. An electric heater coil of negligible thermal capacity is immersed in the water. With 7.0 V across the heater producing a steady current of 1.0 A for exactly 6 minutes, a final temperature of 22 °C was obtained. If the heat loss to the surroundings is negligible, what is the value of the specific heat capacity for water? (See Fig. 17.1.) (SHC copper = $420\,\mathrm{J\,kg^{-1}K^{-1}}$.)

Method

The energy supplied by the heater is

$$Q = VIt = 7.0 \times 1.0 \times 6.0 \times 60 = 2520J$$

If we assume no energy losses to the surroundings then:

Energy Q_w absorbed by the water of mass $m = 50 \times 10^{-3}\,\mathrm{kg}$ in raising its temperature by

$\Delta\theta = (22 - 12) = 10\,\mathrm{K}$ is, assuming it has specific heat capacity c_w,

$$Q_w = mc_w\Delta\theta = 50 \times 10^{-3} \times c_w \times 10 = 0.5\,c_w$$

Energy Q_c absorbed by calorimeter (mass 0.10 kg, specific heat capacity $420\,\mathrm{J\,kg^{-1}\,K^{-1}}$ and temperature rise also 10 K) is

$$Q_c = 0.1 \times 420 \times 10 = 420$$

Equating the terms $Q = Q_w + Q_c$, we get

$$2520 = 0.5\,c_w + 420$$

$$\therefore \quad c_w = 2100/0.5 = 4200$$

Answer

$4.2\,\mathrm{kJ\,kg^{-1}\,K^{-1}}$.

Rate of rise of temperature

Consider a mass m of substance with specific heat capacity c. If the net power supplied is P watts (= power supplied by heater minus heat lost per second to the surroundings) then

$$P = \text{net energy supplied per second}$$
$$= m \times c \times \text{temperature rise per second}$$

and if calculus notation is used (see Chapter 2)

$$P = mc\frac{d\theta}{dt}$$

where θ denotes temperature, t time and $\frac{d\theta}{dt}$ is the rate of change of temperature with time. If power supplied is constant and heat loss to the surroundings is negligible then temperature rises in proportion to time, the temperature versus time graph is linear (i.e. straight) and the slope of the graph equals $\frac{d\theta}{dt}$.

Example 3

Energy is supplied at a rate of 20 W to 40 g of water in a plastic cup. The specific heat capacity of water is $4.2\,\mathrm{kJ\,kg^{-1}\,K^{-1}}$. Calculate the initial rate of rise of temperature.

Method

We have $P = 20$, $m = 40 \times 10^{-3}$ and $c = 4.2 \times 10^3$

$$P = mc\frac{d\theta}{dt} \text{ so that}$$

$$20 = 0.040 \times 4200 \times \frac{d\theta}{dt} \text{ giving}$$

$$\frac{d\theta}{dt} = \frac{20}{168} = 0.12\,\mathrm{K\,s^{-1}}.$$

Answer

$0.12\,\mathrm{K\,s^{-1}}$.

The continuous-flow calorimeter

Fig. 17.2 shows a continuous-flow calorimeter suitable for use with a liquid such as water.

When the temperature θ_2 is steady we know that the glassware has reached steady temperatures throughout.

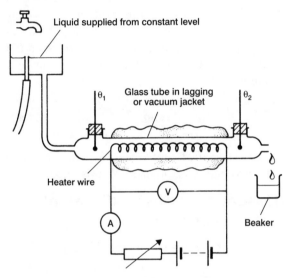

Fig. 17.2   A continuous-flow calorimeter

The energy supplied per second is VI and equals $mc(\theta_2 - \theta_1)$ where m is the mass of liquid per second flowing through the apparatus and collected in the beaker. The specific heat capacity c can be calculated directly from this equation if heat loss to the surroundings is neglected. The rate of heat loss, say q watt, which is usually quite small, can be allowed for as follows. The current is changed to a new value I', using a new voltage V', and the liquid flow is adjusted to m' to make up for this so that θ_2 is the same as before.

$$\therefore \quad VI = mc(\theta_2 - \theta_1) + q \qquad \text{(17.5a)}$$

$$\text{and} \quad V'I' = m'c(\theta_2 - \theta_1) + q \qquad \text{(17.5b)}$$

The heat loss per second is the same for both situations because all parts of the apparatus have the same temperatures as before.

Subtracting these equations we get

$$V'I' - VI = c(\theta_2 - \theta_1)(m' - m)$$

from which an accurate value of c can be calculated.

Example 4

In a continuous-flow calorimeter the readings were: 6.0 V, 2.1 A, $\theta_1 = 17.0\,^\circ$C, $\theta_2 = 22.0\,^\circ$C, 35 g min^{-1} followed by 4.0 V, 1.4 A, $\theta_1 = 17.0\,^\circ$C, $\theta_2 = 22.0\,^\circ$C, 15 g min^{-1}.

Obtain a value for the specific heat capacity of the liquid and the rate of loss of heat to the surroundings.

Method

Using $VI = mc(\theta_2 - \theta_1) + q$, we have, working in grams and seconds,

$$6.0 \times 2.1 = \left(\frac{35}{60} \times c \times 5.0\right) + q$$

and, using $V'I' = m'c(\theta_2 - \theta_1) + q$, we have

$$4.0 \times 1.4 = \left(\frac{15}{60} \times c \times 5.0\right) + q$$

Subtracting these equations (see Chapter 2) gives

$$12.6 - 5.6 = \frac{20}{60} \times c \times 5.0$$

so that

$$c = \frac{7.0}{5.0} \times \frac{60}{20} = 4.2\,\text{J g}^{-1}\,\text{K}^{-1}$$

Substituting this value for c in the first equation gives

$$12.6 = \left(\frac{35}{60} \times 4.2 \times 5.0\right) + q$$

from which

$$q = 12.6 - 12.25 = 0.35\,\text{W}$$

Answer

$4.2\,\text{J g}^{-1}\,\text{K}^{-1}$, 0.35 W.

Mixing hot with cold

If hot solid or liquid is introduced into cold liquid, the heat lost by the hot material cooling down is equal to the heat gained by the cold liquid and calorimeter plus any heat loss to the surroundings.

Example 5

21.0 g of liquid at 60.0 °C is mixed into 100 g of water at 12.5 °C which is already in a metal calorimeter of mass 70.0 g and specific heat capacity 400 J g^{-1} K^{-1}. If heat escape to the surroundings may be neglected, calculate the expected new temperature of the water, given that the specific heat capacity is 4200 J kg^{-1} K^{-1} for water and 4000 J kg^{-1} K^{-1} for the liquid.

Method

The liquid cools from $60\,°C$ to the final temperature which can be called $x\,°C$. The calorimeter and water start at $12.5\,°C$ and rise to $x\,°C$.

(i) The thermal energy given out by the hot liquid is

$$\frac{21}{1000} \times 4000 \times (60 - x)$$

(ii) The thermal energy absorbed by the cold liquid + calorimeter is:

$$\frac{100}{1000} \times 4200 \times (x - 12.5) + \frac{70}{1000} \times 400 \times (x - 12.5)$$

Since (i) and (ii) are equal:

$$21 \times 4 \times (60 - x) = 420 \times (x - 12.5) + 7 \times 4 \times (x - 12.5)$$

$$\therefore \quad 5040 - 84x = 420x - 5250 + 28x - 350$$

$$\therefore \quad x = \frac{5040 + 5250 + 350}{420 + 28 + 84}$$

$$\therefore \quad = 20\,°C$$

Answer

$20\,°C$.

Exercise 17.1

1   A coil of wire of heat capacity $12\,J\,K^{-1}$ has a PD of $5.0\,V$ applied between its ends, so that a current of $0.20\,A$ flows. Calculate the temperature rise produced in 1 minute. Assume the coil to be thermally insulated.

2   Using a continuous-flow calorimeter for measuring the specific heat capacity of a liquid, a PD of $5.0\,V$ was applied to the heating coil. The rate of flow of liquid was then doubled and, by adjusting the applied PD, the same inlet and outlet temperatures were obtained. Assuming heat losses to be negligible, calculate the new value of the applied PD. (It is necessary to assume that the resistance of the heater wire remains constant.)

3   $380\,g$ of a liquid at $12\,°C$ in a copper calorimeter weighing $90\,g$ is heated at a rate of 20 watt for exactly 3 minutes to produce a temperature of $17\,°C$. If the specific heat capacity of copper is $400\,J\,kg^{-1}\,K^{-1}$, the thermal capacity of the heater is negligible, and there is negligible heat loss to the surroundings, obtain a value for the specific heat capacity of the liquid.

4   Calculate the final temperature when $200\,g$ of water at $50\,°C$ is mixed with $80\,g$ of water at $10\,°C$. Heat losses to the calorimeter and surroundings may be neglected.

Specific latent heats

The thermal energy required to melt unit mass of substance at its melting point is called its specific latent heat of fusion. It is usually measured in $J\,kg^{-1}$. Similarly for evaporation at the boiling point we have the specific latent heat of vaporisation. The same latent heats are given out when solidifications or condensations occur. For all latent heats

$$l = \frac{Q}{m} \quad \text{or} \quad Q = ml \qquad (17.6)$$

where $l =$ specific latent heat, Q is heat concerned (in joules) and m is the mass of substance.

Measurement of specific latent heat of vaporisation

A suitable apparatus is illustrated in Fig. 17.3.

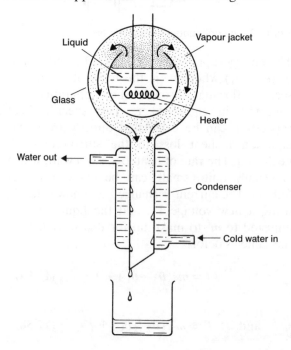

Fig. 17.3   Apparatus for measuring the specific latent heat of vaporisation

The energy equation is

$$V_1 I_1 t = m_1 l + H \qquad (17.7)$$

where V_1 is the PD across the heater and I_1 the current through it. m_1 is the mass of substance evaporated in time t, l the specific latent heat and H the heat loss in time t to the surroundings.

Using a different heater voltage V_2 we get a current I_2 and a mass m_2 is evaporated in time t. Thus

$$V_2I_2t = m_2l + H \qquad (17.8)$$

Eliminating H from the above equations we get

$$V_1I_1t - m_1l = V_2I_2t - m_2l \qquad (17.9)$$

We have assumed H to be the same for both rates of heating. This should be true enough since the liquid is at its boiling point each time and H is already small due to the vapour jacket reducing heat loss.

Example 6

In an experiment to determine the specific latent heat of vaporisation of an alcohol using a self-jacketing vaporiser the following results were taken:

Experiment 1:
$V_1 = 7.40\,\text{V}$, $I_1 = 2.60\,\text{A}$, mass $m_1 = 5.80 \times 10^{-3}\,\text{kg}$ collected in 300 s.

Experiment 2:
$V_2 = 10.0\,\text{V}$, $I_2 = 3.60\,\text{A}$, mass $m_2 = 11.3 \times 10^{-3}\,\text{kg}$ collected in 300 s.

Calculate (a) the specific latent heat of vaporisation of the alcohol, (b) the average rate of heat loss to the surroundings, (c) the power of the heater required to produce a rate of evaporation of 1.50 g per minute.

Method

(a) We arrange Equation 17.9 with $t = 300$:

$$l = \frac{(V_2I_2 - V_1I_1)t}{(m_2 - m_1)}$$

$$= \frac{(10 \times 3.6 - 7.4 \times 2.6) \times 300}{(11.3 - 5.8) \times 10^{-3}}$$

$$= 914 \times 10^3\,\text{J kg}^{-1}$$

(b) To find H we rearrange Equation 17.7 and use the above value for l. So

$$H = V_1I_1t - m_1l$$

$$= (7.4 \times 2.6 \times 300) - (5.8 \times 10^{-3} \times 914 \times 10^3)$$

$$= 471\,\text{J}$$

The average rate of heat loss is

$$H/t = 471/300 = 1.57\,\text{W}$$

(c) We can use Equation 17.7 or 17.8. Referring to Experiment 1 and Equation 17.7 we have power

$$P_1 = V_1I_1 = 7.4 \times 2.6 = 19.2\,\text{W}$$

and $\quad \dfrac{m_1}{t} = \dfrac{5.8 \times 10^{-3}}{300} = 1.93 \times 10^{-5}\,\text{kg s}^{-1}$

We require power P at which $m/t = 1.5\,\text{g}$ per minute or $2.50 \times 10^{-5}\,\text{kg s}^{-1}$. Rearranging Equation 17.9 gives

$$P = P_1 + \left(\frac{m}{t} - \frac{m_1}{t}\right)l$$

$$= 19.2 + (2.5 - 1.93) \times 10^{-5} \times 914 \times 10^3$$

$$= 24.4\,\text{W}$$

Alternatively, we write $V_3I_3t = m_3l + H$. For $t = 300\,\text{s}$ (5 min), $H = 471\,\text{J}$, $m_3 = 5 \times 1.5 \times 10^{-3}$, $l = 914 \times 10^3$ and we calculate V_3I_3.

Answer

(a) $914\,\text{kJ kg}^{-1}$, (b) 1.57 W, (c) 24.4 W.

Exercise 17.2

1 Assuming that heat losses can be neglected, calculate the power of a heater required to boil off water at a rate of 10.0 g per minute. Assume l for water $= 2.26\,\text{MJ kg}^{-1}$

2 An experiment was performed to determine the specific latent heat of vaporisation of a liquid at its boiling point. The following table summarises the results:

Voltage (V)	Current (A)	Mass (g) evaporated in 400 s
10.0	2.00	14.6
15.0	2.50	30.6

Calculate (a) the specific latent heat of vaporisation of the liquid, (b) the energy loss to the surroundings in 400 s, (c) the rate of evaporation of the liquid when a 30.0 W rate of heating is used.

Heat Transfer

The three common processes by which thermal energy is transferred are conduction, convection and radiation. In this section we shall deal with conduction. Radiation is dealt with in Chapter 31.

Thermal conductivity

The thermal conductivity k of a material describes how easy it is for thermal energy to pass through it from a hotter place (temperature θ_1) to a cooler place (temperature θ_2) separated by a distance l. It is defined by the equation for the rate of transfer of thermal energy $\Delta Q/\Delta t$ in which:

$$\frac{\Delta Q}{\Delta t} = kA(\theta_1 - \theta_2)/l \qquad (17.10)$$

where, as shown in Fig. 17.4, A is the area of cross section perpendicular to the direction of thermal energy transfer. $\Delta Q/\Delta t$ has units joule per second, i.e. watt (W).

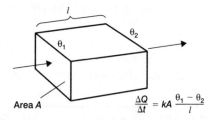

$$\frac{\Delta Q}{\Delta t} = kA\frac{\theta_1 - \theta_2}{l}$$

Area A

Fig 17.4 Transfer of thermal energy by conduction

Thermal conductivity k is analogous to electrical conductivity (see Chapter 20), kA/l is conductance, l/kA is resistance, $\Delta Q/\Delta t$ is analogous to electrical current and $\theta_1 - \theta_2$ is analogous to electrical potential difference. $(\theta_1 - \theta_2)/l$ may be called the temperature gradient and is constant if A, k and $\Delta Q/\Delta t$ are constant (negligible heat loss from the sides). The unit for temperature gradient is $\mathrm{K\,m^{-1}}$ and the unit for k is $\mathrm{W\,m^{-1}\,K^{-1}}$.

Thermal conductors in series

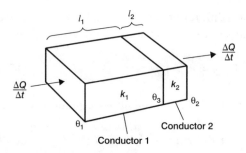

Fig 17.5 Conductors in series

In Fig. 17.5, with no loss from the sides, $\Delta Q/\Delta t = k_1 A(\theta_1 - \theta_3)/l_1$ and here, we emphasise, $\Delta Q/\Delta t$ is the same for conductor 2. Therefore $\Delta Q/\Delta t = k_2 A(\theta_3 - \theta_2)/l_2$.

Most problems can be solved by using these two equations.

From the two equations for $\Delta Q/\Delta t$, if we eliminate θ_3, we can get $\Delta Q/\Delta t = (\theta_1 - \theta_2)/R$. R is the thermal resistance given by $R = R_1 + R_2$ for the two conductors in series where $R_1 = l_1/k_1 A$ and $R_2 = l_2/k_2 A$.

Thermal conduction in buildings

A wall of a building may consist of (say) glass for part of its area and brick elsewhere. In this case the total rate of thermal energy transfer through the wall is the sum of that through the glass and that through the brick.

If the wall, or floor, or ceiling comprises two layers then we have two conductors in series and the calculation is different. This is shown in Example 7.

Example 7

One room in a house has a floor made entirely of concrete which is 200 mm thick. The lower surface of the concrete, in contact with the ground, has a temperature of 10.0 °C and the upper surface, in contact with the living area, has a temperature of 15.0 °C. The floor is square and of sides 10 m × 10 m.

(a) Calculate the rate at which thermal energy is conducted through the concrete. Assume the thermal conductivity of concrete is $0.750\ \mathrm{W\,m^{-1}\,K^{-1}}$.

The house owner decides to cover the concrete with carpet of thickness 15.0 mm. Calculate:

(b) the temperature at the carpet/concrete interface

(c) the rate at which thermal energy is conducted through the two layers.

Assume that the carpet has thermal conductivity $= 0.060 \, \text{W m}^{-1} \text{K}^{-1}$. Assume also that the temperature of the upper surface of the carpet is 15.0 °C and that the temperature of the lower surface of the concrete remains at 10.0 °C.

Method

(a) Almost without exception a thermal conductivity question requires the use of

$$\frac{\Delta Q}{\Delta t} = \frac{kA(\theta_1 - \theta_2)}{l}$$

Using this formula, $\Delta Q/\Delta t$ is the energy per second that must be calculated for part (a) of the question, $k = 0.750 \, \text{W m}^{-1} \text{K}^{-1}$ for the concrete floor, A is $10 \times 10 = 100 \, \text{m}^2$, l is 200 mm $(= 0.200 \, \text{m})$, $\theta_1 - \theta_2 = 15.0 - 10.0 = 5.0 \, \text{K}$.

So $\dfrac{\Delta Q}{\Delta t} = \dfrac{0.75 \times 100 \times 5.0}{0.200} = 1875 \, \text{W} = 1.875 \, \text{kW}$

Since all data used in the calculation were given to 3 significant figures the answer for $\Delta Q/\Delta t$ is 1.88 kW

(b) The rate of conduction of thermal energy $\Delta Q/\Delta t$ through the carpet and then through the concrete floor is the same since the two conductors are in series. Now $\theta_1 = 15.0 \, °\text{C}$ is the temperature of the upper surface of the carpet and $\theta_2 = 10.0 \, °\text{C}$ that of the lower surface of the concrete. Let θ_3 be the temperature of the carpet/concrete interface. Area $A = 10 \times 10 = 100 \, \text{m}^2$.

For the carpet, for which $k_1 = 0.060 \, \text{W m}^{-1} \text{K}^{-1}$:

$$\Delta Q/\Delta t = k_1 A(\theta_1 - \theta_3)/l_1$$

$$= 0.060 \times 100 \times (15.0 - \theta_3)/0.0150$$

$$= 400(15.0 - \theta_3)$$

For the concrete, for which $k_2 = 0.750 \, \text{W m}^{-1} \text{K}^{-1}$:

$$\Delta Q/\Delta t = k_2 A(\theta_3 - \theta_2)/l_2$$

$$= 0.750 \times 100 \times (\theta_3 - 10.0)/0.200$$

$$= 375(\theta_3 - 10.0)$$

Equating the two expressions for $\Delta Q/\Delta t$ gives

$$400(15.0 - \theta_3) = 375(\theta_3 - 10.0)$$

Rearranging gives

$$\theta_3 = 9750/775 = 12.58 \, °\text{C}$$

(c) We can use $\theta_3 = 12.58 \, °\text{C}$ with either of the above expressions, for carpet or concrete, to find $\Delta Q/\Delta t$. For the carpet:

$$\Delta Q/\Delta t = 400(15.0 - 12.58)$$

$$= 968 \, \text{W} \text{ or } 0.968 \, \text{kW}$$

Note that even though the thickness of the carpet is small compared with the concrete, there is a marked reduction (about 50%) in energy transfer as a result of covering the floor with carpet. This is a result of the decrease in temperature gradient across the concrete, since the temperature drop across the concrete reduces from 5.0 K to only 2.6 K.

Answers

(a) 1.88 kW, (b) 12.6 °C, (c) 0.968 kW

U-value of a sheet

Heat-insulating materials can be bought as sheets of various thicknesses and the value of k/l for a sheet is called its *U*-value.

$$\text{So } \Delta Q/\Delta t = \frac{kA(\theta_1 - \theta_2)}{l}$$

$$= UA \times \textbf{temperature difference}$$

$$(17.11)$$

The SI unit for U is watt $\text{m}^{-2} \text{K}^{-1}$.

Exercises 17.3

1 Calculate the rate of energy transfer through a layer of cork of 2.0 mm thickness and 24 cm² area when the temperature difference between its surfaces is 60 K.
(k for cork $= 0.050 \, \text{W m}^{-1} \text{K}^{-1}$.)

2 A sheet of insulating material is of thickness 1.5 mm and the temperature drop across the sheet is 50 K. If the rate at which thermal energy is conducted through the sheet is 8.0 kW m⁻², calculate the thermal conductivity of the material. (Hint: assume a cross-sectional area of 1.0 m².)

3 A 10 cm long brass bar is joined end-on to a copper bar of equal length and diameter, so as to form a compound bar with a cross-section area of 6.0 cm². The join has negligible thermal resistance and the bar is well lagged. The free end of the brass bar is maintained at 100 °C and the far end of the compound bar is kept at 20 °C. Calculate the rate of energy transfer along the bar and also the temperature of the junction.

Assume k for copper $= 400\,\mathrm{W\,m^{-1}\,K^{-1}}$ and for brass $= 100\,\mathrm{W\,m^{-1}\,K^{-1}}$.

4 The base of the loft in a house consists of wooden board which is 15 mm thick and of area 200 m². The thermal conductivity of the board is $0.15\,\mathrm{W\,m^{-1}\,K^{-1}}$. The temperature of the interior of the house is maintained at 20 °C, whilst that of the loft is 0 °C. Calculate:

(a) the rate of thermal energy transfer into the loft through the board.

If the owner now decides to insulate the loft space by covering the board with a layer of insulating material of thickness 10 cm and thermal conductivity $30\,\mathrm{mW\,m^{-1}\,K^{-1}}$, calculate:

(b) the temperature of the board/insulating material interface and

(c) the new rate of thermal energy transfer into the loft.

Assume that the board and insulating material are in good contact. Comment on your answers.

5 (a) State two factors which affect the U value of a material.

(b) A suit made for use in cool climates has a U value of $0.80\,\mathrm{W\,m^{-2}\,K^{-1}}$. It has a total exposed area of 2.0 m² and the skin temperature is 34 °C. Calculate the air temperature at which the heat loss from the suit is 48 W. Assume that the suit is tight fitting and that losses other than conduction can be ignored.

(Hint: use Equation 17.11)

Exercise 17.4: Examination questions

1 (a) Define the **specific heat capacity** of a material.

(b) It is required to determine the specific heat capacity of copper, using an electrical method. Draw a labelled diagram of the **circuit** you would use.

(c) A block of material, of mass 1.75 kg, is heated by a 120 W heater for 5.00 minutes. The block is completely lagged. The initial temperature of the block is 18.0 °C. The specific heat capacity of the material of the block is $435\,\mathrm{J\,kg^{-1}\,°C^{-1}}$.

(i) Calculate the final temperature of the block.

(ii) What is the purpose of having the block completely lagged?

(d) The lagging around the block in (c) is removed and the block is placed in thermal contact with an identical block which is at a temperature of 120 °C. Heat (thermal energy) is transferred from the block at the higher temperature to the one at the lower temperature.

(i) Name the principal method of heat transfer in this situation.

(ii) Describe the mechanism by which energy is transferred in this method.

[CCEA 2001]

2 The following data refer to a dishwasher.

power of heating element	2.5 kW
time to heat water	360 s
mass of water used	3.0 kg
initial temperature of water	20 °C
final temperature of water	60 °C

(a) Taking the specific heat capacity of water to be $4200\,\mathrm{J\,kg^{-1}\,K^{-1}}$, calculate

(i) the energy provided by the heating element,

(ii) the energy required to heat the water.

(b) Give **two** reasons why your answers in part (a) differ from each other. [AQA 2001]

3 A teacher is demonstrating the power used by different devices. She drills a hole in the wall for 30 s with an electric drill connected to the 230 V mains supply. The average current is 0.90 A.

When she puts the drill down, the tip of the steel drill bit melts a hole in a plastic tray.

Assume that all the electrical energy supplied to the drill is transferred to the bit where it produces heating. Calculate the temperature of the bit at the end of the drilling.

Mass of the drill bit = 13 g
Specific heat capacity of steel $= 510\,\mathrm{J\,kg^{-1}\,°C^{-1}}$
Room temperature = 20 °C

Discuss whether this is likely to be the actual temperature of the tip of the drill bit.

[Edexcel S-H 2000]

4 (a) Define the specific heat capacity of a substance.

(b) The energy of foodstuffs may be determined by measuring the thermal energy produced when the substance burns. In such a determination, a sample of food, of mass 15 g, is placed in an atmosphere of oxygen in a sealed, thermally-insulated stainless steel vessel of mass 5.1 kg. The initial equilibrium temperature of the system of food sample and vessel is 14.0 °C. The food is then ignited electrically, and the equilibrium temperature is found to rise to 43.5 °C. No heat energy is lost to the surroundings.

(i) Calculate the heat energy supplied to the stainless steel vessel by burning the food. [Specific heat capacity of stainless steel $= 4.4 \times 10^2 \, \text{J kg}^{-1} \, °\text{C}^{-1}$.]

(ii) On packets of food, the energy content of the foodstuff is often expressed in kJ per 100 g portion.

Neglecting the energy supplied by the electrical ignition system, the energy contained in the 15 g sample of food is equal to the heat energy supplied to the stainless steel vessel when the sample is burnt. Use your answer to (i) to calculate the energy content of the foodstuff. Give your answer in kJ per 100 g portion.

(iii) In (ii), you were told to neglect energy contributed by the electrical ignition system.

In fact, the food is burnt by supplying a current of 0.80 A to a filament of resistance 3.0 Ω for 12.0 minutes. Calculate the true value of the energy content of the foodstuff.

(c) When the specific heat capacity of a gas is measured, the value obtained is less when the gas is kept at a constant volume than when it is allowed to expand against atmospheric pressure. Making reference to the First Law of Thermodynamics, suggest an explanation.

[CCEA 2000]

5 (a) Define
   (i) specific latent heat of vaporization;
   (ii) specific heat capacity.

(b) The electric heating element of an instant hot water shower has a power of 5.0 kW. The volume flow rate of water through the heater is $3.6 \times 10^{-3} \, \text{m}^3 \, \text{min}^{-1}$.
   (i) Determine the mass flow rate in kg s^{-1} given that the density of water is $1.0 \times 10^3 \, \text{kg m}^{-3}$.
   (ii) Calculate the increase in temperature of the water as it flows through the heater. Assume that the specific heat capacity of water is $4.2 \times 10^3 \, \text{J kg}^{-1} \, \text{K}^{-1}$ and that the heat lost to the surroundings is negligible.

[OCR 2001]

6 A piece of aluminium of mass 0.20 kg and specific heat capacity 1.2 kJ kg^{-1} K^{-1} is heated to a steady temperature t and is then quickly but carefully placed in 0.22 kg of water contained in a copper calorimeter of water equivalent 0.020 kg. The temperature of the water rises from 16 °C to 21 °C. Calculate the temperature t, given that the specific heat capacity of water is 4.2 kJ kg^{-1} K^{-1}.

7 An energy conservation leaflet states that using a shower rather than a bath saves energy.

A student takes some measurements to test this.

Shower

The student's shower uses an electrical heater to heat cold water.
The heater is rated at 11 kW.
Time for shower to deliver 1 litre (0.001 m^3) of water = 12 s.
Density of water = 1000 kg m^{-3} (1 kg litre^{-1}).

(a) (i) Show that the mass of water delivered by the shower in one second is about 0.08 kg.
   (ii) The shower lasts for 8 minutes. Calculate the total energy used by the heater to heat the water.

Bath

The student's bath uses a mixture of hot water from a tank heated with an immersion heater and cold water from the main supply.

The bath is run using 30 litres from the cold tap and 42 litres from the hot tap:
Temperature of cold water = 15 °C
Temperature of water from hot tap = 55 °C
Specific heat capacity of water $= 4.2 \times 10^3 \, \text{J g}^{-1} \, \text{K}^{-1}$.

(b) (i) Show that this mixture of hot and cold water reaches a final temperature of about 38 °C for the bath. State *one* assumption you are making.
   (ii) Calculate the energy supplied by the immersion heater for this bath.

In this project, the student assumes that the immersion heater heating her bath water is 100% efficient. Explain whether or not this is a reasonable assumption.

Discuss the accuracy of the statement that 'using a shower rather than a bath saves energy'.

[Edexcel S-H 2000]

8 A kettle rated at 2.00 kW takes 200 s to raise the temperature of 800 g of water by 80.0 °C. If the specific heat capacity of water is 4.20 kJ kg^{-1} K^{-1}, calculate the mean rate at which energy is lost to the surroundings.

9 A block of ice at a temperature of $0\,°C$ and of mass $0.75\,kg$ absorbs thermal energy from its surroundings at a steady rate of $60\,W$. Calculate the minimum time it will take to melt, given that the specific latent heat of fusion of water is $3.2 \times 10^5\,J\,kg^{-1}$.

10 In an experiment to determine the specific latent heat of vaporisation of a liquid, an electrical heater boils the liquid in a well insulated container. The resistance of the heater is $3.00\,\Omega$ and the potential difference across the heater is $8.00\,V$. In a time of $500\,s$, the mass of liquid decreases by $0.110\,kg$. Calculate:

(a) the energy transferred to the liquid

(b) the specific latent heat of vaporisation of the liquid.

11 In a heating experiment, energy is supplied at a constant rate to a liquid in a beaker of negligible heat capacity. The temperature of the liquid rises at $4.0\,K$ per minute just before it begins to boil. After 40 minutes all the liquid has boiled away. For this liquid, what is the ratio

$$\frac{\text{specific heat capacity}}{\text{specific latent heat of vaporisation}}?$$

$\mathbf{A}\ \dfrac{1}{10}\,K^{-1}$ $\mathbf{B}\ \dfrac{1}{40}\,K^{-1}$ $\mathbf{C}\ \dfrac{1}{160}\,K^{-1}$ $\mathbf{D}\ \dfrac{1}{640}\,K^{-1}$

[OCR 2000]

12 An electric kettle with a rating of $3.0\,kW$ contains water that has been brought to the boil. The automatic cutout fails to operate and the electrical supply continues to be maintained. Assuming that all the energy supplied goes to converting the water to steam and that the kettle initially contains $1.20\,kg$ of water, how long will it take before half of the water is boiled off?

The specific latent heat of vaporisation of water is $2.3 \times 10^6\,J\,kg^{-1}$.

13 A thin beaker is filled with $400\,g$ of water at $0\,°C$ and placed on a table in a warm room. A second identical beaker, filled with $400\,g$ of an ice-water mixture, is placed on the same table at the same time. The contents of both beakers are stirred continuously.

The graph below shows how the temperature of the water in the *first* beaker increases with time.

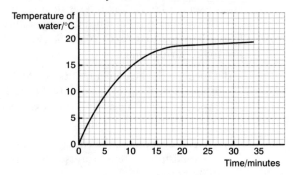

(a) (i) Use the graph to find the initial rate of rise of water temperature. Give your answer in $K\,s^{-1}$.

(ii) The specific heat capacity of water is $4200\,J\,kg^{-1}\,K^{-1}$. Use your value for the rate of rise of temperature to estimate the initial rate at which this beaker of water is taking in heat from the surroundings.

The graph below shows the temperature of the water in the *second* beaker from the moment it is placed on the table.

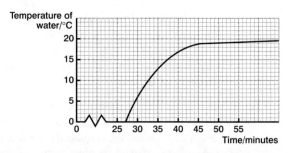

(b) (i) How do you explain the delay of twenty-seven minutes before the ice-water mixture starts to warm up?

(ii) The specific latent heat (enthalpy) of ice is $3.36 \times 10^5\,J\,kg^{-1}$. Estimate the mass of ice initially present in the ice-water mixture. [Edexcel 2000]

14 Ice is commonly used to cool drinks. If an ice cube, at a temperature of $0\,°C$ and of mass $0.015\,kg$ is dropped into a beaker containing $0.15\,kg$ of water with an initial temperature of $18\,°C$, calculate the final temperature of the resulting water. Assume that no heat is exchanged with the surroundings.

Specific heat capacity of water $= 4.2\,kJ\,kg^{-1}\,K^{-1}$
Specific latent heat of ice $= 3.4 \times 10^5\,J\,kg^{-1}$

15 (a) Define the terms *specific latent heat of melting* and *specific heat capacity*. State briefly how each of these quantities can be measured for a substance such as water.

(b) A well-insulated picnic hamper has placed in it twelve 330 ml cans of lemonade, initially at 20 °C, together with 2.0 kg of ice at 0 °C. Use the data below to calculate the final temperature of the lemonade. State your assumptions.
[Specific heat of water (or lemonade) $= 4200 \, \text{J kg}^{-1} \, \text{K}^{-1}$.
Specific latent heat of ice $= 3.3 \times 10^5 \, \text{J kg}^{-1}$; density of water $= 1000 \, \text{kg m}^{-3}$; $1 \, \text{ml} = 10^{-6} \, \text{m}^3$]

(c) In fact the picnic hamper gains heat from its surroundings by thermal conduction through the insulating polystyrene. The energy gain is proportional to the temperature difference ΔT between the outside and inside of the hamper. The rate of energy gain for this hamper is found to equal $0.25 \Delta T \, \text{J s}^{-1}$, where ΔT is measured in °C.
  (i) Show that, when all the ice has melted, the temperature difference ΔT decays exponentially with time.
  (ii) Hence or otherwise estimate the time (in hours) taken for the hamper's internal temperature to rise from 6 °C to 18 °C, when it is kept in the boot of a car at a constant temperature of 30 °C.
    [OCR spec 2001]

16 On a very cold day, the air temperature is −5.0 °C. A pond has a layer of ice of thickness 50 mm and the temperature of the water in the pond is uniform at 0 °C. Calculate:

(a) the magnitude of the temperature gradient across the ice layer

(b) the rate of transfer of thermal energy per m² through the ice layer.

Thermal conductivity of ice $= 2.3 \, \text{W m}^{-1} \, \text{K}^{-1}$
Assume that a steady state has been achieved.

17 A hot-water tank is lagged with a material which allows thermal energy to escape at a rate of 100 W. The owner is dissatisfied with this and replaces the lagging with another material of half the thermal conductivity of the original and twice the thickness. Calculate the rate of thermal energy transfer through the new lagging.

18 A domestic refrigerator can be thought of as a rectangular box of dimensions 0.90 m × 0.50 m × 0.50 m and is lined throughout with a layer of insulation which is 4.0 mm thick and of thermal conductivity $0.040 \, \text{W m}^{-1} \, \text{K}^{-1}$. If the room temperature is 24 °C and the temperature inside the refrigerator is maintained at 4 °C, calculate the rate at which heat flows into the refrigerator from the room.

19 A greenhouse, which may be assumed to be made entirely of glass, needs a 3.00 kW heater to maintain it at a steady temperature. The glass is 3.00 mm thick and has a total area of 5.00 m², and the thermal conductivity of glass is $1.20 \, \text{W m}^{-1} \, \text{K}^{-1}$. Calculate the temperature difference across the glass.

Assume that all other forms of heat loss, other than conduction through the glass, are negligible.

20 The diagram shows the only two external walls of one dwelling in a multi-storey building in a hot country. The average outside temperature is 33 °C. The building is air conditioned and the inside temperature is 22 °C.

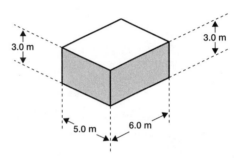

In which direction does energy flow through the walls? Explain your answer.

(a) (i) State, in terms of energy flow, what an air conditioner has to do to keep the inside at 22 °C.
  (ii) The walls have an average U-value of $0.60 \, \text{W m}^{-2} \, \text{K}^{-1}$. Calculate the average power flow through the walls.
  (iii) The walls incorporate a layer of insulation. Without this the U-value would be $1.8 \, \text{W m}^{-2} \, \text{K}^{-1}$. How may times larger or smaller would the power flow be without this layer?

(b) The walls and floors are made of concrete. They have a total mass of 11 tonnes, (1 tonne $= 1000 \, \text{kg}$.) The specific heat capacity of the concrete is $920 \, \text{J kg}^{-1} \, \text{K}^{-1}$. Calculate the average power flow from the concrete to reduce its temperature from 33 °C to an average temperature of 25 °C in the first hour of switching on. [Edexcel 2000]

21 (a) Describe the principal process of thermal conduction in
  (i) a non-metallic solid;
  (ii) a metal.

(b) Fig. 17.6 shows a cross-sectional view of the casing of a domestic freezer. This freezer is operating under steady state conditions.

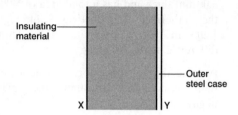

Fig. 17.6

thickness of insulating material = 30 mm
thickness of outer steel case = 0.50 mm

thermal conductivity of steel = $50\,\mathrm{W\,m^{-1}\,K^{-1}}$
thermal conductivity of insulating material = $0.040\,\mathrm{W\,m^{-1}\,K^{-1}}$

(i) If $\Delta\theta_s$ = temperature drop across the outer steel casing and $\Delta\theta_p$ = temperature drop across the insulating material, show that the ratio $\dfrac{\Delta\theta_s}{\Delta\theta_p} = 1.3 \times 10^{-5}$.

(ii) The effective area of each of the surfaces **X** and **Y** of the freezer casing is $2.5\,\mathrm{m^2}$. Calculate the rate P at which thermal energy will be conducted into the freezer when the temperatures of **X** and **Y** are $-15\,^\circ\mathrm{C}$ and $7\,^\circ\mathrm{C}$ respectively.

[OCR 2001]

18
The ideal gas laws and kinetic theory

The gas laws

The laws obeyed by a perfect or ideal gas are as follows (for a fixed mass of gas):

$pV = $ Constant, at constant T (Boyle's law)

$\dfrac{V}{T} = $ Constant, at constant p (Charles' law)

$\dfrac{p}{T} = $ Constant, at constant V (Pressure law)

where p is the pressure, V the volume and T the absolute temperature (K) of the gas.

The ideal gas equation

The three laws above are incorporated in the ideal gas equation:

$$\frac{pV}{T} = \text{Constant} \qquad (18.1)$$

An alternative way of writing this is

$$\frac{p_1 V_1}{T_1} = \frac{p_2 V_2}{T_2} \qquad (18.2)$$

where p_1, V_1, T_1 refer to the initial state and p_2, V_2, T_2 to the final state. Note that pressure and volume may be expressed in any suitable units (see Chapter 3) that we choose, but *temperature must be in kelvin*.

Example 1

A gas cylinder has a volume of $0.040\,\mathrm{m^3}$ and contains air at a pressure of $2.0\,\mathrm{MPa}$. Assuming that temperature remains constant calculate (a) the equivalent volume of air at atmospheric pressure $(1.0 \times 10^5\,\mathrm{Pa})$, (b) the

volume of air, at atmospheric pressure, which escapes from the cylinder when it is opened to the atmosphere.

Method

(a) If temperature is constant then $T_1 = T_2$, and Equation 18.2 reduces to the equation for Boyle's law and becomes

$$p_1 V_1 = p_2 V_2 \qquad (18.3)$$

We have
$p_1 = 2.0 \times 10^6, V_1 = 0.040, p_2 = 1.0 \times 10^5$
and require V_2.

Rearranging Equation 18.3 gives

$$V_2 = \frac{p_1 V_1}{p_2} = \frac{2 \times 10^6 \times 0.04}{1 \times 10^5}$$

$$= 0.80\,\mathrm{m^3}$$

(b) Air escapes from the cylinder until it contains $0.04\,\mathrm{m^3}$ of air at atmospheric pressure. It is then 'empty', so that a volume ΔV will escape where

$$\Delta V = 0.80 - 0.04 = 0.76\,\mathrm{m^3}$$

Note that ΔV is the volume of air, at atmospheric pressure, which would have to be pumped into the 'empty' cylinder to raise its pressure to $2.0\,\mathrm{MPa}$.

Answer

(a) $0.80\,\mathrm{m^3}$, (b) $0.76\,\mathrm{m^3}$.

Example 2

A flask containing air is corked when the atmospheric pressure is $750\,\mathrm{mmHg}$ and the temperature is $17\,^{\circ}\mathrm{C}$. The temperature of the flask is now raised gradually. The cork blows out when the pressure in the flask exceeds atmospheric pressure by $150\,\mathrm{mmHg}$. Calculate the temperature of the flask when this happens.

Method

Note that we have to assume that corking the flask did not change the original pressure of the air inside it, that the atmospheric pressure remains unchanged and that the volume of the flask does not change appreciably during the change of temperature.

If the volume is constant then $V_1 = V_2$ and Equation 18.2 reduces to the equation for the Pressure law and becomes

$$\frac{p_1}{T_1} = \frac{p_2}{T_2} \qquad (18.4)$$

We have

$p_1 = 750\,\text{mmHg}$

$T_1 = 273 + 17 = 290\,\text{K}$

$p_2 = \text{Atmospheric pressure} + \text{Excess pressure}$
$\quad = 750 + 150 = 900\,\text{mmHg}$

To find T_2 we rearrange Equation 18.4. This gives

$$T_2 = \frac{p_2 \times T_1}{p_1} = \frac{900 \times 290}{750}$$

$$= 348\,\text{K} = 75\,°\text{C}$$

Note again that the units can be mmHg for pressure *provided that both p_1 and p_2 are in the same units.*

Answer

The cork blows out at 75° C.

Example 3

A gas cylinder of volume 4.0 litre $(4.0 \times 10^{-3}\,\text{m}^3)$ contains oxygen at a temperature of 15 °C and a pressure of 2.5 MN m^{-2}. Calculate (a) the equivalent volume of oxygen at standard temperature and pressure (STP), (b) the mass of oxygen in the cylinder. The density of oxygen is 1.4 kg m^{-3} at STP.

Method

Standard temperature and pressure (STP) are 0 °C and 1.0×10^5 N m^{-2} respectively.

(a) We use Equation 18.2 in which we have

$p_1 = 2.5 \times 10^6$, $V_1 = 4.0 \times 10^{-3}$

$T_1 = 273 + 15 = 288$

$p_2 = 1.0 \times 10^5$, $V_2 = \text{unknown}$,

$T_2 = 273 + 0 = 273$

Rearranging Equation 18.2 gives

$$V_2 = \frac{p_1 V_1 T_2}{T_1 p_2}$$

$$= \frac{2.5 \times 10^6 \times 4 \times 10^{-3} \times 273}{288 \times 1 \times 10^5}$$

$$= 94.8 \times 10^{-3}\,\text{m}^3$$

(b) The density of oxygen is 1.4 kg m^{-3}. To find the mass of gas:

$$\text{Mass} = \text{Volume} \times \text{Density}$$
$$= 94.8 \times 10^{-3} \times 1.4 = 0.133\,\text{kg}$$

Note that since the density is quoted at STP we must use the volume of gas at STP.

Answer

(a) $95 \times 10^{-3}\,\text{m}^3$, (b) 0.13 kg.

Exercise 18.1

1 Change the following Celsius temperatures into degrees absolute:
(a) 7 °C, (b) 710 °C, (c) −80 °C, (d) −199 °C

2 A fixed mass of gas is held at 27 °C. To what temperature must it be heated so that its volume doubles if its pressure remains constant?

3 A car tyre has a volume of $18 \times 10^{-3}\,\text{m}^3$ and contains air at an *excess* pressure of 2.5×10^5 N m^{-2} above atmospheric pressure $(1.0 \times 10^5$ N m$^{-2})$. Calculate the volume which the air inside would occupy at atmospheric pressure, assuming that its temperature remains unchanged.

4 Inside a sealed container is a fixed mass of gas at a pressure of 1.5×10^5 Pa when the temperature is 17 °C. At what temperature will the pressure inside it be 2.5×10^5 Pa?

5 A fixed mass of gas has a volume of 200 cm^3 at a temperature of 57 °C and a pressure of 780 mm mercury. Find its volume at STP (0 °C and 760 mm mercury).

6 A gas cylinder has a volume of 20 litres $(20 \times 10^{-3}\,\text{m}^3)$. It contains air at a temperature of 17 °C and an *excess* pressure of 3.0×10^5 N m^{-2} above atmospheric pressure $(1.0 \times 10^5$ N m$^{-2})$. Calculate the mass of air in the cylinder, given that the density of air at STP is 1.3 kg m^{-3}.

The equation of state

For a *given amount* of an ideal gas, *the equation of state* is as follows:

$$pV = nRT \qquad (18.5)$$

where p is the pressure (N m^{-2} or Pa), V the volume (m^3), n the number of moles of the gas (mol), R the *universal molar gas constant* (value $8.31\,\mathrm{J\,mol^{-1}\,K^{-1}}$) and T the temperature (K). Note that one mole of a gas is the amount which contains Avogadro's number N_A ($= 6.02 \times 10^{23}$) of molecules.

Equation 18.5 can be rewritten to include the mass M_g (kg) of the gas involved. If M_m (kg) is the molar mass (i.e. the mass of one mole), then the number of moles n is given by

$$n = \frac{\textbf{Mass of gas}}{\textbf{Molar mass}} = \frac{M_g}{M_m} \qquad (18.6)$$

Using Equation 18.6 to substitute for n in Equation 18.5 gives

$$pV = M_g \left(\frac{R}{M_m} \right) T \qquad (18.7)$$

Note that M_m depends on the particular gas. Also, if m is the mass of a molecule of the gas, then

$$M_m = \left(\begin{array}{c} \textbf{Avogadro's} \\ \textbf{number } N_A \end{array} \right) \times \left(\begin{array}{c} \textbf{Mass of} \\ \textbf{molecule } m \end{array} \right)$$
$$(18.8)$$

Example 4

A cylinder of volume $2.00 \times 10^{-3}\,\mathrm{m^3}$ contains a gas at a pressure of $1.50\,\mathrm{MN\,m^{-2}}$ and at a temperature of 300 K. Calculate (a) the number of moles of the gas, (b) the number of molecules of the gas, (c) the mass of gas if its molar mass is $32.0 \times 10^{-3}\,\mathrm{kg}$, (d) the mass of one molecule of the gas.

Assume that the universal gas constant R is $8.31\,\mathrm{J\,mol^{-1}\,K^{-1}}$ and the Avogadro constant N_A is $6.02 \times 10^{23}\,\mathrm{mol^{-1}}$.

Method

(a) We use Equation 18.5 in which $p = 1.5 \times 10^6$, $V = 2 \times 10^{-3}$, $R = 8.31$ and $T = 300$. Rearranging to find n gives us

$$n = \frac{pV}{RT} = \frac{1.5 \times 10^6 \times 2 \times 10^{-3}}{8.31 \times 300} = 1.20$$

(b) One mole contains 6.02×10^{23} molecules, so that 1.20 mol contains $1.20 \times 6.02 \times 10^{23} = 7.22 \times 10^{23}$ molecules.

(c) We have $M_m = 32 \times 10^{-3}$, $n = 1.2$ and require the mass of gas M_g. Rearranging Equation 18.6 gives us

$$M_g = nM_m = 1.2 \times 32 \times 10^{-3}$$
$$= 38.4 \times 10^{-3}\,\mathrm{kg}$$

(d) We use Equation 18.8, in which $M_m = 32 \times 10^{-3}$, $N_A = 6.02 \times 10^{23}$ and we require m. Thus

$$m = \frac{M_m}{N_A} = \frac{32 \times 10^{-3}}{6.02 \times 10^{23}}$$
$$= 5.32 \times 10^{-26}\,\mathrm{kg}$$

Answer

(a) 1.20,   (b) 7.22×10^{23}
(c) $38.4 \times 10^{-3}\,\mathrm{kg}$,   (d) $5.32 \times 10^{-26}\,\mathrm{kg}$.

Example 5

A cylinder contains 2.0 kg of nitrogen at a pressure of $3.0 \times 10^6\,\mathrm{N\,m^{-2}}$ and at a temperature of 17 °C. What mass of nitrogen would a cylinder of the same volume contain at STP (0 °C and $1.0 \times 10^5\,\mathrm{N\,m^{-2}}$)?

Method

We use Equation 18.7, and note that V and M_m are constants for a given volume of a particular gas. At $p_1 = 3.0 \times 10^6$ and $T_1 = 17\,°C = 290\,K$, $M_g = 2.0$. So Equation 18.7 gives

$$3.0 \times 10^6 \times V = 2 \times \left(\frac{R}{M_m} \right) \times 290 \qquad (i)$$

At STP we have $p_2 = 1.0 \times 10^5$, $T_2 = 273\,K$ and require the mass M_g in the cylinder. So

$$1.0 \times 10^5 \times V = M_g \left(\frac{R}{M_m} \right) \times 273 \qquad (ii)$$

Dividing (i) by (ii) to eliminate the constants gives

$$\frac{3.0 \times 10^6}{1.0 \times 10^5} = \frac{2 \times 290}{M_g \times 273}$$
$$\therefore \quad M_g = 7.08 \times 10^{-2}\,\mathrm{kg}$$

Answer

$7.1 \times 10^{-2}\,\mathrm{kg}$ at STP.

Example 6

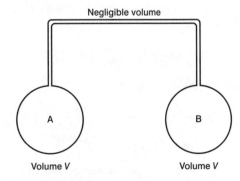

Fig. 18.1   Information for Example 6

Two vessels A and B, of equal volume, are connected by a tube of negligible volume, as shown in Fig. 18.1. The vessels contain a total mass of $2.50 \times 10^{-3}\,\mathrm{kg}$ of air and initially both vessels are at $27\,^\circ$C when the pressure is $1.01 \times 10^5\,\text{N m}^{-2}$. Vessel A is now cooled to $0\,^\circ$C and vessel B heated to $100\,^\circ$C. Calculate (a) the mass of gas now in each vessel, (b) the pressure in the vessels.

Method

(a) Let the volume of each vessel be V (we assume this does not change). Note that *since the vessels are connected, the pressure is equal in the two vessels*; let the final pressure be p. We apply Equation 18.7 to each vessel separately:

Vessel A contains mass M_{gA} of gas at temperature 273 K, so

$$pV = M_{gA}\left(\frac{R}{M_m}\right) \times 273 \qquad \text{(i)}$$

Vessel B contains mass M_{gB} of gas at 373 K, so

$$pV = M_{gB}\left(\frac{R}{M_m}\right) \times 373 \qquad \text{(ii)}$$

Comparing (i) and (ii) we see that

$$M_{gA} \times 273 = M_{gB} \times 373 \qquad \text{(iii)}$$

Now the total mass of gas is $2.5 \times 10^{-3}\,\text{kg}$, so

$$M_{gA} + M_{gB} = 2.5 \times 10^{-3} \qquad \text{(iv)}$$

Substituting $M_{gA} = (373/273)M_{gB}$ from (iii) into (iv) we find

$M_{gA} = 1.44 \times 10^{-3}\,\text{kg}$ and $M_{gB} = 1.06 \times 10^{-3}\,\text{kg}$.

(b) We apply Equation 18.7 to the original whole system at temperature $273 + 27 = 300\,\text{K}$, pressure $1.01 \times 10^5\,\text{N m}^{-2}$, volume $2V$ (since A and B each have volume V) and mass $M_g = 2.5 \times 10^{-3}\,\text{kg}$.

Hence

$$1.01 \times 10^5 \times 2V = 2.5 \times 10^{-3}\left(\frac{R}{M_m}\right) \times 300 \qquad \text{(v)}$$

To find the final pressure p, we make use of (i), in which $M_{gA} = 1.44 \times 10^{-3}$, so

$$pV = 1.44 \times 10^{-3}\left(\frac{R}{M_m}\right) \times 273 \qquad \text{(i)}$$

Dividing (i) by (v) gives

$$\frac{p}{1.01 \times 10^5 \times 2} = \frac{1.44 \times 10^{-3} \times 273}{2.5 \times 10^{-3} \times 300}$$

or $p = 1.06 \times 10^5\,\text{N m}^{-2}$.

Using (ii) should give the same answer for p. Try this as a check.

Answer

(a) $1.44 \times 10^{-3}\,\text{kg (A)}$, $1.06 \times 10^{-3}\,\text{kg (B)}$,
(b) $1.06 \times 10^5\,\text{N m}^{-2}$.

Exercise 18.2

(Assume that the universal molar gas constant R is $8.31\,\text{J mol}^{-1}\,\text{K}^{-1}$ and Avogadro's number N_A is 6.02×10^{23}.)

1 Calculate the volume occupied by one mole of gas at standard temperature $(0\,^\circ\text{C})$ and standard pressure $(1.01 \times 10^5\,\text{N m}^{-2})$.

2 The molar mass of carbon dioxide is $44.0 \times 10^{-3}\,\text{kg}$. Calculate (a) the number of moles and (b) the number of molecules in $1.00\,\text{kg}$ of the gas.

3 The molar mass of nitrogen is $28.0 \times 10^{-3}\,\text{kg}$. A sample of the gas contains 6.02×10^{22} molecules. Calculate (a) the number of moles of the gas, (b) the mass of the gas and (c) the volume occupied by the gas at a pressure of $0.110\,\text{MN m}^{-2}$ and a temperature of $290\,\text{K}$.

4 An oxygen cylinder contains $0.50\,\text{kg}$ of gas at a pressure of $0.50\,\text{MN m}^{-2}$ and a temperature of $7\,^\circ\text{C}$. What mass of oxygen must be pumped into the cylinder to raise its pressure to $3.0\,\text{MN m}^{-2}$ at a temperature of $27\,^\circ\text{C}$. If the molar mass of oxygen is $32 \times 10^{-3}\,\text{kg}$ calculate the volume of the cylinder.

5 Two vessels, one having three times the volume of the other, are connected by a narrow tube of negligible volume. Initially the whole system is filled with a gas at a pressure of $1.05 \times 10^5\,\text{Pa}$ and a temperature of $290\,\text{K}$. The smaller vessel is now cooled to $250\,\text{K}$ and the larger heated to $400\,\text{K}$. Find the final pressure in the system.

Kinetic theory

The pressure exerted by a gas arises as a result of gas molecules bombarding the walls of the container. There are very many molecules in a typical sample of gas, and the molecules have a whole range of speeds. Fig. 18.2 shows the number of molecules having speed c at a given temperature.

The laws of Newtonian mechanics are used to show that the pressure exerted by the gas is given by

$$p = \tfrac{1}{3}\rho <c^2> \qquad \textbf{(18.9)}$$

where ρ is the density of the gas and $<c^2>$ the mean square speed of the molecules of the gas (i.e. the average of all the values of speed squared). Now

THE IDEAL GAS LAWS AND KINETIC THEORY

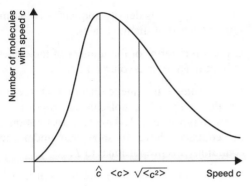

Key $\hat{c}$:   most probable speed

         $<c>$:   average or mean speed

         $\sqrt{<c^2>}$:   root mean square speed

Fig. 18.2   Distribution of molecular speeds in a gas

$$\rho = \frac{\text{Mass of gas}}{\text{Volume}} = \frac{M_g}{V}$$

Substituting ρ in Equation 18.9 gives

$$pV = \tfrac{1}{3}M_g<c^2> = \tfrac{1}{3}Nm<c^2> \qquad \textbf{(18.10)}$$

Since $M_g = Nm$ where N is the number of molecules and m is the mass of a molecule.

By comparing Equations 18.5 and 18.10, for one mole of a gas, we can show that the mean *translational* kinetic energy per molecule of a gas is given by

$$\textbf{mean KE} = \tfrac{1}{2}m<c^2> = \frac{3}{2}\frac{R}{N_A}T \qquad \textbf{(18.11)}$$

where m is the mass of a molecule and $\frac{R}{N_A} = k$ is the Boltzmann constant.

The square root of $<c^2>$ is called the root mean square (RMS) speed ($c_{\text{r.m.s.}}$) and has theoretical significance. Note from Equation 18.11 that, for a particular gas,

$$c_{\text{r.m.s.}} = \sqrt{<c^2>} \propto \sqrt{T} \qquad \textbf{(18.12)}$$

Example 7

At a certain time, the speeds of seven particles are as follows:

Speed/m s^{-1}	2.0	3.0	4.0	5.0	6.0
Number of particles	1	3	1	1	1

Calculate the root mean square speed of the particles.

Method

Table 18.1

Number of particles n	1	3	1	1	1
Speed c	2.0	3.0	4.0	5.0	6.0
c^2	4	9	16	25	36

We first square the speeds (see Table 18.1). The mean square speed $<c^2>$ is the *average of the squares of the speeds*, as follows:

$$\begin{aligned}<c^2> &= \tfrac{1}{7}\{(1 \times 4) + (3 \times 9) + (1 \times 16) + (1 \times 25) \\ &\quad + (1 \times 36)\} \\ &= \tfrac{1}{7}(4 + 27 + 16 + 25 + 36\} \\ &= 15.4 \, \text{m}^2\,\text{s}^{-2}\end{aligned}$$

Note: This is done by adding up the '*speed squared*' *values for each particle* and dividing by the number of particles.

To find the RMS speed we take the square root of $<c^2>$, hence

$$\begin{aligned}\text{RMS speed } c_{\text{r.m.s.}} &= \sqrt{<c^2>} = \sqrt{15.4} \\ &= 3.9 \, \text{m s}^{-1}\end{aligned}$$

Note that the *most probable speed* $\hat{c}$ is $3.0\,\text{m s}^{-1}$ since most (3) particles have this speed. The *average speed* $<c>$ is found from the average of the speeds, as follows:

$$\begin{aligned}<c> &= \tfrac{1}{7}\{(1 \times 2) + (3 \times 3) + (1 \times 4) + (1 \times 5) + (1 \times 6)\} \\ &= 3.7 \, \text{m s}^{-1}\end{aligned}$$

Answer

$3.9 \, \text{m s}^{-1}$.

Example 8

Calculate the RMS speed of air molecules in a container in which the pressure is $1.0 \times 10^5 \, \text{Pa}$ and the density of air is $1.3 \, \text{kg m}^{-3}$.

Method

We have $p = 10^5$ and $\rho = 1.3$. Rearranging Equation 18.9 to find $\sqrt{<c^2>}$ gives

$$\begin{aligned}c_{\text{r.m.s.}} = \sqrt{<c^2>} = \sqrt{\frac{3p}{\rho}} &= \sqrt{\frac{3 \times 10^5}{1.3}} \\ &= 480 \, \text{m s}^{-1}\end{aligned}$$

Answer

$0.48 \, \text{km s}^{-1}$.

Example 9

Calculate the temperature at which the RMS speed of oxygen molecules is twice as great as their RMS speed at $27\,°\text{C}$.

Method

We use Equation 18.12. Thus, since 27 °C is 300 K,

$$\frac{c_{\text{r.m.s.}} \text{ at } T}{c_{\text{r.m.s.}} \text{ at } 300} = \frac{\sqrt{T}}{\sqrt{300}}$$

$$\therefore \qquad 2 = \frac{\sqrt{T}}{\sqrt{300}}$$

Squaring both sides gives

$$T = 4 \times 300 = 1200\,\text{K} = 927\,^\circ\text{C}$$

Answer

927 °C

Exercise 18.3

1   Eight molecules have the following speeds: 300, 400, 400, 500, 600, 600, 700, 900 m s^{-1}. Calculate their RMS speed.

2   The following table shows the distribution of speed of 20 particles:

Speed/m s^{-1}	10	20	30	40	50	60
Number of particles	1	3	8	5	2	1

Find (a) the most probable speed, (b) the average speed, (c) the RMS speed.

3   The RMS speed of helium at STP is 1.30 km s^{-1}. If 1 standard atmosphere is $1.01 \times 10^5\,\text{N m}^{-2}$, calculate the density of helium at STP.

4   The RMS speed of nitrogen molecules at 127 °C is 600 m s^{-1}. Calculate the RMS speed at 1127 °C.

5   If the density of nitrogen at STP ($1.01 \times 10^5\,\text{Pa}$ and 0 °C) is 1.25 kg m^{-3}, calculate the RMS speed of nitrogen at 227 °C.

Exercise 18.4: Examination questions

(Assume Avogadro's number $N_A = 6.02 \times 10^{23}$, and universal molar gas constant $R = 8.31\,\text{J mol}^{-1}\,\text{K}^{-1}$ unless otherwise stated.)

1   A rigid gas-tight container holds 150 cm^3 of air at a temperature of 100 °C and a pressure of 1.00×10^5 Pa. The temperature of the air is raised to 150 °C. Calculate the new pressure.

2   According to kinetic theory, the pressure p of an ideal gas is given by the equation

$$p = \tfrac{1}{3}\rho\langle c^2 \rangle$$

where ρ is the gas density and $\langle c^2 \rangle$ is the mean squared speed of the molecules.

Express ρ in terms of the number of molecules N, each of mass m, in a volume V.

It is assumed in kinetic theory that the mean kinetic energy of a molecule is proportional to kelvin temperature T. Use this assumption, and the equation above, to show that under certain conditions p is proportional to T.

State the conditions under which p is proportional to T.

A bottle of gas has a pressure of 303 kPa above atmospheric pressure at a temperature of 0 °C. The bottle is left outside on a very sunny day and the temperature rises to 35 °C. Given that atmospheric pressure is 101 kPa, calculate the new pressure of the gas inside the bottle.

[Edexcel 2001]

3   A flask of volume $9.0 \times 10^{-4}\,\text{m}^3$ contains air. A vacuum pump reduces the pressure in the flask to 150 Pa at a temperature of 300 K.

Avogadro constant $= 6.0 \times 10^{23}\,\text{mol}^{-1}$
molar gas constant $= 8.3\,\text{J mol}^{-1}\,\text{K}^{-1}$
molar mass of air $= 0.029\,\text{kg mol}^{-1}$

For the air remaining in the flask, calculate

(a)   its density;

(b)   the number of molecules present.

[OCR 2001]

4   Fig. 18.3 shows a balloon being prepared at ground-level for a long-distance flight. The envelope of the balloon is being filled with helium.

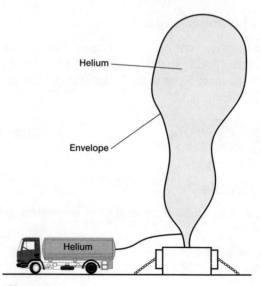

Helium

Envelope

Helium

Fig. 18.3

(a) The envelope is made of a thin plastic material with a silvered outer surface. State and explain why the temperature variations of the gas in the balloon will be less during a 24-hour period than if the surface of the material were of a darker colour.

(b) The envelope, when fully inflated, has an internal volume of $10\,000\,m^3$. For take-off, it is partially inflated with $5000\,m^3$ of helium at a pressure of $105\,kPa$ and a temperature of $293\,K$. Both pressure and temperature change as the balloon rises into the cool upper atmosphere. The result of these changes is an increase in volume of the helium.

   (i) The envelope first becomes fully inflated when the temperature of the helium is $243\,K$. What is the pressure of the helium at this time?

   (ii) Suggest why it is necessary to release helium from the envelope as the balloon continues to rise.

   (iii) The balloon reaches a height where the fully-inflated envelope contains helium at a temperature of $217\,K$ and a pressure of $7.5\,kPa$. Calculate the percentage of the number of moles of helium supplied at ground level now remaining in the envelope. [OCR 2000]

5 In the diagram the volume of bulb X is twice that of bulb Y. The system is filled with an ideal gas and a steady state is established with the bulbs held at $200\,K$ and $400\,K$.

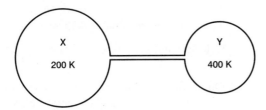

There are x moles of gas in X.
How many moles of gas are in Y?

A $\dfrac{x}{4}$ B $\dfrac{x}{2}$ C x D $2x$

[OCR 2000]

6 The pressure in a car tyre is adjusted to the manufacturer's recommended value before setting out on a journey. The temperature of the air in the tyre is then $15\,°C$. After driving some distance, it is found that the temperature of the air in the tyre is $41\,°C$. Assume that the air in the tyre behaves as an ideal gas, and that the volume of the air within the tyre remains constant.

(a) By what percentage of the recommended value has the pressure in the tyre increased?

(b) The driver reduces the tyre pressure to the recommended value by letting some air escape through the valve. The temperature of the air in the tyre remains at $41\,°C$. What percentage of the mass of air originally in the tyre is released?

(Hint: the mass of air in the tyre is proportional to the number of moles of air in the tyre.) [CCEA 2000, part]

7 A balloon has volume $5.50 \times 10^{-2}\,m^3$. It is filled with helium to a pressure of $1.10 \times 10^5\,Pa$ at a temperature of $20\,°C$. Calculate:

(a) the number of moles of helium inside the balloon

(b) the number of helium atoms inside the balloon

(c) the net force acting on one square centimetre of the material of the balloon if atmospheric pressure is $1.01 \times 10^5\,Pa$.

8 (a) State **two** quantities which increase when the temperature of a given mass of gas is increased at constant volume.

(b) A car tyre of volume $1.0 \times 10^{-2}\,m^3$ contains air at a pressure of $300\,kPa$ and a temperature of $290\,K$. The mass of one mole of air is $2.9 \times 10^{-2}\,kg$.

Assuming that the air behaves as an ideal gas, calculate

   (i) n, the amount, in mol, of air,
   (ii) the mass of the air,
   (iii) the density of the air.

(c) Air contains oxygen and nitrogen molecules. State, with a reason, whether the following are the same for oxygen and nitrogen molecules in air at a given temperature.
   (i) The average kinetic energy per molecule
   (ii) The r.m.s. speed [AQA 2001]

9 (a) Give non-mathematical explanations, in terms of molecules, for the following:
   (i) A gas exerts a pressure on the walls of its container.
   (ii) The gas pressure increases as the temperature increases.

(b) A cylinder of volume $30 \times 10^{-3}\,m^3$ contains $0.20\,kg$ of oxygen gas at a temperature of $300\,K$. Calculate
   (i) the number of molecules of gas in the container.
   [The mass of a mole of oxygen molecules is $0.032\,kg$.]
   (ii) the pressure exerted by the gas.

    (iii) the root-mean square speed of the molecules. [WJEC 2000]

10 A small dust particle suspended in a gas is struck successively by five gas molecules whose speeds are $300\,\mathrm{m\,s^{-1}}$, $500\,\mathrm{m\,s^{-1}}$, $400\,\mathrm{m\,s^{-1}}$, $600\,\mathrm{m\,s^{-1}}$ and $200\,\mathrm{m\,s^{-1}}$.

Calculate the root mean square speed of these five molecules.

Kinetic theory leads to the equation

$$pV = \tfrac{1}{3}Nm{<}c^2{>}$$

The ideal gas equation is written

$$pV = nRT$$

The second equation may be derived from the first equation with the aid of an additional assumption which involves ideal gas temperature T. State this assumption in algebraic form.

Show how, with this assumption, the ideal gas equation may be derived from the kinetic theory equation. [Edexcel 2000]

11 (a) State an algebraic relation between the molar mass M of an ideal gas and the mass m of one of its molecules. Identify any other symbol used.

(b) The Earth's atmosphere at ground level consists principally of oxygen (molar mass $0.032\,\mathrm{kg}$) and nitrogen (molar mass $0.028\,\mathrm{kg}$), both gases being at the same temperature. Calculate the ratio of the r.m.s. speeds of the molecules of these gases.

(c) (i) At ground level the mean density of atmospheric gases at a temperature of $288\,\mathrm{K}$ and pressure $101\,\mathrm{kPa}$ is $1.22\,\mathrm{kg\,m^{-3}}$. Calculate the mean density of the atmosphere at a height of $10.0\,\mathrm{km}$, where the temperature is $223\,\mathrm{K}$ and the pressure is $26.4\,\mathrm{kPa}$. Assume the composition of the atmosphere is the same at both levels.

    (ii) Comment on the assumption in (i) in the light of the existence of the greenhouse effect. [OCR 2001]

12 (a) Write down an equation relating the pressure p and volume V of an ideal gas and the mean-square speed $<c^2>$ of the molecules of the gas. Define any other terms which appear in the equation.

(b) A volume of $71\,200\,\mathrm{cm^3}$ of a certain ideal gas contains 1.03×10^{24} atoms. The gas has density $0.800\,\mathrm{kg\,m^{-3}}$.
  (i) Calculate the mass of one atom of the gas.
  (ii) The pressure exerted by this gas is measured, and is found to be $80.0\,\mathrm{kPa}$.
    1. Calculate the root-mean-square speed of the atoms of the gas.
    2. Calculate the temperature of the gas. [CCEA 2000]

13 Two moles of argon have a mass of $0.036\,\mathrm{kg}$ and occupy a rigid container of volume $4.0 \times 10^{-2}\,\mathrm{m^3}$ at a pressure of $1.0 \times 10^5\,\mathrm{Pa}$. Calculate:

(a) the root mean square speed of an argon atom

(b) the temperature of the argon gas

(c) the total internal energy of the gas atoms.

The safety valve in the container will open if the pressure of the gas inside it exceeds $1.5 \times 10^5\,\mathrm{Pa}$. If the gas is now heated calculate:

(d) the temperature at which the safety valve will open.

14 (a) Describe how the concept of the absolute zero of temperature is explained in terms of
  (i) the ideal gas laws,
  (ii) the kinetic theory of gases.

(b) A flask of volume $2.0 \times 10^{-3}\,\mathrm{m^3}$, containing an ideal gas at room temperature ($290\,\mathrm{K}$) and atmospheric pressure ($100\,\mathrm{kPa}$), is sealed with a rubber stopper.

The Avogadro constant is $6.0 \times 10^{23}\,\mathrm{mol^{-1}}$.
The molar gas constant is $8.3\,\mathrm{J\,mol^{-1}\,K^{-1}}$.

Calculate the number of gas molecules in the flask.

(c) On heating the flask in (b) the rubber stopper is forced out when the temperature exceeds $400\,\mathrm{K}$.
  (i) The area of the lower surface of the stopper is $4.0 \times 10^{-4}\,\mathrm{m^2}$. Calculate the force exerted on this area at $400\,\mathrm{K}$.
  (ii) Calculate, for ideal gas molecules, the ratio

$$\frac{\text{r.m.s. speed at } 400\,\mathrm{K}}{\text{r.m.s. speed at } 290\,\mathrm{K}}$$

[OCR 2000]

19
Ideal gases and thermodynamics

The first law of thermodynamics

In mathematical terms the first law is written as

$$\Delta Q = \Delta U + \Delta W \qquad \text{(19.1)}$$

where ΔQ is the thermal energy supplied *to* the system, ΔU the *increase* in internal energy of the system and ΔW the work done *by* the system on the surroundings.

Thus if 5 J (ΔQ) of energy was given to a sample of gas by heating it, and if the gas then expanded and did 3 J (ΔW) of work (e.g. by pushing a piston), Equation 19.1 tells us that 2 J (ΔU) of energy would remain inside the gas. For an ideal gas this would correspond to a rise in kinetic energy, only, of the molecules – so there would be an increase in RMS speed and temperature (see Equation 18.11). Note that no change in potential energy is possible since the interatomic forces are zero.

Work done by an expanding gas

Fig. 19.1 shows a gas enclosed in a cylinder by a frictionless piston. If the gas expands and moves the piston outwards, the gas does work against the external force. The external work ΔW is given by

$$\Delta W = \int_{V_1}^{V_2} p \, \mathrm{d}V \qquad \text{(19.2)}$$

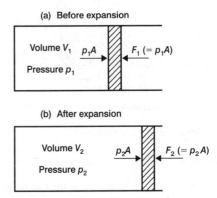

(a) Before expansion

(b) After expansion

Fig. 19.1   A gas expanding in a cylinder

This mathematical operation needs to be carried out if the pressure p of the gas changes as it expands. If the pressure remains constant, so that $p_1 = p_2 = p$, then Equation 19.2 becomes

$$\Delta W = p(V_2 - V_1) \qquad \text{(19.3)}$$

where ΔW is in joules when p is in pascals and $(V_2 - V_1)$ is in m^3.

Example 1

Figure 19.2 shows a sample of gas enclosed in a cylinder by a frictionless piston of area 100 cm^2. The cylinder is now heated, so that 250 J of energy is transferred to the gas, which then expands against atmospheric pressure $(1.00 \times 10^5 \, \mathrm{N\,m^{-2}})$ and pushes the piston 15.0 cm along the cylinder as shown. Calculate (a) the external work done by the gas, (b) the increase in internal energy of the gas.

Method

(a) Referring to Fig. 19.2, we see that the force F exerted by the atmosphere on the piston is given by

$$F = H \times A = 1 \times 10^5 \times 1 \times 10^{-2}$$

$$= 1 \times 10^3 \, \mathrm{N}$$

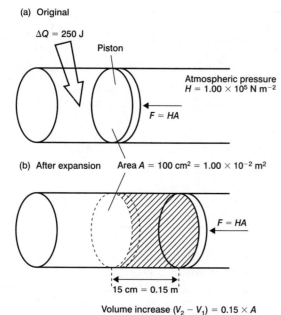

(a) Original

$\Delta Q = 250$ J

Piston

Atmospheric pressure
$H = 1.00 \times 10^5$ N m^{-2}

$F = HA$

(b) After expansion   Area $A = 100$ cm$^2 = 1.00 \times 10^{-2}$ m^2

$F = HA$

15 cm = 0.15 m

Volume increase $(V_2 - V_1) = 0.15 \times A$

Fig. 19.2   Information for Example 1

Thus the work done ΔW during expansion is

$$\Delta W = \text{Force } F \times \text{Distance moved by piston}$$

$$= 10^3 \times 0.15$$

$$= 150 \, \text{J}$$

We could use Equation 19.3 to calculate ΔW to get the same answer, as follows. The pressure p of the gas is equal to atmospheric pressure during the expansion. Thus, since $(V_2 - V_1)$ is $0.15 \times A$,

$$\Delta W = p(V_2 - V_1)$$

$$= 1 \times 10^5 \times 0.15 \times 1 \times 10^{-2}$$

$$= 150 \, \text{J}$$

(b) We have $\Delta Q = 250$ and $\Delta W = 150$. Rearranging Equation 19.1 gives

$$\Delta U = \Delta Q - \Delta W = 250 - 150$$

$$= 100 \, \text{J}$$

Thus as heat is supplied to the gas the speed of the molecules increases. This would increase the pressure in the container, if it were not for the fact that the piston is pushed out. This decreases the density of the gas, and thus (see Equation 18.9) the pressure of the gas can remain at atmospheric pressure. The net effect is one of heat input being used to do work in pushing back the atmosphere, and to increase the internal energy (and so increase molecular speeds and temperature) of the gas.

Answer

(a) 150 J, (b) 100 J.

Example 2

When 1.50 kg of water is converted to steam (at 100 °C) at standard atmospheric pressure $(1.01 \times 10^5 \, \text{N m}^{-2})$, 3.39 MJ of heat are required. During the transformation from liquid to vapour state, the increase in volume of the water is 2.50 m^3. Calculate the work done against the external pressure during the process of vaporisation. Explain what happens to the rest of the energy.

Method

When the liquid is converted into steam, the molecules have to push back the atmosphere during the accompanying increase in volume. We use Equation 19.3 with $p = 1.01 \times 10^5$ and $(V_2 - V_1) = 2.50$. So,

$$\Delta W = p(V_2 - V_1) = 1.01 \times 10^5 \times 2.50$$

$$= 0.253 \times 10^6 \, \text{J}$$

The external work done $\Delta W = 0.253$ MJ.

The rest of the energy goes to an increase in internal energy ΔU of the water molecules and is given by Equation 19.1:

$$\Delta U = \Delta Q - \Delta W = 3.39 - 0.253$$

$$= 3.14 \, \text{MJ}$$

This is needed to do work in separating the water molecules during the liquid–vapour transition. It thus becomes potential energy. No kinetic energy change occurs because there is no increase in temperature.

Answer

External work done is 0.253 MJ.

Exercise 19.1

1   A fixed mass of gas is cooled, so that its volume decreases from 4.0 litres to 2.5 litres at a constant pressure of 1.0×10^5 Pa. Calculate the external work done by the gas. Note: 1 litre = 10^{-3} m^3.

2   Referring to Fig. 19.2a, suppose that the sample of gas is cooled down so that 120 J of heat is extracted from it. If as a result the piston moves inwards 5.0 cm along the cylinder, calculate (a) the external work done by the gas, (b) the increase in internal energy of the gas.

3   The specific latent heat of vaporisation of steam is 2.26 MJ kg^{-1}. When 50 cm^3 of water is boiled at standard atmospheric pressure of 1.01×10^5 Pa, 83×10^3 cm^3 of steam are formed. Calculate (a) the mass of water boiled, (b) the heat input needed, (c) the external work done during vaporisation, (d) the increase in internal energy. (Density of water = 1000 kg m^{-3}; 1 cm^3 = 10^{-6} m^3.)

Isothermal and adiabatic changes

An *isothermal change* is one which takes place in such a way that the *temperature remains constant*. Thus for an isothermal change, Equation 18.2 reduces to Equation 18.3:

$$p_1 V_1 = p_2 V_2 \qquad (18.3)$$

where p_1 and V_1 are the initial pressure and volume and p_2 and V_2 are pressure and volume after the isothermal change.

An *adiabatic change* is one which takes place in such a way that *no heat can enter or leave the system* during the process. This means that, from Equation 19.1, since $\Delta Q = 0$, any external work done by the gas must lead to a corresponding decrease in internal energy (and hence a temperature drop). Similarly an adiabatic compression leads to an increase in internal energy and hence a temperature rise. For an *adiabatic change* it can be shown that (for a fixed mass of gas)

$$p_1 V_1^\gamma = p_2 V_2^\gamma \qquad (19.4)$$

where p_1 and V_1 refer to initial pressure and volume, p_2 and V_2 to pressure and volume after the adiabatic change and γ is a constant which depends upon the number of atoms per molecule of the gas. Any suitable units may be used for pressure and volume.

Note that *for any change, Equations 18.2 and 18.5 can be used*.

By combining Equations 19.4 and 18.2 we can eliminate pressure to get, for an adiabatic change,

$$T_1 V_1^{(\gamma-1)} = T_2 V_2^{(\gamma-1)} \qquad (19.5)$$

where T_1 and T_2 refer to initial and final temperature respectively.

Example 3

A gas at an initial pressure of 760 mm mercury is expanded adiabatically until its volume is doubled. Calculate the final pressure of the gas if γ is 1.40.

Method

We have $p_1 = 760$ and $\gamma = 1.4$. Let $V_1 = V$, so $V_2 = 2V$.

Rearranging Equation 19.4 gives

$$p_2 = p_1 \left(\frac{V_1}{V_2}\right)^\gamma = 760 \times \left(\frac{V}{2V}\right)^{1.4} = \frac{760}{2^{1.4}} = \frac{760}{2.64}$$
$$= 288$$

Answer

Final pressure is 288 mm mercury.

Example 4

The piston of a bicycle pump is slowly moved in until the volume of air enclosed is one-fifth of the total volume of the pump and is at room temperature (290 K). The outlet is then sealed and the piston suddenly drawn out to full extension. No air passes the piston. Find the temperature of the air in the pump immediately after withdrawing the piston, assuming that air is a perfect gas with $\gamma = 1.4$. [WJEC, part]

Method

The pushing-in of the piston results in some air remaining trapped in the body of the pump. Its initial temperature is $T_1 = 290$; let its initial volume $V_1 = V$. The act of *suddenly* drawing out the piston indicates an adiabatic expansion and, since no air passes the piston, a fixed mass of gas. The final volume $V_2 = 5V$ and we require the final temperature T_2. Rearranging Equation 19.5 with $\gamma - 1 = 0.4$:

$$T_2 = T_1 \left(\frac{V_1}{V_2}\right)^{(\gamma-1)} = 290 \left(\frac{V}{5V}\right)^{0.4} = 152\,\text{K}$$

Note that we could have used Equation 19.4 to find p_2 in terms of p_1 and then used Equation 18.2 to find T_2. It is worth checking the answer using this method, which is equivalent to proving Equation 19.5.

The final temperature is less than the initial value because the external work done by the gas, on expansion, results in a corresponding decrease in internal energy, hence temperature.

Answer

Final temperature is 152 K.

Example 5

A fixed mass of gas, initially at 7 °C and a pressure of $1.00 \times 10^5\,\text{N m}^{-2}$, is compressed isothermally to one-third of its original volume. It is then expanded adiabatically to its original volume. Calculate the final temperature and pressure, assuming $\gamma = 1.40$.

Method

We must treat the two processes *separately and in order*. For the *isothermal* change we have $p_1 = 1 \times 10^5$; let $V_1 = V$ and $V_2 = V/3$. We rearrange Equation 18.3 to find p_2:

$$p_2 = \frac{p_1 V_1}{V_2} = \frac{1 \times 10^5 \times V}{V/3}$$

$$= 3 \times 10^5 \, \text{N m}^{-2}$$

For the *adiabatic* change our initial temperature is still $7\,°C$, so

initial state: $p_2 = 3.00 \times 10^5, V_2 = V/3,$

$$T_2 = 273 + 7 = 280\,\text{K}$$

final state: $p_3 = ?, \qquad V_3 = V, \qquad T_3 = ?$

Note that we have the initial state with suffix 2 and the final state with suffix 3, so Equation 19.4 becomes $p_2 V_2^\gamma = p_3 V_3^\gamma$. Rearranging to find p_3:

$$p_3 = p_2 \left(\frac{V_2}{V_3}\right)^\gamma = 3 \times 10^5 \left(\frac{V/3}{V}\right)^{1.4} = \frac{3 \times 10^5}{3^{1.4}}$$

$$= 0.644 \times 10^5 \, \text{N m}^{-2}$$

To find T_3 we can use Equation 18.2 (or, alternatively, Equation 19.5):

$$T_3 = \frac{p_3 V_3 T_2}{p_2 V_2} = \frac{0.644 \times 10^5 \times V \times 280}{3 \times 10^5 \times V/3}$$

$$= 180\,\text{K}$$

Answer

Final temperature is $180\,\text{K}$, final pressure is $0.644 \times 10^5 \, \text{N m}^{-2}$.

Exercise 19.2

(Assume $\gamma = 1.40$ for air.)

1 2.00 litre of air initially at a pressure of $1.01 \times 10^5 \, \text{N m}^{-2}$ and a temperature of $17\,°C$ is compressed to volume of 0.30 litre (a) under isothermal conditions, (b) under adiabatic conditions. Calculate in each case the final pressure and temperature.

2 $3.00 \times 10^{-4}\,\text{m}^3$ of air at $7\,°C$ and a pressure of $5.00 \times 10^5 \, \text{N m}^{-2}$ is allowed to expand until the pressure falls to $1.00 \times 10^5 \, \text{N m}^{-2}$. Calculate the final volume and temperature in each case if the expansion takes place under (a) isothermal, (b) adiabatic conditions.

3 A fixed mass of air at an initial pressure of 760 mm mercury and $0\,°C$ is expanded adiabatically to 1.50 times its volume and then compressed isothermally to 0.50 times its *original* volume. Calculate its final temperature and pressure.

Heat engines

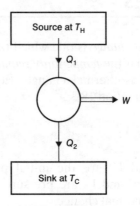

Fig. 19.3   Operation of a heat engine

As shown in Fig. 19.3, a heat *engine* takes heat Q_1 from a source at a (high) temperature T_H, does useful work W and gives out heat Q_2 to its surroundings, the sink, at a (low) temperature T_C. Note that:

$$Q_1 = W + Q_2 \qquad (19.6)$$

The efficiency E of the engine is defined by (see also Equation 6.2):

$$E(\%) = \frac{\text{Useful work done by engine } (W)}{\text{heat supplied by source } (Q_1)} \times 100 \qquad (19.7)$$

The *maximum* possible efficiency E_{max} is that of an ideal, or Carnot, engine. Theoretically this is given by:

$$E_{\text{max}} = \frac{(T_H - T_C)}{T_H} \times 100 \qquad (19.8)$$

Example 6

(a) A heat engine operates between a source at $227\,°C$ and a sink at $27\,°C$. Calculate the (theoretical) maximum efficiency of this engine.

(b) In practice the engine accepts heat at a rate of $9.0\,\text{kW}$ and does useful work at a rate of $2.5\,\text{kW}$. Calculate:
    (i) the actual efficiency of the engine;
    (ii) the rate at which 'waste' heat passes to the sink.

Method

(a) We have $T_H = 273 + 227 = 500\,\text{K}$ and $T_C = 273 + 27 = 300\,\text{K}$. From Equation 19.8 we have:

$$E_{max} = \left(\frac{T_H - T_C}{T_H}\right) \times 100 = \left(\frac{500 - 300}{500}\right) \times 100$$

$$= 40\%$$

(Note that if the source temperature T_H is increased relative to the sink temperature T_C then the theoretical maximum efficiency will increase but it can never reach 100%.)

(b) We have, *in one second*:

> heat supplied by source $Q_1 = 9.0$ kJ and
> useful work done by engine $W = 2.5$ kJ

(i) From Equation 19.7:

> Efficiency $= \frac{2.5}{9.0} \times 100 = 27.8\%$

(ii) We require Q_2. From Equation 19.6:

> $Q_2 = Q_1 - W = 9.0 - 2.5 = 6.5$ kJ

(Note that use can be made of the 'waste' heat, for example to give a supply of warm water.)

Answer

(a) 40%,

(b) (i) 28%; (ii) 6.5 kJ per second or 6.5 kW.

Heat pumps

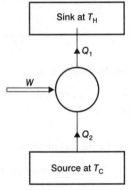

Fig. 19.4 Operation of a heat pump

A heat *pump* is a heat engine 'in reverse'. Heat Q_2 is taken from a source at (low) temperature T_C and heat Q_1 is released into a 'reservoir' at (high) temperature T_H. Energy W must be provided in order to operate the heat pump as shown in Fig. 19.4. Note that once again:

$$Q_1 = W + Q_2 \qquad (19.6)$$

The *coefficient of performance CP* of a heat pump is a useful measure of its efficiency.

For a refrigerator:

$$CP = \frac{\text{heat taken from cool box}}{\text{work done } W}$$

$$\qquad\qquad (19.7)$$

$$= \frac{Q_2}{W} = \frac{Q_2}{(Q_1 - Q_2)}$$

For a heat pump used to heat the inside of a building:

$$CP = \frac{\text{heat supplied to inside of building}}{\text{work done } W}$$

$$\qquad\qquad (19.8)$$

$$= \frac{Q_1}{W} = \frac{Q_1}{(Q_1 - Q_2)}$$

Example 7

A heat pump in a refrigerator has a coefficient of performance of 4.0. If 60 W of heat must be transferred from inside the refrigerator in order to keep its contents cool, calculate:

(a) the rate at which the heat pump operates

(b) the rate at which heat is discharged into the area surrounding the refrigerator.

Method

(a) We have $CP = 4.0$, $Q_2 = 60$ J (per second) and require the rate W at which the heat pump operates. From Equation 19.9:

$$W = \frac{Q_2}{CP} = \frac{60}{4.0} = 15 \text{ J (per second)}$$

(b) We require Q_1. From Equation 19.6:

$$Q_1 = W + Q_2 = 15 + 60 = 75 \text{ J (per second)}$$

Answer

(a) 15 W, (b) 75 W.

Exercise 19.3

1   A modified car engine uses a mixture of air and natural gas as its energy source. The temperature of the spark ignited cylinder is 2.20×10^3 K and the exhaust temperature is 920 K. The difference between the rate at which heat is supplied to the engine and the work done by the engine is 5.0 MW. Calculate:

(a) the maximum (Carnot) efficiency of the engine

(b) the rate at which the engine does useful work if 8.0 MW is input via the engine source

(c) the actual efficiency of the engine.

2 A heat pump is used to transfer heat from the outside of a building to the inside. If 1.5 kW is needed to operate the heat pump in order to heat the interior at a rate of 7.5 kW, calculate the coefficient of performance of the heat pump.

Work done during a cycle

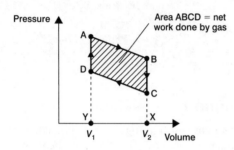

Fig. 19.5   A pressure-volume cycle

The work done by a gas during expansion at constant pressure has been covered previously (Example 1). In Fig. 19.5 a gas expands from state A (volume V_1) to state B (volume V_2) and the pressure is not constant. In general *the work done is equal to the area ABXY under the curve*. If the gas is taken through a cycle of events ABCD as shown in Fig. 19.5 then work (equal to area ABXY) is done *by* the gas as it expands from A to B – and work (equal to area CDYX) is done *on* the gas as it contracts from C to D.

The *net work done by the gas during the cycle is thus equal to the enclosed area ABCD.*

This underlies the principle by which energy is transferred during the operation of an engine. The air–fuel mixture is taken through a cycle of events and work is done *by* the gaseous mixture which results in energy transfer to moving parts.

Example 8

A fixed mass of gas is taken through the closed cycle ABCD as shown in Fig. 19.6. Calculate the work done by the gas during this cycle of events.

Method

The net work done by the gas is equal to the enclosed area ABCD. Now:

area ABCD = AB × BC

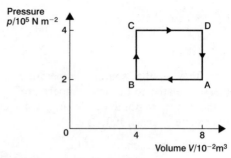

Fig. 19.6   Diagram for Example 8

where $AB = (8 - 4) \times 10^{-2} \, \text{m}^3$
and $BC = (4 - 2) \times 10^{+5} \, \text{Nm}^{-2}$
Thus:

$$\text{area ABCD} = 4 \times 10^{-2} \times 2 \times 10^{+5}$$
$$= 8 \times 10^{+3} \, \text{J}$$

Answer

$8 \times 10^{+3}$ J.

Example 9

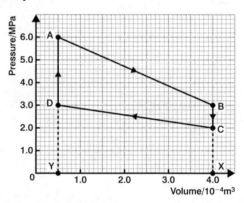

Fig. 19.7   Diagram for Example 9

Fig. 19.7 shows a simplified indicator diagram (pressure-volume cycle) for one cylinder of an engine. Calculate:

(a) the work done by the gas on expansion from A to B

(b) the work done by the gas on contraction from C to D

(c) the net work done by the gas during one cycle ABCD.

If the engine rotates at 50 cycles per second and it has four cylinders, calculate:

(d) the power generated by the engine.

Method

(a) Since the gas expands it does work on its surroundings. Thus:

work done *by* the gas = area ABXY
$$= \tfrac{1}{2}(\text{BD} \times \text{AD}) + (\text{BX} \times \text{XY})$$
$$\text{or } \tfrac{1}{2} \times (\text{AY} + \text{BX}) \times \text{XY}$$

where $AY = 6.0 \times 10^{+6}$, $BX = 3.0 \times 10^{+6}$ and $XY = (4.0 - 0.50) \times 10^{-4}$. Thus

area $ABXY = 4.5 \times 10^{+6} \times 3.50 \times 10^{-4} = 1575\,J$

(b) Since the gas contracts then it has work done on it. Thus:

work done = area CDYX

$$= \tfrac{1}{2} \times (DY + CX) \times XY$$

where $DY = 3.0 \times 10^{+6}$, $CX = 2.0 \times 10^{+6}$ and $XY = 3.50 \times 10^{-4}$. Thus

area $CDYX = 2.5 \times 10^{+6} \times 3.50 \times 10^{-4} = 875\,J$

Thus the work done by the gas $= -875\,J$ (note the minus sign signifying that work is done *on* the gas).

(c) Work done *by* the gas during the cycle ABCD is equal to the enclosed area ABCD.

Area $ABCD =$ area $ABXY -$ area $CDYX$

$$= 1575 - 875 = 700\,J$$

(d) In one second each cylinder is taken through 50 cycles and there are 4 cylinders. Therefore the power generated is $50 \times 4 = 200$ times the work done by the gas in one cycle. Thus:

power generated $= 200 \times 700 = 140 \times 10^3\,W$

The power generated is partly used to overcome friction within the engine/car system and partly to provide a driving force.

Answer

(a) 1.6 kJ, (b) −0.88 kJ, (c) 0.70 kJ, (d) 0.14 MW.

Exercise 19.4

1

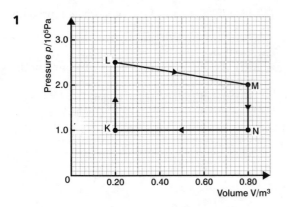

Fig. 19.8   Diagram for Question 1

A fixed mass of gas is subjected to the cycle of pressure and volume changes KLMN as shown in Fig. 19.8. Calculate the work done by the gas during this cycle.

2

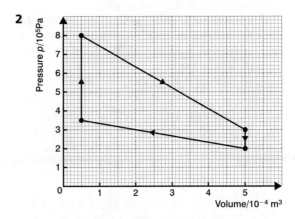

Fig. 19.9   Diagram for Question 2

Fig. 19.9 shows a simplified indicator diagram for one cylinder of a high compression petrol engine. Calculate:

(a) the net work done by the fuel–air mixture during one cycle

(b) the power generated by the engine if it has 4 cylinders rotating at 3600 revolutions per *minute*.

Exercise 19.5: Examination questions

1 A fixed mass of gas is heated, so that its volume increases from $0.5\,m^3$ to $0.8\,m^3$, at a constant pressure of $1.0 \times 10^5\,Pa$. Calculate the external work done by the gas.

2 A fixed mass of an ideal gas is sealed in a container by a frictionless piston which is free to move. 400 J of heat is supplied to the gas which expands under a constant pressure of 25 kPa from a volume of $5.0 \times 10^{-3}\,m^3$ to a volume of $15 \times 10^{-3}\,m^3$. Calculate the change in internal energy of the gas.

3 (a) An electric kettle has a power of 2.4 kW. It contains boiling water at 100 °C. Calculate how long it takes to boil away 0.50 kg of water. (The specific latent heat of vaporisation of water is $2.2\,MJ\,kg^{-1}$.)

　　(b) (i) 0.50 kg of water contains 27.8 mol of water and occupies a volume of $0.00050\,m^3$. Show that the volume of the water vapour it produces at 100 °C is approximately $0.9\,m^3$.
(Atmospheric pressure is $1.01 \times 10^5\,Pa$.)

　　　　(ii) Calculate the work done by the water pushing the atmosphere back as it turns from liquid into vapour.

(c) The equation* $\Delta U = \Delta Q + \Delta W$ is applied to the 0.50 kg of water during the process of converting it to vapour. What are the values of each of the three terms?
(Assume $R = 8.31\,\mathrm{J\,mol^{-1}\,K^{-1}}$)

[Edexcel 2001]

4 A fixed mass of an ideal gas at atmospheric pressure is compressed adiabatically to one third of its original volume, at which point it has a pressure of 5 times atmospheric. If the original temperature was 27 °C, calculate its final temperature.

5 A fixed mass of an ideal gas at a temperature of 27 °C is adiabatically compressed to half of its original volume and then cooled, at this volume, until the pressure is restored to its original value. Calculate the new temperature of the gas.

6 A fixed mass of an ideal gas (with $\gamma = 1.67$) at a temperature of 280 K, is subject to an adiabatic expansion in which its volume is trebled. Calculate its new temperature.

7 Calculate the theoretical maximum efficiency of a steam engine which exhausts into the atmosphere, at a temperature of 15 °C, if the engine utilises high pressure steam at a temperature of 170 °C.

8 A fixed mass of gas is taken around a cycle of changes ABCD as shown in Fig. 19.10.

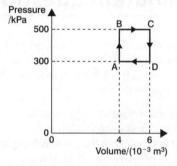

Fig. 19.10   Information for Question 8

Calculate the net work done by the gas during one cycle.

9 Fig. 19.11 shows the indicator diagram for one cycle of an engine. Calculate the net work done by the engine per cycle.

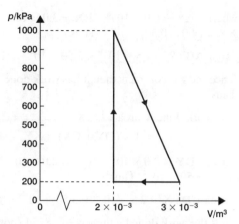

Fig. 19.11   Information for Question 9

10 Fig. 19.12 shows an idealised indicator diagram for a petrol engine.

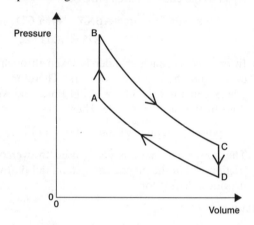

Fig. 19.12

In one particular cycle, 380 J of energy is supplied when the fuel is burned and 180 J is lost in the exhaust gases.

By reference to Fig. 19.12,

(a) identify that part of the cycle which represents the burning of the fuel,

(b) calculate
  (i) the energy represented by the area of the loop ABCD,
  (ii) the efficiency of the engine.

[OCR 2000, part]

*Authors' note: Use $\Delta Q = \Delta U + \Delta W$ for our purposes.)

Section G
Electricity and magnetism

20
Direct current circuits

Electric charge

All solids, liquids and gases are made of electrons, protons and neutrons. Electrons repel each other and we explain this effect by saying that electrons possess an electric charge or are 'charged'. Similarly protons repel each other so they are also charged. But an electron and proton attract each other, so the charge on a proton is not the same as that on an electron; we describe the charges as positive (+) and negative (−) respectively.

The charges on these particles are all equally strong, although + and − charges have opposite effects. Normally the number of electrons in an object equals the number of protons so that the object neither attracts nor repels any nearby charge. A surplus of electrons in an object means that it is negatively charged (− sign) while a deficiency of electrons (a surplus of protons) is a positive charge (+ sign).

Two well-charged objects having the same signs repel each other. Opposite signs attract. A charged object may also show a weak attraction on an uncharged object.

The size of charge called a coulomb (abbreviation C) is a surplus or deficiency of approximately 6 thousand million electrons.

Electric current

A current is a flow of charge. In a metal wire many electrons are free to move, so that a current can flow in a metal wire as a flow of electrons, i.e. the current carriers (or charge carriers) are electrons.

The unit for current is the ampere (A), defined in Chapter 23. Current size I is related to charge q moving through (entering and leaving) a wire in time t seconds by

$$I = \frac{q}{t} \text{ and } 1\,\text{A} = 1\,\text{C}\,\text{s}^{-1} \qquad (20.1)$$

Equation 20.1 defines the coulomb as $1\,\text{A}\,\text{s}$.

The direction of current flow is taken to be that of positive charge flow, i.e. opposite to that of electron flow.

Carrier velocity

If carriers, e.g. electrons in a metal wire, are moving with an average drift velocity along the wire of v metre per second, then the current is

$$I = nAqv \qquad (20.2)$$

where n is the carrier density (number per m³), A is the cross-section area of the wire (so that nA is the carriers per metre length of wire) and q is the charge of each carrier.

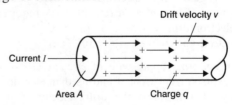

Fig. 20.1 $I = nAqv$

Example 1

How many electrons are passing through a wire per second if the current is 1.00 mA, given that the charge carried by each electron is 1.6×10^{-19} C?

Method

$I = 10^{-3}$, $q = 1.6 \times 10^{-19}$ C; let time $t = 1$ s and the number of electrons be n. Using $I = Q/t$ (Equation 20.1) we have

$$1.00 \times 10^{-3} = \frac{n \times 1.6 \times 10^{-19}}{1}$$

$$\therefore \qquad n = \frac{10^{-3}}{1.6 \times 10^{-19}}$$

$$\therefore \qquad n = 6.25 \times 10^{15}$$

Answer

6.2×10^{15}.

Example 2

Calculate the mean velocity of electron flow (the drift velocity) in a wire where the free electron density is 5.0×10^{28} m^{-3} if the current is 1.0 A and the wire has a uniform cross-section area of 1.0 mm^2. (Electron charge $= -1.6 \times 10^{-19}$ C.)

Method

$I = nqvA$ (Equation 20.2) and $I = 1$ A, $n = 5 \times 10^{28}$ m^{-3}, $q = 1.6 \times 10^{-19}$ C and $A = 10^{-6}$ m^2

$$v = \frac{I}{nqA} = \frac{1}{5 \times 10^{28} \times 1.6 \times 10^{-19} \times 10^{-6}}$$

$$= \frac{1}{8} \times 10^{-3} = 0.125 \times 10^{-3} \text{ m s}^{-1}$$

Answer

1.2×10^{-4} m s^{-1} (if we assume an accuracy of two significant figures).

Exercise 20.1

1   In a certain semiconducting material the current carriers each have a charge of 1.6×10^{-19} C. How many are entering the semiconductor per second when the current is 2.0 μA?

2   How many free electrons are there per metre length of wire if a current of 2.0 A requires the electron drift velocity to be 10^{-3} m s^{-1}?
    (Electronic charge $= 1.6 \times 10^{-19}$ C.)

3   A uniform copper wire of circular cross-section has its current trebled and its diameter doubled. By what factor is the drift velocity of its free electrons multiplied as a result?

Potential and potential difference

The potential of a place may be thought of as its attractiveness for electrons or unattractiveness for positive charges. A place where there is a high concentration of electrons or which has a lot of electrons near it will have a low potential.

The difference of potential (PD) V between two places is defined as the work done per coulomb of charge moved from the one place to the other.

$$V = \frac{W}{q} \qquad (20.3)$$

where W is the work done (e.g. if positive charge q moves from lower potential ($-$) to higher potential ($+$)) or energy obtainable from the movement (e.g. if negative charge q goes from $-$ to $+$ place).

The unit for PD is the volt (V).

The potential of a place measured in volts is the PD between the place concerned and some reference point, usually taken to be a place far away from any electric charges (i.e. at infinity), or otherwise the Earth. In other words, either of these places may be taken as zero potential.

Electric current flows spontaneously from a higher potential place ($+$) to a lower potential place ($-$) if the two places are joined by a conducting path.

Ohm's law

This law states that the current I through a given conductor is proportional to the PD between its ends, provided that its temperature does not change.

$$I \propto V \text{ or } \frac{V}{I} = \text{Constant} \qquad (20.4)$$

This law applies to metallic conductors and many others.

Resistance *R* of a conductor

This is the opposition of the conductor to current flow through it, and it is defined as the PD needed across it (between its ends) per ampere of current:

$$R = \frac{V}{I} \tag{20.5}$$

The unit for resistance is the ohm (Ω).

Resistors

These are devices for providing resistance to the flow of current. Some variable resistors are called rheostats.

A thermistor is a temperature sensitive resistor. An LDR is a light dependent resistor (photoconductor).

Electric circuits

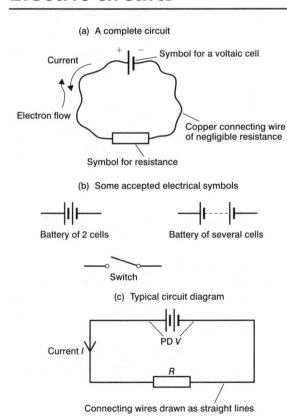

(a) A complete circuit

Current
+ − Symbol for a voltaic cell
Electron flow
Copper connecting wire of negligible resistance
Symbol for resistance

(b) Some accepted electrical symbols

Battery of 2 cells
Battery of several cells
Switch

(c) Typical circuit diagram

Current *I*
PD *V*
R
Connecting wires drawn as straight lines

Fig. 20.2

Often a current is produced by use of a voltaic cell or battery (two or more cells joined together). The cell creates and maintains a PD between its terminals. A current is obtained if these two terminals are joined by a conducting path, i.e. when a complete circuit is formed. (Fig. 20.2)

The current obtained from a voltaic cell is direct current (DC) because its direction is constant.

Resistors in series

When two resistances R_1 and R_2 ohm are connected as shown in Fig. 20.3a they are in series and the total resistance is R, where

$$R = R_1 + R_2 \tag{20.6}$$

R_1 and R_2 carry the same current.

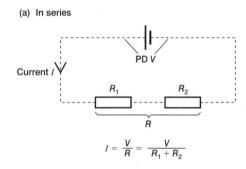

(a) In series

PD *V*
Current *I*
R_1
R_2
R

$$I = \frac{V}{R} = \frac{V}{R_1 + R_2}$$

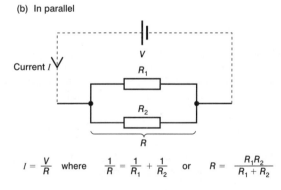

(b) In parallel

V
Current *I*
R_1
R_2
R

$$I = \frac{V}{R} \quad \text{where} \quad \frac{1}{R} = \frac{1}{R_1} + \frac{1}{R_2} \quad \text{or} \quad R = \frac{R_1 R_2}{R_1 + R_2}$$

Fig. 20.3 Resistors in series and parallel

Resistances in parallel

In this arrangement the resistance of the combination is given by

$$\frac{1}{R} = \frac{1}{R_1} + \frac{1}{R_2} \quad \text{or} \quad R = \frac{R_1 R_2}{R_1 + R_2} \tag{20.7}$$

In a parallel combination the PD across one resistor is the same as that across the other, but the total circuit current I in Fig. 20.3b is shared between the resistors.

Example 3

Calculate the current through, and PD across, each of the resistors in the circuit shown (Fig. 20.4).

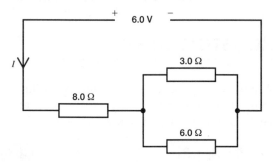

Fig. 20.4   Circuit diagram for Example 3

Method

The resistance of $3\,\Omega$ in parallel with $6\,\Omega$ is

$$R = \frac{R_1 R_2}{R_1 + R_2} = \frac{3.0 \times 6.0}{3.0 + 6.0} = 2.0\,\Omega$$

or $\quad \dfrac{1}{R} = \dfrac{1}{R_1} + \dfrac{1}{R_2} = \dfrac{1}{3.0} + \dfrac{1}{6.0} = 0.50^*$

$$\therefore \quad R = 2.0\,\Omega$$

We see that the circuit can be regarded as $8.0\,\Omega$ in series with $2.0\,\Omega$. Circuit resistance is

$$R = R_1 + R_2 = 8.0 + 2.0 = 10\,\Omega$$

$$\therefore \quad I = \frac{V}{R} = \frac{6.0}{10} = 0.60\,\text{A}$$

Note that we know only one PD, namely $6.0\,\text{V}$, and to use $I = V/R$ we must use $V = 6.0$ with the correct resistance. It is the $10\,\Omega$ across which the PD is $6.0\,\text{V}$.

The current through the $8.0\,\Omega$ resistor is I, which is $0.60\,\text{A}$.

PD across the $8.0\,\Omega$ (using $V = IR$ for this resistor now that its current is known) is given by

$$V = 0.6 \times 8.0 = 4.8\,\text{V}$$

To obtain answers for the $3.0\,\Omega$ and $6.0\,\Omega$ we can say either:

PD across the $3.0\,\Omega$ and $6.0\,\Omega$ is $6.0\,\text{V} - 4.8\,\text{V} = 1.2\,\text{V}$. The current I_3 through the $3.0\,\Omega$ is

$$I_3 = 1.2/3.0 = 0.40\,\text{A}$$

and for the $6.0\,\Omega$ the current I_6 is $1.2/6.0$ or $0.20\,\text{A}$; *or* (in view of the simple values of 3.0 and 6.0 for the parallel resistors) we can say:

*A common error is to forget that this is 1/R, not R.

The $3.0\,\Omega$ and $6.0\,\Omega$ are in the ratio of $1:2$, so that the easier route for the current $(3.0\,\Omega)$ will carry two parts of the $0.60\,\text{A}$ while the $6.0\,\Omega$ route will carry one part. The $6.0\,\Omega$ carries one-third of the $0.60\,\text{A}$, namely $0.20\,\text{A}$; the $3.0\,\Omega$ carries two-thirds, namely $0.40\,\text{A}$.

Answer

$0.60\,\text{A}, 4.8\,\text{V}; 0.20\,\text{A}, 1.2\,\text{V}; 0.40\,\text{A}, 1.2\,\text{V}.$

Exercise 20.2

1   A PD of $6.0\,\text{V}$ is maintained across a series combination of two resistors A and B. A is $20\,\Omega$ and B is $40\,\Omega$. Calculate

   (a) the current that should flow and

   (b) the expected PD across resistor A.

2   A PD of $3.0\,\text{V}$ is maintained across a parallel combination of $2.0\,\Omega$ and $3.0\,\Omega$. Calculate

   (a) the current that the voltage supply must be providing and

   (b) the current through the $2.0\,\Omega$ resistor.

3   Calculate the current through each resistor and the PD across each in the circuit shown in Fig. 20.5.

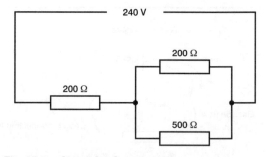

Fig. 20.5   Circuit for Question 3

Resistivity ρ of a material

The resistance R of a conductor is proportional to its length l, inversely proportional to its area of cross-section A, and dependent upon the nature of the material, described by its resistivity ρ, which is defined by the following equation:

$$R = \rho\frac{l}{A} \qquad (20.8)$$

The unit for ρ (which is given by $\rho = RA/l$) is $\Omega\,\text{m}$.

The term 'conductivity' σ of a material is used for the reciprocal of ρ so that

$$\sigma = \frac{1}{\rho} \qquad (20.9)$$

Temperature coefficient of resistance or resistivity

This quantity is denoted by α.

For many materials, e.g. a metal, a conductor's resistance increases steadily with increase of temperature in accordance with the equation

$$\alpha = \frac{R - R_0}{R_0 \theta} \text{ or } R = R_0(1 + \alpha \theta) \qquad (20.10)$$

where R is the resistance at Celsius temperature θ, R_0 is the resistance at $0\,^\circ\text{C}$ and α is the temperature coefficient of resistance of the material.

The unit for α is K^{-1}.

We can also write

$$\rho = \rho_0(1 + \alpha \theta)$$

where ρ and ρ_0 are resistivities at temperature θ and $0\,^\circ\text{C}$.

Example 4

Calculate the length of wire of 1.0 mm diameter and $5.0 \times 10^{-6}\,\Omega\,\text{m}$ resistivity that would have a resistance of $5.0\,\Omega$.

Method

$$R = \frac{\rho l}{A} \quad A = \pi \times \text{radius}^2 = \frac{\pi d^2}{4} \quad R = 5.0,$$

$$\rho = 5.0 \times 10^{-6}, \quad d = 1.0 \times 10^{-3}$$

$$\therefore \quad A = \frac{\pi \times 10^{-6}}{4} = 7.9 \times 10^{-7}$$

$$\therefore \quad l = \frac{RA}{\rho} = \frac{5.0 \times 7.9 \times 10^{-7}}{5.0 \times 10^{-6}} = 0.79\,\text{m}$$

Answer

0.79 m

Example 5

A coil of wire has resistance $6.00\,\Omega$ at $60\,^\circ\text{C}$ and $5.25\,\Omega$ at $15\,^\circ\text{C}$. What is its temperature coefficient of resistance?

Method

$$R = R_0(1 + \alpha \theta) \quad \therefore \quad 6.00 = R_0(1 + \alpha 60) \text{ and}$$
$$5.25 = R_0(1 + \alpha 15)$$
$$\therefore \quad \frac{5.25}{6.00} = \frac{R_0(1 + 15\alpha)}{R_0(1 + 60\alpha)}$$

Cancelling the R_0 factor and cross multiplying gives

$$5.25 + 315\alpha = 6.00 + 90\alpha$$
$$\therefore \quad 315\alpha - 90\alpha = 6.00 - 5.25$$
$$\therefore \qquad \alpha = \frac{0.75}{225} = 0.0033\,\text{K}^{-1}$$

Answer

$0.0033\,\text{K}^{-1}$.

Exercise 20.3

1   The electrical resistivity of manganin is $45 \times 10^{-8}\,\Omega\,\text{m}$ and is affected very little by temperature change. Calculate the resistance of 2.0 m of manganin wire of 1.0 mm diameter.

2   The resistivity of mild steel is $15 \times 10^{-8}\,\Omega\,\text{m}$ at $20\,^\circ\text{C}$ and its temperature coefficient is $50 \times 10^{-4}\,\text{K}^{-1}$. Calculate the resistivity at $60\,^\circ\text{C}$.

3   A certain coil of wire has an electrical resistance of $24\,\Omega$ at $10\,^\circ\text{C}$ and at $20\,^\circ\text{C}$ the resistance increases to $28\,\Omega$. Calculate the temperature coefficient of resistance for the metal of which the coil is made.

Electrical heating in a resistance

When current flows through a resistance there is a PD V across the resistance and, for Q coulombs passing through, electrical potential energy is lost (work W is done), this becoming internal energy of the resisting material (its temperature has risen). Since $V = W/Q$ and $Q = It$ (Equation 20.3 and 20.1) we have

$$W = VIt \qquad (20.11)$$

i.e. the heat produced is VIt where t is the time for which current flows.

Using $R = V/I$ (Equation 20.5), we can also write

$$W = \frac{V^2 t}{R} \qquad (20.12)$$

or

$$W = I^2 R t \qquad (20.13)$$

The work done per second or heat produced per second is the power P and

$$P = VI \left(\text{or} \ \frac{V^2}{R} \ \text{or} \ I^2 R \right) \qquad (20.14)$$

The unit for power is watt (W). $1\,\text{W} = 1\,\text{J}\,\text{s}^{-1}$.

The expression 'power dissipated' (in a resistance) is often used. It means 'heat produced (per second)'; but reminds us that the heat normally spreads and escapes from the place where it is produced.

The kilowatt–hour (product of kW and hour) is a unit for energy and is the quantity of energy converted in 1 hour when the power is one kilowatt. 1 kilowatt–hour = 1000 watt $\times$ 60 $\times$ 60 seconds = $3600 \times 10^3\,\text{J} = 3.6\,\text{MJ}$.

Exercise 20.4

1 Calculate the heat produced in a $10\,\Omega$ resistor when a current of 2.0 A flows through it for 1 minute exactly.

2 Calculate the energy dissipated by a 100 watt lamp working for 1 day. Give the answer

    (a) in killowatt–hours and

    (b) in joules.

3 Calculate the heat produced in 5 minutes in a pair of $10\,\Omega$ resistors connected in parallel with a PD of 2.0 V across the combination.

Electromotive force and internal resistance

The PD between the terminals of a cell is caused by a chemical action which stops when the PD reaches a value characteristic of the type of cell, called the EMF of the cell. EMF stands for electromotive force, although it is a voltage not a

force. When the cell is producing no current, i.e. it is on open circuit, the terminal PD V equals the EMF E:

$$V = E, \text{ on open circuit} \qquad (20.15)$$

When a current is being produced, the PD falls from the EMF value E, the chemical action starts again and the terminal PD V that is maintained is less than E by an amount called the 'lost volts'. This drop $E - V$ is a consequence of internal resistance r in the cell that hinders the cell's working. The lost volts equals $I \times r$, so that

$$E - V = Ir \qquad (20.16)$$

Either of the statements $E = V$ when $I = 0$ or $E - V = Ir$ may be used to define E, but a more satisfactory definition is

$$E = \frac{P}{I} \qquad (20.17)$$

where P is the total power $(I^2 R + I^2 r)$ dissipated in the circuit resistance R and the internal resistance r. This means that

$$E = \frac{I^2 R + I^2 r}{I} = IR + Ir$$

$$\therefore \quad E = V + Ir$$

which agrees with Equation 20.16 and gives $E = V$ when $I = 0$.

For calculations a cell or other voltage source can be regarded as a cell of zero internal resistance with a separate resistance r in series with it (Fig. 20.6a).

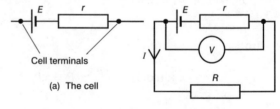

(a) The cell

(b) In the complete circuit

Fig. 20.6   Cell with EMF E and internal resistance r

A cell represented in this way is seen in the circuit of Fig. 20.6b. This circuit agrees with $E - V = Ir$ and $E = P/I$ and it is also seen that

$$I = \frac{E}{R + r} \qquad (20.18)$$

Example 6

A cell of EMF 1.5 V and internal resistance $1.0\,\Omega$ is connected to a $5.0\,\Omega$ resistor to form a complete circuit. Calculate the current expected, the terminal PD and the power dissipated in the external circuit and in the cell.

Method

A suitable diagram is shown in Fig. 20.7.

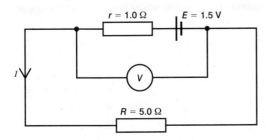

Fig. 20.7   Circuit diagram for Example 6

$$I = \frac{E}{R+r} \text{ with } E = 1.5, R = 5.0 \text{ and } r = 1.0.$$

$$\therefore \quad I = \frac{1.5}{5.0+1.0} = 0.25\,\text{A}$$

Using $E - V = Ir$,

$$V = E - Ir = 1.5 - 0.25 \times 1.0 = 1.25\,\text{V}$$

The power in the $5.0\,\Omega$ is $I^2R = 0.25^2 \times 5.0 = 0.31\,\text{W}$.

Alternatively, this power equals

PD across $R \times$ Current $= VI = 1.25 \times 0.25 = 0.31\,\text{W}$

The power in the $1.0\,\Omega$ internal resistance is

$$I^2r = 0.25^2 \times 1.0$$

$$= 0.0625\,\text{W}$$

Alternatively, this power equals lost volts squared $\times$ internal resistance. Also the total power $I^2R + I^2r$ can be equated to $E \times I$.

Answer

0.25 A, 1.2 V, 0.31 W, 0.062 W.

Cells in series and parallel

When cells are joined in series, each cell adds its EMF to the total EMF if its + terminal connects to the − terminal of the next cell. It subtracts if − joins on to −. The internal resistances add.

For identical cells (same E and r) connected in parallel the EMF of the combination equals E, while the internal resistance of the combination is that of equal resistors in parallel (see Equation 20.7).

Maximum power

When a cell or other voltage source, having internal resistance r, is connected to a 'load' resistance (R in Fig. 20.6b), the current through R is given by $E/(R+r)$, the PD across it is $ER/(R+r)$ and the power dissipated in it is equal to the product of these. The current is at its largest when $R \ll r$; the PD is large when $R \gg r$ and, it can be shown, the power is greatest when $R = r$.

Example 7

With reference to Fig. 20.7:

(a) What value would be needed for resistance R in order that maximum power should be drawn from the cell?

(b) Calculate the maximum power value.

Method

(a) For maximum power dissipation in resistance R this resistance must equal the internal resistance, which is $1.0\,\Omega$.

(b) The total resistance of the circuit will then be $2.0\,\Omega$ and the current will be EMF/2.0 or 1.5/2.0 or 0.75 A.

$$\therefore \text{ power in } R \text{ is } P = I^2R = 0.75^2 \times 1.0 = 0.5625\,\text{W}$$

Answer

(a) $1.0\,\Omega$ (b) 0.56 W

Exercise 20.5

1 A 3.0 V battery having an internal resistance of $2.0\,\Omega$ is connected across a $4.0\,\Omega$ resistor. Calculate the PD between the terminals of the battery.

2 A 3.0 V battery is connected across a parallel combination of two resistors with resistance values of $10\,\Omega$ and $40\,\Omega$. The total current provided by the battery is measured as 0.25 A. Obtain a value for the internal resistance of the battery.

3 A certain large 6.0 V battery is used to produce a current of 60 A. (a) If this current is obtained when the load resistance is $0.08\,\Omega$, what is the internal resistance of the battery? (b) What would the maximum current be that could be drawn from the battery? (c) How much heat would be produced per second in the battery when this maximum current is flowing, if the internal resistance is assumed to remain constant?

Kirchhoff's circuit laws

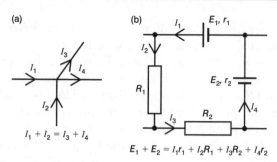

Fig. 20.8   Kirchhoff's laws

$$I_1 + I_2 = I_3 + I_4$$

$$E_1 + E_2 = I_1r_1 + I_2R_1 + I_3R_2 + I_4r_2$$

The two Kirchhoff laws are:

(1) At any point in a circuit where conductors join, the total current towards the point equals the total current flowing away from it (Fig. 20.8a).

(2) For any path that forms a complete loop the total of the EMFs equals the sum of the products of current and resistance, allowing for polarity: i.e. the algebraic sum of the EMFs equals the algebraic sum of the IR products ($\Sigma E = \Sigma IR$, the Σ denoting 'the sum of the values of'). This law is illustrated in Fig. 20.8b.

Example 8

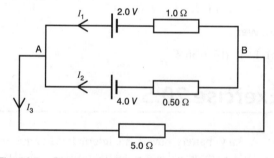

Fig. 20.9   Diagram for Example 8

Calculate the current passing through the 2.0 V cell in Fig. 20.9.

Method

For the currents at A,

$$I_3 = I_1 + I_2 \qquad \text{Equation 1}$$

For the top loop, the sum of the EMFs is $2.0 - 4.0$, so that

$$-2.0 = 1.0I_1 - 0.50I_2 \qquad \text{Equation 2}$$

For the outer loop containing the $1.0\,\Omega$ and $5.0\,\Omega$ resistors the single EMF is 2.0 V, so

$$2.0 = 1.0I_1 + 5.0I_3 \qquad \text{Equation 3}$$

From Equation 2,

$$I_2 = \frac{I_1 + 2.0}{0.5}$$

and from Equation 3,

$$I_3 = \frac{2.0 - I_1}{5.0}$$

Substituting for I_2 and I_3 in Equation 1 gives

$$\frac{2.0 - I_1}{5.0} = I_1 + \frac{I_1 + 2.0}{0.5}$$

$\therefore$ if both sides of the equation are multiplied by 5 we get

$$2.0 - I_1 = 5I_1 + 10I_1 + 20$$

$$\therefore \qquad 16I_1 = -18$$

$$\therefore \qquad I_1 = \frac{-18}{16} = -1.125 \text{ or } -1.1\,\text{A to 2 sig figs}$$

Answer

The current through the 2.0 V cell is 1.1 A in the direction towards the right in the diagram.

Comments

For the top loop the 2.0 V is trying to produce an anticlockwise current. So if the 2.0 V is regarded as positive the anticlockwise currents are positive and this is seen in Equation 2. The 4.0 V cell is opposing the 2.0 V cell.

The lower loop which contains the $5.0\,\Omega$ resistance gives the equation

$$4.0 = 5.0I_3 + 0.5I_2$$

It does not provide any information additional to that provided by the three equations used. Neither was it needed to get the answer. It could have been used in place of one of the equations used. However it is less convenient than Equation 2 or 3 because it contains two unknown currents, neither of which is the current to be calculated.

Exercise 20.6

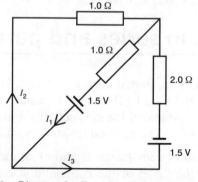

Fig. 20.10   Diagram for Exercise 20.6

Calculate the current flowing in the diagonal path in Fig. 20.10.

The moving-coil meter

The commonest type of meter is the 'moving-coil' design. Its action is explained in detail in Chapter 23. This kind of meter can be very sensitive and is made so that the pointer deflection is proportional to the current.

A galvanometer is a sensitive instrument that is suitable for detecting the presence of a current.

Conversion of a sensitive current-measuring meter to measure large currents

This range multiplication is common practice with sensitive moving-coil meters. A resistance of suitable value is fitted in parallel with the sensitive meter. This resistance is called a 'shunt'.

Only a fraction of the current to be measured passes through the sensitive meter. How the shunt achieves the required conversion is best explained by an example, as follows.

Example 9

Calculate the shunt resistance required to convert a 0–10 mA moving-coil meter whose resistance is $5.0\,\Omega$ into a 0–2.0 A meter.

Method

Fig. 20.11 shows the position of the shunt and illustrates the situation when the current to be measured is at its highest value, namely 2.0 A (2000 mA). The meter must then give full scale deflection, i.e. *10 mA flows through it*.

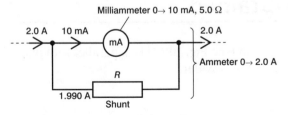

Fig. 20.11   Circuit diagram for Example 9

The current through the shunt resistance R must be 2000 mA minus 10 mA, i.e. 1990 mA.

We know that the meter has resistance $5.0\,\Omega$ and has 10 mA through it. Therefore the PD across it is $5.0 \times 10/1000$ volt, i.e. 50 mV. Because this is also the PD across R and we know the current through R, we can deduce R from $R = V/I$.

$$R\left(=\frac{V}{I}\right) = \frac{50 \times 10^{-3}}{1990 \times 10^{-3}}$$

$$= \frac{50}{1990}$$

$$= 0.025\,125\,\Omega \text{ or } 0.025\,\Omega$$

Answer

$0.025\,\Omega$ or $25 \times 10^{-3}\,\Omega$.

Meter resistance

The resistance of a current-measuring meter should be so small that the current to be measured is not changed when the meter is fitted into the circuit. A shunted milliammeter usually satisfies this requirement. In contrast a voltmeter should have as high a resistance as possible.

Voltmeters

The common type of voltmeter is the moving-coil design.

The moving-coil voltmeter works on the principle that a PD can be measured by allowing it to produce a current, which is measured. A larger PD gives a larger current. For example, the 0–10 mA meter mentioned in Example 9 could be used as a 0–50 mV meter but it would be a very poor voltmeter because its resistance is only $5\,\Omega$. When it is connected to a circuit, perhaps to measure the PD between the ends of a certain resistor, the $5\,\Omega$ would be in parallel with the resistor and would completely change the current through, and therefore the PD across, the resistor before the measurement is made. A good voltmeter should have a resistance that is high compared with the circuitry under test. Satisfactory voltmeters can be obtained by having a high resistance fitted in series with a sensitive moving-coil meter. This is shown in Fig. 20.12.

The series resistor is often called a *multiplier*.

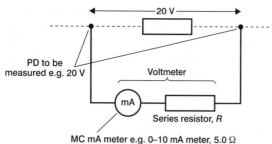

PD to be measured e.g. 20 V

Voltmeter

mA

Series resistor, R

MC mA meter e.g. 0–10 mA meter, 5.0 Ω

Fig. 20.12   Using an MC meter to make a voltmeter

Example 10

Calculate the value required for the series resistor when a 0–10 mA, 5.0 Ω moving-coil meter is converted into a 0–20 V meter.

Method

The diagram required is shown in Fig. 20.12.

We consider the situation where the meter reading is at its maximum, i.e. 20 V is being measured, and the current is then 10 mA. Using the fact that $I = V/R$ we have

$$10 \times 10^{-3} = \frac{20}{R + 5}$$

$$\therefore \qquad R + 5 = \frac{20}{10 \times 10^{-3}} = 2000 \, \Omega$$

$$\therefore \qquad R = 1995 \, \Omega$$

Answer

$2.0 \, \text{k}\Omega$.

Exercise 20.7

1   A microammeter gives full scale deflection (FSD) with $100 \, \mu\text{A}$ and it has a resistance of $100 \, \Omega$.

    (a) What PD is needed across it for FSD?

    (b) If a $1.00 \, \Omega$ shunt were fitted across the meter, what current would flow through the shunt when the meter gives FSD?

2   How exactly could a $10 \, \Omega$, 10 mA FSD moving-coil meter be converted to read (a) 0 to 2.5 A, (b) 0 to 20 V? (c) If this meter were converted into a 0–200 mV meter by fitting a series resistor, what would be the resistance of the voltmeter?

The voltage divider

Fig. 20.13 shows circuits each of which provides a PD V_{AC} which is a fraction $R_1/(R_1 + R_2)$ of the supply voltage V. These circuits are often called

'potential dividers' although 'potential difference dividers' would be better. 'Voltage dividers' avoids the difficulty.

$$V_1 \text{ or } V_{AC} = \frac{R_1}{R_1 + R_2} \times V \text{ supply} \qquad (20.19)$$

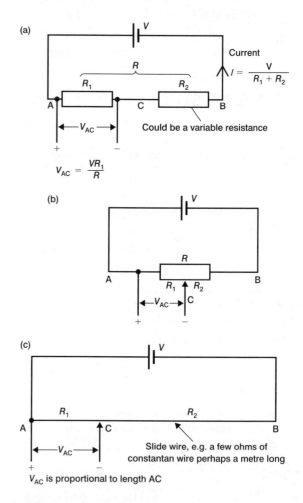

Fig. 20.13   Voltage dividers

Effect of load on PD obtained

The load (see Fig. 20.14) is in parallel with part of the voltage divider.

Example 11

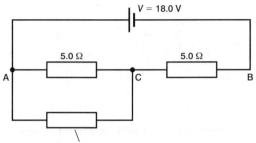

20 Ω load connected to terminals AC of voltage divider

Fig. 20.14   Circuit for Example 11

Calculate the PD between A and C in the circuit of Fig. 20.14.

Method

We have $20\,\Omega$ in parallel with the $5.0\,\Omega$. Using $R = R_1R_2/(R_1 + R_2)$ (Equation 20.7) we get

$$R_{AC} = \frac{20 \times 5.0}{20 + 5.0} = 4.0\,\Omega$$

$$V_{AC} = \frac{V \times R_1}{R_1 + R_2}$$

(see Fig. 20.13) and in this equation $R_1 = R_{AC}$, $V = 18\,V$ and $R_2 = 5.0\,\Omega$.

$$V_{AC} = \frac{18 \times 4.0}{4.0 + 5.0} = 8.0\,V$$

(This compares with 9.0 V when the load is a very high resistance or open circuit, i.e. when $R_{AC} = 5.0\,\Omega$.)

Answer

8.0 V.

Exercise 20.8

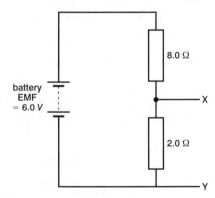

Fig. 20.15

1   With reference to Fig. 20.15 calculate the potential difference between points X and Y (a) if the battery has a negligible internal resistance and (b) if the battery's internal resistance is $2.0\,\Omega$.

2   A 12 V battery of negligible internal resistance is connected to a $5.0\,\Omega$ and a $10\,\Omega$ resistor in series. What is the PD across the $5.0\,\Omega$ resistor (a) when measured by a high-resistance voltmeter, (b) when apparatus with a resistance of $20\,\Omega$ is connected in parallel with the $5.0\,\Omega$ resistor?

Exercise 20.9: Examination questions

1   The current I through a metal wire of cross-sectional area A is given by the formula

$$I = nAve$$

where e is the electronic charge on the electron. Define the symbols n and v.

Two pieces of copper wire, X and Y, are joined end-to-end and connected to a battery by wires which are shown as dotted lines in the diagram. The cross-sectional area of X is double that of Y.

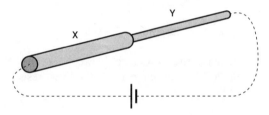

In the table below, n_X and n_Y denote the values of n in X and Y, and similarly for the other quantities. Write in the table the value of each ratio, and alongside it explain your answer.

Ratio	Value	Explanation
$\dfrac{n_Y}{n_X}$		
$\dfrac{I_Y}{I_X}$		
$\dfrac{v_Y}{v_X}$		

[Edexcel 2001]

2   An electric shower is connected to the mains supply by a copper cable 20 m long. The two conductors inside the cable each have a cross-sectional area of $4.0\,mm^2$. The resistivity of copper is $1.7 \times 10^{-8}\,\Omega\,m$. Show that the resistance of each of the conductors is $0.085\,\Omega$.

The operating current of the shower is 37 A. Calculate the total voltage drop caused by the cable supplying the shower.   [Edexcel 2001, part]

3 Fig. 20.16 shows a network of nine identical resistors. Each resistor has resistance $6\,\Omega$. The maximum safe current in a single resistor is $0.3\,\text{A}$.

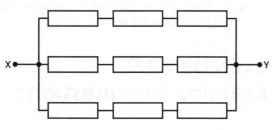

Fig. 20.16

(a) Find the total resistance of the network between the terminals **X** and **Y**.

(b) Find the maximum safe current which can be supplied to the network between **X** and **Y**.
[CCEA]

4

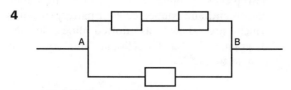

Fig. 20.17   Diagram for Question 4

Calculate the resistance between A and B in Fig. 20.17 given that each of the three resistances is $3.0\,\Omega$.

5 A coil of copper wire is heated slowly in an oil bath. A constant potential difference of $2.0\,\text{V}$ is maintained across the coil. Readings of current and temperature are taken and the graph plotted as shown.

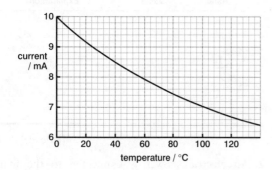

(a) Explain, in terms of the motion of free electrons, why the current decreases as the temperature increases.

(b) (i) Find the **resistance** of the coil at $0\,^{\circ}\text{C}$ and at $100\,^{\circ}\text{C}$.

(ii) Calculate the temperature coefficient of resistance of copper.   [WJEC 2000]

6 Four resistors are connected as shown.

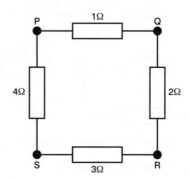

Between which two points is the resistance of the combination a maximum?

A P and Q   **B** Q and S
C R and S   **D** S and P
[OCR 2000]

7 The graph shows how the resistance R of a thermistor depends on temperature θ.

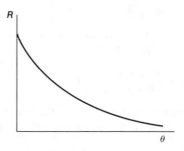

In terms of the behaviour of the material of the thermistor, explain qualitatively the variation shown on the graph.

A student connects the thermistor in series with a $330\,\Omega$ resistor and applies a potential difference of $2.0\,\text{V}$. A high resistance voltmeter connected in parallel with the resistor reads $0.80\,\text{V}$.

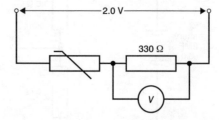

Calculate the resistance of the thermistor

The student now increases the applied p.d. from $2.0\,\text{V}$ to $20\,\text{V}$. She expects the voltmeter reading to increase from $0.80\,\text{V}$ to $8.0\,\text{V}$ but is surprised to find that it is greater. Explain this.

[Edexcel S. H. 2000]

8 The emf of a battery is 3.0 V.

(a) Explain, in energy terms, what this statement means.

(b) When a resistor of 6.0 Ω is connected across the battery the potential difference across its terminals is measured to be 2.4 V.

(i) Draw a diagram of the circuit. Include a second resistor, suitably placed to represent the internal resistance of the battery. Include, also, a voltmeter to measure the potential difference across the battery terminals.

(ii) Calculate the current through the 6.0 Ω resistor.

(iii) Calculate the internal resistance of the battery.

(iv) Write down the potential difference you would have found across the battery terminals before the 6.0 Ω resistor was connected.

(v) When the 6.0 Ω resistor is replaced by one of smaller value the current increases. Explain why the potential difference across the battery terminals now falls. [WJEC 2001]

9 A set of 40 lights for a Christmas tree consists of 5 rows of lights in parallel with 8 lights in each row. Each bulb is rated at 0.5 watt at 1.5 V.

(a) What voltage is required to operate the set with 1.5 V across each bulb?

(b) Calculate the current that will be drawn from the supply.

10

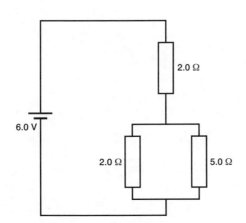

Fig. 20.18 Diagram for Question 10

Calculate the power dissipated in the 2.0 Ω resistance in Fig. 20.18.

11 Fig. 20.19 shows an electrical circuit in which the internal resistance of the battery is negligible.

Complete Fig. 20.20 by giving the electrical quantities for each of the components in the circuit. You are advised to start by completing the column for component A.

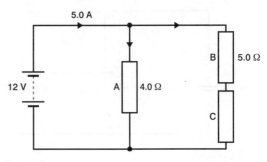

Fig. 20.19

circuit component	A	B	C	whole circuit
potential difference/V				12
current/A				5.0
power/W				
resistance/Ω	4.0	5.0		

Fig. 20.20 [OCR 2000]

12

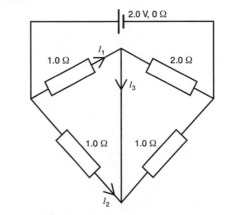

Fig. 20.21 Diagram for Question 12

Calculate the current I_3 in Fig. 20.21.

13 A milliammeter has a resistance of 80 Ω and gives a full-scale deflection for a current of 2.00 mA. Fig. 20.22 is the symbol for the milliammeter.

Fig. 20.22

The milliammeter is to be converted to an ammeter which can be used to measure currents up to a maximum of 1.00 A.

Make any necessary calculations to determine exactly what component(s) is/are required to make the conversion. On Fig. 20.22, show how the component(s) would be connected. Label the terminals by which the ammeter is connected into a circuit in order to measure the current in the circuit. [CCEA 2001, part]

14 A d.c. power supply of e.m.f. 6.0 V and negligible internal resistance is used with a potential divider to generate an output voltage of 4.4 V. The circuit is shown in Fig. 20.23.

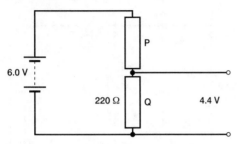

Fig. 20.23

The resistor Q has resistance 220 Ω. The output voltage is obtained across Q.

(i) Calculate the resistance of resistor P.

(ii) A voltmeter of resistance 2000 Ω is now connected across resistor Q. What is the reading on the voltmeter?

[CCEA 2001, part]

15 A technician is asked to construct a potential divider circuit to deliver an output voltage of 1.2 V, using a battery of e.m.f. 3.0 V and negligible internal resistance. To conserve the life of the battery, it is desirable that the current drawn from it should be about 10 μA.

(i) Draw a diagram of a suitable circuit in which the current drawn from the battery is 10 μA. Calculate the values of any resistors used. Show where connections would be made to obtain the 1.2 V output. Label the output terminals **T+** and **T−** to indicate their polarity.

(ii) A resistor of resistance 1.0 kΩ is now connected across the output terminals. Explain why the output voltage and the current drawn from the battery are affected by making this connection. Determine the new values of output voltage and current drawn.   [CCEA 2001, part]

21 Electrostatics

Electric charges

Charges have already been discussed in Chapter 20.

The SI unit for charge is the coulomb (C).

Force between charges

The force F between two small conducting spheres with charges Q_1 and Q_2 is given by

$$F = \frac{Q_1 Q_2}{4\pi\varepsilon r^2} \qquad (21.1)$$

where r is the distance between the centres of the spheres and ε is the permittivity of the medium in which the spheres lie. ε for vacuum is denoted by ε_0 and ε for air is so close to ε_0 that we take it as equal to ε_0. The SI unit for ε is farad per metre $(\mathrm{F\,m^{-1}})$ (see p. 191).

The above formula applies also to the forces between any charged objects provided that their sizes are small compared to the separation r, i.e. they are 'point charges'. The fact that F is proportional to $1/r^2$ is called the inverse square law of electrostatics.

Example 1

Calculate the force between two small metal spheres with charges $+1.0 \times 10^{-9}\,\mathrm{C}$ and $+9.0 \times 10^{-9}\,\mathrm{C}$ whose centres are $30\,\mathrm{cm}$ apart in air, for which the permittivity is $8.9 \times 10^{-12}\,\mathrm{F\,m^{-1}}$. Is the force attractive or repulsive?

Method

The force is

$$F = \frac{Q_1 Q_2}{4\pi\varepsilon r^2} = \frac{1.0 \times 10^{-9} \times 9.0 \times 10^{-9}}{4\pi \times 8.9 \times 10^{-12} \times (0.3)^2}$$

Note the conversion from centimetres to SI units, i.e. metres

$$\therefore \quad F = \frac{9.0 \times 10^{-18}}{4\pi \times 8.9 \times 10^{-12} \times 9 \times 10^{-2}}$$

$$= \frac{1}{4\pi \times 8.9} \times 10^{-4}$$

$$= 8.94 \times 10^{-7}\,\mathrm{N}$$

$$= 0.89\,\mu\mathrm{N}$$

Note too that it may be found helpful to collect together the tens to various powers, as shown in the equation above.

The force is repulsive because both charges are positive.

Answer

$0.89\,\mu\mathrm{N}$. The force is repulsive.

Electric intensity

In the vicinity of any charge Q there is a region within which other charges may be attracted or repelled by it. This region is called the 'field' of the charge Q. We can describe the field strength at any point in an electric field by the value of F/q, where q is the size of a small charge placed at the point concerned and F is the force it experiences due to the presence of Q. This ratio is called the electric intensity E of the field:

$$E = \frac{F}{q} \qquad (21.2)$$

The unit for E could be $\mathrm{N\,C^{-1}}$ but volt per metre (see p. 185) is preferred.

Intensity *E* due to an isolated charged conducting sphere

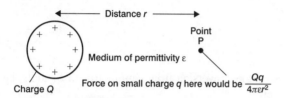

Fig. 21.1   Intensity due to a charged sphere

In Fig. 21.1

$$E = \frac{F}{q} = \frac{Qq}{4\pi\varepsilon r^2 q}$$

$$\therefore \quad E = \frac{Q}{4\pi\varepsilon r^2} \qquad (21.3)$$

The same formula applies if *Q* is a point charge.

Electric lines of force

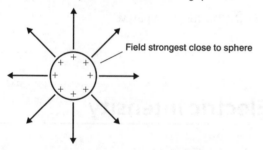

(a) Due to a positive isolated conducting sphere

(b) Between parallel positive and negative plates

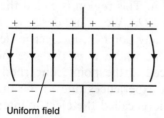

Uniform field

Fig. 21.2   Electric lines of force

Intensity has direction. The direction is that of the force experienced by a small positive charge. Lines of force are lines which show the directions of *E* in an electric field. Two examples are shown in Fig. 21.2.

Example 2

Point charges are located in air at points A and B as shown in Fig. 21.3. Calculate the magnitude of the intensity at P and the direction of the intensity. (Take $1/4\pi\varepsilon_0$ as $9.0 \times 10^9 \, \text{m F}^{-1}$.

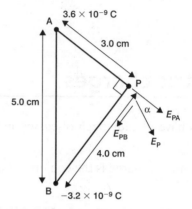

Fig. 21.3   Diagram for Example 2

Method

The intensity E_{PA} at P due to the charge at A is given by

$$E_{\text{PA}} = \frac{Q}{4\pi\varepsilon_0 r^2} = \frac{9.0 \times 10^9 \times 3.6 \times 10^{-9}}{0.03^2}$$

This gives

$$E_{\text{PA}} = 36\,000 \, \text{V m}^{-1}$$

The intensity E_{PB} at P due to the charge at B works out by the same method to be $18\,000 \, \text{V m}^{-1}$.

The directions of E_{PA} and E_{PB} are shown by the arrows in the diagram and the combined effect (intensity E_p) at P is found by vector addition (parallelogram rule, see page 20). Since E_{PA} and E_{PB} are perpendicular this addition can be done by use of Pythagoras' equation.

$$E_\text{p} = 36\,000^2 + 18\,000^2 = 1620 \times 10^6$$

whence

$$E_\text{p} = 40.2 \times 10^3 \text{V m}^{-1}$$

To find the direction of *E* we have $\tan \alpha = \dfrac{E_{\text{PB}}}{E_{\text{PA}}} = 0.5$.

This gives $\alpha = 26.6°$.

Answer

$40 \, \text{kV m}^{-1}$, 27° to direction AP, 63° to PB.

A relationship between intensity and potential

Consider first a small charge $+q$ being moved from close to the negative plate in Fig. 21.2b up to the positive plate through distance d. Let the PD between the plates be V and the intensity E. The work done is $W = Fd$ (see p. 45) and equals Fqd. Also, by definition of PD (see p. 170) $W = Vq$ (Equation 20.3). Hence $Eqd = Vq$ or $E = V/d$.

> **Intensity** $E = \dfrac{V}{d}$ **(21.4)**

This is an important result. It also justifies our measuring E in volt per metre. Example 3 illustrates the use of this formula.

Work done when a charge moves

The work done (W) when a charge moves in any electric field can be deduced from Equation 20.3 in Chapter 20, which shows that $W = qV$, where q is the charge moved in coulombs. This is a very useful equation when you have an electron accelerated in the electric field between parallel plates, starting at rest at the negative plate. The same result would be obtained by using the formula

$$W = F \times d \text{ (Equation 6.1 in chapter 6)}$$

with $\quad E = \dfrac{F}{q}$ (Equation 21.2)

and $\quad E = \dfrac{V}{d}$ (Equation 21.4).

Using e for the charge of an electron the work done on the electron, and therefore the kinetic energy it gains, is:

> **Work done** $= eV$ **(21.5)**

Example 3

A uniform electric field is obtained between two parallel plates by using a PD of 10 V and a plate separation of 20 mm.

An electron initially at rest close to the negative plate is moved by the field to the positive plate. Calculate:

(a) the intensity of the field
(b) the force acting on the electron
(c) the speed of the electron as it arrives at the positive plate.

(Electron charge $e = 1.6 \times 10^{-19}$ C, electron mass $= 9.11 \times 10^{-31}$ kg.)

Method

(a) $E = \dfrac{V}{d} = \dfrac{10}{20 \times 10^{-3}} = 5.0 \times 10^2 \, \text{V m}^{-1}$

(b) Force $F = Eq = Ee = 5.0 \times 10^2 \times 1.6 \times 10^{-19}$
$\qquad = 8.0 \times 10^{-17} \, \text{N}$

(c) Work done is $W = eV$ (Equation 21.5)
$\qquad = 1.6 \times 10^{-19} \times 10 = 1.6 \times 10^{-18} \, \text{J}$, and this equals
the kinetic energy $\frac{1}{2}mv^2$ (see Chapter 6).

Therefore

$$v^2 = \frac{2W}{m} = \frac{2 \times 1.6 \times 10^{-18}}{9.11 \times 10^{-31}} = 3.513 \times 10^{12}$$

and $v = 1.87 \times 10^6 \, \text{m s}^{-1}$.
An alternative way of calculating v is to get the acceleration a from the equation $F = ma$ (Equation 5.5 in Chapter 5) and hence v from $v^2 = u^2 + 2as$ (Equation 5.3).

Answers

(a) $5.0 \times 10^2 \, \text{V m}^{-1}$ (b) $8.0 \times 10^{-17} \, \text{N}$ (c) $1.9 \times 10^6 \, \text{m s}^{-1}$.

Potential at a distance R from a charged sphere or point charge

It is common, in GCE work particularly, to take as zero for potential measurements the potential at a large distance away from any charge, i.e. at infinity.

The potential difference between infinity (e.g. at far right of Fig. 21.1) and position P can be shown to equal $Q/4\pi\varepsilon R$, i.e.

> **Potential at P is** $V = \dfrac{Q}{4\pi\varepsilon r}$ **(21.6)**

Note the r (not r^2).

Example 4

Point charges of -2.0×10^{-10} C and -3.0×10^{-10} C are located in air at A and B which are 4.0 cm apart. Calculate the electric intensity and potential midway between A and B. ($1/4\pi\varepsilon_0$ may be taken as $9.0 \times 10^9 \, \text{m F}^{-1}$.

Method

We are interested in a point which is 2.0 cm or 2.0×10^{-2} m from A and from B.

The intensity there caused by the -2.0×10^{-10} C is given by $E = Q/4\pi\varepsilon_0 r^2$ and so equals

$$\frac{2.0 \times 10^{-10} \times 9.0 \times 10^9}{(2.0 \times 10^{-2})^2} \text{ or } 4.5 \times 10^3 \text{ V m}^{-1}$$

The direction of this intensity, because of the negative charge at A, is from B to A.

The intensity at the midpoint due to the -3.0×10^{10} C charge at B is similarly

$$\frac{3.0 \times 10^{-10} \times 9.0 \times 10^9}{(2.0 \times 10^{-2})^2} \text{ or } 6.75 \times 10^3 \text{ V m}^{-1}$$

The direction of this intensity, because the charge at B is negative, is towards B.

The total intensity at the midpoint caused by the two charges is obtained by adding the two intensities vectorially, i.e. with consideration of their directions.

Total intensity is, in the direction towards B, $6.75 \times 10^3 - 4.5 \times 10^3$ V m^{-1} or 2.25×10^3 V m^{-1}. To two significant figures we have 2.2 kV m^{-1}.

The potential at the midpoint due to the charge -2.0×10^{-10} C at A is given by $V = Q/4\pi\varepsilon_0 r$ and so equals

$$\frac{-2.0 \times 10^{-10} \times 9.0 \times 10^9}{2.0 \times 10^{-2}} \text{ or } -90 \text{ V}$$

The potential due to the -3.0×10^{-10} C at B is

$$\frac{-3.0 \times 10^{-10} \times 9.0 \times 10^9}{2.0 \times 10^{-2}} \text{ or } -135 \text{ V}$$

Since potential is a scalar quantity (no direction) we add the two contributions to the potential algebraically to get

$$-90 + -135 \text{ or } -225 \text{ V}$$

To two significant figures this is -0.22 kV.

Answer

2.2 kV m^{-1}, -0.22 kV.

Exercise 21.1

(Take ε for air to be 8.9×10^{-12} F m^{-1} unless otherwise stated.)

1 Calculate (a) the force between two charges of $+1.4$ nC and $+1.6$ nC on point conductors 40 cm apart in air. (b) What size of charge on a third point conductor placed midway between the first

two conductors would result in doubling of the magnitude of the force on the 1.4 nC charge?

2 (a) Calculate (i) the electric intensity and (ii) the potential at a point midway between two point charges of $+10^{-9}$ C and -10^{-9} C which are 20 cm apart in air.

   (b) To produce an equally large electric intensity midway between two large-area, parallel plates 2.0 cm apart in air, what PD would be needed between the plates?

3 When a charge of 50 μC is moved between two points P and Q in a uniform electric field, 100 μJ of work is done. What is the potential difference between P and Q?

4 Calculate the potential at the surface of an isolated metal sphere carrying a negative charge of 2.0×10^{-8} C and surrounded by air, if the sphere's radius is 2.0 cm.

   How much work would be done in moving a positive charge of 1.6×10^{-6} C from the sphere's surface to a point 3.0 cm further from the centre?

5 Calculate the electric potential and electric field strength (or intensity) at C in Fig. 21.4.

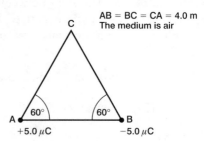

AB = BC = CA = 4.0 m
The medium is air

A +5.0 μC   B −5.0 μC

Fig. 21.4   Diagram for Question 5

6 An electron is initially at rest in a uniform electric field of intensity 0.50×10^3 V m^{-1}.

   This field causes the electron to have an acceleration a and to reach a speed v after it has travelled a distance of 50 mm. Obtain values for a and v.

   (Electron charge $e = 1.6 \times 10^{-19}$ C, electron mass $= 9.11 \times 10^{-31}$ kg.)

Exercise 21.2: Examination questions

Where necessary use the following values:

Permittivity of free space $(\varepsilon_0) = 8.85 \times 10^{-12}$ F m^{-1}
Electronic charge $(e) = 1.60 \times 10^{-19}$ C
Electron mass $(m) = 9.11 \times 10^{-31}$ kg
$\pi = 3.142$ (value also obtainable from calculator).

1 This question is about the deflection of an electron beam near a charged sphere in a vacuum.

(a) The voltage between the anode and cathode of an electron gun is 2500 V. Show that the electrons are emitted from the gun at about $3 \times 10^7 \mathrm{m\,s}^{-1}$.

electronic charge, $e = 1.6 \times 10^{-19}\,\mathrm{C}$
mass of electron, $m_e = 9.1 \times 10^{-31}\,\mathrm{kg}$

(b) A charged sphere is moved towards the electron gun along a line perpendicular to the direction in which electrons leave the gun (Fig. 21.5). When the centre of the sphere is about 0.34 m from the gun, the path of the beam is an arc of a circle.

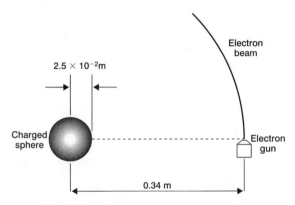

Fig. 21.5

(i) State whether the sphere is positively or negatively charged. Explain your reasoning.
(ii) Explain why the speed of each electron remains constant while it is following a circular path.
(iii) Show that the centripetal force on each electron is about $2.4 \times 10^{-15}\,\mathrm{N}$.
(iv) Hence calculate the strength of the electric field 0.34 m from the centre of the sphere.
(v) Hence calculate the charge on the sphere.
$$\varepsilon_0 = 8.9 \times 10^{-12}\,\mathrm{F\,m}^{-1}.$$
[OCR 2001]

2 A beam of electrons is directed at a target. They are accelerated from rest through 12 cm in a uniform electric field of strength $7.5 \times 10^5\,\mathrm{N\,C}^{-1}$.

Calculate the potential difference through which the electrons are accelerated.

Calculate the maximum kinetic energy in joules of one of these electrons.

Calculate the maximum speed of one of these electrons.   [Edexcel 2001, part]

3 (a) An electric field may be produced in the region between two charged parallel plates. Fig. 21.6 shows two such plates.

Fig. 21.6

On Fig. 21.6 sketch the pattern of field lines between the plates.

(b) An isolated point charge of magnitude Q is situated in a vacuum. At a distance of $1.0 \times 10^{-10}\,\mathrm{m}$ from this charge, the electric potential is $+14.3\,\mathrm{J\,C}^{-1}$.

(i) Explain what is meant by **electrical potential** at a point in an electric field.
(ii) Electric potential may be a positive or a negative quantity. Explain the significance of the **positive** value of potential in this case.
(iii) Calculate the magnitude of Q.
(iv) Complete Table 21.1, showing the electric potential V at various distances r from the isolated point charge of magnitude Q.

$r/10^{-10}\mathrm{m}$	$V/\mathrm{J\,C}^{-1}$
1.0	+14.4
2.0	
3.0	

(v) An electron of charge $-1.6 \times 10^{-19}\,\mathrm{C}$ is moved from a distance of $3.0 \times 10^{-10}\,\mathrm{m}$ to a distance of $1.0 \times 10^{-10}\,\mathrm{m}$ from the isolated point charge of magnitude Q. Making use of your answer to (iv), determine the work done in moving the electron.

Is this work performed against the field of the point charge of magnitude Q, or does the field do the work? Give a reason for your answer.   [CCEA 2000]

4 A simple model of a hydrogen atom consists of an electron moving at constant speed in a circular path around a central nucleus (proton).

(a) Write down an expression for the electrostatic force on the electron in its orbit.

(b) If the speed of the electron is $1.1 \times 10^6\,\mathrm{m\,s}^{-1}$, calculate the radius of the electron's orbit.
[WJEC 2000]

5 (a) Point charges of $+2.0\,\mu C$ and $+4.0\,\mu C$ are fixed at the points **W** and **X** respectively, as shown in Fig. 21.7. The distance between the charges is 2.0 mm.

Fig. 21.7

(i) Explain why the $+4.0\,\mu C$ charge experiences a force.

(ii) Find the magnitude and direction of the force on the $+4.0\,\mu C$ charge.

(iii) What is the force on the $+2.0\,\mu C$ charge?

(b) (i) Define **electric potential** at a point in an electric field. State the relationship between the electric potential energy W of a charge Q placed at a point in a field and the electric potential V at that point.

(ii) For the arrangement of charges shown in Fig. 21.7, calculate the electric potential energy possessed by the $+4.0\,\mu C$ charge.

(iii) With the $+2.0\,\mu C$ charge still fixed at **W**, the $+4.0\,\mu C$ charge is now moved to point **Y**, 3.0 mm to the right of **X**, and is fixed there (Fig. 21.8).

Fig. 21.8

Calculate the change in electric potential energy of the $+4.0\,\mu C$ charge.

(iv) With the $+2.0\,\mu C$ charge still fixed at **W**, the $+4.0\,\mu C$ charge is now moved along the arc of a circle of radius 5.0 mm from **Y** to point **Z**, as shown in Fig. 21.9. **WZ** is at right angles to **WY**.

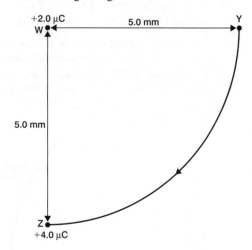

Fig. 21.9

Find the work done in this operation. Explain your answer. [CCEA 2001, part]

22 Capacitors

Capacitance

(a) Circuit symbol for capacitor

(b) Capacitor charging

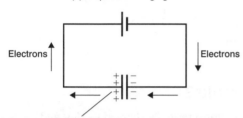

Electrons ↑   ↓ Electrons

Capacitor will finally be charged with charges +Q and −Q on the conductors and PD (V) will equal supply PD

(c) Capacitor discharging

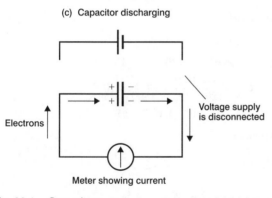

Electrons ↑

Voltage supply is disconnected

Meter showing current

Fig. 22.1   Capacitors

A capacitor consists of two conducting surfaces close together, two metal sheets for example, as in Fig. 22.1. The surfaces are often described as 'plates'. When the capacitor is charged as in Fig. 22.1b it has equal + and − charges on its plates. The electrostatic attraction between the opposite charges makes it easier to build up large charges on the plates so that charges of useful sizes are stored in the capacitor. This 'ability to store charges' is called the 'capacitance' C of the capacitor and is defined as (or measured by) the ratio of charge Q stored (on each plate) to the potential difference across it.

$$C = \frac{Q}{V} \qquad (22.1)$$

The S.I unit for capacitance is the farad (F).

Capacitances are mostly met in microfarad (μF) and smaller sizes.

A capacitor continues to charge until the p.d. between its plates equals the applied p.d., for example, of a battery.

Energy stored in a charged capacitor

This equals $\frac{1}{2}QV$ because, during charging, Q coulombs of electrons have in effect been taken from one conductor of the capacitor to the other through a PD which was initially zero, is finally V, and has an average value of $\frac{1}{2}V$. Note that because $Q = CV$ we can also write $\frac{1}{2}QV$ as $\frac{1}{2}Q^2/C$.

$$\text{Energy stored} = \tfrac{1}{2}QV \qquad (22.2)$$

Example 1

A capacitor is charged by a 20 V DC supply and when it is discharged through a charge meter it is found to have carried a charge of $5.0\,\mu$C. What is its capacitance, and how much energy was stored in it?

Method

$$C = \frac{Q}{V}$$

$$\therefore \quad C = \frac{5.0 \times 10^{-6}}{20} = 0.25 \times 10^{-6}\,\text{F}$$

$$= 0.25\,\mu\text{F}$$

Energy stored $= \frac{1}{2}QV$

$$= \tfrac{1}{2} \times 5.0 \times 10^{-6} \times 20$$
$$= 5.0 \times 10^{-5}\,\text{J}$$

Or, using the formula $\tfrac{1}{2}CV^2$, the energy is

$$\tfrac{1}{2} \times 0.25 \times 10^{-6} \times 20^2$$

This equals 50×10^{-6} or $5.0 \times 10^{-5}\,\text{J}$.

Answer

$0.25\,\mu\text{F}$, $5.0 \times 10^{-5}\,\text{J}$

Experimental measurements of capacitance

Electric charge meters are now available so that it is convenient to charge a capacitor using a known PD V and then discharge it through the meter to measure the charge Q. Then C can be calculated from $C = Q/V$.

The repeated discharge method is illustrated in Fig. 22.2.

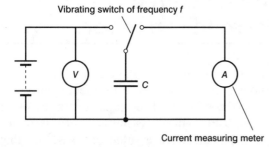

Vibrating switch of frequency f

Current measuring meter

Fig. 22.2 Measurement of capacitance by the repeated discharge method

In this method a switch, usually a reed switch operated by an alternating current of frequency f, causes the capacitor to be charged to a PD V, and then the capacitor is connected to the current meter through which it discharges its charge Q. This cycle is repeated f times per second so that the charge per second (the current I) through the meter is fQ or fCV.

$$C = \frac{I}{fV} \qquad (22.3)$$

Example 2

A capacitor of capacitance C_1 is connected to a 2 V supply and is discharged through a charge meter. A deflection of 10 divisions is observed. A second capacitor of capacitance C_2 is connected to give capacitance $C_1 + C_2$ and the experiment is repeated. This time the deflection is 15 divisions. Calculate the ratio C_1/C_2.

Method

The first charge measured Q_1 is 10 units compared with 15 units for the second charge Q_2, the actual size of the unit being unimportant.

$$\frac{Q_1}{Q_2} = \frac{10}{15} = \frac{2}{3}$$

But $Q_1 = C_1 V$ and $Q_2 = (C_1 + C_2)V$

$$\therefore \quad \frac{Q_1}{Q_2} = \frac{C_1 V}{C_2 V + C_1 V} = \frac{C_1}{C_1 + C_2}$$

$$\therefore \quad \frac{C_1}{C_1 + C_2} = \frac{2}{3}$$

$$\therefore \quad 3C_1 = 2C_1 + 2C_2$$

$$\therefore \quad C_1 = 2C_2$$

$$\frac{C_1}{C_2} = 2$$

Answer

2.

Example 3

A capacitor repeatedly charged to 15 V and discharged through a milliammeter by use of a reed switch working at 120 cycles per second causes a meter reading of 3.6 mA. Calculate the capacitance of the capacitor.

Method

$$I = fCV$$

$$\therefore \quad 3.6 \times 10^{-3} = 120 \times C \times 15$$

$$\therefore \quad C = \frac{3.6 \times 10^{-3}}{120 \times 15} = \frac{3.6}{1800} \times 10^{-3}$$

$$= 2.0 \times 10^{-6}\,\text{F or } 2.0\,u\text{F}$$

Answer

$2.0\,\mu\text{F}$.

Formula for the capacitance of a parallel-plate capacitor

When the two conductors of a capacitor are parallel as in a 'parallel-plate' capacitor or a waxed-paper capacitor the capacitance C is given by the formula

$$C = \frac{\epsilon A}{d} \qquad (22.4)$$

ε is the permittivity of the medium, called the dielectric, separating the conductors ('plates') of the capacitor, A is the area of the plates and d is their separation, i.e. the dielectric thickness.

The SI unit for ε

The SI unit for ε is farad per metre ($\mathrm{F\,m^{-1}}$). This is seen to be appropriate from the formula $\varepsilon = Cd/A$ (Equation 22.4).

Dielectric constant

ε denotes the permittivity of a medium, ε_0 is the permittivity of air or vacuum. The relative permittivity of a medium is the ratio of its permittivity to ε_0, i.e. $\varepsilon/\varepsilon_0$. When the medium is being used as a dielectric, its relative permittivity is often described as the dielectric constant.

Example 4

A $0.10\,\mu\mathrm{F}$ capacitor is to be constructed with metal foil and waxed paper (dielectric constant 2.0). The width of the foil is to be 4.0 cm, and the length no more than 5.0 m. What is the maximum thickness of the waxed paper? ($\varepsilon_0 = 8.9 \times 10^{-12}\,\mathrm{F\,m^{-1}}$.)

Method

$$C = \frac{\varepsilon A}{d}$$

$$\therefore \quad 0.1 \times 10^{-6} = \frac{2.0 \times 8.9 \times 10^{-12} \times 0.04 \times 5.0}{d}$$

$$\therefore \quad d = 2.0 \times 8.9 \times 0.04 \times 5.0 \times 10^{-5}$$

which gives 3.6×10^{-5} m or 0.036 mm.

Answer

0.036 mm.

Exercise 22.1

1 Calculate the capacitance of a pair of semicircular brass discs, of 3.5 cm radius, with their planes parallel and 0.50 mm apart in air (a) with the maximum overlap of the plates, (b) when the area of overlap is halved. ($\varepsilon_{\mathrm{air}} = 8.9 \times 10^{-12}\,\mathrm{F\,m^{-1}}$.)

2 A $2.0\,\mu\mathrm{F}$ capacitor is charged using 0.40 V from a cell and voltage divider. It is then allowed to discharge through a meter. The meter gives a charge deflection of 6.4 cm. What is the sensitivity of the meter in $\mathrm{mm\,C^{-1}}$?

A second capacitor charged to the same PD gives a deflection of 9.5 cm. What is its capacitance?

A third capacitor charged to 40 V gives a deflection of 3.3 cm. What is its capacitance?

3 A simple capacitor is constructed from a pair of metal plates separated by insulating spacers so as to leave a 1.5 mm air space between them. The plates, each 20 cm square, are placed to achieve maximum capacitance. The capacitance is then measured by repeatedly charging it to 100 V and discharging it through a calibrated microammeter 200 times per second. The current recorded is $4.8\,\mu\mathrm{A}$. Obtain a value for the capacitance of this capacitor and for the permittivity of air.

4 A $5.0\,\mu\mathrm{F}$ capacitor is alternately connected to a 20 V DC supply and then to a milliammeter by a vibrating switch working at a frequency of 80 Hz. What reading is expected on the meter?

Capacitors in series and parallel

Two capacitors C_1 and C_2 can be joined together either in series or parallel, as shown in Figs 22.3a and b, where the capacitance, charge and PD are labelled for each capacitor and for the combined capacitance C.

(a) In parallel

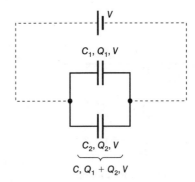

(b) In series

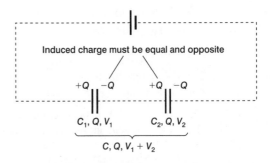

Fig. 22.3   Combinations of capacitors

191

For C_1 and C_2 in parallel

$$C = C_1 + C_2 \qquad (22.5)$$

$$\left(\text{because } C = \frac{Q_1 + Q_2}{V} = C_1 + C_2 \right).$$

For C_1 and C_2 in series

$$\frac{1}{C} = \frac{1}{C_1} + \frac{1}{C_2} \quad \text{or} \quad C = \frac{C_1 C_2}{C_1 + C_2} \qquad (22.6)$$

$$\left(\text{because } \frac{1}{C} = \frac{V_1 + V_2}{Q} = \frac{V_1}{Q} + \frac{V_2}{Q} = \frac{1}{C_1} + \frac{1}{C_2} \right).$$

Example 5

A $2.0\,\mu\text{F}$ capacitor is charged to $12\,\text{V}$. The voltage supply is removed and then a $4.0\,\mu\text{F}$ capacitor is fitted in parallel with the $2.0\,\mu\text{F}$ one. Calculate the charge stored in the $2.0\,\mu\text{F}$ capacitor (a) initially, (b) finally.

Method

(a)

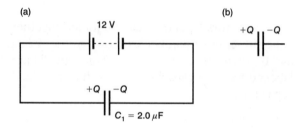

(c)

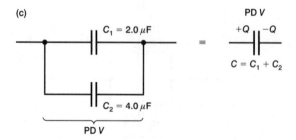

Fig. 22.4   Diagrams for Example 5

(a) In Fig. 22.4a Q is given by

$$Q = CV = 2.0 \times 10^{-6} \times 12$$
$$= 24 \times 10^{-6}\,\text{C}$$

In Fig. 22.4b the battery has been removed and we have $+Q$ on the left and $-Q$ on the right.

(b) In Fig. 22.4c the $4.0\,\mu\text{F}$ has been connected in parallel. The total capacitance is, from Equation 22.5,

$$C = C_1 + C_2 = 2.0 + 4.0 = 6.0\,\mu\text{F}$$

Now we must realise that the charge on the final $6.0\,\mu\text{F}$ combined capacitor is still $+Q$ on the left and $-Q$ on the right, i.e. $24\,\mu\text{C}$. The charge on the left in Fig. 22.4b is now shared by C_1 and C_2, but it cannot escape from the left or be added to. Thus the PD V across the combined capacitor is

$$V = \frac{Q}{C} = \frac{24 \times 10^{-6}}{6.0 \times 10^{-6}} = 4.0\,\text{V}$$

This is the PD across C_1 and across C_2 in Fig. 22.4c, so that the new charge on C_1 is given by

$$\text{Charge} = C_1 \times \text{PD}$$

i.e. $2.0 \times 10^{-6} \times 4.0$ or $8.0 \times 10^{-6}\,\text{C}$.

(The $8.0\,\mu\text{C}$ on the $2.0\,\mu\text{F}$, and similarly the $16\,\mu\text{C}$ on the $4.0\,\mu\text{F}$ illustrate that, in a parallel combination, the charge is shared in proportion to the capacitances.)

Answer

(a) $24\,\mu\text{C}$, (b) $8.0\,\mu\text{C}$.

Example 6

(a) Calculate the charge stored in a $3.0\,\mu\text{F}$ capacitor and a $6.0\,\mu\text{F}$ capacitor joined in series and then connected across the terminals of an $18\,\text{V}$ battery.

(b) What is the PD across each of these capacitors?

Method

(a) A diagram should be sketched (see Fig. 22.3b).

The combined capacitance C is given by Equation 22.6:

$$\frac{1}{C} = \frac{1}{C_1} + \frac{1}{C_2}$$

$$\therefore \quad \text{in } \mu\text{F}, \frac{1}{C} = \frac{1}{3.0} + \frac{1}{6.0} = \frac{1}{2}$$

So that $C = 2.0\,\mu\text{F}$.

Therefore the charge stored Q is

$$Q = CV = 2.0 \times 10^{-6} \times 18$$
$$= 36 \times 10^{-6}\,\text{C}$$

and, for capacitors in series, this is the same for both capacitors.

(b) The PD across the $3.0\,\mu\text{F}$ is given by charge divided by capacitance and equals $36 \times 10^{-6}/3.0 \times 10^{-6}$ or $12\,\text{V}$. For the $6.0\,\mu\text{F}$, we have PD $= 36 \times 10^{-6}/6.0 \times 10^{-6}$ or $6.0\,\text{V}$. (We note that the total PD, $18\,\text{V}$ here, is shared by capacitors in series in *inverse proportion* to the capacitances.)

Answer

(a) $36\,\mu\text{C}$, (b) $12.0\,\text{V}$ and $6.0\,\text{V}$.

Example 7

(a) A $5.0\,\mu F$ capacitor is charged to $4.0\,V$ and is removed from the voltage supply. How much energy is stored?

(b) If the $5.0\,\mu F$ capacitor is connected in parallel with a $3.0\,\mu F$ capacitor, what is the new energy stored in the capacitor combination, and how much energy was converted to heat by the movement of charge through the wires between the two capacitors?

Method

(a) From Equation 22.2

$$\text{Energy} = \tfrac{1}{2}C_1V^2$$

$$= \tfrac{1}{2} \times 5.0 \times 10^{-6} \times 4.0^2 = 40 \times 10^{-6}\,J.$$

(b) The new capacitance $(C = C_1 + C_2)$ is $5.0 + 3.0\,\mu F$ or $8.0\,\mu F$. We do not know the new PD, but we know that the charge is the same as in (a). This charge is given by $Q = C_1V$, so that it equals $5.0 \times 10^{-6} \times 4.0\,C$ or $20\mu C$.

The energy now in the $8\,\mu F$ is given by $\tfrac{1}{2}Q^2/C$ as

$$\tfrac{1}{2} \times \frac{(20 \times 10^{-6})^2}{8 \times 10^{-6}} = \frac{200}{8} \times 10^{-6}\,J \text{ or } 25\,\mu J.$$

The electrical potential energy has fallen from $40\,\mu J$ to $25\,\mu J$, i.e. $15\,\mu J$ has become heat in the connecting wires.

Answer

(a) $40\,\mu J$, (b) $25\,\mu J$, $15\,\mu J$.

Time constant

When a charged capacitor C is connected into a circuit of resistance R, as in Fig. 22.5, the current $I = V/R$, or since $C = Q/V$, $I = Q/CR$. This means that the rate of reduction of the charge is proportional to the charge Q. Hence the discharge is exponential and $Q = Q_0 e^{-t/RC}$ (see Chapter 2). Using $Q = CV$ this last equation becomes $V = V_0 e^{-t/RC}$.

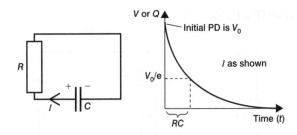

Fig. 22.5   Discharge of capacitor ($V = V_0 e^{-t/RC}$)

The 'time constant' for the discharge is the time for Q or V to fall to $1/e$ of the initial value and is given by

$$e^{-t/RC} = 1/e$$

or $\quad -\dfrac{t}{RC} = \ln\dfrac{1}{e}$

or $\quad -t = -1 \times RC$

so $\quad t = RC.$

| Time constant $= RC$ | (22.7) |

The time required for Q or V to fall to half the initial value is the 'half-life' time and is given by

$$e^{-t/RC} = \tfrac{1}{2}$$

or $\quad -t/RC = \ln(1/2)$

or $\quad -t = -RC \times \ln 2$

or $\quad t = 0.693RC$

| Half-life of capacitor discharge $= RC \times \ln 2$ | (22.8) |

The time constant and half-life values are not affected by the initial Q or V value and so apply starting at any stage of the discharge. The time constant RC also affects the time taken for a capacitor to charge. This is seen in Fig. 22.6.

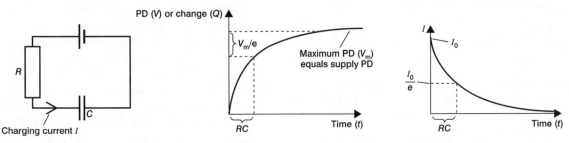

Fig. 22.6   Time taken for charging a capacitor

193

Example 8

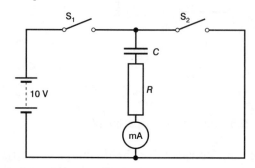

Fig. 22.7   Circuit diagram for Example 8

In the circuit in Fig. 22.7 the resistance R is $10\,\text{k}\Omega$, C is a capacitance of $1000\,\mu\text{F}$ and the resistances of the battery and milliammeter are negligible.

Calculate the milliammeter reading expected (a) immediately after the switch S_1 is closed, (b) $10\,\text{s}$ later, (c) after several minutes

If after this time the switch S_1 is opened and the switch S_2 is closed instead, what is the expected milliammeter reading (d) immediately after S_2 is closed, (e) $10\,\text{s}$ later, (f) after several minutes?

Method

(a) At the first instant of charging, the PD across C is zero, so that the current I is decided only by the supply PD and, of course, the circuit resistance R.

$$I = \frac{10}{10 \times 10^3}$$
$$= 10^{-3}\,\text{A or } 1\,\text{mA}$$

(b) The time constant $= RC$
$$= (10 \times 10^3) \times (1000 \times 10^{-6})$$
$$= 10\,\text{s}$$

The time elapsed is exactly equal to the time constant so that the capacitor PD V is $V_m - V_m/e$, i.e. $10 - 10/2.718$, which equals $10 - 3.7$ or $6.3\,\text{V}$. Consequently $I = (10 - 6.3)/R = 3.7/(10 \times 10^3)$, which is $0.37\,\text{mA}$.

(c) After several minutes, i.e. many times RC, I will be effectively zero because battery EMF and capacitor PD will then be equal and opposite.

(d) During the first instant of discharge the capacitor PD is $10\,\text{V}$, and this causes the current to be $10/R$, i.e. $10/(10 \times 10^3)$ or $10^{-3}\,\text{A}$.

(e) After a time equal to CR the capacitor PD will have fallen to V_0/e, i.e. to $10/2.718$ or $3.7\,\text{V}$, which gives a discharge current of $3.7/(10 \times 10^3)$, i.e. $0.37\,\text{mA}$.

(f) For $t \gg CR$, the discharge current is effectively zero because the capacitor PD is then zero.

Answer

(a) $1.0\,\text{mA}$, (b) $0.37\,\text{mA}$, (c) zero, (d) $1.0\,\text{mA}$, (e) $0.37\,\text{mA}$, (f) zero.

Exercise 22.2

1

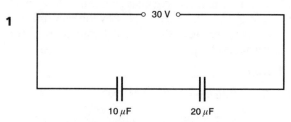

Fig. 22.8   Circuit diagram for Question 1

In the circuit shown in Fig. 22.8, $10\,\mu\text{F}$ and $20\,\mu\text{F}$ capacitors are connected in series with a $30\,\text{V}$ DC supply. What is the charge on each capacitor?

A $0.30\,\text{mC}$ **B** $0.20\,\text{mC}$ **C** $1.0\,\mu\text{C}$
D $0.90\,\text{mC}$ **E** $4.5\,\mu\text{C}$

2 A $2.0\,\mu\text{F}$ capacitor is charged by connecting it across the terminals of a cell whose EMF is $1.5\,\text{V}$.

(a) What is the charge Q stored in this capacitor, the energy E stored in it and the PD V across it?

(b) If the cell remains connected, and a second $2.0\,\mu\text{F}$ capacitor is connected in parallel with the first one, what are Q, E and V for the second capacitor?

(c) The cell is removed without discharging the capacitors, and a third $2.0\,\mu\text{F}$ capacitor is fitted in parallel with the other two. What are Q, E and V for this third capacitor?

3 A capacitor A of capacitance $4.0\,\mu\text{F}$ is charged to a potential difference of $20\,\text{V}$. An uncharged capacitor B of capacitance $2.0\,\mu\text{F}$ is then connected in parallel with A. What is (a) the energy initially stored in A, (b) the potential difference across A after B has been connected, (c) the energy finally stored in A and B?

4 A simple parallel-plate capacitor with a $2\,\text{mm}$-thick air dielectric has a capacitance of $5 \times 10^{-10}\,\text{F}$. A uniform sheet of material whose dielectric constant is 2 and thickness is $1\,\text{mm}$ is now inserted between the plates throughout the capacitor area, the plates remaining $2\,\text{mm}$ apart. What will the new capacitance be?
(Hint: Treat as two capacitors in series.)

5 A $30\,\mu\text{F}$ capacitor is initially charged with $15\,\text{mC}$. It is then discharged through a $200\,\Omega$ resistor. What is the maximum current during the discharge?

A $2.5\,\text{A}$ **B** $10\,\mu\text{A}$ **C** $2.5\,\text{mA}$
D $2.2 \times 10^{-12}\,\text{A}$ **E** $2.5 \times 10^{-12}\,\text{A}$

6 A $2.0\,\mu F$ capacitor initially charged to $20\,V$ is discharged through a $500\,k\Omega$ resistance. Calculate the rate of fall of the capacitor PD (a) at the first instant of discharge, (b) after 1 second. ($e = 2.718$.)

Exercise 22.3: Examination questions

(Where necessary use $\varepsilon_0 = 8.85 \times 10^{-12}\,F\,m^{-1}$.)

1 (a) A parallel plate air capacitor is made of two horizontal metal plates each having an area of $4.0 \times 10^{-2}\,m^2$ and separated by a distance of $1.5\,mm$. The potential difference between the plates is $500\,V$.

Calculate
  (i) the capacitance of the capacitor,
  (ii) the charge on a plate,
  (iii) the energy stored in the capacitor.

(b) The plates are now disconnected from the supply and electrically isolated, the original charges remaining on them. The upper plate is then raised until the separation of the plates is $6.0\,mm$.
  (i) Calculate the increase in energy stored in the capacitor.
  (ii) Explain how this extra energy is supplied.
[WJEC 2000]

2

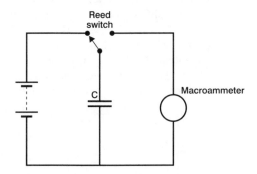

Fig. 22.9   Diagram for Question 2

Fig. 22.9 shows the circuit for measuring a capacitance using a reed switch.

Calculate the capacitance that produces a $50\,\mu A$ current when the charging voltage is $80\,V$ and the reed switch frequency is $200\,Hz$.

Using a different capacitor with the $80\,V$ supply and $200\,Hz$ switch frequency a current of $40\,\mu A$ is obtained. What current is expected if the experiment is repeated for this capacitor, a voltage supply of $100\,V$ and a frequency of $240\,Hz$?

3 Most new cars have an interior light which comes on whenever one of the doors is opened. In some cars the light stays on for a short time after the door is closed. The circuit shown below controls the timing delay.

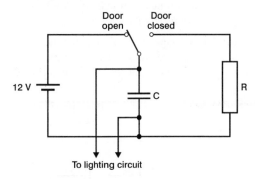

The circuit has a capacitor C which is connected to the car battery when the door is opened. When the door is closed, the capacitor is disconnected from the battery, and connected across a resistor R.

On the axes below, sketch a graph to show how the voltage across the capacitor varies with time.

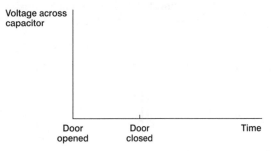

Calculate the time constant of the circuit if $C = 220\,\mu F$ and $R = 100\,k\Omega$.

In order for the light to be lit, there must be at least $6\,V$ across the capacitor. Calculate how long the light will stay on after the door is shut.

Explain the effect on the light if the manufacturer increases the value of R.   [Edexcel S.H. 2000]

4 (a) (i) Define the capacitance of a capacitor.
        (ii) Define the farad, the unit of capacitance.
        (iii) State one function of a capacitor.

(b) Fig. 22.10 shows an arrangement of six identical capacitors. Each capacitor has capacitance $22\,\mu F$. The maximum safe potential difference across a single capacitor is $50\,V$.

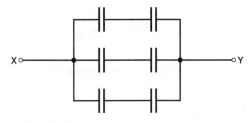

Fig. 22.10

(i) Find the total capacitance of the network between the terminals **X** and **Y**.

(ii) Find the maximum safe potential difference that can be applied between the terminals **X** and **Y**.

(c) (i) A capacitor of capacitance $22\,\mu F$ is charged by a battery of e.m.f. $9.0\,V$ and negligible internal resistance (Fig. 22.11).

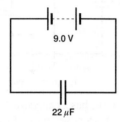

$9.0\,V$

$22\,\mu F$

Fig. 22.11

Calculate the charge on the capacitor.

(ii) The charged capacitor is then disconnected from the battery, and reconnected to an initially uncharged capacitor **Z** (Fig. 22.12).

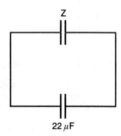

Z

$22\,\mu F$

Fig. 22.12

After this reconnection, the potential difference across the $22\,\mu F$ capacitor falls to $2.9\,V$.

(1) What is then the charge on the $22\,\mu F$ capacitor?

(2) What is the potential difference across capacitor **Z**?

(3) Calculate the capacitance of capacitor **Z**. [CCEA 2001]

5

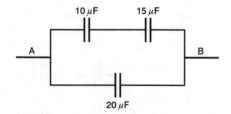

$10\,\mu F$ $15\,\mu F$

A   B

$20\,\mu F$

Fig. 22.13 Diagram for Question 5

(a) The three capacitors shown in Fig. 22.13 behave as a single capacitance between A and B. Calculate this capacitance.

(b) When a PD is applied between A and B, which of the three capacitors will store the greatest charge?

(c) Which will have the greatest PD across it?

6 (a) Two parallel-plate capacitors X and Y have equal plate areas. The capacitance of X is three times the capacitance of Y. Suggest **two** possible reasons for this difference.

(b) In the circuit of Fig. 22.14, switch S_1 is closed and switch S_2 is open.

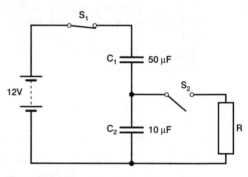

S_1

C_1 $50\,\mu F$

S_2

$12V$

C_2 $10\,\mu F$

R

Fig. 22.14

(i) Calculate the combined capacitance of C_1 and C_2.

(ii) The potential difference V across C_2 is $10\,V$. Explain, by reference to relevant formulae, how this value is obtained.

(c) Switch S_1 in the circuit of Fig. 22.14 is now opened and switch S_2 closed so that C_2 discharges through resistor R.

(i) Fig. 22.15 is to show the variation of V, the potential difference across R, with time t in ms. Two points have been plotted. Plot **two** further points and draw the graph.

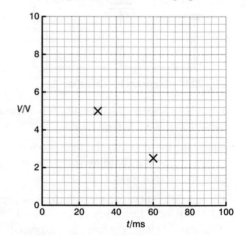

V/V

t/ms

Fig. 22.15

(ii) Suggest how the variation of V with t may be monitored experimentally.

[OCR 2001]

23
Magnetic forces

Force between parallel wires carrying currents

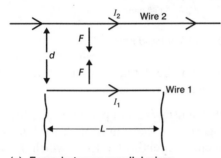

(a) Force between parallel wires

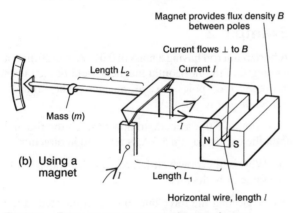

(b) Using a magnet

Fig 23.1   Force on a current-carrying conductor

In Fig 23.1 A straight wire of length L lies parallel to a very long straight wire 2. The wires carry currents I_1 and I_2. Experiments show that wire 1 experiences a force per unit length F/L given by

$$\frac{F}{L} = \frac{\mu I_1 I_2}{2\pi d} \qquad (23.1)$$

where μ is a constant for the medium in which the conductors lie. It is called the permeability of the medium. Its unit is $H\,m^{-1}$ (see p. 205).

For air or vacuum the permeability is denoted by μ_0 and the ampere is defined to make its value $4\pi \times 10^{-7}\,H\,m^{-1}$.

If both wires are very long, i.e. extend well away from the place of interest, then the force per metre is the same for both wires.

Note that if I_1 and I_2 are in the same direction, then the force is an attraction, i.e. like currents attract and opposite currents repel (quite opposite to the rule for poles). The currents act like two magnets attracting or repelling.

Since the currents act like magnets the same force F on wire 1 can be obtained with wire 2 suitably replaced by a magnet, as in Fig 23.1b where L_2 is made to balance the "see-saw" and F can be calculated.

Example 1

Currents of 5.0 A and 15 A flow down two parallel straight wires. The wires are 10 cm apart in air. Calculate the magnitude (size) of the force per unit length acting on each wire due to the currents and state the directions of these forces. (Take the magnetic permeability μ for air as $4\pi \times 10^{-7}\,H\,m^{-1}$.)

Method

The size of the force per metre on each wire is given by Equation 23.1 as

$$\frac{F}{L} = \frac{\mu I_1 I_2}{2\pi d} = \frac{4\pi \times 10^{-7} \times 5 \times 15}{2\pi \times 0.1}$$

$$= 1.5 \times 10^{-4}\,N\,m^{-1}$$

The *forces are inwards*, from one wire towards the other, because *currents in the same direction attract*.

Answer

$1.5 \times 10^{-4}\,N\,m^{-1}$.

Force on a straight wire related to field applied to it

The equation $F/L = \mu I_1 I_2/2\pi d$ can be written as $F = BI_1 L$, if $B = \mu I_2/2\pi d$. B is a characteristic of the place where wire 1 lies. B is decided by the current I_2 and the distance away d of the other conductor 2, and by the permeability of the air between.

So, forgetting wire 1, there is a region around the long wire in which all places have a noticeable B value. This is the 'magnetic field' of the wire and B is a measure of the field strength at the place concerned. B is described as the 'magnetic flux density' and is measured in tesla (T).

So for a long wire

$$B = \frac{\mu I}{2\pi d} \qquad (23.2)$$

where I is the current through the wire and d the distance from the wire to the place concerned.

As regards any wire responding to a field (such as our wire 1) the force on it is

$$F = BIL \qquad (23.3)$$

The current I here being the current through this wire and the length of this wire being L.

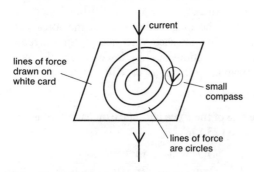

Fig 23.2   Lines of force around a current

Fig. 23.2 shows how lines can be drawn, e.g., on a piece of white card, to show the directions in which a small compass needle will point when a current flows though a long straight wire. These lines are called 'lines of force'. They show the directions of the magnetic flux density B for places near the wire. The direction of the BIL force, whether B is due to our long wire or to a magnet, is related to the direction of the B by the 'left-hand rule' (see Fig. 23.3a).

According to this the direction of the magnetic **F**ield (flux density B) is indicated by the **F**irst finger, **C**urrent by the se**C**ond finger and the force F, or the **M**otion produced by it, by the

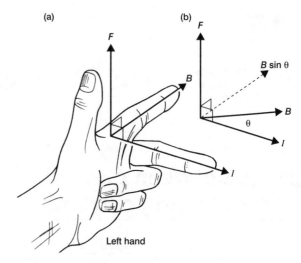

Fig 23.3   The left-hand rule

thu**M**b when these fingers of the left hand are held mutually at right angles.

If the flux density B is not perpendicular to I but is at an angle θ to it, then our formula for F needs the component perpendicular to I, which is $B\sin\theta$ (Fig. 23.3b). The formula in then $F = BIL\sin\theta$.

Example 2

A horizontal rod having a mass of 0.010 kg has 20 mm of its length in the horizontal magnetic field of 0.20 tesla between the poles of a magnet. The rod is perpendicular to this field.

A current of 5.0 A is made to flow through the rod and then the current, again 5.0 A, is reversed in direction.

(a) Why is the downward force on the rod not the same in both cases?

(b) Neglecting the force due to wires connected to the rod, calculate the size of the larger of these two forces. (Acceleration due to gravity or gravitational field strength, $g = 9.81\,\mathrm{m\,s^{-2}}$.)

Method

(a) According to the left-hand rule the BIL force is reversed when the current is reversed and so it will be upward in one case and downward in the other, opposing the rod's weight and adding to it respectively.

(b) The downward force

$$\begin{aligned}
&= \text{weight} + BIL \text{ force} = mg + BIL \\
&= (0.010 \times 9.81) + (0.20 \times 5.0 \times 20 \times 10^{-3}) \\
&= 1.18 \times 10^{-1}\,\mathrm{N} \text{ or } 0.12\,\mathrm{N}
\end{aligned}$$

Answer

(b) 0.12 N.

Example 3

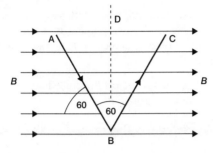

Fig. 23.4   Diagram for Example 3

Fig. 23.4 shows two straight conductors AB and BC, joined at B, carrying a current of 2.0 A and subjected to a uniform magnetic field of flux density 0.01 T whose direction lies in the plane ABC at 60° to AB. Both AB and BC are 5.0 cm long. The angle ABC is 60°. Calculate the forces on AB and BC. What movement do the two forces together try to produce?

Method

The component of B perpendicular to AB (namely $B \cos 30$ or $B \sin 60$)

$$= 0.01 \times \cos 30 = 0.01 \times 0.866 = 0.0087 \text{ T}$$

The force on AB is

$$F = 0.0087 \times 2.0 \times 5 \times 10^{-2} = 0.00087 \text{ N}$$

The force on BC is the same but, while the force on AB is upwards out of the diagram, the left-hand rule gives the force on BC to be downwards. These forces therefore produce a couple about the line BD shown in the diagram, and rotation about this line is expected.

Answer

8.7×10^{-4} N, Rotation about BD.

Force on a charged particle moving through a magnetic field

The formula for this is

$$F = Bqv \qquad (23.4)$$

Here q is the charge carried by the particle, v is the particle's velocity and B is the magnetic flux density perpendicular to v.

We are assuming B to be perpendicular to v. Otherwise B must be replaced in the formula by its component perpendicular to v, since only this component is effective. The direction of F is of course given by the left-hand rule (and note that this rule considers current direction, which for the movement of negative particles such as electrons will be opposite to that of the particle velocity).

Example 4

An electron is moving with a speed of $1.5 \times 10^7 \text{ m s}^{-1}$ perpendicular to a magnetic field having a uniform flux density of 0.0012 T.

(a) Calculate the force on the electron.

(b) Calculate the radius of the circular path followed by the electron. (Electron charge $e = 1.6 \times 10^{-19}$ C, electron mass $m = 9.0 \times 10^{-31}$ kg.)

Note: A 'uniform' field is a constant field, i.e., it has the same B value for all parts of the electron's path. Such fields can be obtained with magnets, coils or solenoids (see Chapter 25).

Method

(a) $F = Bqv$ or Bev when e is used to denote the electron's charge,

$$= 0.0012 \times 1.6 \times 10^{-19} \times 1.5 \times 10^7$$
$$= 2.8 \times 10^{-15} \text{ newton}$$

(b) The force Bev is providing the necessary inwards force mv^2/R for circular motion (see Chapter 8).

$$\therefore \quad Bev = \frac{mv^2}{r}$$

$$\therefore \quad r = \frac{mv}{Be} = \frac{9.0 \times 10^{-31} \times 1.5 \times 10^7}{0.0012 \times 1.6 \times 10^{-19}}$$

$$= 7.0 \times 10^{-2} \text{ m or } 7.0 \text{ cm}$$

Answers

(a) 2.8×10^{-15} N   (b) 7.0 cm

Couple on a coil

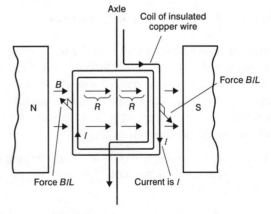

Fig. 23.5   Couple on a coil

In Fig. 23.5 a force BIL acts on each vertical wire. If there are n turns of wire on the coil, the total force is $BILn$ on each side of the coil.

The torque due to this pair of forces (couple) is $C = 2BILnR$. However, the coil area A equals $2RL$, so that

$$C = BAIn \qquad (23.5)$$

If the magnetic field is radial (see Fig. 23.6a), then B is always parallel to the plane of the coil even when the coil is allowed to rotate, and $C = BAIn$ still. If instead the field is uniform (see Fig. 23.6b and c), then the component of B which is effective is $B \cos \theta$ (see diagram), and the torque is $C = BAIn \cos \theta$.

(a) Radial field (seen from above)

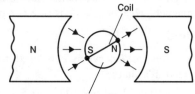

Poles are induced in soft iron cylinder

(b) Uniform field

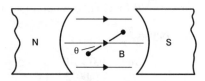

(c) Uniform field with large soft iron cylinder

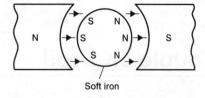

Soft iron

Fig. 23.6 Radial and uniform magnetic fields

In an ordinary moving-coil meter the current to be measured flows through the coil, the field is radial, and the torque $BAIn$ turns the coil. The turning tightens a spring which therefore produces an opposing torque of k newton metre per unit angle of rotation.

The coil comes to rest when $BAIn = k\theta$, where the angle of rotation of the coil θ is in degrees or radians.

$$BAIn = k\theta \qquad (23.6)$$

The torque $BAIn$ is also used to produce rotation in simple electric motors.

Example 5

A moving coil meter has a coil with 40 turns, each with an area of $20 \, \text{cm}^2$. It is suspended in a vertical plane and its sides are perpendicular to a radial magnetic field of 0.30 T.

(a) Calculate the torque on the coil when a current of $100 \, \mu\text{A}$ flows through it.

(b) If the coil has a resistance of $5.0 \, \Omega$ and a sensitivity of $100 \, \mu\text{A}$ full scale deflection, what series resistance is needed to convert the meter to a $10 \, \text{mV}$ FSD meter?

Method

(a) Torque $= BAIn = 0.30 \times 20 \times 10^{-4} \times 100 \times 10^{-6} \times 40 = 2.4 \times 10^{-6} \, \text{N m}$.

(b) $10 \, \text{mV}$ must produce $100 \, \mu\text{A}$, so the resistance of meter plus series resistance must be

$$R(= V/I) = \frac{10 \times 10^{-3}}{100 \times 10^{-6}} \text{ or } 100 \, \Omega$$

and the series resistance $= 100 - 5.0 = 95 \, \Omega$

Answer

(a) $2.4 \times 10^{-6} \, \text{N m}$, (b) $95 \, \Omega$.

The Earth's magnetism

In the United Kingdom the direction of the magnetic flux density due to the Earth's magnetism makes an angle θ (called the 'angle of dip') of about $70°$ to the horizontal. The horizontal component of this flux density B is $B_0 = B \cos \theta$, and the vertical component is $B_V = B \sin \theta$.

Note also that

$$B_V(= B \sin \theta) = \frac{B_0 \sin \theta}{\cos \theta} = B_0 \tan \theta,$$

where θ is the angle of dip as shown in Fig. 23.7.

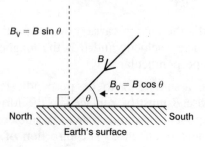

Fig. 23.7 The angle of dip

Example 6

Calculate the size and direction of the force per metre length on a straight, horizontal wire lying with 2.0 A flowing through it in direction north to south. (Earth's horizontal field component $= 1.6 \times 10^{-5}$ T. Angle of dip $= 70°$.)

Method

$B_V = B_0 \tan \theta = 1.6 \times 10^{-5} \times \tan 70 = 4.4 \times 10^{-5}$ T

$F = B_V IL = 4.4 \times 10^{-5} \times 2.0 \times 1 = 8.8 \times 10^{-5}$ N

By the left-hand rule, with B_V downwards, F is eastwards.

Answer

88 μN, eastwards.

Exercise 23.1

(Where necessary take μ_0 to be $4\pi \times 10^{-7}$ H m^{-1}.

1 Two very long parallel wires 0.4 m apart in air each carry a current of 5.0 A. What is the force, in newtons, on each metre length of wire?

2 A horizontal wire of length 4.0 cm is moving vertically downwards, with a current of 1.0 A flowing through it. If the plane in which the wire moves is perpendicular to a magnetic flux density of 0.1 T, calculate the force on the wire due to the current.

3 A moving coil meter has a 50-turn coil measuring 1.0 cm by 2.0 cm. It is held in a radial magnetic field of flux density 0.15 T and its suspension has a torsional constant of 3.0×10^{-6} N m rad^{-1}. What current is required to give a deflection of 0.5 rad?

4 In Fig. 23.8 a flat, rectangular coil is fitted symmetrically on an axle and lies in a horizontal plane. The coil is made of 10 turns of insulated wire and its dimensions are as shown in the figure.
If a current of 2.0 A flows round the coil

(a) What size is the vertical force on side BC caused by interaction between the current and the Earth's magnetic field? (Take the horizontal component of this field to be 1.6×10^{-5} T.)

(b) Calculate the total moment about the axle due to this force and to the similar force on side DA.

(c) Calculate the total moment that would be experienced by the coil if its plane were at an angle of 20° to the horizontal.

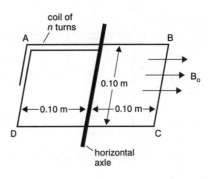

Fig. 23.8   Diagram for Question 4

5 An electron moving at a steady speed of 0.50×10^6 m s^{-1} passes between two flat, parallel metal plates 2.0 cm apart with a PD of 100 V between them. The electron is kept travelling in a straight line perpendicular to the electric field between the plates by applying a magnetic field perpendicular to the electron's path and to the electric field.
Calculate:

(a) the intensity of the electric field

(b) the magnetic flux density needed.

(Hint: the electron charge is not needed. It cancels.)

Exercise 23.2: Examination questions

1 Two long, straight, parallel wires in a vacuum are 0.25 m apart.
    (i) The wires each carry a current of 2.40 A in the same direction. Calculate the force between the wires per metre of their length. Draw a sketch showing clearly the direction of the force on each wire.
    (ii) The current in one of the wires is reduced to 0.64 A. Calculate the current needed in the second wire to maintain the same force between the wires per metre of their length as in (i).

(Take $\mu_0 = 4\pi \times 10^{-7}$ H m^{-1})
[CCEA 2000, part]

2 (a) In Fig. 23.9, PQRS is a rectangular coil consisting of N turns of wire and carrying current I. The plane of PQRS is parallel to a uniform magnetic field of flux density B. The length of PQ is L and the length of QR is b.

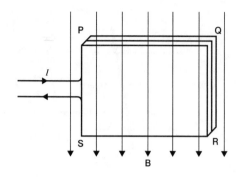

Fig. 23.9

(i) Write down an expression for the force experienced by the side PQ of the coil.
(ii) Show that the torque T experienced by PQRS is given by the expression $T = NBIA$ where $A = Lb$.

(b) The electric motor in a model railway engine is powered by a 6.0 V battery. Within this motor, a coil of resistance $1.2\,\Omega$ rotates in the field of a permanent magnet. With the engine pulling a moderate load, a back e.m.f. of 5.6 V is induced in the coil.
 (i) Calculate the current in the coil.
 (ii) Suggest and explain an undesirable consequence of allowing the engine to pull a heavy load for a long period of time.　　　　　　　 [OCR 2001]

3 Two parallel metal sheets are separated by 25 mm in a vacuum, the lower plate being earthed. A narrow beam of electrons enters symmetrically

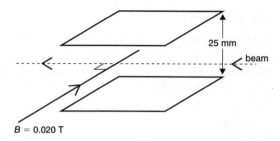

Fig. 23.10

between the plates, as shown in Fig. 23.10. There is a uniform magnetic field of flux density 0.020 T, which is perpendicular to the beam and parallel to the plates, acting in the direction shown. When a potential difference of 3500 V is applied to the plates the electron beam is undeviated.

(a) Calculate the speed of an electron, assuming that the electric field between the plates is uniform.

(b) When the magnetic field is removed, the electron beam is found to deflect downwards (in relation to the diagram). What is the **potential** of the upper plate?　　 [WJEC 2000]

4 The ampere is defined as that current which, flowing in two infinitely long, parallel straight wires 1.0 m apart in vacuum, causes a force per metre length of $2.0 \times 10^{-7}\,\mathrm{N\,m^{-1}}$ to act on each wire. Use this definition to obtain a value for the magnetic permeability of a vacuum.

24
Electromagnetic induction

Straight conductors

Induced EMF in a straight wire

Consider a straight wire of length L moving with a velocity v perpendicular to its length in a magnetic field of flux density B which is perpendicular to both the wire's length and the velocity (Fig. 24.1a).

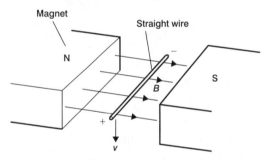

(a) Wire not part of the complete circuit

Magnet

Straight wire

N

S

B

v

(b) Wire is part of a complete circuit

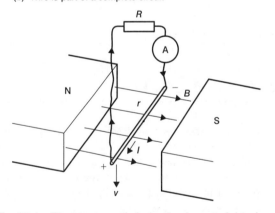

R

A

N

r

B

S

I

v

Fig. 24.1 Electromagnetic induction in a straight wire

The metal wire contains free electrons, so that movement of the wire means movement of these electrons at a velocity v. We can use the formula $F = Bev$ (Chapter 23) and deduce that each free electron is moved by this force along the wire until a PD is established between the two ends of the wire sufficient to stop any further movement of the electrons. This PD is produced almost instantly and is given by

$$E = BLv \qquad (24.1)$$

This is the induced EMF. If the ends of the wire were joined by conductors of resistance R to form a complete circuit of resistance $R + r$, where r is the resistance of the straight wire itself (Fig. 24.1b) then the current I resulting from the electromagnetic induction is

$$I = \frac{BLv}{R + r}$$

and the terminal potential difference between the wire's ends is

$$V = \frac{BLvR}{R + r}$$

The direction in which the induced current flows (Fig. 24.1b) can be deduced by use of Lenz's law together with the left-hand rule (Chapter 23) or, alternatively, the right-hand rule may be used with the first finger for the B direction, thumb for movement direction and the second finger for the induced current.

Magnetic flux Φ

If an area A lies perpendicular to a magnetic flux density B then the product BA is called the magnetic flux and is usually denoted by the symbol Φ. The direction of the flux is the same as the direction of the magnetic field.

The unit for Φ is the weber, and 1 tesla = 1 weber per metre^2.

$$\Phi = BA \qquad (24.2)$$

Induced EMF in a straight wire in terms of magnetic flux

In the formula $E = BLv$ the product Lv is the area cut through per second by the wire moving perpendicular to B, so that $B \times Lv$ is $B \times$ the area per second. Therefore $E =$ flux cut per second by the moving wire. Thus $E = \Phi/t$ where

Φ is the flux cut in time t, and we are assuming that E is constant. If E is not constant then its value at any instant is $d\Phi/dt$. For calculus see page 12:

$$E = \frac{d\Phi}{dt} \qquad (24.3)$$

B not perpendicular to the area

If the direction of B is inclined to the area at an angle θ, then the effective value of B is $B\cos(90 - \theta)$ or $B\sin\theta$. (The same result is obtained if we say that there is an area $A\sin\theta$ perpendicular to B. Either way $E = BLv\sin\theta$.)

Example 1

An aeroplane is travelling at $100\,\mathrm{m\,s^{-1}}$ in a direction which is horizontal and northwards. Calculate the EMF induced between the tips of its wings, which have a span of $20\,\mathrm{m}$. Take the Earth's magnetic flux density to be $5.0 \times 10^{-5}\,\mathrm{T}$ and the angle of dip $71°$ at the place concerned.

Method

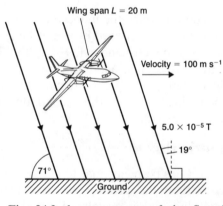

Wing span L = 20 m

Velocity = 100 m s^{-1}

5.0 × 10^{-5} T

19°

71°

Ground

From Fig. 24.2 the component of the flux density perpendicular to the aeroplane wing's movement is $5.0 \times 10^{-5}\cos 19°$. Using the formula $E = BLv$:

$$E = (5.0 \times 10^{-5}\cos 19) \times 20 \times 100$$

$$= 10^{-1} \times \cos 19 = 10^{-1} \times 0.946 = 0.0946\,\mathrm{V}$$

or $95\,\mathrm{mV}$ to two significant figures.

Answer

$95\,\mathrm{mV}$.

Fig. 24.2 Suitable diagram for Example 1

Example 2

A wheel with metal spokes is turning through a steady 2 revolutions per second and it has a radius of $50\,\mathrm{cm}$. Its plane is perpendicular to the horizontal component of the Earth's magnetic field which is $1.6 \times 10^{-5}\,\mathrm{T}$.

Calculate the flux through the wheel.

Hence calculate the induced EMF in a spoke.

Show that this EMF can be calculated from the formula $E = BLv$ if v is the mean speed of rotation of the spoke (half the speed of its outer end).

Method

(a) The wheel has an area of πR^2 and the flux through the wheel is its area $\times$ perpendicular flux density

$$= \pi R^2 \times 1.6 \times 10^{-5}\,\text{weber}$$

or $\quad 3.142 \times 0.50^2 \times 1.6 \times 10^{-5}\,\mathrm{Wb}$

$\therefore \quad$ the flux $= 1.26 \times 10^{-5}\,\mathrm{Wb}$

(b) The spoke cuts through this flux twice per second so that 1 revolution takes $0.50\,\mathrm{s}$ and

$$\text{EMF } E\left(= \frac{d\Phi}{dt}\right) = \frac{1.26 \times 10^{-5}}{0.50} =$$

$$2.52 \times 10^{-5}\,\text{volt or } 2.5 \times 10^{-5}\,\mathrm{V}$$

(c) $E = \dfrac{d\Phi}{dt} = \dfrac{d(BA)}{dt} = \dfrac{B \times \pi R^2}{0.50} = B \times 2\pi R^2$

but mean speed v of spoke $= \frac{1}{2} \times$ circumference/time for a revolution

$$= \frac{1}{2} \times \frac{2\pi R}{0.50} = 2\pi R$$

and spoke length $L = R$ so that

$$E = B \times 2\pi R^2 = B \times R \times 2\pi R = BLv.$$

Answer

(a) $1.3 \times 10^{-5}\,\mathrm{Wb}$, (b) $25\,\mu\mathrm{V}$

Exercise 24.1

1 Calculate the induced EMF in a straight wire when it is moving at $5.0\,\mathrm{m\,s^{-1}}$ perpendicular to its length in a magnetic field of flux density $0.10\,\mathrm{T}$ if the field direction is (a) perpendicular to the plane of movement, (b) parallel to it, (c) at $60°$ to it. The wire length is $1.0\,\mathrm{cm}$.

2 (a) Calculate the EMF induced between the axle and the rim of a spoked metal wheel if the wheel radius is $20\,\mathrm{cm}$ and the uniform field in which it lies is $0.020\,\mathrm{T}$ perpendicular to the plane of the wheel, the speed of rotation being 10 revolutions per second.

(b) What is the expected current size through a $10\,\Omega$ resistor connected between the axle and the rim if the wheel's resistance is negligible?

3 Calculate the flux cut through in 1.0 ms by a straight wire 3.0 cm long moving at $2.0\,\text{m s}^{-1}$ perpendicular to its length and to a magnetic field of flux density 10 mT.

Coils

Induced EMF in a coil

If a flat coil lies with its plane, of area A, perpendicular to a magnetic field whose flux density is B, then the flux Φ 'passing through'* the coil is $B \times A$.

Flux $\Phi = B \times A$	(24.4)

Φ can be changed – e.g. by changing B or by rotating the coil so that less flux passes through it. Now the change of this flux Φ through the coil is also the flux cut through by the wires of the coil. So we can use the formula (Equation 24.3) obtained earlier for the induced EMF, namely $E = \mathrm{d}\Phi/\mathrm{d}t$.

However, for the coil it is appropriate to describe $\mathrm{d}\Phi/\mathrm{d}t$ as the *rate of change of flux through the coil*.

For a coil of n turns the induced EMF is n times greater.

$E = n\dfrac{\mathrm{d}\Phi}{\mathrm{d}t}$	(24.5)

where Φ is the flux through the coil.

Alternatively we write

$E = \dfrac{\mathrm{d}\Phi}{\mathrm{d}t}$	(24.6)

even though the coil has n turns and Φ now represents the 'effective flux' through the coil, called the flux linkage, this quantity being the product of flux through coil × number of turns.

Flux linkage $= n \times$ Flux

The SI unit for flux linkage is also the weber (Wb).

'Passing through' as if flux were a flow of something through the coil, along the lines of force of the magnetic field.

Note that, since both flux and flux linkage are usually denoted by the same symbol Φ and have the same unit, it will sometimes be necessary to distinguish between them, e.g. by writing 'flux Φ' or 'flux linkage Φ'.

Frequently Equation 24.6 is written as

$$E = -\frac{\mathrm{d}\Phi}{\mathrm{d}t}$$

so that, using a suitable sign convention for $\mathrm{d}\Phi/\mathrm{d}t$, the polarity of E is obtained. You are not expected to know this convention.

For a coil of n turns and area A, perpendicular to a uniform flux density B the flux Φ is BA (see above) so the flux linkage is

$\Phi = BAn$	(24.7)

A typical example of induced EMF in such a coil is the steady reduction to zero in time t of the flux density B. The flux linkage change is $BAn - 0$ so that $E = BAn/t$.

Example 3

The flux passing through a coil of 80 turns is reduced quickly but steadily from 2.0 mWb to 0.5 mWb in a time of 4.0 s. Calculate the induced EMF.

Method

$E = \dfrac{\mathrm{d}\Phi}{\mathrm{d}t}$ where Φ is the flux linkage.

$\therefore \quad E = \dfrac{\mathrm{d}(n\Phi)}{\mathrm{d}t} = \dfrac{80(2.0 - 0.50) \times 10^{-3}}{4.0}$

$\therefore \quad E = 3.0 \times 10^{-2}\,\text{V}$

Answer

30 mV.

Self-induction

If the current I in a coil changes, then the magnetic flux density B within the coil changes (as well as the field around the coil of course), and this causes electromagnetic induction in the coil. The coil is an inductor and the induced EMF is

$E = L\dfrac{\mathrm{d}I}{\mathrm{d}t}$	(24.8)

where L is called the self-inductance of the coil. The SI unit for self-inductance is the henry (H). L is decided by the coil's geometry and number of turns and also by the presence of magnetic material (permeability μ) within or around the coil. Hence the unit for μ is H m^{-1}.

Back EMF

The self-induced EMF is often called a 'back EMF' because it opposes the voltage that produced the current I.

Similarly a rotating coil in a motor experiences an induced EMF due to its movement between the poles of its magnet and this voltage (back EMF) opposes the voltage driving the motor.

At the instant when a circuit is connected to a voltage supply the current (I) is zero and the rate of growth of current (dI/dt) will be such that the back EMF equals the supply voltage. Subsequently the induced EMF and the PD ($V = IR$) across resistance in the circuit together equal the supply PD. Finally IR = supply PD as I becomes steady.

Mutual induction

When two coils are close so that a change of current I_1 in one of the coils causes a change in the flux density inside the second coil, an EMF is induced in the second coil.

This fact explains how a transformer works (see page 218).

Example 4

If a 2.0 V DC voltage supply is connected to an inductor of 0.50 H inductance and $100 \, \Omega$ resistance what is the rate of rise of current

(a) at the instant when the connection is made (current = zero)

(b) when the current has risen to 0.010 A

(c) when the current is 0.020 A?

Method

(a) $E = L \dfrac{dI}{dt}$ and equals the supply voltage.

$\therefore \quad 2.0 = 0.5 \times \dfrac{dI}{dt}$ and $\dfrac{dI}{dt} = 4.0 \, \text{A s}^{-1}$

(b) When $I = 0.010$ A, the PD due to the resistance is $V = IR$ and equals 0.010×100 or 1.0 V. But the PD across the inductance and resistance (think of these as in series) must equal the supply PD, so the further 1.0 V is the induced voltage due to the inductance.

$$E = L \frac{dI}{dt}$$

$\therefore \quad 1.0 = 0.5 \times \dfrac{dI}{dt}$

$\therefore \quad \dfrac{dI}{dt} = 2.0 \, \text{A s}^{-1}$

(c) $\quad I = 0.020$ A

$\therefore \quad V = IR = 0.020 \times 100 = 2.0$ V.

This means that the voltage due to self inductance is zero, $\dfrac{dI}{dt}$ is zero and the current is no longer rising.

Answer

(a) $4.0 \, \text{A s}^{-1}$, (b) $2.0 \, \text{A s}^{-1}$, (c) Zero.

Exercise 24.2

1 A flat coil having an area of $8.0 \, \text{cm}^2$ and 50 turns lies perpendicular to a magnetic field of 0.20 T. If the flux density is steadily reduced to zero, taking 0.50 second, what is (a) the initial flux through the coil, (b) the initial flux linkage, (c) the induced EMF?

2 Calculate the self-inductance of a coil that experiences an induced EMF of 20 mV when the current through it changes at a rate of $2.0 \, \text{A s}^{-1}$.

Rotating coils

Induced EMF in a rotating coil in a uniform field

When a coil rotates as in Fig. 24.3 the formula for the induced EMF can be obtained by applying the equation $E = BLv$ to each of the vertical sides of the coil (see Fig. 24.3). The formula is

$$E = 2\pi f BAn \sin 2\pi ft \text{ or } 2\pi f \, BAn \, \sin \omega \, t$$

$$(24.9)$$

Otherwise the formula $E = \dfrac{d\Phi}{dt}$ can be used. In this formula Φ is the flux linkage and equals $BAn \cos \theta$ where θ is the angle between the coil axis and B. Also θ equals ωt where t is the time which started when θ was zero.

Now if you look at the simple harmonic motion formulae for displacement ($y = r \sin \omega t$ or $r \sin \omega t$) and velocity ($v = r\omega \cos \omega t$) you conclude that the rate of change of $\sin \omega t$ with t

(a) Coil rotating in a uniform field

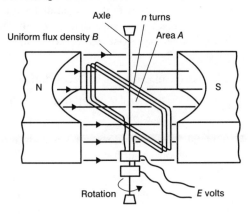

(b) The EMF produced

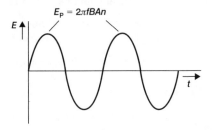

Fig. 24.3   Induced EMF in a rotating coil (a simple generator)

equals $\omega \cos \omega t$. Similarly comparison of the SHM formulae for velocity $(v = r\omega \cos \omega t)$ and acceleration $(a = -\omega^2 y = -\omega^2 r \sin \omega t)$ shows that the rate of change of $\cos \omega t$ with time equals $-\omega \sin \omega t$.

$$\frac{d(\sin \omega t)}{dt} = \omega \cos \omega t$$

$$\text{and } \frac{d(\cos \omega t)}{dt} = -\omega \sin \omega t \qquad (24.10)$$

So $E = -\dfrac{d\Phi}{dt} = -\dfrac{d(BAn \cos \omega t)}{dt}$

$= -BAn(-\omega \sin \omega t) = 2\pi f BAn \sin \omega t$

in which f is the number of revolutions per second (i.e. the frequency of rotation), A is the coil area, n the number of turns of wire on the coil and t is the time. The magnetic flux density B is assumed to be uniform (the same everywhere) and perpendicular to the axis of rotation.

As t increases, $\sin 2\pi f t$ will reach a maximum value of unity $(= 1)$, so that the maximum, or peak, value of E is $2\pi f BAn$ and we can write $E = E_P \sin \omega t$ where E_p is the peak value and ω

is the angular frequency $(2\pi f)$ of the EMF or the angular velocity of the rotating coil (Fig. 24.3b). In these equations t is zero when the plane of the coil is perpendicular to B, and ωt is the angle between the coil axis (not rotation axis) and B. The graph shape is sinusoidal.

Example 5

A coil of 200 turns and $12 \, \text{cm}^2$ area is rotating at 20 revolutions per second in a uniform magnetic field of flux density $0.020 \, \text{T}$. Calculate the induced EMF when the coil's plane is momentarily (i) parallel to B, (ii) at $20°$ to B.

Method

(i) The induced EMF is $E = 2\pi f BAn \sin 2\pi f t$ (Equation 24.9).

We have

$f = 20 \, \text{s}^{-1}, B = 0.020 \, \text{T}, A = 12 \times 10^{-4} \, \text{m}^2, n = 200$ and $\sin 2\pi f t = 1$ when the coil's plane is parallel to B.

$$\therefore \quad E = 2\pi \times 20 \times 0.020 \times 12 \times 10^{-4} \times 200 \times 1$$

$$= 0.603 \, \text{V or } 0.60 \, \text{V}$$

(ii) The angle $2\pi f t$ equals $90°$ when the coil's plane is parallel to B, and movement through $20°$ from that position means that the angle $2\pi f t = 70°$ (or $110°$).

$$E = 2\pi \times 20 \times 0.02 \times 12 \times 10^{-4} \times 200 \times \sin 70°$$

$$= 0.603 \, \text{V} \times 0.94$$

$$= 0.57 \, \text{V}$$

Answer

(i) $0.60 \, \text{V}$, (ii) $0.57 \, \text{V}$.

Exercise 24.3

1   A flat coil of area $4.5 \, \text{cm}^2$ having 200 turns of resistance $20 \, \Omega$ lies with its area perpendicular to a field for which $B = 0.60 \, \text{T}$. If the coil is turned through $90°$ in $0.50 \, \text{s}$ what is the average induced current if the external circuit resistance is zero?

2   A coil is rotating in a uniform field of $0.01 \, \text{T}$ perpendicular to the axis of rotation (as in Fig. 24.3). The coil area is $2.0 \, \text{cm}^2$, the number of turns is 50 and the steady speed of rotation is 20 revolutions per second. Calculate (a) the maximum induced EMF, (b) the induced EMF at the instant when the plane of the coil lies at $40°$ to the field direction.

Exercise 24.4: Examination questions

1 At the beginning of a horse-race, a horizontal straight wire of length 20 m is raised vertically through a height of 3.0 m in 0.20 s.

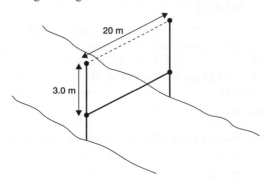

The horizontal component of the Earth's magnetic field strength perpendicular to the wire is 2.0×10^{-5} T.

What is the average e.m.f. induced across the ends of the wire?

A zero   **B** 0.24 mV   **C** 1.2 mV   **D** 6.0 mV

[OCR 2001]

2

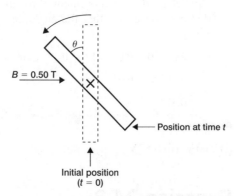

(a) A flat, circular coil of wire of 30 turns, each of area 0.025 m², is initially placed with its plane at right angles to a uniform magnetic field of flux density 0.50 T, as shown. Calculate the flux linking the coil.

(b) The coil is now rotated steadily at 60 rad s⁻¹ about a diameter which is perpendicular to the magnetic field. At time t the coil is in the position shown.
   (i) Give an expression for the flux linking the coil at time t.
   (ii) Hence show that the induced e.m.f. E at time t is given by

$$E = 22.5 \sin 60t.$$

[WJEC 2000]

3 A flat circular coil of 120 turns, each of area 0.070 m², is placed with its axis parallel to a uniform magnetic field. The flux density of the field is changed steadily from 80 mT to 20 mT over a period of 4.0 s.

What is the e.m.f. induced in the coil during this time?

A 0   **B** 130 mV   **C** 170 mV   **D** 500 mV

[OCR 2000]

4 A metal framed window is 1.3 m high and 0.7 m wide. It pivots about a vertical edge and faces due south.

Calculate the magnetic flux through the closed window.

(Horizontal component of the Earth's magnetic field = 20 μT. Vertical component = 50 μT)

The window is opened through an angle of 90° in a time of 0.80 s. Calculate the average e.m.f. induced.

State and explain the effect on the induced e.m.f. of converting the window to a sliding mechanism for opening. [Edexcel 2001]

5 Fig. 24.4 shows a series circuit containing a 2.0 V cell, a switch S, a 0.25 Ω resistor R, and an inductor L. The internal resistance of the cell and the resistance of L are negligible.

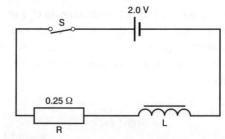

Fig. 24.4

(a) After closing S, the current in the circuit rises, eventually becoming steady. While the current is increasing from zero to 0.20 A, the rate of change of current can be assumed to be constant at 40 A s⁻¹.
   (i) Calculate, for the instant when the current is 0.20 A, the potential difference (p.d.)
       1 across R;
       2 across L.
   (ii) Use your result from (a) (i) 2 in calculating the inductance of L.

(b) The current in the circuit eventually becomes steady.
   (i) Calculate the magnitude of the steady current.
   (ii) Explain why the inductor L plays no part in determining the magnitude of this steady current. [OCR 2000]

25
Magnetic field calculations

Field due to current in a long straight wire

As shown in Chapter 23 (p. 198) the field strength,* called the magnetic flux density, is given by

$$B = \frac{\mu I}{2\pi d} \qquad (25.1)$$

where I is the current through the straight wire and B is the resulting flux density at a point distance d from the wire. μ equals permeability of the medium. The lines of force of this field are circles centred upon the wire (as stated in Chapter 23), and this is shown in Fig. 23.2. The directions of the lines of force are given by the 'corkscrew rule' according to which these directions are clockwise when one looks along the wire in the direction of the current.

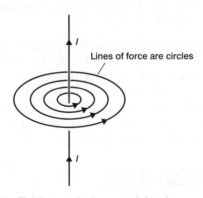

Fig. 25.1   Field around a long straight wire

*Magnetic intensity H is a different quantity that is also used to describe field strength.

Neutral points in magnetic fields

If a magnetic field results from more than one current-carrying conductor or magnet then at a certain place in the field the flux densities may be equal in magnitude and opposite in direction so that their effects cancel, i.e., the resultant flux density is zero. Such a place is called a 'neutral point'.

Example 1

A long, straight, vertical wire carries a downward current of 4.0 A. The earth's magnetic field in which this wire is placed has a horizontal component of 1.6×10^{-5} T. Calculate:

(a) the resultant horizontal magnetic flux density at a point 10 cm to the west of the wire

(b) the distance from the wire of the neutral point.

(Take μ as $4 \times 10^{-7}\,\mathrm{H\,m^{-1}}$.)

Method

(a) The flux density due to the wire at a distance d of 10 cm (0.10 m) is given by

$$\frac{\mu I}{2\pi d} = \frac{4\pi \times 10^{-7} \times 4.0}{2\pi \times 0.10} = 8.0 \times 10^{-6}\,\mathrm{T}$$

Due west of the wire this flux density is, according to the corkscrew rule, directed northwards. It therefore adds to the earth's horizontal flux density. So the resultant flux density is $1.6 \times 10^{-5}\,\mathrm{T} + 0.80 \times 10^{-5}\,\mathrm{T}$ or 24 μT.

(b) At the neutral point the flux density due to the wire is equal in magnitude to the 1.6×10^{-5} T of the earth's field.

$$\therefore \quad \frac{\mu I}{2\pi d} = 1.6 \times 10^{-5}$$

$$\therefore \quad \frac{4\pi \times 10^{-7} \times 4.0}{2\pi d} = 1.6 \times 10^{-5}$$

$$\therefore \quad d = \frac{2 \times 10^{-7} \times 4.0}{1.6 \times 10^{-5}} = 5.0 \times 10^{-2}\,\mathrm{m} \text{ or } 5.0\,\mathrm{cm}$$

Answer

(a) 24 μT, (b) 5.0 cm

Magnetic field at a point within a toroid or well inside a solenoid

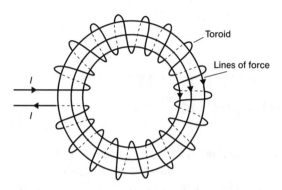

Fig. 25.2   Field within a toroid

Within a toroid (an endless coil, see Fig. 25.2) the magnetic flux density is given by

$$B = \mu I \times \text{number of turns per metre}$$

$$\text{or} \quad B = \mu I \times \frac{n}{L} \qquad (25.2)$$

A solenoid is a long coil, i.e. its length is considerably greater than its diameter, as shown in Fig. 25.3.

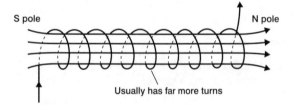

Fig. 25.3   A solenoid

A solenoid can be thought of as part of a large toroid, and the turns of the remainder of the toroid are too far from the middle of the solenoid to affect the flux density there. Hence the same formula (25.2) applies to a solenoid.

Example 2

Calculate the flux density in the middle of a solenoid having 10 turns per centimetre and carrying a current of 0.50 A. The medium within the solenoid is air, for which the permeability is $4\pi \times 10^{-7}\,\mathrm{H\,m^{-1}}$.

Method

The flux density is given by Equation 25.2:

$$B = \frac{\mu I n}{L}$$

$\mu = 4\pi \times 10^{-7}$, $I = 0.50$, $n/L = 10$ per cm, i.e. $1000\,\mathrm{m^{-1}}$.

$$\therefore \quad B = 4\pi \times 10^{-7} \times 0.5 \times 1000$$
$$= 2\pi \times 10^{-4}$$
$$= 6.3 \times 10^{-4}\,\mathrm{T}$$

Answer

$6.3 \times 10^{-4}\,\mathrm{T}$

Exercise 25.1

(The permeability of air may be taken as $4\pi \times 10^{-7}\,\mathrm{H\,m^{-1}}$.)

1   A vertical wire carries a downward current of 5.0 A, and 12 cm east of this there is another vertical wire carrying an equal downward current. The earth's horizontal component is $1.6 \times 10^{-5}\,\mathrm{T}$. What is the flux density at a distance 2.0 cm from the first wire and 10.0 cm from the other?

2   Two long, parallel, straight wires are 10 cm apart. One wire carries a current of 2.0 A and the other carries 3.0 A. In the resulting magnetic field there is a neutral point. Calculate its distance from the 2.0 A wire

(a)   when the currents are in the same direction

(b)   when the currents are in opposite directions.

3   A solenoid having 200 turns per metre and carrying a current of 0.050 A lies with its axis east–west. Well inside the solenoid is a small compass whose needle points 37° west of north. Calculate the Earth's horizontal magnetic field component B_0.

4   An air-cored toroid has 200 turns and a length of 15 cm. Around its centre is wound a coil of radius 3.0 cm with 20 turns. If the current in the toroid is initially 20 mA and is reduced steadily to zero in a time of 0.10 s, what EMF will be induced in the 20-turn coil during this time. (Take permeability of air to be $4\pi \times 10^{-7}\,\mathrm{H\,m^{-1}}$.)

Exercise 25.2: Examination questions

1   The magnetic flux density at a certain point P close to a long, straight wire carrying a current I is 3.0 mT. A line perpendicular to the wire and passing through P meets a point Q which is twice as far from the wire as P. What is the flux density at Q when the current in the wire is reduced to 0.5 I?

2 A slinky spring of 180 turns is stretched uniformly along a horizontal bench-top. When a current of 1.20 A is passed through the spring, it acts as a solenoid.

    (a) Calculate the magnetic flux density at the centre of this solenoid when the tension in the spring is such that its length is 2.00 m.

    (b) The tension in the spring is reduced so that its length becomes 1.50 m. Find the new flux density at the centre of the solenoid.

                         [CCEA 2000]

3 Use the equations $F = BIL$, $B = \dfrac{\mu I}{2\pi d}$ and $E = -L\dfrac{dI}{dt}$ to show that henry per metre ($\mathrm{H\,m^{-1}}$) is an appropriate unit for μ. (The henry is the unit for self-inductance L.) L in BIL denotes length.

4 (a) Fig. 25.4 shows a rear-view cross-section of the body of a railway carriage and of an electric cable under the floor of the carriage. The cable carries a current of 80 A towards the front of the carriage. A magnetic compass is held horizontally at P, 1.5 m above the cable.

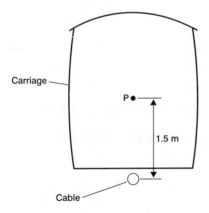

Fig. 25.4

      (i) Calculate the flux density B_C of the magnetic field at P due to the current in the cable. Take the permeability of air to be $1.3 \times 10^{-6}\ \mathrm{H\,m^{-1}}$.

      (ii) On Fig. 25.4, draw an arrow at P to show the direction of B_C.

    (b) The flux density B_H of the horizontal component of the Earth's magnetic field is $1.8 \times 10^{-5}\ \mathrm{T}$. Assume that this acts in the direction of true North, and that there are no other magnetic fields apart from that of the current in the cable.

    Calculate the resultant horizontal magnetic flux density B at P, and state the direction in which the compass points, when the carriage is oriented with its front:

    (i) towards the east;

    (ii) towards the north.   [OCR 2001]

26 Alternating currents

Variation of voltage with time

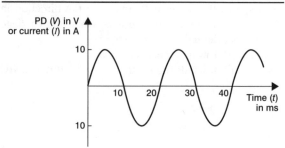

Fig. 26.1   Sinusoidal variation of voltage and current with time

Fig. 26.1 shows a graph for an alternating voltage or current that is sinusoidal. Mains AC supply is like this.

Unless otherwise stated 'alternating current' or 'alternating voltage' means sinusoidal current or PD. As shown in Chapter 24 a uniform-field generator produces a sinusoidal voltage.

The variation of voltage with time is described by the formula

$$V = V_p \sin 2\pi ft \qquad (26.1)$$

where f is the number of cycles (i.e. repeats) per second and is the frequency, t is the time measured from an instant when $V = 0$, and V_p is

the maximum or peak value of the voltage. Note that $2\pi f$ may be written as ω, known as the angular frequency. If the voltage is produced by a rotating-coil generator, then ω may be identified with the angular frequency of the coil's rotation and $\omega = \theta/t$, where θ is the angle through which the coil rotates. The unit for ω is rad s^{-1}.

$$V = V_p \sin \omega t \quad \text{or} \quad V_p \sin\theta \qquad (26.2)$$

However, regardless of the cause of the voltage, the value of θ ($=2\pi ft$) is important for describing the stage reached by the voltage variation and is called the phase angle, as explained in Chapter 11.

Fig. 26.2 shows how the variation of voltage is described by a rotating radius (see also Fig. 11.2, describing simple harmonic motion). It is a phasor because it has size and phase.

Size of current in a purely resistive circuit

If the circuit concerned contains no significant capacitance or inductance, only resistance R, then at all times $I = V/R$ so that

$$I = \frac{V_p \sin 2\pi ft}{R}$$

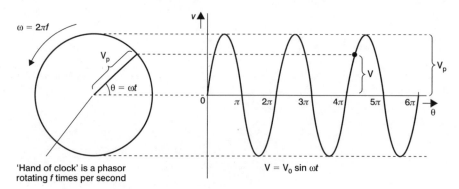

Fig. 26.2   Use of a phasor for voltage or current ($V = V_0 \sin \omega t$)

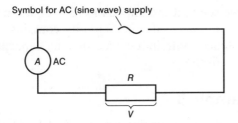

Symbol for AC (sine wave) supply

Fig. 26.3   A purely resistive circuit

$$\text{or} \quad I = I_p \sin 2\pi ft \qquad (26.3)$$

where $I_p = V_p/R$.

The current rises and falls in step with the voltage, i.e. I and V are in phase. $I = I_p$ when $V = V_p$.

Average and RMS values

The effects produced by alternating currents will often depend on some kind of average of the current. A simple average over a half-cycle is known as the average (or mean) value of the current or voltage. Heating by a current is decided by the mean value of I^2R or V^2/R, and the square root of mean I^2 or V^2 is the root mean square (RMS) value. The sizes quoted for alternating voltages and currents, unless otherwise stated, are always RMS values.

For a sine wave variation the mean value equals $(2/\pi) \times$ peak value and the RMS value equals $(1/\sqrt{2}) \times$ peak value.

$$\text{i.e.} \quad I_{RMS} = \frac{1}{\sqrt{2}}I_p \quad \text{and} \quad \bar{I} = \frac{2}{\pi}I_p \qquad (26.4)$$

In Fig. 26.1 for example the peak voltage is 10 V and for the sine wave $V_{RMS} = 7.1\,\text{V}$, $\bar{V} = 6.3\,\text{V}$.

Example 1

A sinusoidal alternating voltage displayed on a cathode ray oscilloscope is seen to have a peak value of 75 V. What reading should be obtained with a voltmeter indicating RMS voltage?

Method

$V_p = 75\,\text{V}$ but $V_{RMS} = V_p/\sqrt{2}$.
Therefore $V_{RMS} = 75/1.414 = 53\,\text{V}$.

Answer

53 V.

Example 2

Calculate the value of a sinusoidal voltage having a peak value of 30 V at a time of one-tenth of a cycle after a peak has been reached. What current will be present at this instant if the total resistance of the circuit is $9.0\,\Omega$?

Method

$V = V_p \sin\theta$ (Equation 26.2).

One-tenth of a cycle is 360/10 degrees, i.e. 36°. Therefore we need V when θ is 36° greater than 90°, i.e. $\theta = 126°$. However, we should realise that V will have the same value at 36° less than 90°, namely $\theta = 54°$.

$$V = 30 \times \sin 54 \text{ or } 30 \times \sin 126$$

Hence $\quad V = 30 \times 0.81 = 24.3\,\text{V}$

$$\text{Current} = \frac{V}{R} = \frac{24.3}{9.0} = 2.7\,\text{A}$$

Answer

24 V, 2.7 A.

Impedance

This is the opposition of a circuit to the flow of alternating current. It is denoted by Z and is defined by

$$Z = \frac{V_{RMS}}{I_{RMS}} \qquad (26.5)$$

where V_{RMS} is the RMS supply voltage and I_{RMS} the resulting current. Clearly we could use peak values or mean values in place of RMS in the above equation. Z is decided not only by the resistance R of the circuit but, as we shall soon see, by the presence of inductance or capacitance in the circuit also. In a purely resistive circuit Z equals R because $V_{RMS}/I_{RMS} = R$.

Inductive reactance

Suppose that an alternating voltage is applied to a copper coil of appreciable inductance L (see Chapter 24) but negligible resistance i.e. an "inductor" (Fig. 26.4). The continual changes of current I cause induced voltages that oppose every rise and fall of current. Consequently there

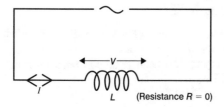

Fig. 26.4   A purely inductive circuit

is opposition to the flow of the alternating current. This opposition due to inductance is called inductive reactance X_L.

It is defined as the ratio $\dfrac{V_p}{I_p}$ or $\dfrac{V_{RMS}}{V_{RMS}}$ and is, of course, measured in ohms. Its magnitude is given by

$$X_L = \omega L \qquad (26.6)$$

where ω is the angular frequency ($=2\pi f$) of the alternating current.

Capacitive reactance

When an alternating voltage is applied to a capacitor C, it repeatedly charges, discharges and recharges the capacitor with opposite polarity for each successive charging. Thus alternating current is flowing in the circuit (see Fig. 26.5).

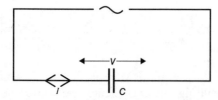

Fig. 26.5   AC circuit containing capacitance but negligible inductance or resistance

The extent of each charging of the capacitor, and hence the size of the current obtained, is limited by the PD that builds up across the capacitor. The current is greater if C is large and the process is rapid (i.e. the frequency is high).

The opposition to alternating current flow due to the presence of capacitance is called 'capacitive reactance' (X_C) defined as $\dfrac{V_{Cp}}{I_p}$ or $\dfrac{V_{C\,RMS}}{I_{RMS}}$.

Its size is given by

$$X_C = \frac{1}{\omega C} \qquad (26.7)$$

So we see that the impedance Z is equal to R or ωL or $1/\omega C$ if the circuit contains only resistance, only inductance or only capacitance respectively.

Example 3

A sinusoidal alternating voltage of 6.0 V RMS and frequency 1000 Hz is applied to a coil of 0.5 H inductance and negligible resistance. What is the expected value for the RMS current?

Method

$$\frac{V_{RMS}}{I_{RMS}} = Z = X_L = \omega L = 2\pi f L$$

$$\therefore \quad I_{RMS} = \frac{V_{RMS}}{2\pi f L} = \frac{6.0}{2\pi \times 1000 \times 0.5}$$

$$= 0.0019 = 1.9 \times 10^{-3}\,\text{A}$$

Answer

1.9 mA.

Example 4

A 25 V peak, 50 Hz sinusoidal voltage is applied to a capacitor. If the peak current is 15.7 mA, what is the value of the capacitance?

Method

$$\frac{V_p}{I_p} = \frac{V_{RMS}}{I_{RMS}} = Z = X_C = \frac{1}{\omega C} = \frac{1}{2\pi f C}$$

$$\therefore \quad \frac{25}{15.7 \times 10^{-3}} = \frac{1}{2\pi \times 50 \times C}$$

$$\therefore \qquad C = \frac{15.7 \times 10^{-3}}{25 \times 2\pi \times 50}$$

$$= 2.0 \times 10^{-6}\,\text{F or } 2.0\,\mu\text{F}$$

Answer

2.0 μF.

Series *LCR* circuits

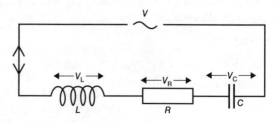

Fig. 26.6   The series *LCR* circuit

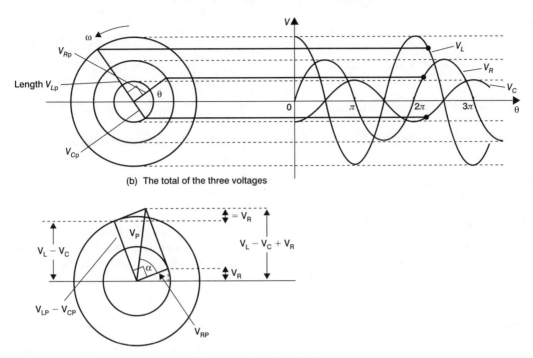

(a) A 'clock with three hands' for V_{Lp}, V_{Rp} and V_{Cp}

(b) The total of the three voltages

Fig. 26.7 Use of rotating phasors with an *LCR* circuit (see Fig. 26.6)

An AC circuit may contain a combination of resistances, inductances and capacitances. We will deal only with the case of these all being in series as shown in Fig. 26.6. Unfortunately the values of resistance R, inductive reactance ωL and capacitive reactance $1/\omega C$ cannot simply be added to find the impedance Z of the circuit.

In fact Z is less than what would result from simple addition of these ohms because the voltages V_L and V_C are not in phase. V_L reaches its peak value (V_{Lp}) a quarter of a cycle before the current peaks and V_C peaks a quarter cycle after the current peaks.

V_R peaks when the current peaks, as you expect. These facts can be illustrated using the rotating phasor method, as in Fig. 26.7a.

In Fig. 26.7 V_{Lp} is shown greater than V_{Rp}, and V_{Cp} is smallest. As a result the total voltage V leads V_R, and so leads the current I (by the phase angle α). If the capacitive reactance played a larger part in the circuit, α would be negative, i.e. the current would reach its peak *before* the total voltage (or supply voltage). α should be remembered as the lag of current behind the supply PD.

At any instant the PDs V_R, V_L and V_C must simply add algebraically. It can be shown that the

phasors V_{Rp}, V_{Lp} and V_{Cp} in Fig. 26.7b agree with this requirement and their resultant, obtained by applying the parallelogram rule (see Chapter 2) is V_p, given by

$$V_p{}^2 = V_{Rp}{}^2 + (V_{Lp} - V_{Cp})^2$$

It follows that, since I_p is the same throughout, that

$$Z^2 = R^2 + \left(\omega L - \frac{1}{\omega C}\right)^2 \qquad (26.8)$$

Also we can see from the triangle containing α in Fig. 26.7b that

$$\tan \alpha = \frac{V_{Lp} - V_{Cp}}{V_{Rp}}$$

or, dividing top and bottom of the fraction by I_p,

$$\tan \alpha = \frac{\omega L - \dfrac{1}{\omega C}}{R} \qquad (26.9)$$

where α is the angle by which the current lags on the supply PD.

By Pythagoras

$$z^2 = R^2 + \left(\omega L - \frac{1}{\omega C}\right)^2$$

and

$$\tan a = \frac{\omega L - 1/\omega C}{R}$$

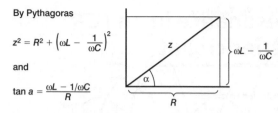

Fig. 26.8 Combining R, ωL and $1/\omega C$ in series

The facts described by Equations 26.8 and 26.9 are summarised in Fig. 26.8.

It is useful to note that

$$\omega L = \frac{V_{LRMS}}{I_{RMS}}, \qquad \frac{I}{\omega C} = \frac{V_{CRMS}}{I_{RMS}}$$

$$R = \frac{V_{RRMS}}{I_{RMS}}, \qquad Z = \frac{V_{RMS}}{I_{RMS}}$$

$$(26.10)$$

Example 5

Calculate the current expected when a 0.30 H coil having 55 Ω resistance is connected to a 22 V RMS, 70 Hz voltage supply.

Method

$1/\omega C = 0$ here because, where a capacitor might have been, we have a low resistance connecting wire instead. Equation 26.8 becomes

$$Z^2 = R^2 + (\omega L)^2,$$

and $Z = V_{RMS}/I_{RMS}$, where V_{RMS} and I_{RMS} are the voltage supply and current.

Also $\omega = 2\pi f$

$L = 0.3$, $\omega = 2\pi f = 2\pi \times 70$, $R = 55$, $V_{RMS} = 22$.

$$\therefore \quad Z^2 = 55^2 + (2 \times \pi \times 70 \times 0.3)^2 = 20\,435$$

$$\therefore \quad Z = 143\,\Omega$$

$I_{RMS}\ (=V_{RMS}/Z) = 22/143 = 0.154\,\text{A RMS}$

Answer

0.15 A RMS.

Example 6

A 16 μF capacitor and an inductive coil of 300 Ω resistance are connected in series across a 20 V, 50 Hz AC supply. The current obtained is 40 mA RMS. What is the inductance of the coil?

Method

$$Z^2 = R^2 + \left(\omega L - \frac{1}{\omega C}\right)^2$$

$$\omega = 2\pi f \quad \text{and} \quad Z = \frac{V_{RMS}}{I_{RMS}}$$

$V_{RMS} = 20$, $I_{RMS} = 40 \times 10^{-3}$, $f = 50$, $R = 300$, $C = 16 \times 10^{-6}$.

$$Z = \frac{20}{40 \times 10^{-3}} = 500\,\Omega, \quad \omega = 2\pi \times 50 = 314.2$$

$$\therefore \quad Z^2 = 500^2 = 300^2 + \left(314.2L - \frac{10^6}{314.2 \times 16}\right)^2$$

$$(314.2L - 199)^2 = 500^2 - 300^2 = 160\,000$$

$$314.2L - 199 = 400 \quad \text{and} \quad L = \frac{599}{314} = 1.9\,\text{H}$$

Answer

1.9 H

Example 7

Calculate the time interval by which the current lags on the 50 Hz supply voltage for a circuit in which a 10 H, 1000 Ω coil only is connected to the supply. This supply has negligible internal resistance and reactance.

Method

$$\tan \alpha = \frac{\omega L - 1/\omega C}{R} \quad \text{(Equation 26.9)}$$

and $\omega = 2\pi f$, $f = 50$, $L = 10$, $R = 1000$, $1/\omega C = 0$.

$$\tan \alpha = \frac{2\pi \times 50 \times 10}{1000} = 3.14$$

$$\therefore \quad \alpha = 72.3°$$

But 360° is a whole cycle, i.e. one-fiftieth of a second.

Therefore the lag is

$$\frac{1}{50} \times \frac{72.3}{360} = 4.0 \times 10^{-3}\,\text{s}$$

Answer

$4.0 \times 10^{-3}\,\text{s}$

Heating by an alternating current

In a resistance R the heat produced per second (i.e. the electrical energy per second or power converted into internal energy within the resistance) is the mean value of I^2R, i.e. $I_{RMS}^2 R$.

In a pure inductance or capacitance there is no production of heat. So the power dissipated in an LCR circuit is

$$P = I_{RMS}^2\, R \quad \text{or} \quad I_{RMS}^2\, Z \cos\alpha$$
$$\text{or} \quad V_{RMS}\, I_{RMS} \cos\alpha \qquad (26.11)$$

($R = Z \cos \alpha$, as shown in Fig. 26.8.)

The product $V_{RMS} I_{RMS}$ is often called the 'apparent power' and $\cos \alpha$, called the power factor, tells us the ratio of true to apparent power.

If $\cos \alpha = 1$, i.e. $P = V_{RMS} I_{RMS}$, then the circuit or device (across which the PD is V_{RMS}) is acting as a pure resistance (a series LCR circuit at resonance for example (see this page)).

Example 8

Calculate the true power and the apparent power in Example 6.

Method

The true power is $I_{RMS}^2 R$. Using $I_{RMS} = 40 \times 10^{-3}$ A and $R = 300 \,\Omega$ we get $(40 \times 10^{-3})^2 \times 300$ which equals 0.48 W.

The apparent power is $V_{RMS} \times I_{RMS}$. Using $V_{RMS} = 20$ V and $I_{RMS} = 40 \times 10^{-3}$ A we get $20 \times 40 \times 10^{-3}$, which equals 0.80 W.

Answer

0.48 W, 0.80 W.

Exercise 26.1

1 A sinusoidal alternating voltage supply has an RMS value of 2.0 V. Calculate (a) the peak voltage, (b) the expected peak current if the circuit's resistance is $20 \,\Omega$.

2 What is the shortest time it takes for a 100 Hz alternating current to change from zero to (a) its peak value, (b) half of its peak value?

3 A sinusoidal voltage supply having an angular frequency ω of $200 \,\mathrm{rad\,s^{-1}}$ and a peak voltage of 100 V is connected to an inductor of 0.50 H and negligible resistance. Calculate (a) the inductive reactance of the inductor, (b) the peak current, (c) the PD at a time of one-sixth of a cycle after the PD was zero, (d) the current at this time.

4 A coil having inductance 0.040 H and resistance X is connected in series to a $25 \,\Omega$ resistor and a sinusoidal voltage supply with a frequency of 50 Hz. If the RMS PD across the coil equals that across the resistor, calculate (a) the impedance of the coil, (b) the value of X.

5 A 6.0 V RMS alternating voltage supply with a frequency of 700 Hz and negligible impedance is connected to a $35 \,\Omega$ resistor and a $3.5 \,\mu F$ capacitor in series. Calculate (a) the impedance of the circuit, (b) the peak PD across the resistor.

6 A 25 W, 100 V heater is to be run from a 250 V 50 Hz sinusoidal AC supply. Calculate the inductance to be included in the circuit.

Resonance in an LCR series circuit

In the formula $Z^2 = R^2 + (\omega L - 1/\omega C)^2$ it can be seen that $Z = R$ if $\omega L = 1/\omega C$, but under all other circumstances Z is greater. Thus for a given PD applied to an LCR series circuit the current is exceptionally high when $\omega L = 1/\omega C$. This condition usually arises as a result of the supply's frequency being varied until $\omega^2 = 1/LC$ (or $\omega = 1/\sqrt{LC}$) or

$$f = \frac{1}{2\pi\sqrt{(LC)}} \qquad (26.12)$$

This phenomenon is called *resonance*. It is the result of the applied frequency matching the circuit's own (or natural) frequency of $\frac{1}{2\pi\sqrt{(LC)}}$. The high current occurs because V_L becomes equal to V_C, so that the would-be opposition to current flow due to L is cancelled by that due to C.

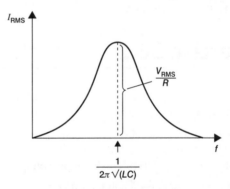

Fig. 26.9   Current in an LCR series circuit, showing resonance

At resonance

$$\tan \alpha \left(= \frac{\omega L - 1/\omega C}{R} \right) = 0$$

i.e. α is zero and I is in phase with the supply PD V. Resonance is illustrated in Fig. 26.9.

Example 9

(a) Calculate the resonant frequency for a series LCR circuit in which $L = 0.010$ H, $C = 1.0 \,\mu F$ and $R = 20 \,\Omega$.

(b) If the voltage supply is 12 V RMS, what current flows at resonance?

(c) What is the RMS PD across L and across C at resonance?

Method

(a) We use Equation 26.12:
$$f = 1/2\pi\sqrt{(LC)}, \quad L = 0.01, \quad C = 1.0 \times 10^{-6}.$$

$$f = \frac{1}{2\pi\sqrt{(0.01 \times 1.0 \times 10^{-6})}}$$

$$= \frac{1}{2\pi\sqrt{10^{-8}}} = \frac{10^4}{2\pi} = 1591 \text{ Hz}$$

(b) $I_{RMS} = \dfrac{V_{RMS}}{Z} = \dfrac{V_{RMS}}{R} = \dfrac{12}{20} = 0.60 \text{ A RMS}$

(c) $\dfrac{V_{LRMS}}{I_{RMS}} = \omega L$

$\therefore \quad V_{LRMS} = 2\pi f \times 0.01 \times I_{RMS}$

$$= 2\pi \times 1591 \times 0.01 \times 0.6$$

$$= 60 \text{ V RMS}$$

$\dfrac{V_{CRMS}}{I_{RMS}} = \dfrac{1}{\omega C}$ gives 60 V also for V_{CRMS}

(not surprisingly since, at resonance, V_C is equal and opposite in polarity to V_L at all times).

Answer

1.6 kHz, 0.60 A RMS, 60 V, 60 V.

Exercise 26.2

1 Calculate the resonant frequency for a 0.20 H inductor in series with a 2.0 μF capacitor.

2 What size of capacitor is needed in series with a 2.0 H, 100 Ω coil in order to get the current in phase with the 50 Hz, 240 V supply voltage?

What size of current will then flow?

3 A 100 Ω resistor, a 1.0 μF capacitor and a 0.20 H inductor of negligible resistance are connected in series with a supply of sinusoidal alternating EMF of 20 V RMS whose frequency f can be varied (see Fig. 26.10). Calculate (a) the resonant frequency, (b) the value of the maximum RMS current, (c) the RMS voltage across each of the components at this frequency.

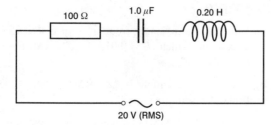

100 Ω   1.0 μF   0.20 H

20 V (RMS)

Fig. 26.10   Circuit for Question 3

The transformer

A transformer consists of a primary coil to which an alternating voltage V_1 is applied, and a secondary coil from which the required alternating voltage V_2 is obtained as the result of mutual induction between the two coils. The coils are often wound on a core of magnetic material such as iron.

In an ideal transformer the coil resistances are negligible and eddy-current heating in the core is negligible (so that there is no energy wastage as heat). Also the flux Φ passes through ('links with') all the turns of the coils. For this ideal transformer V_2 and V_1 (RMS values) are related by

$$\frac{V_2}{V_1} = \left(\frac{n_2 d\Phi/dt}{n_1 d\Phi/dt}\right) = \frac{n_2}{n_1} = \text{the turns ratio } n \quad \text{(26.13)}$$

where n_2 and n_1 are the number of turns on the secondary coil and the primary coil respectively. A step-up transformer produces a secondary voltage greater than the primary voltage: the opposite is step-down.

Assuming a small resistive load on the secondary (Fig. 26.11), the currents I_1 and I_2 are approximately in phase with their respective voltages so that the power dissipated in the primary is $V_1 I_1$ and the power dissipated in the load is $V_2 I_2$. Since we are assuming no losses these powers can be equated, giving $V_1 I_1 = V_2 I_2$, from which

$$\frac{I_2}{I_1} = \frac{V_1}{V_2} = \frac{n_1}{n_2} = \frac{1}{n} \quad \text{(26.14)}$$

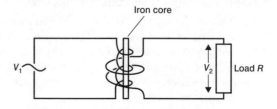

Iron core

V_1 V_2 Load R

Fig. 26.11   Principle of the transformer

If the load on the secondary is resistance R, then $V_2/I_2 = R$, and so

$$\frac{V_1}{I_1} \left(= \frac{V_2/n}{nI_2}\right) = \frac{R}{n^2} \quad \text{(26.15)}$$

Transformers used for high-voltage power distribution

The product of RMS voltage and RMS current is the power. (We will assume no phase difference.) Consequently the same power is obtainable with high voltage and low current or with lower voltage and higher current. A generator in a power station produces its power for the consumer by means of a large current at low voltage, but for transmission of this power it is more convenient and economical to use small currents. Transformers are employed to step up the voltage and subsequently to step down the voltage again before it reaches the consumer.

Example 10

(a) What turns ratio would be needed for an ideal transformer to provide 12 V RMS when connected to 240 V RMS mains supply?

(b) If the transformer were loaded with a non-inductive 12 V, 60 W heater, what current would flow in the mains supply lead?

(c) If the transformer in practice gives 11.8 V RMS and 4.5 A RMS when the primary current is 0.25 A RMS, what is the efficiency of energy conversion by the transformer?

Method

(a) According to Equation 26.13,

$$\frac{V_2}{V_1} = \frac{n_2}{n_1} = n$$

$V_2 = 12$, $V_1 = 240$.

$$\therefore \quad n = \frac{12}{240} = \frac{1}{20}$$

(The primary has the greater number of turns.)

(b) From Equation 26.14

$$\frac{I_2}{I_1} = \frac{1}{n}$$

I_2 is known from the fact that, for a power of 60 W using 12 V, the current must be 60 W divided by 12 V, i.e. 5.0 A. $n = \frac{1}{20}$

$$\frac{I_2}{I_1} = \frac{5.0}{I_1}$$

but $\quad \frac{I_2}{I_1} = \frac{1}{n} = \frac{1}{1/20}$

$$= 20$$

$$\therefore \quad \frac{5.0}{I_1} = 20 \text{ or } I_1 = 0.25 \text{ A}$$

(c) The input power is $V_1 I_1 = 240 \times 0.25 = 60$ W.

The output power is $V_2 I_2 = 11.8 \times 4.5 = 53.1$ W.

$$\text{Efficiency} = \frac{\text{Useful power output}}{\text{Power input}}$$

$$= \frac{53.1}{60} = 0.88 \quad \text{or} \quad 88\%$$

Answer

(a) $\frac{1}{20}$, (b) 0.25 A, (c) 88%.

Exercise 26.3

1 A transformer with a 100-turn primary winding and a 500-turn secondary winding is connected to a 2.0 V RMS supply. Calculate values for the output voltage from the secondary, and the maximum secondary current if the primary winding is to be limited to 0.10 A. State the assumptions made.

2 10 kW of electrical power at 100 V RMS is to be delivered by use of a 5.0 Ω cable. What is the rate of heat production in the cable?

If, instead, the power were transmitted, again through 5.0 Ω, but at 30 kV stepped down to 100 V at the user-end of the cable, what would be the new heat dissipation in the cable? Assume the transformer to be ideal, i.e. no energy loss occurs.

Exercise 26.4: Examination questions

1 A simple generator has a 300-turn rectangular coil of dimensions 20 mm × 35 mm. The coil rotates in a uniform magnetic field of flux density 0.25 T. How many revolutions must the coil make per second in order to produce a peak output of 12 V?
[CCEA 2001, part]

2 A capacitor having a capacitance of 1.0 μF is connected to a 10 V RMS, 700 Hz sinusoidal voltage supply. Calculate the RMS value of the expected current. (Take π as 3.142.)

3 A source of sinusoidal alternating voltage is connected to a component which is concealed inside a box. The graph of Fig. 26.12(a) shows the variation of the voltage V across the component with time t. The simultaneous current I in the component is shown in Fig. 26.12(b).

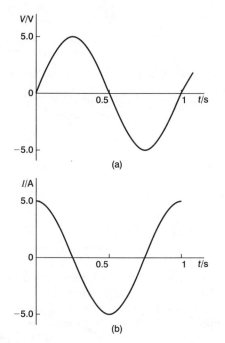

Fig. 26.12

(a) Identify the component, giving the reason for your answer.

(b) Using data from the graphs, calculate:
   (i) the frequency of the voltage across the component;
   (ii) the root-mean-square value of this voltage;
   (iii) the reactance of the component.

[OCR 2001]

4 A coil has an inductance of 0.20 H and a resistance of 100 Ω. Find the peak voltage across the coil when a 50 Hz alternating current of 1.5 A rms flows through it. What is the phase difference between voltage and current, and which leads?

[WJEC 2000, spec]

5

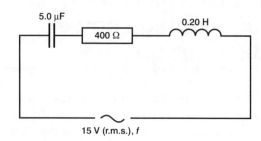

Fig. 26.13

An a.c. source of 15 V (r.m.s.) and variable frequency f is used as shown in the above circuit.

(a) At a certain value of the frequency the p.d. across the resistor is equal to that of the source and the current then has its maximum value. Calculate the value of this frequency.

(b) The frequency is now doubled.

Calculate

   (i) the new value of the r.m.s. current,
   (ii) the power dissipated in the circuit.

[WJEC 2000]

6 A transformer, assumed to be 100% efficient, is used with a supply voltage of 120 V. The primary winding has 50 turns. The required output voltage is 3000 V. The output power is 200 W.

(a) Name this type of transformer.

(b) Calculate the number of turns in the secondary winding.

(c) Calculate the current supplied to the primary winding.

[CCEA 2001, part]

Section H
Atomic and nuclear physics

27
Photoelectric emission and atomic structure

Photoelectric emission from the surface of a solid

Electromagnetic radiation is made up of separate ('discrete') quantities of energy which we may describe as light particles (photons). Each photon consists of energy hf joules, where f is the frequency of the light and h is the Planck constant (or hc/λ because f = velocity of light/wavelength $= c/\lambda$). For an electron to escape from a solid by photo-electric emission it must acquire the energy of an incident photon and use this energy to (1) 'get to the solid's surface' and (2) get through the surface (the energy needed is called the work function energy, WFE), leaving it with (3) some kinetic energy $\frac{1}{2}mv^2$. Thus

$$\begin{array}{c} hf \\ (\text{or } hc/\lambda) \end{array} = \left(\begin{array}{c} \textbf{Energy} \\ \textbf{to get to} \\ \textbf{surface} \end{array} \right) + \text{WFE} + \tfrac{1}{2}mv^2 \qquad (27.1)$$

where c is velocity of the light, λ the wavelength, m mass of the electron, v velocity of the escaped electron (photoelectron).

If $hf <$ WFE then no electron emission occurs.

Of all the electrons escaping the *fastest* will be those which did not have to use energy to reach the surface so that, for them,

$$hf \left(\text{or } \frac{hc}{\lambda} \right) = \text{WFE} + \tfrac{1}{2}mv^2 \qquad (27.2)$$

If an electrode is placed near the emitting surface and is made negative by V volts, then the photo-electrons can be repelled back to the surface. Even the fastest electrons that aim directly at the negative electrode will be prevented from reaching it if the retarding PD V equals or exceeds the value given by

$$eV = \tfrac{1}{2}mv^2 \text{ which equals } \frac{hc}{\lambda} - \text{WFE} \qquad (27.3)$$

where eV (the work to be done in reaching the electrode) is the electron charge $\times$ PD.

Work function energy can be quoted in joules or electron-volts. This $\frac{1}{2}mv^2$ is the highest kinetic energy that an escaping electron can have (KE_{max}). The work function *voltage* is the PD needed to accelerate electrons to such an energy.

The electron-volt (eV)

This is a unit of energy which is particularly useful in particle physics (e.g. atomic and nuclear calculations). It is the energy acquired by an electron freely accelerated (i.e. in vacuum) through a PD of 1 volt. Therefore, since work $W = qV$,

$$1\,eV = e\ \text{joule} \qquad (27.4)$$

where e is the electronic charge (1.6×10^{-19} when working in SI units).

$$\therefore \quad 1\,eV = 1.6 \times 10^{-19}\,\text{J}.$$

Example 1

Electromagnetic radiation of frequency 0.88×10^{15} Hz falls upon a surface whose work function is 2.5 V.

(a) Calculate the maximum kinetic energy of photo-electrons released from the surface.

(b) If a nearby electrode is made negative with respect to the first surface using a PD V, what value is required for V if it is to be just sufficient to stop any of the photoelectrons from reaching the negative electrode?

(Planck constant $h = 6.6 \times 10^{-34}\,\text{J s}$, electron charge $e = -1.6 \times 10^{-19}\,\text{C}$.)

Method

(a) Using Equation 27.1 or 27.2,

$$hf = \text{WFE} + \left(\begin{array}{c}\text{Kinetic energy of}\\\text{fastest photoelectrons}\end{array}\right)$$

we have

$$6.6 \times 10^{-34} \times 0.88 \times 10^{15}$$
$$= 2.5 \times 1.6 \times 10^{-19} + E_{\text{max}}$$

(2.5 is multiplied by 1.6×10^{-19} here in order to convert the 2.5 eV energy to joules.)

So

$$E_{\text{max}} = 5.8 \times 10^{-19} - 4.0 \times 10^{-19}$$
$$= 1.8 \times 10^{-19}\,\text{J}$$

In electron-volts,

$$E_{\text{max}} = \frac{1.8 \times 10^{-19}}{1.6 \times 10^{-19}} = 1.125\,\text{eV} \quad \text{or} \quad 1.1\,\text{eV}$$

(b) Working in joules again (our equations are all written for SI units) we have, from Equation 27.3

$$eV = \tfrac{1}{2}mv^2$$

or

$$eV = E_{\text{max}} = 1.8 \times 10^{-19}\,\text{J}$$

$$\therefore \quad V = \frac{1.8 \times 10^{-19}}{1.6 \times 10^{-19}} = \frac{1.8}{1.6} = 1.125 \quad \text{or} \quad 1.1\,\text{V}$$

More simply, $E_{\text{max}} = 1.1\,\text{eV}$ and retarding PD $= 1.1\,\text{V}$.

Answer

(a) 1.1 eV, (b) 1.1 V.

De Broglie wavelength for a particle of matter

Light and other electromagnetic radiations must be regarded as waves but also as particles (quanta of energy), i.e. photons. Each photon then has a mass $m = E/c^2$ where E is the energy of the photon and c is the velocity of light (see also Chapter 29). Using $E = hc/\lambda$ for the photon:

$$m = \frac{hc/\lambda}{c^2} \quad \text{or} \quad mc = \frac{h}{\lambda}$$

$$\text{Momentum } mc = \frac{h}{\lambda} \qquad (27.5)$$

De Broglie proposed that any particle of *matter*, e.g. an electron or proton, has, like a photon, both wave and particle properties, so that it has a wavelength given by

$$\lambda = \frac{h}{\text{momentum}} = \frac{h}{mv} \qquad (27.6)$$

where v is the particle's velocity, m its mass.

Note that the electron wave's velocity is not equal to the velocity of light c and so $E = \dfrac{hc}{\lambda}$ does not apply.

Example 2

Calculate the wavelength of electrons that have been accelerated from rest through a PD of 100 V. What kind of electromagnetic radiation has wavelengths similar to this value?

(Electron mass $m = 9.1 \times 10^{-31}$ kg, electron charge $e = -1.6 \times 10^{-19}$ C, Planck constant $h = 6.6 \times 10^{-34}$ J s.)

Method

From Equation 27.6, wavelength $= h/\text{momentum}$.

To find the electron's momentum:

$$\tfrac{1}{2}mv^2 = eV \quad \text{(see Equation 21.5)}$$

$$\therefore \quad (mv)^2 = 2meV$$

$$\therefore \quad \text{Momentum} = mv = \sqrt{(2meV)}$$

$$\text{so} \quad \lambda = \frac{h}{\text{Momentum}} = \frac{h}{\sqrt{(2meV)}}$$

$$h = 6.6 \times 10^{-34}, \quad m = 9.1 \times 10^{-31}, \quad e = 1.6 \times 10^{-19},$$
$$V = 100.$$

$$\therefore \quad \lambda = \frac{6.6 \times 10^{-34}}{\sqrt{(2 \times 9.1 \times 10^{-31} \times 1.6 \times 10^{-19} \times 100)}}$$

$$= 1.22 \times 10^{-10} \text{ m}$$

In the electromagnetic spectrum this would be X-radiation.

Answer

1.2×10^{-10} m, X-radiation.

Example 3

A monochromatic source emits a narrow, parallel beam of light of wavelength 546 nm, the power in the beam being 0.080 W. How many photons leave the source per second? If this beam falls on the cathode of a photocell, what is the photocell current, assuming that 1.5% of the photons incident on the cathode liberate electrons?

(Planck constant $= 6.6 \times 10^{-34}$ J s,
velocity of light in vacuum $= 3.0 \times 10^8$ m s^{-1},
electronic charge $= 1.6 \times 10^{-19}$ C.)

Method

At 546 nm, i.e. 546×10^{-9} m wavelength the photon energy is hc/λ and equals

$$\frac{6.6 \times 10^{-34} \times 3 \times 10^8}{546 \times 10^{-9}} \quad \text{or} \quad 3.626 \times 10^{-19} \text{ J}$$

The number of photons per second

$$= \frac{\text{Joules per second}}{\text{Photon energy}} = \frac{0.08}{3.626 \times 10^{-19}}$$

$$= 2.2 \times 10^{17}$$

The number of electrons liberated per second

$$= \frac{1.5}{100} \times 2.2 \times 10^{17} = 3.3 \times 10^{15}$$

The current

$$= \text{Electrons per second} \times \text{electronic charge}$$

$$= 3.3 \times 10^{15} \times 1.6 \times 10^{-19}$$

$$= 5.28 \times 10^{-4} \text{ A} \quad \text{or} \quad 0.53 \text{ mA}$$

Answer

2.2×10^{17}, 0.53 mA.

Circular orbits

For an electron (mass m, charge $-e$) in circular orbit, radius r, about a nucleus containing Z protons (each of charge $+e$) we have

Electrostatic force of attraction (Equation 21.1) is

$$F = \frac{q_1 q_2}{4\pi\varepsilon_0 r^2} = \frac{Ze \times e}{4\pi\varepsilon_0 r^2}$$

and this must equal mv^2/r (see Chapter 8) where v is the electron's speed.

$$\frac{mv^2}{r} = \frac{Ze^2}{4\pi\epsilon_0 r^2} \tag{27.7}$$

In addition it is found that the distance round the orbit $2\pi r$, must equal a whole number n of electron wavelengths, so that

$$2\pi r = n\lambda = \frac{nh}{mv}$$

(using the de Broglie relation, 27.6)

$$\therefore \quad 2\pi mvr = nh \tag{27.8}$$

where n may be 1, 2, 3, ... The above two equations can be solved to find r for each n value. For each of these allowed orbits we can calculate the electron's energy E. These allowed energy values are called energy levels.

The circular orbit closest to the nucleus has $n = 1$, and the energy of an electron here is lower than for $n = 2, 3, \ldots$ The hydrogen atom has one electron only and this will normally reside in this innermost orbit, i.e. this is its ground state. The hydrogen atom is simple also because the energies for its elliptical orbits are near enough the same as for the circular ones.

Atomic electrons may be classified in groups called shells. In the hydrogen atom all electrons of the same shell have the same energy and same n value.

Excitation

An atomic electron if it acquires sufficient energy, e.g. from a colliding particle or from incident electromagnetic radiation (a photon), can move within the atom to a higher energy level. Typically it will stay at this 'excited' level for only a minute fraction of a second and then 'fall back' to its ground state or other lower energy level, giving out energy as a photon of electromagnetic radiation (often visible light).

$$E - E' = hf \qquad (27.9)$$

where E is electron energy in the higher state and E' electron energy in the lower state; h is the Planck constant, f the frequency of radiation emitted as a photon (energy hf). This process is called *excitation*.

Example 4

The three lowest energy levels of the electron in the hydrogen atom have energies

$$E_1 = -21.8 \times 10^{-19}\,\text{J}$$
$$E_2 = -5.45 \times 10^{-19}\,\text{J}$$
$$E_3 = -2.43 \times 10^{-19}\,\text{J}$$

The energies are measured so that the electron would have zero energy if it were completely free of the atom and at rest.

(a) What is the wavelength of the H_α spectral line due to transition between levels E_3 and E_2?

(b) Through what potential difference must an electron be accelerated to enable it to
  (i) ionise a normal hydrogen atom
  (ii) cause emission of the H_α line.

(Planck constant $= 6.6 \times 10^{-34}\,\text{J s}$,
speed of light in vacuum $= 3.0 \times 10^8\,\text{m s}^{-1}$,
electron charge $= 1.6 \times 10^{-19}\,\text{C}$.)

Method

(a) (Note that the energy values given are negative because of the zero chosen.)

For the E_3 to E_2 transition, using Equation 27.9:

$$(-2.43 \times 10^{-19}) - (-5.45 \times 10^{-19})$$

$$= hf = \frac{hc}{\lambda}$$

$$3.02 \times 10^{-19} = \frac{6.6 \times 10^{-34} \times 3.0 \times 10^8}{\lambda}$$

$$\therefore \qquad \lambda = \frac{6.6 \times 3}{3.02} \times 10^{-7}\,\text{m}$$

$$= 6.6 \times 10^{-7}\,\text{m}$$

Accelerating PD needed:

(b) (i) For ionisation of normal atom, transition is from ground state (E_1) to outside of the atom (our zero of energy).

$$eV\left(= \tfrac{1}{2}mv^2\right) = (0) - (-21.8 \times 10^{-19})$$

$$V = \frac{21.8 \times 10^{-19}}{1.6 \times 10^{-19}} = 13.6\,\text{V}$$

(ii) For emission of H_α radiation an electron must be moved from ground state (E_1) to E_3 so that it can fall to E_2.

$$\therefore \quad eV = (-2.43 \times 10^{-19}) - (-21.8 \times 10^{-19})$$

$$= 19.4 \times 10^{-19}$$

$$\therefore \quad V = \frac{19.4 \times 10^{-19}}{1.6 \times 10^{-19}} = 12.1\,\text{V}$$

Answer

(a) $6.6 \times 10^{-7}\,\text{m}$,   (b) (i) $13.6\,\text{V}$,   (ii) $12.1\,\text{V}$.

Exercise 27.1

(Where necessary take
velocity of light in vacuum $c = 3.0 \times 10^8\,\text{m s}^{-1}$,
Planck constant $h = 6.6 \times 10^{-34}\,\text{J s}$.)

1 Calculate the frequency and the photon energy for blue light of wavelength $4.0 \times 10^{-7}\,\text{m}$.

2 Through what potential difference must electrons be accelerated to be able to produce visible light of wavelength $6.0 \times 10^{-7}\,\text{m}$?
(Electron charge $e = -1.6 \times 10^{-19}\,\text{C}$.)

3 A metal surface is illuminated with monochromatic light and it becomes charged to a steady positive potential of $1.0\,\text{V}$ relative to its surroundings. The work function energy of the metal surface is $3.0\,\text{eV}$, and the electron charge e is $1.6 \times 10^{-19}\,\text{C}$. Calculate the frequency of the light.

4 A clean surface is irradiated with light of wavelength $5.5 \times 10^{-7}\,\text{m}$ and electrons are just able to escape from the surface. When light of wavelength $5.0 \times 10^{-7}\,\text{m}$ is used, electrons emerge with energies of up to $3.6 \times 10^{-20}\,\text{J}$. Obtain a value for the Planck constant h.

5 The beam of light from a certain laser has a power of $1.0\,\text{mW}$ and a wavelength of $633\,\text{nm}$. How many photons are emitted per second by this laser?

6 Calculate the de Broglie wavelength of $300\,\text{V}$ electrons.
(Mass of electron $= 9.1 \times 10^{-31}\,\text{kg}$,
electron charge $= -1.6 \times 10^{-19}\,\text{C}$)

7 The first excitation energy of the hydrogen atom is $10.2\,\text{eV}$. Calculate the speed of the slowest electron that can excite a hydrogen atom.
(Electron charge/mass ratio e/m
$= 1.7 \times 10^{11}\,\text{C kg}^{-1}$.)

Exercise 27.2:
Examination questions

Where necessary use

electronic charge $(e) = 1.60 \times 10^{-19}\,\text{C}$
electronic mass $(m) = 9.11 \times 10^{-31}\,\text{kg}$
velocity of light in vacuum $(c) = 3.00 \times 10^{8}\,\text{m s}^{-1}$
Planck constant $(h) = 6.63 \times 10^{-34}\,\text{J s}$
1 electron-volt $(\text{eV}) = 1.60 \times 10^{-19}\,\text{J}$

1 An electron travelling at $8.0 \times 10^{6}\,\text{m s}^{-1}$ in a vacuum enters a region of uniform magnetic field of flux density 30 mT, as shown in Fig. 27.1.

Fig. 27.1

 (i) On Fig. 27.1, mark the direction of the force on the electron when it enters the magnetic field at **P**.
 (ii) Calculate the magnitude of the force on the electron.
 (iii) Explain why, when the electron is moving in the magnetic field, it follows part of a circular path.
 (iv) Calculate the radius of this circular path.
[CCEA 2000, part]

2 Ultraviolet light of wavelength 12.2 nm is shone on to a metal surface. The work function of the metal is 6.20 eV.

Calculate the maximum kinetic energy of the emitted photoelectrons.

Show that the maximum speed of these photoelectrons is approximately $6 \times 10^{6}\,\text{m s}^{-1}$.

Calculate the de Broglie wavelength of photoelectrons with this speed.

Explain why these photoelectrons would be suitable for studying the crystal structure of a molecular compound. [Edexcel 2001]

3 The diagram (Fig. 27.2) shows some of the energy levels for atomic hydrogen.

Add arrows to the diagram showing all the single transitions which could ionise the atom.

Why is the level labelled −13.6 eV called the ground state?

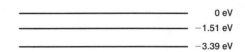

Fig. 27.2

Identify the transition which would result in the emission of light of wavelength 660 nm.
[Edexcel 2000, part]

4 In a simple model of the hydrogen atom, an electron of mass m_{e} and charge $-e$ is supposed to move in a circular orbit of radius r about a proton of charge $+e$. The mass of the proton is very much greater than that of the electron. The linear speed of the electron in orbit is v.
 (i) Write down an expression for the electrical force between the electron and the proton. This electrical force provides the centripetal force required to make the electron move in circular orbit. Hence obtain an expression for v in terms of ε_{o}, e, m_{e} and r.
 (ii) In this model, only certain values $r_1, r_2, \ldots r_n, \ldots$ of the orbital radius are allowed. The corresponding values of the orbital speed are $v_1, v_2, \ldots v_n, \ldots$. The relation fixing the values of r_n and v_n is

$$m_{\text{e}}v_n r_n = \frac{nh}{2\pi}, \text{ where } n = 1, 2, 3, \ldots, \quad (27.10)$$

and h is the Planck constant.

 1. Use Equation 27.10 and your answer to (i) to show that

$$r_n = An^2,$$

 where A is a constant. Obtain an expression for A in terms of ε_{o}, e, h and m_{e}.
 2. Hence calculate the radius of the **smallest** electron orbit.
 3. Draw a sketch showing the proton, the smallest electron orbit, and the next three orbits.
 4. According to the de Broglie theory, a moving particle has an associated wavelength. Use the de Broglie relation $(\lambda = h/p)$ and Equation 27.10 to show that the de Broglie wavelength of the electron in the smallest orbit is equal to the circumference of that orbit. Deduce how the de Broglie wavelengths of the electron in the next three orbits are related to the circumference of these orbits. [CCEA 2000]

28
Radioactivity and X-rays

The nucleus

The nucleus of an atom consists of a number Z of protons and a number N of neutrons. Z is called the atomic number or proton number. The sum of the proton number and the neutron number is the total number of nuclear particles (or nucleons). It is usually denoted by A and is often called the mass number. For example $^{238}_{92}U$ denotes a uranium atom for which $Z = 92$ and $Z + N = 238$.

Radioactivity

Some nuclei, because of the particular numbers of protons and neutrons they contain, are unstable and may change (i.e. decay or disintegrate) at any time. These decays usually cause emission of an α or β particle, often followed immediately by emission of a photon of γ radiation.

A typical radioactive source consists of a large number n of such nuclei, all of the same kind (i.e. all the same nuclide), and the number of decays per second, or (what is the same thing) the reduction in the value of n occurring in a second is called the 'activity' of the source. The unit for activity is the becquerel (Bq). It means 1 decay per second.

In a small time δt the reduction in n is $-\delta n$, so the activity is $A = -\dfrac{\delta n}{\delta t}$.

Now A is found to be proportional to n and we write $A = \lambda n$ and λ is called the 'radioactive decay constant'. This λ is characteristic of the particular nuclide concerned. Its SI unit is s^{-1}.

This means that

$$A = \lambda n \text{ and } -\frac{\delta n}{\delta t} = \lambda n \qquad (28.1)$$

Writing the last equation as $\dfrac{dn}{d(\lambda t)} = -n$ shows that we have a decay that is exponential, i.e. has the form $\dfrac{dy}{dx} = -y$, so that $y = y_0 e^{-x}$ (see Chapter 2). Hence with $x = \lambda t$,

$$n = n_0 e^{-\lambda t} \qquad (28.2a)$$

Multiplying both sides of this equation by λ gives

$$A = A_0 e^{-\lambda t} \qquad (28.2b)$$

e is the exponential function (=2.718 on your calculator).

The decay graph relating n and t is shown in Fig. 28.1. The use of a log-linear graph for describing radioactive decay is explained in Chapter 30.

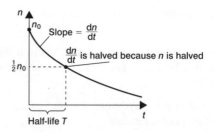

Fig. 28.1   Graph for radioactive decay

The time T required for n to fall from n_0 to $\frac{1}{2}n_0$ is called the half-life of the nuclide, and from Equation 28.2a we can get

$$0.5 = e^{-\lambda T} \text{ or } -\ln 2 = -\lambda T$$

$$\text{or } \quad T = \frac{0.693}{\lambda} \qquad (28.3)$$

An alternative and more obvious formula for n is

$$n = \frac{n_0}{2^{\frac{t}{T}}} \quad \text{or} \quad n = n_0 2^{-\frac{t}{T}} \qquad (28.4a)$$

and $\quad A = A_0 2^{-t/T}$ (28.4b)

where $Y = t/T$ = the number of half-lives. Thus if the time elapsed is T, then $Y = 1$ and $n = \frac{1}{2}n_0$; if $2T$, then the answer is halved again to give $\frac{1}{4}n_0$, because $Y = 2$.

Example 1

(a) Calculate the decay constant in s^{-1} of a radioactive source that has a half-life of 23 days.

(b) If this source has an initial activity of 40 kBq what will its activity be 10 days later?

Method

(a) $\lambda = \dfrac{\ln 2}{T} = \dfrac{0.6931}{T}$

and $T = 23$ days $= 23 \times 24$ hours $= 23 \times 24 \times 60 \times 60$ s
$= 1.987 \times 10^6$ s

$\lambda = \dfrac{0.6931}{1.987 \times 10^6} = 3.488 \times 10^{-7}\,s^{-1}$

or $3.5 \times 10^{-7}\,s^{-1}$.

(b) $A = A_0 2^{-t/T}$

$= 40 \times 10^3 \times 2^{-\frac{10}{23}} = 40 \times 2^{-0.4348} \times 10^3$

$= 40 \times 0.7398 \times 10^3 = 29.58 \times 10^3$ Bq

Answer

(a) $3.5 \times 10^{-7}\,s^{-1}$, (b) 30 kBq

Comment

In part (b) t was given in days and so it is was easier to use the equation containing T for which we had a value in days rather than the $A = A_0 e^{-\lambda t}$ equation in which λ involved seconds.

The value inserted for A_0 in the calculation was converted to SI units by introducing 10^3, but subsequently the answer was divided by 10^3 to get an answer in kBq. If this cancelling of the conversion factor is foreseen, then A_0 could have been left in kBq.

Exercise 28.1

1 The half-life of radium is 1620 years. How long will it take for the initial activity of a radium compound to fall to a fifth of its original value? (Remember $\log a^b = b \log a$.)

2 The activity of a radioactive source decreases by seven-eighths of its initial value in 30 hours. Calculate the half-life and the decay constant for this source. ($\log_e 2 = 0.693$.)

3 Calculate the number of phosphorus-32 atoms required to give an activity of 30 kBq. (Decay constant for P-32 is $5.6 \times 10^{-7}\,s^{-1}$.)

Mass of nuclide related to activity

The mass of a nuclide needed to provide a certain activity A can be calculated as follows:

$$A = \lambda n = \lambda \times \frac{\text{Mass of sample}}{\text{Mass of atom}} \quad (28.5)$$

The mass of the atom may be given in kg or in unified atomic mass units (u). Since the mass of a nucleon is approximately 1 u, the mass of an atom in u is approximately equal to its nucleon number (mass number).

Also, since the molar mass for hydrogen atoms is one gram, the mass of one-atom (and so for one nucleon approximately) is 1 gram/Avagadro number (see also Chapter 2).

Example 2

Calculate the activity of 1 g of pure $^{238}_{92}U$ given that its half-life is 4.5×10^9 year. ($1\,u = 1.66 \times 10^{-27}$ kg.)

Method

$$A = \lambda n = \lambda \times \frac{\text{Mass of sample}}{\text{Mass of atom}}$$

$\lambda = \ln 2/T$, $T = 4.5 \times 10^9$ year, mass of atom $= 238\,u$, mass of sample $= 10^{-3}$ kg

$\therefore\; A = \dfrac{0.693}{4.5 \times 10^9} \times \dfrac{10^{-3}}{238 \times 1.66 \times 10^{-27}}$

$= 3.9 \times 10^{11}$ year^{-1}

or $\dfrac{3.9 \times 10^{11}}{365 \times 24 \times 60 \times 60}\,s^{-1}$, which is 12 kBq.

Answer

12 kBq.

Isotopes

These are different nuclides of the same element. For example $^{238}_{92}U$ and $^{235}_{92}U$ are isotopes. They differ only in the number of neutrons they contain.

Example 3

Calculate the atomic mass for natural uranium which consists of 99.3% ^{238}U and 0.7% ^{235}U. The atomic masses of ^{238}U and ^{235}U are 238.051 u and 235.044 u.

Method

The atomic mass required is the mean for all its atoms.

Out of every 100 atoms we would have 99.3 weighing 238.051 u each and 0.7 weighing 235.044 u each. Therefore the average mass is

$$\frac{(99.3 \times 238.051) + (0.7 \times 235.044)}{100}$$

and equals 238.03 u.

Note that we can say the *relative* atomic mass is 238.03 or 238.03 g per mole.

Answer

238.03 u.

Exercise 28.2

1 The atomic mass of ordinary chlorine gas is 35.5 u, while it is known that this gas is made up of two isotopes whose atomic masses are 35 u and 37 u. Calculate the ratio of the number of the heavier atoms to the number of lighter ones.

2 The radioactive nuclide radium-221 has a half-life of 30 s.

 (a) What approximately is its atomic mass
   (i) in unified atomic mass units (u) and
   (ii) in grams given that the mass of a normal carbon-12 atom is 2.0×10^{-26} kg?

 (b) How many atoms are there in 0.10 g of the pure element?

 (c) Calculate the activity of 0.10 g of the pure element.
   (ln 2 = 0.693)

3 In a certain radioactive source 16% of the atoms are radioactive with a half-life of 3 days. What percentage of the atoms in the source 9 days later will be radioactive?

Nuclear changes

An alpha particle consists of 2 protons and 2 neutrons holding together, like a helium nucleus, and travelling very fast. When an α-emission occurs, the proton number falls by 2 and the nucleon number falls by 4. For β^--emission the nucleus changes one of its neutrons into a proton plus a negative electron. This electron is emitted from the nucleus at extremely high speed as the β-particle. Thus the nucleon number does not change, but the proton number increases by one. β^+-emission causes decrease of the proton number by one, a β^+ particle being a positive electron.

Example 4

A radioactive isotope of cobalt $^{60}_{27}$Co decays by emission of a beta particle followed by emission of one or more gamma photons. Thus we may write

$$^{60}_{27}\text{Co} \rightarrow \text{P} + \beta + \gamma$$

where P is the product nuclide, β is the beta particle and γ represents one or more gamma photons. The atomic number of P is

A 29 **B** 28 **C** 27 **D** 26 **E** 25

Method

Beta emission increases the atomic number (i.e. proton number) by 1, so that the atomic number of P is $27 + 1$ or 28. Note that β means β^- unless otherwise stated.

Answer

B.

Radioactive dating

A good example of this concerns the decay of the carbon $^{14}_{6}$C isotope which is a β^--emitter and is found in a small, known concentration in atmospheric carbon dioxide. Plant or other material contains carbon taken from the atmosphere at the time of the material's formation. When the living material dies it will subsequently contain a decreasing proportion of $^{14}_{6}$C to normal $^{12}_{6}$C atoms as the carbon-14 slowly decays.

Example 5

The half-life of $^{14}_{6}$C is 5570 years. (a) What is its decay constant? (b) How many disintegrations per second are obtained from 1 g of carbon if 1 carbon atom in 10^{12} is of the radioactive $^{14}_{6}$C type? (c) After what time will the activity per gram have fallen to 3 disintegrations per minute? ($1 \text{ u} = 1.66 \times 10^{-27}$ kg.)

Method

(a) $\lambda = \dfrac{\ln 2}{T}$ (Equation 28.3) $= \dfrac{0.693}{5570}$ year^{-1}

$\therefore \quad \lambda = \dfrac{0.693}{5570 \times 365 \times 24 \times 60 \times 60}$ s^{-1}

$\therefore \quad \lambda = 3.94 \times 10^{-12}$ s^{-1}

(b) For 1 g carbon:

Number of carbon atoms $= \dfrac{\text{Mass}}{\text{Mass of atom}}$

$= \dfrac{10^{-3}\,\text{kg}}{12 \times 1.66 \times 10^{-27}}$

(assuming all atoms are carbon $^{12}_{6}$C – almost true)

$= 5.02 \times 10^{22}$

$\therefore$ number of carbon-14 atoms

$= 5.02 \times 10^{22} \times \dfrac{1}{10^{12}}$

$= 5.02 \times 10^{10}$

The (carbon-14) activity is

$A = \lambda n = 3.94 \times 10^{-12} \times 5.02 \times 10^{10}$

$= 0.198$ s^{-1}

Note: This means 12 disintegrations per minute.

(c) Using $A = A_0/2^Y$ (Equation 28.4) we get

$2^Y = \dfrac{12}{3} = 4 \quad \therefore \quad Y = 2$

$\therefore$ time of decay $t = 2 \times$ half-life, i.e. 2×5570 year

Answer

(a) 3.94×10^{-12} s^{-1}, (b) 0.198 s^{-1},
(c) 11.1×10^3 years.

Inverse square law

For a small γ-source in a low density medium, e.g. air, there will be little loss of the emitted photons, but they spread as they travel so that the number of particles per second (the particle flux) received per unit area at distance r is proportional to $1/r^2$. Hence the term 'inverse square law'. This follows from the fact that, at distance r, the total flux F is spread over a spherical area $4\pi r^2$ so that the flux per unit area is

$$\dfrac{F}{\text{area}} = \dfrac{\text{Activity}}{4\pi r^2} \qquad (28.6)$$

assuming that each disintegration of the source produces one particle (i.e. one emergent γ-photon from the γ-source). Consequently a detector should give a count rate proportional to $1/r^2$. For α- and β-particles the flux per unit area is also reduced by intervening air.

Fig. 28.2 illustrates this law.

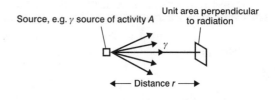

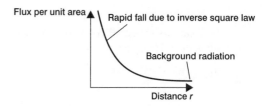

Fig. 28.2   The inverse square law

Absorption

When γ-photons pass into any medium their number will begin to decrease because of the ionising of atoms and other processes. (A photon is used up when an ionisation is produced.)

The half-thickness of a medium is that thickness which halves the flux. We can denote it by T. The emergent flux F is given by

$$F = \dfrac{F_0}{2^Y} \qquad (28.7)$$

where $Y = x/T$, x being the thickness of the absorbing medium, T the half-thickness; F_0 is the initial flux entering. We can also write

$$F = F_0 e^{-\mu x} \qquad (28.8)$$

where μ is the linear absorption coefficient (attenuation coefficient) for the medium. Also

$$\mu = \dfrac{\ln 2}{T} \qquad (28.9)$$

Note that Equations 28.7, 8 and 9 are analogous to Equations 28.4, 2 and 3 for decay; this makes learning the equations easier. The same formulae happen to apply to absorption of β-radiation from radioactive sources.

Example 6

A γ-emitting nuclide in a small source has a half-life of 60 minutes. Its initial γ count rate, recorded by a counter placed 1.0 m from the source, is $320\,s^{-1}$. The distance between the counter and the source is changed. After 2.0 h the count rate recorded is $125\,s^{-1}$: What is the new distance between the counter and the source?

Method

Here the count rate changes due to both decay and the change of distance.

As regards decay the expected count rate R, say, after 2.0 h can be deduced from Equation 28.6:

$$A = \frac{A_0}{2^Y}$$

Since the count rate R is proportional to the activity we can now write

$$R = \frac{320}{2^Y}$$

and $\quad Y = \dfrac{t}{T} = \dfrac{2.0\,h}{1.0\,h} = 2$

$$\therefore \quad R = \frac{320}{2^2} = 80\,s^{-1}$$

The change from $80\,s^{-1}$ to $125\,s^{-1}$ must be due to decreased distance, and since flux and therefore count rate are proportional to $1/r^2$ (the inverse square law).

$$\frac{\text{First count rate}}{\text{Second count rate}} = \frac{r_2{}^2}{r_1{}^2}$$

$$\therefore \quad \frac{80}{125} = \frac{r_2{}^2}{1.0^2}$$

$$\therefore \quad r_2{}^2 = 0.64 \text{ and } r_2 = 0.8\,m$$

Answer

0.8 m.

Example 7

The half-thickness of a certain material for β-radiation, from a particular source is 3.0 mm. What thickness is needed to reduce the β-radiation flux by 90%?

A 15 mm	B 5.4 mm	C 10 mm
D 0.6 mm	E 0.15 mm	

Method

An approximate value for the thickness can be obtained as follows:

One half-thickness gives a reduction to 50% of the initial flux; two half-thicknesses gives a further reduction to half of this, i.e. 25% of the initial flux would emerge; three gives $12\frac{1}{2}\%$ and four gives 6.25%. Therefore 90% reduction, i.e. 10% emerging, requires a thickness between three and four times the half-thickness 3 mm, i.e. between 9 mm and 12 mm. In fact this argument does tell us that the only answer which can be correct is C.

The alternative approach is to use Equation 28.7, as follows:

$$F = \frac{F_0}{2^Y} \quad \text{and} \quad Y = \frac{x}{T}$$

F is the flux surviving and is to be 10%, F_0 is the original flux, i.e. 100%, half-thickness $T = 3.0$ mm and x is the thickness to be found.

$$\therefore \quad 10 = \frac{100}{2^Y} \quad \therefore \quad 2^Y = 10$$

$$\therefore \quad \log_{10} 2^Y = \log_{10} 10 = 1$$

But $\log 2^Y = Y \log 2$ so that

$$Y \log 2 = 1 \quad \therefore \quad Y = \frac{1}{\log 2} = \frac{1}{0.3010} = 3.322$$

$$\therefore \quad \frac{x}{T}(= Y) = 3.322$$

$$\therefore \quad x = 3.322 \times 3.0\,mm$$

$$= 9.966\,mm \quad \text{or} \quad 10\,mm$$

Answer

C.

Exercise 28.3

1   A certain radioactive nucleus $^{222}_{86}X$ decays by alpha emission. What are the nucleon number and the proton number of the daughter nucleus?

2   If the γ-radiation flux per unit area is acceptably low at a distance of 1.0 m from a certain small source, at what distance from the source will the same radiation level be obtained when the source is enclosed in a container whose walls have a thickness equal to 2 half-thicknesses?

3   A very old wooden tool is found to have a carbon-14 activity of 0.07 Bq per gram of total carbon content. The half-life for carbon-14 is 5570 years and the carbon dioxide of the atmosphere, from which the wood's carbon was obtained, gives 19 disintegrations per minute per gram of carbon. What is the age of the tool?

X-radiation

X-radiation is, like gamma radiation, an electromagnetic radiation but its wavelength is generally longer than that of gamma rays. However, the main distinction between these radiations is the method of production.

X-rays are usually produced by accelerating electrons to high speeds in an evacuated glass tube and letting them strike a target or anode (Fig. 28.3). The X-radiation results from two processes. One is the rapid slowing down of the electrons as they enter the target's surface. The other is excitation of the target atoms.

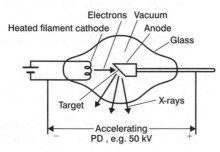

Fig. 28.3   An X-ray tube

The slowing process gives X-rays over a range of wavelengths, i.e. a continuous spectrum. This radiation is called 'white'. The shortest wavelength (λ_{min}) obtained (highest photon energy) is due to incident electrons stopping abruptly, so that all the electron's kinetic energy becomes one photon of X-radiation. Thus

$$\frac{hc}{\lambda_{min}} = hf_{max} = \tfrac{1}{2}mv^2 = eV \qquad (28.10)$$

where λ_{min} is the shortest wavelength of X-radiation produced, f_{max} is its frequency, v is the velocity of the accelerated electrons.

V is the potential difference used for the acceleration. The excitation process gives X-rays at a few definite wavelengths called 'characteristic X-rays' and decided by the allowed transitions in the target atoms, according to the equation below:

$$E - E' = hf = \frac{hc}{\lambda} \qquad (28.11)$$

In this process an accelerated electron removes an inner electron of the atom at energy level E'. Then an electron at higher energy level E can fall to level E' and so emit X-radiation.

Fig. 28.4 shows a typical X-radiation spectrum for such an X-ray tube.

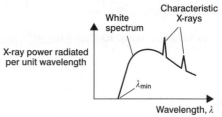

Fig. 28.4   Typical X-ray spectrum

Example 8

Calculate the maximum frequency of X-rays emitted by an X-ray tube using an accelerating voltage of 33.0 kV. (Planck constant $= 6.6 \times 10^{-34}$ J s, charge e on an electron $= 1.6 \times 10^{-19}$ C.)

Method

We use Equation 28.10:

$$\frac{hc}{\lambda_{min}} = eV \quad \text{or} \quad hf_{max} = eV$$

$h = 6.6 \times 10^{-34}, e = 1.6 \times 10^{-19}, V = 33 \times 10^3$.

$$\therefore \quad f_{max} = \frac{eV}{h} = \frac{1.6 \times 10^{-19} \times 33 \times 10^3}{6.6 \times 10^{-34}}$$

$$= 8.0 \times 10^{18} \text{ Hz}$$

The whole of the accelerated electron's energy has been used to produce a photon of this frequency. A higher frequency is not possible.

Answer

8.0×10^{18} Hz.

Example 9

Characteristic X-radiation is described as K_α if it is due to an electron transition to the $n = 1$ shell (K shell) from the $n = 2$ (L shell). K_β radiation is due to $n = 3$ (M shell) to K shell transitions. For molybdenum the K_α wavelength is 0.071 nm and K_β is 0.063 nm. Calculate the difference between (a) the K and L energy levels and (b) the L and M energy levels.

(Planck constant $h = 6.6 \times 10^{-34}$ J s, velocity of electromagnetic radiation in vacuum $c = 3.0 \times 10^8$ m s^{-1}.)

Method

We use Equation 28.11.

$$E - E' = \frac{hc}{\lambda}$$

(a) For K_α, $\lambda = 0.071 \times 10^{-9}$ m; $h = 6.6 \times 10^{-34}$ J s, $c = 3.0 \times 10^8$ m s^{-1}.

$$\therefore \quad E_L - E_K = \frac{6.6 \times 10^{-34} \times 3.0 \times 10^8}{0.071 \times 10^{-9}}$$

$$= 278.9 \times 10^{-17} \text{ J or } 2.8 \times 10^{-15} \text{ J}$$

(b) For K_β, $\lambda = 0.063 \times 10^{-9}$ m;

$$\therefore \quad E_M - E_K = \frac{6.6 \times 10^{-34} \times 3.0 \times 10^8}{0.063 \times 10^{-9}}$$

$$= 314.3 \times 10^{-17} \text{ J}$$

and

$$E_M - E_L = (E_M - E_K) - (E_L - E_K)$$

$$= 314.3 \times 10^{-17} - 278.9 \times 10^{-17}$$

$$= 35.4 \times 10^{-17} \text{ J or } 3.5 \times 10^{-16} \text{ J}$$

Answer

(a) 2.8×10^{-15} J, (b) 3.5×10^{-16} J.

Exercise 28.4

1 Calculate the highest frequency of X-radiation that can be obtained from an X-ray tube operated with a PD of 17 kV. If the target of such a tube is made of molybdenum, will the K_α (0.071 nm) and K_β (0.063 nm) characteristic radiations be obtained?
(Electron charge $e = -1.6 \times 10^{-19}$ C; Planck constant $h = 6.6 \times 10^{-34}$ J s; velocity of electromagnetic radiation in vacuum $c = 3.0 \times 10^8$ m s^{-1}.)

2 An X-ray tube has an electron beam current of 10 mA, and the accelerating voltage is 50 kV. The efficiency (i.e. percentage of input power converted into X-ray power) is 0.5%. Calculate (a) the input power, (b) the power lost in the tube as heat, (c) the minimum wavelength of X-rays produced.
(Electron charge $e = -1.6 \times 10^{-19}$ C; Planck constant $h = 6.6 \times 10^{-34}$ J s; velocity of electromagnetic radiation in vacuum $c = 3.0 \times 10^8$ m s^{-1}.)

3 The K_α and K_β characteristic X-radiations from a copper target have wavelengths 0.154 nm and 0.139 nm, and are due to electron transitions to the $n = 1$ shell from the $n = 2$ and $n = 3$ shells respectively. Calculate a value for the energy difference between (a) the $n = 1$ and $n = 3$ shells, (b) the $n = 2$ and $n = 3$ shells.
($h = 6.6 \times 10^{-34}$ J s; $c = 3.0 \times 10^8$ m s^{-1}.)

Exercise 28.5:
Examination questions

(Note that $\ln 2 = 0.6931$.)

1 The half-life of protactinium is 73 seconds. A freshly-prepared sample has an activity of 30 kBq.
  (i) Calculate the decay constant of protactinium.
  (ii) How many atoms of protactinium are initially present in the sample?
  (iii) What would be the activity of the sample 2.0 minutes after preparation?
[CCEA 2000, part]

2 (a) (i) The radioactive decay law may be written as

$$A = A_0 e^{-\lambda t}$$

  Identify the following terms in this relationship: A; A_0; λ.

  (ii) Define the half-life of a radioactive substance.

(b) A newly-prepared radioactive source contains 5.0 μg of Strontium-90, which has a half-life of 28 years.
  (i) Calculate the mass of Strontium-90 present in the source after it has been used for 16 years.
  (ii) How much longer will it take for the mass of Strontium-90 in the source to be reduced to 1.0 μg? [CCEA 2000]

3 The graph shows the decay of a radioactive nuclide.

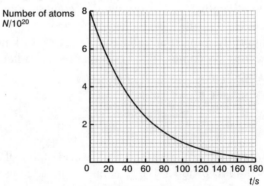

Determine the half-life of this radionuclide.

Use your value of half-life to calculate the decay constant λ of this radionuclide.

Use the graph to determine the rate of decay dN/dt when $N = 3.0 \times 10^{20}$.

Use your value of the rate of decay to calculate the decay constant λ of this radionuclide.

Explain which method of determining the decay constant you consider to be more reliable.
[Edexcel 2000]

4 Carbon-14, decays by β-emission, with a half-life of 5730 years.

A sample of wood found in a bog has a mean activity of 0.20 Bq after correction for background radiation.

(a) Define the term decay constant.

(b) Show that the sample contains 5.3×10^{10} carbon-14 atoms.
1 year $= 3.2 \times 10^7$ s.

(c) An identical sample of living wood is taken and found to have a mean activity of 0.25 Bq after correction for background radiation. Find the age of the wood taken from the bog. [OCR 2000]

5 A certain X-ray tube operates at 110 kV. Calculate the shortest wavelength of X-rays produced.

(Electronic charge $(e) = 1.60 \times 10^{-19}$ C
velocity of light in vacuum $(c) = 3.00 \times 10^8\,\mathrm{m\,s^{-1}}$
Planck constant $(h) = 6.63 \times 10^{-34}$ J s)
[CCEA 2000, part]

29
Nuclear reactions

The Einstein mass–energy relationship

The mass m of any body, defined by the equation $F = ma$ (see Chapter 5), and the total energy E of the body are related by the equation

$$E = mc^2 \qquad (29.1)$$

where c = velocity of light in vacuum.

For a body at rest m is the rest mass and the corresponding energy $E \, (= mc^2)$ is the rest mass energy. If the body were then to move, E would increase on account of the body's acquiring kinetic energy and so m increases also.

In nuclear reactions the energy changes are sufficient for the mass changes to be significant.

If a nuclear change, i.e. reaction, occurs with no supply of energy from outside, then we have a spontaneous reaction such as a radioactive decay. The potential energy of the nucleus must fall, and so the rest mass must decrease. The energy lost escapes from the nucleus usually as the kinetic energy of an emitted particle or as a γ-photon, or both.

The loss of rest mass or the energy released in a reaction is denoted by Q (the 'Q value' of the reaction) and the loss of rest mass when a nucleus is formed from its component particles can be called the 'mass defect' of the nucleus.

If Q is negative then the reaction cannot occur without a supply of energy.

Example 1

Beta particle emission from $^{210}_{83}\mathrm{Bi}$ can be described by the equation

$$^{210}_{83}\mathrm{Bi} = {}^{210}_{84}\mathrm{Po} + {}^{0}_{-1}e + v + Q$$

where e denotes the electron which is the β^- particle, v denotes a neutrino and Q is the energy that becomes the kinetic energy of the particles produced.

The masses of the atoms concerned are 209.984110 u for the bismuth 210 and 209.982866 u for the polonium 210.

Calculate the value of Q (a) in joules and (b) in electron-volts.

Take $1\,\mathrm{u} = 1.7 \times 10^{-27}\,\mathrm{kg}$, the speed of light $c = 3.0 \times 10^8\,\mathrm{m\,s^{-1}}$ and the electron charge $e = 1.6 \times 10^{-19}\,\mathrm{C}$. The rest mass of the neutrino is zero.

Method

(Using atomic masses rather than nuclear masses means that the masses of 83 electrons are included in the bismuth atom on the left of the equation. However the mass of the polonium atom includes 84 electrons, 83 of which will balance the 83 on the left and the remaining one will allow the mass of the beta particle electron to be neglected. In A-level calculations you will assume that electron masses can be overlooked.)

(a) The total rest mass on the left of the equation is 209.984 110 u and on the right it is 209.982 866 u.

The loss of rest mass which is the mass of the energy Q is

$$209.984\,110 - 209.982\,866 = 0.001\,244\,\mathrm{u}$$

In kilograms this is $0.001244 \times 1.7 \times 10^{-27}\,\mathrm{kg}$ or $2.1148 \times 10^{-30}\,\mathrm{kg}$.

In joules it is, using $E = mc^2$,

$$E = 2.1148 \times 10^{-30} \times (3.0 \times 10^8)^2$$
$$= 19.03 \times 10^{-14}\,\mathrm{J}$$

(b) In electron-volts, using $1\,\mathrm{eV} = e$ joules, we get

$$E = \frac{19.03 \times 10^{-14}}{1.6 \times 10^{-19}}$$

$$= 11.90 \times 10^5\,\mathrm{eV} \quad \text{or} \quad 1.19\,\mathrm{MeV}$$

Answer

(a) $19 \times 10^{-14}\,\mathrm{J}$ (b) $1.2\,\mathrm{MeV}$

Equations for nuclear reactions

As in the above example, the right-hand side of the equation contains the reaction products. On the left is the particle or particles at the start. The equation can also include Q, the net loss of rest mass energy or gain in other energy.

Note too that the sum of the mass numbers on left and right must be equal (there is no way in which nucleons can be created or destroyed) and similarly the proton numbers must add up to the same total on each side because charge is conserved.

To conform with these rules we write the negative electron, the positive electron and the neutron as $_{-1}^{0}\text{e}$, $_{1}^{0}\text{e}$ and $_{0}^{1}\text{n}$ respectively.

Example 2

$_{15}^{32}\text{P}$ decays by a β^- emission to a nucleus $_{Z}^{A}\text{X}$. What are the values of A and Z?

Method

$$_{15}^{32}\text{P} = _{Z}^{A}\text{X} + _{-1}^{0}\text{e}$$

Equating nucleon numbers gives $32 = A + 0$ so that $A = 32$. Equating atomic numbers (proton numbers) gives $15 = Z - 1$ so that $Z = 16$.

Answer

$A = 32$, $Z = 16$.

Induced nuclear reactions

Some nuclei can be made to change as a result of their interacting with a bombarding particle, e.g. by a proton or an alpha particle, and the incident particle may have a useful amount of kinetic energy. The reaction is an 'induced reaction'.

Example 3

The nuclear reaction between a deuteron and a lithium-6 nucleus does not require the deuteron to have any kinetic energy and the reaction results in the production of two alpha particles according to the equation

$$_{1}^{2}\text{H} + _{3}^{6}\text{Li} \rightarrow 2_{2}^{4}\text{He} + Q$$

The mass of the deuteron is 2.0141 u, of the lithium nucleus 6.0151 u and of each alpha particle 4.0026 u.

Calculate the kinetic energy of each of the alpha particles immediately following the reaction.

(Speed of light in vacuum is $3.0 \times 10^8 \, \text{m s}^{-1}$.)

Method

The total rest mass of the reacting particles is

$$2.0141 + 6.0151 \quad \text{or} \quad 8.0292 \, \text{u}$$

The total rest mass of the products is

$$2 \times 4.0026 \quad \text{or} \quad 8.0052 \, \text{u}$$

The loss of rest mass is

$$8.0292 - 8.0052 = 0.0240 \, \text{u}$$
$$= 0.0240 \times 1.66 \times 10^{-27} = 4.0 \times 10^{-29} \, \text{kg}$$

and, using $E = mc^2$,

$$E \text{ or } Q = 4.0 \times 10^{-29} \times (3.00 \times 10^8)^2$$
$$= 3.6 \times 10^{-12} \, \text{J}$$

The energy *per α-particle* is half of 3.6×10^{-12} J, i.e. 1.8×10^{-12} J.

Answer

1.8×10^{-12} J.

Nuclear binding energy

Nucleons hold together in a nucleus only because of attractive *nuclear forces*. The potential energy of the nucleons is lower when they are in a nucleus close together than when they are separated from each other. When this potential energy decreases, the rest mass energy decreases and energy Q is released in the form of γ-radiation or kinetic energy. Conversely, to break up a nucleus into separate nucleons requires an equal energy to be provided, and this is called the binding energy of the nucleus. The binding energy of a nucleus divided by its nucleon number gives the 'binding energy per nucleon'.

Example 4

The ^3H isotope of hydrogen is known as tritium and the mass of the $_{1}^{3}\text{H}$ atom is 3.0160 u. Calculate the binding energy of this atom. (Neutron mass = 1.009 u, mass of $_{1}^{1}\text{H}$ atom 1.008 u, 1 u = 931 MeV.)

Method

The $_{1}^{3}\text{H}$ atom contains a proton and an orbiting electron (i.e. a $_{1}^{1}\text{H}$ atom) and two neutrons. The total mass of these particles is

$$1.008 + (2 \times 1.009) \quad \text{or} \quad 3.026 \, \text{u}$$

The Q value (or energy released during formation) is the difference between the tritium mass and the mass of its constituents.

$$Q = 3.026 - 3.016 \quad \text{or} \quad 0.010\,u$$

Converted to MeV we have 0.010×931 or 9.31 MeV.

Answer

9.3 MeV.

Nuclear fission

The binding energy per nucleon is smaller (potential energy is bigger) for nucleons in large nuclei such as uranium nuclei and is greatest for nucleons in medium sized nuclei. Consequently splitting, or fission, can occur where a large nucleus divides into two approximately equal-size nuclei, and energy is released.

Example 5

Fission of a ^{235}uranium nucleus can be induced according to the following equation

$$^{235}_{92}U + ^{1}_{0}n = ^{95}_{42}Mo + ^{139}_{57}La + 2^{1}_{0}n + Q$$

The atomic masses of the nuclei concerned are:

    U235   235.042 77 u
    Mo95   94.905 53 u
    La139   138.905 34 u

and the mass of a neutron is 1.008 66 u.

Calculate the energy released in this reaction in joules. Any kinetic energy of the interacting particles can be neglected.

Method

For the left side of the equation the sum of the masses is

$$235.042\,77 + 1.008\,66 \quad \text{or} \quad 236.051\,u$$

For the right side the sum is

$$94.905\,53 + 138.905\,34 + (2 \times 1.008\,66)$$
$$\text{or } 235.828\,u$$

$\therefore \quad Q = 236.051 - 235.828 = 0.223\,u$

$\therefore \quad$ But $1\,u = 1.66 \times 10^{-27}$ kg

$\therefore \quad Q = 0.223 \times 1.66 \times 10^{-27}$ kg or, using $E = mc^2$,

$$Q = 0.223 \times 1.66 \times 10^{-27} \times (3.0 \times 10^8)^2$$
$$= 3.33 \times 10^{-11}\,J$$

Answer

3.3×10^{-11} J.

Comment

In electron volts $Q = \dfrac{3.33 \times 10^{-11}}{1.6 \times 10^{-19}}$ eV or 2.08×10^8 eV or 208 MeV.

Values needed but not given in the question are given in lists of data provided with exam questions.

The additions and subtraction can be done without the use of a calculator. If a calculator is used then it should be made to give answers to 6 sig. figs. (see Use of calculators in Chapter 2). Proton numbers do not balance in this equation because the fission is accompanied by β-emitting decays.

Nuclear fusion

This is the joining together of two small nuclei to produce a nucleus for which the binding energy per nucleon is larger. Energy is released as kinetic energy of a product particle (if any), or otherwise as gamma radiation.

Example 6

Calculate the energy in MeV released by fusing two protons and two neutrons to form a helium nucleus. (This reaction is difficult to achieve.)

The atomic masses of hydrogen ($^{1}_{1}H$) and helium ($^{4}_{2}He$) are 1.007 825 u and 4.002 604 u respectively. The mass of the neutron is 1.008 665 u. $1\,u = 931$ MeV.

Method

The reaction is

$$2^{1}_{1}H + 2^{1}_{0}n = ^{4}_{2}He + Q$$

The sum of the masses for the left-hand side is

$$(2 \times 1.007\,825) + (2 \times 1.008\,665)$$
$$\text{or} \quad 4.032\,98\,u$$

(In addition to the protons, neutrons and helium nucleus, two electrons are included in each side of the equation, one in each hydrogen atom and two in the helium atom. These electron masses cancel.)

Subtracting the mass on the right from that on the left gives Q.

$\therefore \quad Q = 4.032\,98 - 4.002\,604 = 0.030\,376\,u$

and converted to MeV,

$$Q = 0.030\,376 \times 931 = 28.3\,MeV$$

per fusion.

Answer

28.3 MeV.

Exercise 29.1

1 A possible induced fission reaction is shown by the following equation

$$^{235}_{92}U + ^{1}_{0}n = ^{87}_{35}Br + ^{x}_{57}La + 3\,^{1}_{0}n$$

What number is represented by the x?

2 Use the following data to show that the binding energy of U235 (i.e. $^{235}_{92}$Uranium) is approximately 1.7×10^3 MeV.

Mass of U235 atom $= 390.295 \times 10^{-27}$ kg

Mass of neutron $= 1.675 \times 10^{-27}$ kg

Mass of proton $= 1.672 \times 10^{-27}$ kg

Velocity of light (in vacuum) $= 3.0 \times 10^8\,\mathrm{m\,s^{-1}}$

$1\,\mathrm{eV} = 1.60 \times 10^{-19}$ J

3 Calculate the energy in MeV released in the fusion reaction

$$^{2}_{1}H + ^{2}_{1}H \rightarrow ^{3}_{1}H + ^{1}_{1}H + Q$$

The atomic masses

deuterium $^{2}_{1}H$, $2.014\,102\,u$

tritium $^{3}_{1}H$, $3.016\,049\,u$

hydrogen $^{1}_{1}H$, $1.007\,825\,u$

$(1\,u = 931\,\mathrm{MeV}.)$

Exercise 29.2: Examination questions

Where necessary use
    electronic charge $(e) = 1.60 \times 10^{-19}$ C
    unified atomic mass unit $(u) = 1.66 \times 10^{-27}$ kg
    velocity of light in vacuum $(c) = 3.00 \times 10^8\,\mathrm{m\,s^{-1}}$
    Avogadro's number $(N_A) = 6.02 \times 10^{23}\,\mathrm{mol^{-1}}$

1 (a) Part of a series of radioactive decay processes is shown below.

Fig. 29.1

In a particular sample, the number of thallium (Tl) nuclei present remains constant. For this sample, calculate the ratio:

$$\frac{\text{number of Bi nuclei}}{\text{number of Tl nuclei}}$$

(b) Give the nuclear equation for the thallium-207 decay, including any other particles that are produced. [OCR 2001, part]

2 In one fusion reaction, two deuterium ($^{2}_{1}$H) nuclei combine to form a helium nucleus ($^{3}_{2}$He). Write an equation for this reaction, including nucleon and proton numbers.

The masses involved are:

	mass/u
$^{2}_{1}$H nucleus	2.014 10
$^{3}_{2}$He nucleus	3.016 03
neutron	1.008 67

$1\,u = 1.66 \times 10^{-27}$ kg

(a) Calculate the energy released in this reaction.

(b) Hence calculate the energy released when 1.0 kg of deuterium nuclei fuse to form $^{3}_{2}$He. 1.0 kg of deuterium contains 3.0×10^{26} deuterium nuclei. [Edexcel S-H 2000, part]

3 (a) A typical nuclear fission event is represented by the equation

$$^{1}_{0}n + ^{235}_{92}U \rightarrow ^{92}_{36}Kr + ^{A}_{Z}Ba + 3\,^{1}_{0}n$$

(i) Calculate the number of protons and the number of neutrons in the $^{A}_{Z}$Ba nucleus.

(ii) Calculate the energy released in one of these events from the following data.

mass of $^{1}_{0}n = 1.67 \times 10^{-27}$ kg
mass of $^{235}_{92}U = 390.19 \times 10^{-27}$ kg
mass of $^{92}_{36}Kr = 152.57 \times 10^{-27}$ kg
mass of $^{A}_{Z}Ba = 233.92 \times 10^{-27}$ kg

(b) Estimate the useful power output of a nuclear power station which has an efficiency of 33% and uses up $^{235}_{92}U$ at a rate of $4.4 \times 10^{-5}\,\mathrm{kg\,s^{-1}}$. [WJEC 2000, part]

4 Explain what is meant by the **binding energy** of a nucleus.

Use the data below to calculate the binding energy per nucleon for an alpha-particle. Give your answer in MeV per nucleon.

Proton mass $m_p = 1.0073\,u$;

neutron mass $m_n = 1.0087\,u$;

mass of ^{4}He$^{++} = 4.0015\,u$. [CCEA 2000, part]

Section I
Calculations involving graphs

30
Graphs and oscilloscope traces

Introduction

The way in which the variation of one quantity affects another can be expressed as a graph as an alternative to an equation. One advantage of a graph is the quickness with which its information can be grasped.

Unless other symbols are preferred, we use x to denote the quantity plotted on the horizontal axis (abscissa) and y for the other quantity (ordinate).

The origin is the place where the axes meet.

Plotting graphs

If you are required to plot a graph using numerical data provided by the exam question you should use scales that give a graph that fills the available space, that is convenient for easy plotting of points and that produces a graph not more than twice as wide as it is high (or vice versa).

Unless you have good reason to do otherwise you should start both axes at zero.

For accurate graphs (as opposed to sketches) it is assumed that proper graph paper is used. A typical A4-size graph sheet has about 24 squares upwards, each 1 cm by 1 cm, and about 16 across. Each of these is subdivided with fainter small squares, 5 up and 5 across, inside each large square. A voltage ranging from zero up to 10 volts, for example, could be plotted very easily on the y axis if 2 large squares, i.e. 2 cm, along the scale used were allocated for each volt. Then each of the ten small squares (0.2 cm of the paper) would represent 0.1 V. The plotted values would use most of the paper height.

The quantities being plotted should be marked against their respective axes, together with the units being used, with the quantity divided by unit.

Along each axis the numbers of units, e.g. 1.0, 2.0, etc for 1.0 V, 2.0 V, etc., should be marked at intervals normally no closer than 5 mm and no greater than 2 cm.

Example 1

The data below describe the stretching of a spring. Plot a graph of the applied force as abscissa and the extension as ordinate.

Force / N	Extension / mm
2.0	6
3.0	9
4.0	12
5.0	16
6.0	19
7.0	20
8.0	24
9.0	28
10	31
11	33

Method

As shown in Fig. 30.1, each 2 mm of extension has been represented by one large square along the y axis, so that the results are accommodated within 17 out of the available 24 squares. The force values have used 11 out of the 16 large squares along the x axis. The graph could have been made wider by using, for example, 6 scale divisions (6 times 2 mm of paper) for each newton so that 11 N would be 132 mm of paper along the x axis, i.e. more than 13 large squares. This scale of 6 divisions per newton would make plotting more difficult.

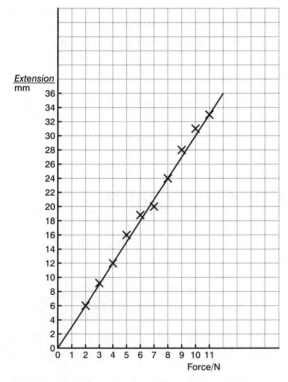

Fig 30.1 Plotting a graph (Large squares only shown. Paper size 16 cm × 24 cm)

The best straight line has been drawn through the plotted points with, as far as possible, equal numbers of points above and below all parts of the line.

Plotting a graph from a formula

Suppose the formula is $P = I^2R$, where R is a constant resistance of 5.0 ohm, P denotes power in watts and I denotes electric current in amperes. You are required to plot and so discover the shape of the graph.

You consider simple values of one of the variables, e.g. the current, and calculate the corresponding P values. So, if we choose 1.0 A, 2.0 A, 3.0 A, 4.0 A and 5.0 A, the P value for $I = 2.0$ A is $2.0^2 \times 5.0$, which is 20 W, and the set of P values is as shown below.

I/A	1.0	2.0	3.0	4.0	5.0
P/W	5.0	20	45	80	125

These are the values to plot. The shape of the graph is as shown in Fig. 30.2d.

If only a quick sketch is required rather than an accurate graph, the x and y scales can be drawn reasonably straight on plain paper and a few scale divisions marked at approximately equal spacings.

Exercise 30.1

1 Plot a graph of the following data, with PD (V) on the x axis and current (I) on the y axis, and read from the graph the the current expected at 4.7 V.

$PD\,/\,V$	0.0	1.0	2.0	3.0	4.0	5.0	6.0
$I\,/\,A$	0.0	0.55	1.0	1.5	1.9	2.5	3.1

2 Sketch the graph for the formula $\rho = \dfrac{M}{V}$ where ρ is a fixed density of $10 \times 10^3 \,\text{kg m}^{-3}$, M is the mass in kg, and V is volume in m^3. Put M on the y axis.

Some common graphs

If $y \propto x$, i.e. $y = mx$ where m is a constant (not affected by variation of x and y), then we get a straight line which passes through the point $x = 0$, $y = 0$ (Fig. 30.2a). Other common examples are also shown in Fig. 30.2.

The most important of these examples is

$$y = mx + C \qquad\qquad (30.1)$$

When $y = mx + C$ the graph is a straight line and if $C = 0$ the line passes through $x = 0$, $y = 0$. Then y is proportional to x.

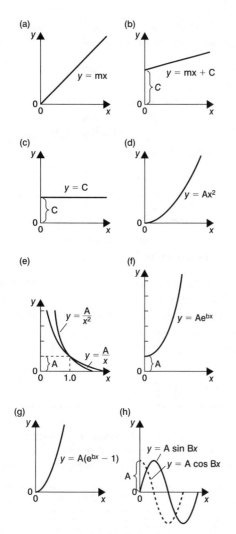

Fig. 30.2   Some common graphs

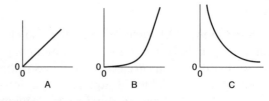

Fig. 30.3   Graphs for Example 2

Answer

(i) C, (ii) A, (iii) B

Small changes

If symbols x and y are used to denote the values of two related quantities, then it is usual to denote small corresponding increases in x and y by δx and δy (where δ is pronounced 'delta'). Small decreases would be $-\delta x$ and $-\delta y$.

If now x were made extremely small (too small to measure it), i.e. $\delta x \to 0$, we denote the value of $\delta y/\delta x$ when $\delta x \to 0$ by dy/dx (pronounced 'dee y by dee x'). (See also page 12.)

Slope of a graph

The gradient or slope s of a graph at a chosen point is

$$s = \frac{\text{Small change in } y}{\text{Corresponding small change in } x}$$
$$= \frac{\delta y}{\delta x} \qquad (30.2)$$

For a straight line graph s is the same for all parts of the graph, and so

$$s = \frac{\text{Any change in } y}{\text{Corresponding change in } x}$$
$$= \frac{a}{b} \text{ in Fig. 30.4a} \qquad (30.3)$$

Note that for $y \propto x$, we can write

$$s \text{ (or } m) = \frac{y}{x}$$

and for any straight line $y = mx + C$ with the gradient m. To measure s at a point on a curve it is not too difficult to draw a tangent to the curve and measure its slope, which equals the required slope (Fig. 30.4b).

Example 2

Select from the three graphs labelled A, B and C in Fig. 30.3 the graph that describes the relationship (i) pressure versus volume of a fixed mass of perfect gas at a constant temperature (ii) acceleration versus applied force for a constant mass (iii) power radiated versus absolute temperature of a hot surface.

Method

(i) The equation relating pressure p to volume V is
$pV = \text{constant}$ or $p = \dfrac{\text{constant}}{V}$ or $p \propto \dfrac{1}{V}$ This
agrees with Fig. 30.2e, so the answer to select is C.

(ii) The relationship between acceleration a and force
F is $a = \dfrac{F}{\text{mass}}$ or $a = \dfrac{1}{M} \times F$ which agrees with
$y = mx$ and Fig. 30.2a, so the answer to select is A.

(iii) The relation between power P radiated and absolute temperature T is $P = \sigma A T^4$ and the P versus T graph is like $y = Ax^2$ and Fig. 30.2d, but rises more steeply. The answer to select is B.

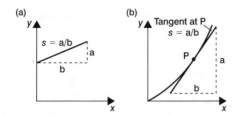

Fig. 30.4   Slope of a graph

Note that for measuring the slope of a graph the choice of origin is not important.

The intercepts of a graph

The y intercept is the value of y when $x = 0$, and the x intercept is x when $y = 0$. Often we need to make use of only one of the two intercepts and we normally use the y intercept. The graph $y = mx + C$ has a y intercept C (Fig. 30.2b). For a straight line (i.e. linear) graph the x intercept and y intercept are related by

$$\text{Slope } s \text{ (or } m) = \frac{\textbf{Magnitude of } y \textbf{ intercept}}{\textbf{Magnitude of } x \textbf{ intercept}}$$

$$(30.4)$$

(see Fig. 30.5)

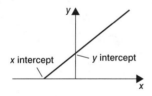

Fig. 30.5   Intercepts

Example 3

The following values of resistance R and corresponding Celsius temperature θ conform to the equation $R = R_0(1 + \alpha\theta)$ where R_0 and α are constants. Plot a straight line graph from these results and hence determine R_0 and α.

$\theta/°C$	10	30	60	90
R/Ω	10.3	11.0	12.0	13.0

Method

The graph is plotted as shown in Fig. 30.6.

To find R_0 we first note the resemblance between $R = R_0(1 + \alpha\theta)$ and $y = C + mx$. If we rewrite the R equation as $R = R_0 + R_0\alpha\theta$, it is seen that R_0 is the intercept on the R axis and $R_0\alpha$ is the slope.

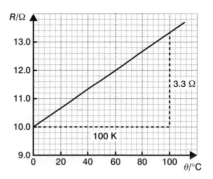

Fig. 30.6   Graph for Example 3

From the graph the slope is 3.3/100 or $0.033\ \Omega\,\text{K}^{-1}$ and the intercept is $10.0\ \Omega$. So $R_0 = 10.0\ \Omega$ and $R_0\alpha = 0.033\ \Omega\,\text{K}^{-1}$, whence α $(= 0.033/R_0)$ $= 0.0033\ \text{K}^{-1}$.

Answer

$R_0 = 10.0\ \Omega, \alpha = 0.0033\ \text{K}^{-1}$

Advantage of a straight line graph

If a graph is obtained with a straight line, we can easily determine the mathematical relationship described by the graph. If the line passes through point $(0, 0)$ then it agrees with $y = mx$ and m is obtained from the gradient. If it has an intercept, then $y = mx + C$ and m is the slope and C is the intercept.

Slope of a graph as a method of averaging

In Fig. 30.7 the resistance of the conductor is $R = V/I$.

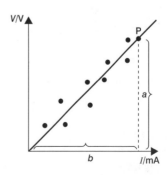

Fig. 30.7   Graph of V versus I for an ohmic conductor

No particular pair of results can be regarded as the correct one for calculating R but the line drawn is an average for the results. For any point on the line such as P the x and y values (i.e. the coordinates of P) can be regarded as average values and R is calculated as $R = a/b$, which is the slope of the graph. P does not need to be a plotted point.

Exercise 30.2

1 Plot the graph of $y = 2x^2 + 4$, where y is in kg and x is in cm and measure the slope at $y = 12.0$ kg.

2 The following readings were taken of power P supplied to evaporate a liquid and the resulting mass per second of liquid evaporated m.

P/W	2.0	4.0	6.0	8.0	10
$m/10^{-5}$ kg s^{-1}	0.3	1.0	1.65	2.3	3.0

Plot a graph to check agreement of these results with the equation $P = mL + h$, where h is a constant rate of heat loss and L is constant. Hence evaluate L and h.

Plotting log y versus log x

Logarithms were explained in Chapter 2. Equations 2.12 and 2.13 are used here.

If $y = Ax^n$ then

$$\log_{10} y = \log_{10} A + \log_{10} x^n$$
$$= \log_{10} A + n \log_{10} x \qquad (30.5)$$

The same is true if we use $\log_e$ (i.e. $\ln y$, etc.) instead.

We can plot $\log_{10} y$ versus $\log_{10} x$ to obtain a straight line graph whose slope equals the power n. The intercept is $\log_{10} A$ from which A can be calculated.

Logarithms are numbers (without units) but x, y and A may have units, e.g. $p = 100$ kPa in Example 4. The units can be indicated by writing 'log p with p in kPa.' However, units are more easily displayed if we plot logs of numerical values, e.g. $\log(p/\text{kPa})$, $\log(V/\text{cm}^3)$. This practice does not affect the method or results because

physical quantities and their numerical values are always related by the same equation.

Example 4

Corresponding values of volume (V) and pressure (p) are given below for a fixed mass of air at constant temperature. Given that $p = AV^n$, where A and n are constants, deduce the values of n and A.

p/kPa	100	200	300	400	500	600
V/cm^3	90	50	32	25	20	16

Method

Plotting V versus p gives a graph like Fig. 30.2e and the power n would not be determined. Now $\log p = \log A + n \log V$ and n is the slope of the $\log p$ versus $\log V$ graph. So we need log values:

$\log_{10}(p/\text{kPa})$	2.0	2.3	2.5	2.6	2.7	2.8
$\log_{10}(V/\text{cm}^3)$	1.95	1.7	1.5	1.4	1.3	1.2

The graph is plotted in Fig. 30.8.

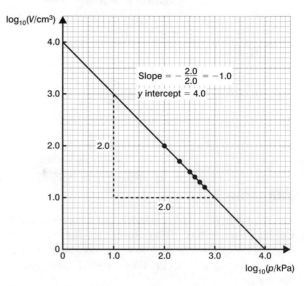

Fig. 30.8 Graph for Example 4

The slope is found to be -1.0, i.e. $n = -1$ (Boyle's law of course).

The y intercept is $\log_{10} A$ and equals 4.0, i.e. $A = 10^4$.

The unit for A is the same as for p/V^n, i.e. kilopascal/cm^{3n} or kPa cm^{-3n}. Here $n = -1$, so that $A = 10^4$ kPa cm^3.

Answer

$n = -1$, $A = 1.0 \times 10^4$ kPa cm^3.

Plotting log y or ln y versus x

If an equation is of the form $y = Ab^{cx}$ where A, b and c are constants then, taking logarithms with base b of each side of the equation, using Equations 2.12 and 2.13, and remembering that $\log_b b = 1$,

$$\log_b y = \log_b A + cx \qquad (30.6)$$

For example if $N = N_0 e^{-\lambda t}$ (Equation 28.2a) and we take logs to the base e (or ln) of each side of the equation we get

$$\ln N = \ln N_0 - \lambda t \qquad (30.7)$$

This equation has the form $y = mx + C$ and so plotting ln y versus x gives a straight line with a slope of $-\lambda$ and an intercept on the ln N (or y) axis of ln N_0.

Note that while natural logs (ln) are preferred for an equation involving e, other logs e.g. to base 10 (ordinary logs), may be used, for example

$$\log_{10} N = \log_{10} A + \log_{10} e^{-\lambda t}$$
$$\text{and} \quad \log_{10} N = \log_{10} A + -\lambda t \log_{10} e \qquad (30.8)$$
$$\log_{10} N = \log_{10} A - 2.303 \lambda t$$

Example 5

For a hot object cooling in a draught the excess temperatures θ recorded at times t were as follows:

t/min	0	5	10	15	20	25	30
θ/°C	60	50	40.5	31.5	25.8	20.1	16.5

Show that the results agree with

$$\frac{\mathrm{d}\theta}{\mathrm{d}t} = -A\theta \text{ or } \theta = \theta_0 e^{-At}$$

and deduce the values of the constants A and θ_0.

Method

A graph of ln θ versus t should give a straight line because $\ln \theta = \ln \theta_0 - At$.

t/min	0	5	10	15	20	25	30
ln (θ/°C)	4.09	3.91	3.70	3.45	3.25	3.00	2.80

The graph obtained is shown in Fig. 30.9.

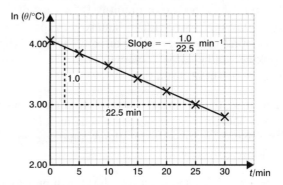

Fig. 30.9   Graph for Example 5

The slope is found to be $-0.044\,\text{min}^{-1}$ so $A = 0.044\,\text{min}^{-1}$.

When $t = 0$, $\theta = \theta_0$ and is equal to 60 °C.

Answer

$A = 0.044\,\text{min}^{-1}$, $\theta_0 = 60\,°\text{C}$.

Note that logarithms of numbers less than 1 are negative. So do not be put off by suddenly getting negative values to plot in log graphs.

Exercise 30.3

1 The following values were obtained for the electric current I and potential difference V for a certain electronic device.

V/V	1.0	2.0	3.0	4.0	5.0
I/A	3.5	20	55	112	196

It is suggested that $I = AV^p$ where A and p are constants. Plot a graph to show that this suggestion is correct and evaluate A and p.

2 The activity A of a radioactive sample decreases with time t giving the following results:

t/day	0	20	40	60	80	100
A/Bq	1000	830	665	545	446	368

If $A = A_0 e^{-\lambda t}$ where A_0 and λ are constants, plot a suitable straight line graph and from it determine λ.

Area under a graph

In Fig. 30.10 velocity v is plotted against time t. The area under line A is 120 squares. Each square represents (in a sense 'measures') $0.5 \, \text{m s}^{-1}$ by $0.5 \, \text{s}$, i.e. a quantity $0.5 \, \text{s} \times 0.5 \, \text{m s}^{-1}$ or $0.25 \, \text{m}$. So 120 squares represents $120 \times 0.25 \, \text{m}$ or $30 \, \text{m}$. More directly the 'area' (by looking at the large rectangle below line A) is $6 \, \text{m s}^{-1} \times 5 \, \text{m}$ or $30 \, \text{m}$.

The area under line B (area of triangle = half base × height) is $0.5 \times 6 \times 4 = 12 \, \text{m}$. This is average value of velocity times time i.e. gives distance correctly.

The area under a graph always has the dimensions of the PRODUCT of the x and y axis quantities.

The area under a current against PD graph gives average power for example.

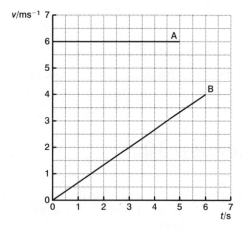

Fig. 30.10   Graph of velocity versus time

Example 6

Fig. 30.11 shows a graph of intensity versus distance from a point charge. The area under this graph equals the potential difference across the distance concerned. Obtain a value for the potential $0.15 \, \text{m}$ from the charge.

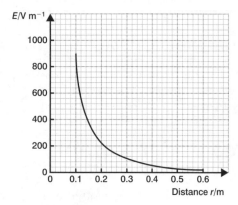

Fig. 30.11   Graph for Example 6

Method

The number of squares of the graph paper is counted for the area under the graph between $r = 0.15 \, \text{m}$ and $r = 0.6 \, \text{m}$ or more. The answer is about 66. The area of each of these squares is $40 \, \text{V m}^{-1}$ by $0.02 \, \text{m}$, i.e. $0.8 \, \text{V}$.

$\therefore$ Total area is $66 \times 0.8 = 53 \, \text{V}$

Answer

53 V approximately.

Exercise 30.4

1 An object travels in a straight line with its velocity v related to time t as shown in Fig. 30.12. How far from its start is the object after the 8 seconds?

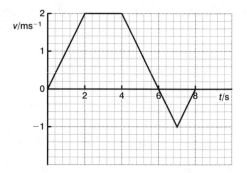

Fig. 30.12   Graph for Question 1

2 With reference to Example 1 in this chapter, calculate the work done in stretching the spring to an extension of $24 \, \text{mm}$.

The cathode ray oscillope

The cathode ray oscilloscope (CRO) can be regarded as a very special voltmeter. A PD V to be measured is amplified and then applied to metal plates above and below the electron beam (Y-plates, Fig. 30.13).

The beam is deflected up or down depending upon the polarity of V. The size of the deflection is proportional to the size of V.

When a voltage produced by the 'time base' section of the CRO is correctly applied to the X-plates, the beam moves steadily and repeatedly

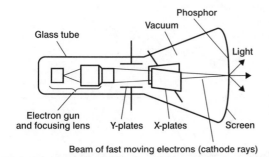

Fig. 30.13   A cathode ray oscilloscope tube

across the screen. If now an alternating voltage is applied to the Y-plates, the trace seen on the screen is a wave. It is in fact a voltage–time graph.

Measurement of current using a CRO

The technique for this is to let the current flow through a small resistance R and measure the PD V across it with the CRO. Then $I = V/R$.

Frequency measurement

A cathode-ray oscilloscope is convenient for measuring the frequency of an alternating voltage. You need to be familiar with the following method.

If the number of millimetres in the X-direction occupied by one cycle of a trace is recorded from the screen, then the period can usually be deduced from the time base velocity (marked in seconds per centimetre on the time base selector or control).

So if one of the waves seen in Fig. 30.14 has adjacent peaks separated by 20 mm and the time base is set at 1 ms per mm, then the period of the wave is 20 ms or 0.020 s and its frequency is 1/0.020 or 50 Hz.

Example 7

A 5.0 V RMS, 50 Hz voltage is obtained from a transformer connected to the mains supply and is fed to the Y-plates of a CRO. If the Y sensitivity is set at $10\,\mathrm{V\,cm^{-1}}$ and the time base at $10\,\mathrm{ms\,cm^{-1}}$, what will be

the total peak-to-peak height of the trace and how many complete cycles of the voltage will be displayed if the trace is 4.0 cm wide?

Method

If the RMS voltage is 5.0 V, then we find the peak value by multiplying this by $\sqrt{2}$ (see Chapter 26), and to obtain the peak-to-peak value we multiply the peak value by 2.

Peak-to-peak voltage is $5.0 \times \sqrt{2} \times 2 = 14.14\,\mathrm{V}$

Since 10 V is represented by 1.0 cm, then 14.14 V will be represented by $1.0 \times 14.14/10$, i.e. 1.414 cm. (1.4 cm to two significant figures.)

The time occupied by one cycle is one-fiftieth of a second or 20 ms. At 10 ms per cm time base velocity one cycle occupies 2.0 cm. Thus 4.0 cm of screen width will accommodate 2 cycles exactly.

Answer

1.4 cm, 2.0 cycles.

Measuring phase difference

To discover or display the extent to which two periodic signals are in phase the Lissajous' figures technique may be used. Alternatively the signals may be fed to the two beams of a double-beam oscilloscope using the Y_1 and Y_2 terminals. The time base is employed and it controls both beams to produce identical X movements.

Example 8

What is the phase relationship between the two signals shown in Fig. 30.14?

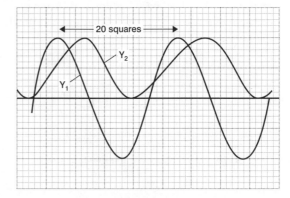

Fig. 30.14   Graph for Example 8

Method

Each peak of trace Y_2 occurs 4 small squares after a peak in Y_1 and this delay ('lag') can be compared with the period of the traces which amounts to 20 small squares. Thus the lag is 1/5 of a cycle or 60 degrees (out of a cycle of 360 degrees).

Answer

Y_2 lags behind Y_1 by 60°.

Exercise 30.5

1

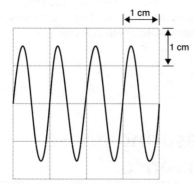

Fig. 30.15   Graph for Question 1

The screen of a cathode ray oscilloscope displays the trace shown in Fig. 30.15. The Y sensitivity is set at 10 mV/cm, and the time base is set at 0.20 ms/cm. Obtain values for (a) the peak voltage and (b) the frequency of the alternating signal.

2   An oscilloscope is used to measure the time it takes to send a pulse of sound along a 70 cm length of metal rod and back again. Fig. 30.16 shows the appearance of the oscilloscope screen. A indicates the original pulse and B the reflected pulse. If the time base speed is 0.10 mm μs^{-1}, what is the speed of travel of the pulse through the rod?

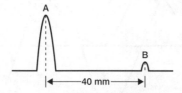

Fig. 30.16   Oscilloscope trace for Question 2

3   By how many degrees are the signals out of phase in Fig. 30.17?

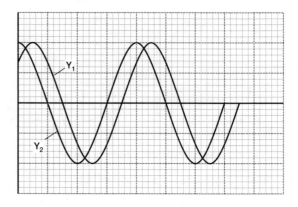

Fig. 30.17   Diagram for Question 3

Exercise 30.6: Examination questions

1   The diagram, scale 1:1, shows some equipotentials in the region of a positive point charge, $+q$.

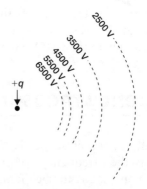

(a)   Add two field lines to a copy of the diagram.

(b)   Plot a graph of electric potential against distance from the point charge.

(c)   Write down the expression for electric potential in a radial field.

(d)   Show that the plotted values are consistent with this expression.

(e)   Calculate the magnitude of the point charge q.

(Permittivity at vacuum $(\varepsilon_0) = 8.85 \times 10^{-12}\,F\,m^{-1}$
[Edexcel 2000]

2   An electric toaster is labelled 780 W 230 V $\sim$ 50 Hz.

(a)   On a copy of the axes below sketch a graph to show how the potential difference across the toaster varies with time. Add a scale to both axes.

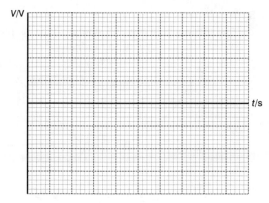

(b) Calculate the peak current in the toaster.

[Edexcel 2001]

3 A sample of cobalt 60 is found to have an activity of 8000 disintegrations per second. Make use of the graph grid to find the number of disintegrations per second after a time of 8.0 years. The half-life of cobalt 60 is 5.3 years.

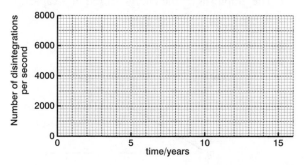

[WJEC 2000, part]

4 (a) The circuit shown is to be used to investigate the photoelectric effect. Monochromatic light of known frequency is shone onto the emissive surface. Describe how you would find the maximum kinetic energy, KE_{max}, of the emitted electrons.

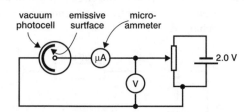

(b) A graph of KE_{max} against frequency f of light shone onto the surface is given below:

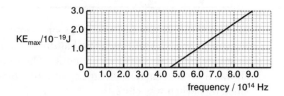

State Einstein's photoelectric equation and hence determine from the graph a value for the Planck constant. [WJEC 2000]

5 A lorry accelerates from rest.

The graph below shows how the momentum of this lorry varies over the first minute.

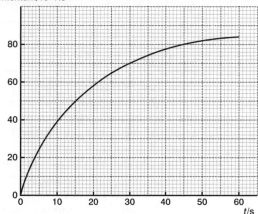

Define momentum.

State the physical quantity represented by the slope of this graph.

Determine the magnitude of this quantity at $t = 20\,\text{s}$.

Explain the shape of this graph.

[Edexcel S-H 2001]

6 The graph shows how the resistance R of a thin film of platinum, connected to two terminals, varies with the Celsius temperature θ in the range $0\,°\text{C} - 100\,°\text{C}$.

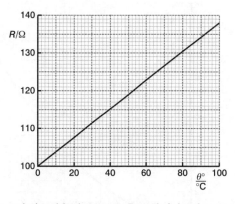

The relationship between R and θ in the range $0\,°\text{C}–100\,°\text{C}$ is given by

$$R = R_0 + k\theta$$

where R_0 is the resistance of the platinum film at $0\,°\text{C}$ and k is a constant.

(a) (i) Calculate the value of k.

(ii) Find the value of θ at which R is zero. Comment on this value.

(b) (i) Assuming that the relationship between R and θ holds up to $500\,°C$, draw up a table of values for R at $100\,°C$ intervals of θ from $0\,°C$ up to $500\,°C$.

(ii) The measured values of the resistance R_m of the platinum film at these values of θ are

$\theta/°C$	0	100	200	300	400	500
R_m/Ω	100	138	175.5	212	247	281

Show how R (assumed linear) differs from R_m (measured experimentally) over the range $0\,°C–500\,°C$ by plotting a graph of the difference $\Delta R = R - R_m$ against θ.

(iii) Calculate values for ΔR as a percentage of R_m at values of θ equal to $200\,°C$, $300\,°C$ and $400\,°C$. Hence estimate the value of θ above which this percentage is greater than 1.0%. [Edexcel 2000, part]

7 (i) A battery has an e.m.f. of $12.0\,V$ and an internal resistance of $3.0\,\Omega$. Calculate the p.d. across the battery when it is delivering a current of $3.0\,A$.

(ii) The same battery is now connected to a filament lamp. The graph shows how the p.d. across the lamp would depend on the current through it.

Use your answer to part (i) to help you draw, on the same axes, a line showing how the p.d. across the battery would depend on the current through it.

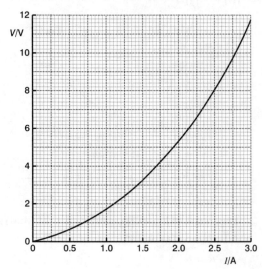

What current will the battery drive through the lamp? [Edexcel 2000, part]

8 Which one of the graphs best represents the relationship between the energy W of a photon and the frequency f of the radiation?

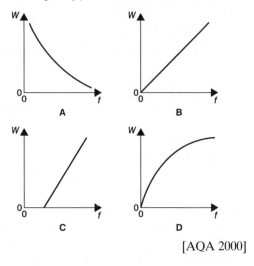

[AQA 2000]

9 The graph shows the charge stored in a capacitor as the voltage across it is varied.

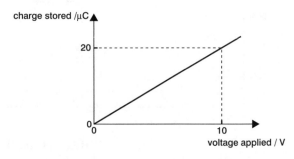

The energy stored, in μJ, when the potential difference across the capcitor is $5\,V$, is

A 25   B 50   C 100   D 200

[AQA 2000]

10 A student assembles the circuit shown in which the switch is initially open and the capacitor uncharged.

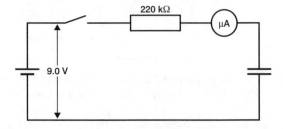

He closes the switch and reads the microammeter at regular intervals of time. The battery maintains a steady p.d. of $9.0\,V$ throughout. The graph shows how the current I varies with the time t since the switch was closed.

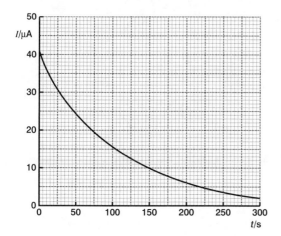

Use the graph to estimate the total charge delivered to the capacitor.

Estimate its capacitance.   [Edexcel 2001]

11 It is suggested that the turn-on time, T_{on}, for a liquid crystal display is given by the equation

$$T_{on} = \frac{k\eta d^2}{V^2}$$

where η is the viscosity,
  V the voltage applied,
  d the thickness of crystal, and
  k is a constant.

Data showing how the turn-on time T_{on} depends on the voltage V is provided in the table below.

Turn-on time T_{on}/ms	Voltage V/V
5	2.01
10	1.42
15	1.16
20	1.00
27	0.86

(a) On the grid opposite*, plot a suitable graph to test the relationship suggested between T_{on} and V. Record the results of any calculations that you perform by adding to the table above.

(b) Discuss whether or not your graph confirms the suggested relationship between T_{on} and V.
[Edexcel S-H 2001]

12 A cathode-ray oscilloscope has its amplifier sensitivity control set at $10\,\text{V cm}^{-1}$. (The calibration of both amplifier sensitivity and timebase controls on this c.r.o. is accurate.)

An a.c. voltage of frequency 10 kHz is applied to the input of the amplifier. Fig. 30.18 shows the trace obtained on the screen.

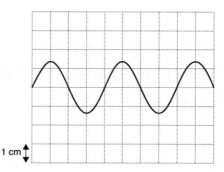

1 cm

Fig. 30.18

(1) Calculate the amplitude of the input signal.
(2) What is the setting of the timebase control?   [CCEA 2001, part]

*An A4 graph sheet was provided with this question.

Section J
Special topics

31
Astronomy

Looking at stars from the Earth

The Earth approximates to a sphere. It spins and the axis of spin meets the Earth's surface at the north pole and the south pole. The equator of the Earth is an imaginary line on its surface, a circle round the Earth's middle whose plane is perpendicular to the axis. Where a person is on the Earth's surface affects what stars may be seen. How near this observer is to the north pole can be described by the angle called the latitude ϕ, as shown in Fig. 31.1.

A plane drawn tangential to the Earth's surface at the observer's position (a line in Fig. 31.1) is the 'horizon' line because the observer can only see what is above it. The observer's 'zenith' is a point that lies straight up above the observer's head on a line perpendicular to the Earth's surface. A line from the observer to a star being viewed makes an angle A with the horizon which is called the star's 'altitude'. The angle z between a star's direction from the observer and the observer's zenith is oddly called the 'zenith DISTANCE'.

The bright star known as 'polaris' lies on the Earth's axis line at a vast distance north of the Earth and Fig. 31.2 shows a line from our observer directed towards Polaris.

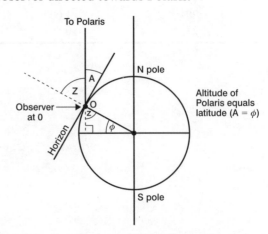

Fig. 31.2   An observer in the northern hemisphere

The zenith distance for Polaris is vertically opposite an angle in the right-angled triangle shown, and so $\phi = 90 - z$. But z and A are adjacent and add to $90°$, so $A = 90 - z$ also, which means that A for Polaris $= \phi$.

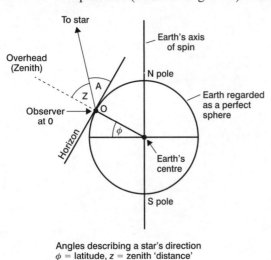

Angles describing a star's direction
ϕ = latitude, z = zenith 'distance'
A = altitude

Fig. 31.1   An observer on the Earth

Altitude of Polaris = latitude of observer

(31.1)

How far east or west an observer is from some reference point is the 'longitude'. The reference point is Greenwich (near London in England) or, you may say, a reference line drawn through Greenwich from N pole to S pole, the Greenwich 'meridian'. The longitude of a place is the number of degrees the place lies to the east or west of this reference and can be up to 180 °E or W of Greenwich.

The celestial sphere

Stars are at all sorts of large distances from the Earth. Except for the Sun which is comparatively close, the stars give the impression of being fixed in position on a spherical surface that you might call the sky but is known as the 'celestial sphere'. How far from the Earth you imagine this to be does not matter. The equator and poles of this sphere are as shown in Fig. 31.3. Polaris coincides with the north celestial pole.

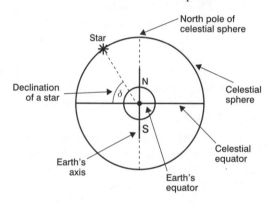

Fig. 31.3   The celestial sphere

The position of a star is partly described by the angle (δ) its direction makes with the celestial equator. This angle is the star's 'declination' and is analogous to an observer's latitude on the Earth.

Just as there is a longitude angle for an observer on Earth, so for a star's position on the celestial sphere we measure an angle from a reference point which is the 'first point of Aries'. There is a choice here between quoting this angle in degrees or, what can be more useful, describing the angle by the time that the Earth needs to turn to move through the angle concerned. If the angle is measured in degrees westwards from the first point of Aries (FPA) then we call the angle the 'sidereal hour angle' (SHA) and this can be up to 360°.

The SHA of a star is the number of degrees it is west of the FPA

If we measure the angle from the same reference but eastwards, and specify it as a time, then the angle is called the 'right ascension' (RA). Since one revolution of the Earth takes 24 hours, each hour corresponds to a rotation of 360/24 degrees, i.e. 15°.

An RA of 1 hour corresponds to 15°

Declination related to altitude

Fig. 31.4 shows how the declination of a star is related to the altitude seen by an observer. The diagram applies when the star is passing over the observer's meridian, i.e. when the star 'culminates' (is seen at its highest position). Remember that, in spite of the impression given by such diagrams, the Earth is of negligible size compared with the celestial sphere. Allowing for this, the declination $\delta = \phi + z = \phi + 90 - A$.

Declination = latitude plus 90 minus altitude

(31.2)

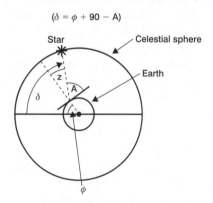

$(\delta = \phi + 90 - A)$

Fig. 31.4   Declination related to altitude

Example 1

The star Deneb (Alpha Cygni) has a declination of 45°. What is its altitude when it culminates for an observer at Newcastle, which has a latitude of 55°?

Method

$$\delta = \phi + 90 - A$$

$$\therefore \quad 45 = 55 + 90 - A$$

$$\therefore \quad A = 55 + 90 - 45 = 100°$$

(The fact that the altitude exceeds 90° by 10° means that the star is seen 10° to the south of the observer's zenith.)

Answer

100°.

Circumpolar stars

In Fig. 31.5 an observer, perhaps in England, is initially at position O_1, but due to the spin of the earth will be at O_2 at a later time.

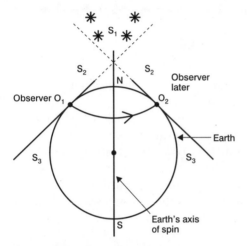

Fig. 31.5   Circumpolar stars

The observer cannot see below his horizon, so a star anywhere in the region marked by S_3 is not seen from O_1 or from O_2. It is never seen. A star in the region marked by S_1 is seen from O_1 and from O_2. It is seen by the observer at all times unless daylight makes seeing stars difficult. Such stars are known as 'circumpolar stars'. They are close to the north celestial pole on the celestial sphere and appear to rotate about the pole as the earth spins. A star in the region S_2 is visible sometimes for this observer.

In Fig. 31.6 a star is considered that is just circumpolar, being visible at position O_1 at altitude A and lying on the observer's horizon when subsequently he is at position O_2.

The two directions aiming at the (distant) star must be parallel. So angle $A = 2\phi$ (alternate angles), and using the $\delta = \phi + 90 - A$ equation obtained earlier gives $\delta = 90 - \phi$. If a star has a smaller δ than this value (making A larger), then the star is not circumpolar.

> **For a circumpolar star $\delta \geqslant 90 - \phi$**
>
> (31.3)

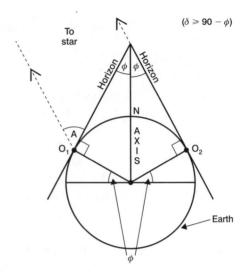

Fig. 31.6   A circumpolar star

Example 2

Vega is a star in the constellation Lyra (The Lyre) and has a declination of $38°45'$ and an RA of 18 hours 36 minutes.

(a) Show that this star is circumpolar for an observer at latitude 55° north of the equator.

(b) Calculate its sidereal hour angle (SHA).

Method

(a) $\delta = 38°45'$, $\phi = 55°$ and for the star to be circumpolar δ must equal or exceed $90 - \phi$, which is $90 - 55$ or $35°$. So the star is circumpolar for $\phi = 55°$ north.

(b) An RA of 18 h 36 min (or 18.6 h) means that the star is at an angle of 18.6×15 or $279°$ east of the FPA. It is therefore $360 - 279$ or $81°$ west of the FPA. So its SHA is 81° west.

Answer

(b) 81° west.

The Earth's orbit

The Earth is moving in an orbit around the Sun. This path is elliptical (as suggested by Fig. 31.7)

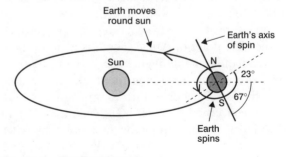

Fig. 31.7   The Earth's orbit

but almost a circle. The Earth is spinning at the same time as moving along the orbit, and the axis of spin is perpendicular to the path but tilts at an angle of 67° to the plane of the orbit.

The time taken for a complete orbit of the Sun is a year and during this time the Earth spins 365 times (more precisely 365.24), giving 365 nights and daytimes, i.e. 365 'solar days'.

Sidereal and solar times

A solar day is the time from mid-day at a place (culmination of the Sun there) through a night to next mid-day. There are therefore 365 of these days in a year.

It is also convenient for some purposes to define a day as the time for an observer to observe the appearance of the stars in the sky, see this change (a result of the Earth's turning) and return to its original appearance. This is a 'star' day, i.e. a 'sidereal day'.

Now if the Earth were not spinning but its movement was only the orbiting around the Sun, an observer, e.g. in London, would get a changing view of the stars until after a complete orbit he would get his original view again. With the 365 cycles of change in the sky's appearance due to the Earth's spinning the total number of cycles is 365 + 1 or 366 sidereal days in a year. This means that a sidereal day is slightly shorter than a solar day. Like the solar day we divide this sidereal day into 24 hours and then into minutes, the difference (known as the 'equation of time') being about 4 minutes (either solar minutes or sidereal).

One sidereal day = 23 h 56 min of solar time

The equation of time is 4 min

The time indicated by our clocks, radios and TV programmes is 'Greenwich Mean Time', which uses solar hours and counts from midnight at Greenwich. The 'mean' in its title arises because the angular velocity of the Earth's movement along its elliptical orbit is not constant. We have a 'mean solar time' and the sidereal day referred to above is really a mean sidereal day.

Greenwich sidereal time (GST) uses sidereal hours and starts at a time when the first point of Aries culminates for Greenwich.

A local sidereal time (LST) applies to any place and starts when the FPA culminates at that place. The time at which a particular star culminates at this place is the time difference (or angle measured in hours) between this star and the FPA, i.e. the star's RA.

Example 3

The dog star, Sirius, has an RA of 6 h 44 min. At what GST will it culminate for an observer (a) in Greenwich (b) in Swansea, South Wales, at a longitude of 4° west?

Method

(a) The culmination occurs when the local sidereal time equals the RA. But this LST, in the special case of Greenwich, is also the GST, so the answer is 6 h 44 min GST. In Fig. 31.8 the Earth must turn for 6 h 44 min to take G round to S.

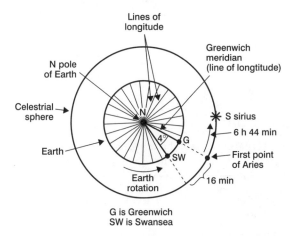

Fig. 31.8   Earth and celestial sphere seen from north at time GST = 0

(b) 4° corresponds to $\frac{4}{15}$ hours or 16 min and the Earth will have to turn an extra 16 min to make Swansea's meridian coincide with the celestial meridian of Sirius, as illustrated in Fig. 31.8.

Answer

(a) 6 h 44 min, (b) 7 h 00 min.

The Sun as a star on the celestial sphere

As the Earth spins, an observer on the Earth sees the Sun move across the sky and therefore over the celestial sphere. So the Sun appears to orbit the Earth, and its path on the celestial sphere is a circle, known as the 'ecliptic'. Because the Earth's equator is tilted at 23° to the Earth's orbit the ecliptic makes an angle of 23° with the celestial equator.

As the Sun follows the ecliptic it crosses the equator in two places. At these times, one in the Spring and one in the Autumn, night and day last 12 hours each. The time in the Spring (the 'Vernal equinox') occurs (for the northern hemisphere) when the Sun is at the first point of Aries. Consequently the FPA is often called the Vernal Equinox (see Fig. 31.9).

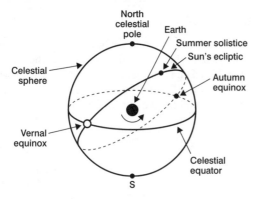

Fig. 31.9   The Sun's ecliptic

Example 4

The vernal equinox in the year 2000 occurred on March 20. At GST of 00 hours on that day which of the following times was the GMT approximately?

A 00 hours   B 00 hours 04 minutes
C 12 hours   D Some other value

Explain your answer.

Method

At the vernal equinox the Sun and first point of Aries coincide on the celestial sphere. At GST = 00 hours the FPA is above the Greenwich meridian (i.e. the FPA culminates for Greenwich). However the Sun also culminates for Greenwich at the same moment, so this is mid-day (12 hours GMT).

Answer

C

The planets and their orbits

Planets are bodies that orbit a star and so the Earth is a planet of the Sun. The orbits are generally elliptical, similar to an egg shape. The graph having the shape of an ellipse must obey

the formula $\dfrac{x^2}{a^2} + \dfrac{y^2}{b^2} = 1$, but if a and b are equal the width and length of the ellipse become equal so that the ellipse becomes a circle with $a = b = r$, giving $x^2 + y^2 = r^2$ which is the equation for a circle.

In the case of a circular orbit the inwards force required to keep a planet in orbit is mv^2/r as explained in Chapter 8. This force is provided by the gravitational force F between the Sun and the planet concerned. This force (see also Chapter 9) can be calculated from

$$F = \frac{GMm}{r^2} \qquad (31.4)$$

where G is the 'universal gravitational constant', M the mass of the Sun and m the planet's mass. r is the distance between the centres of Sun and planet. So for a circular orbit

$$\frac{GMm}{r^2} = \frac{mv^2}{r} \text{ (and cancelling is possible)}$$
$$(31.5)$$

For an elliptical orbit the planet's movement satisfies some rules discovered by Kepler. One of these rules tells us that the Sun is at one of two special positions called the 'foci' of the ellipse, as shown in Fig. 31.10.

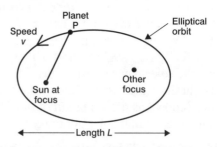

Fig. 31.10   Foci of the earth's orbit

Another of his rules says that for any elliptical orbit (including the special case of a circle) the square of the orbit period is proportional to the cube of the ellipse length.

$$T^2 \propto L^3 \qquad (31.6)$$

Example 5

The Sun's mass is 2.0×10^{30} kg. The distance between the centre of the Sun and that of the Earth is 1.5×10^{11} m. Given that the gravitational constant G

equals $6.7 \times 10^{-11}\,\mathrm{N\,m^2\,kg^{-2}}$, obtain a value for the speed of travel of the Earth along its orbit.

Method

The force F on the Earth is given by $F = \dfrac{GMm}{r^2}$ where M is the Sun's mass, m the Earth's mass and r the distance of the Earth from the Sun. But this is the force that keeps the Earth on its circular path, so $F = \dfrac{mv^2}{r}$ where r is the radius of orbit and v the speed to be determined.

$$\therefore \quad \frac{GMm}{(r)^2} = \frac{mv^2}{r}$$

$$\therefore \quad v^2 = \frac{GM}{r} = \frac{6.7 \times 10^{-11} \times 2.0 \times 10^{30}}{1.5 \times 10^{11}}$$

$$= 9.0 \times 10^8\,\mathrm{m^2\,s^{-2}}$$

$$\therefore \quad v = 3.0 \times 10^4\,\mathrm{m\,s^{-1}}$$

Answer

$3.0 \times 10^4\,\mathrm{m\,s^{-1}}$

Example 6

Jupiter has a period of about 12 Earth years. If the radius of the Earth's orbit is 150×10^9 m what approximately is the length of the elliptical orbit of Jupiter?

Method

Using the Kepler law $T^2 \propto L^3$ or $\dfrac{T_2^2}{T_1^2} = \dfrac{L_2^3}{L_1^3}$ we have

$$\frac{12^2}{1^2} = \frac{L_2^3}{(150 \times 10^9)^3}$$

$$\therefore \quad L_2^3 = 12^2 \times 150^3 \times 10^{27}$$

$$= 144 \times 3.375 \times 10^6 \times 10^{27} = 4.86 \times 10^{35}$$

$$\therefore \quad L_2 = 7.9 \times 10^{11}\,\mathrm{m}$$

Answer

7.9×10^{11} m

Distances of stars

It is often convenient to measure distances of stars from the Earth in special units. For example

    1 'astronomical unit' (AU) equals the radius of the Earth's orbit
    1 'light-year' equals the distance travelled in 1 year by light in vacuum
    1 'parsec' (pc) is the distance at which the Earth's orbit radius subtends an angle of 1 second of arc (i.e. $\frac{1}{3600}$ of a degree angle)

One method of measuring the distance of a star from the Earth is known as the 'parallax' method.

This method requires that the altitude of the star is accurately measured from the Earth, then measured again from the same place on the Earth six months later when the Earth has moved a distance equal to the diameter of its orbit.

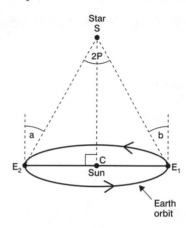

Fig. 31.11   Diagram concerning the parallax method

a and b in Fig. 31.11 are calculated and hence the parallax angle p. Then, bearing in mind that p will be extremely small because of the star's great distance, $2p$ in radians equals the distance $E_1 E_2$ divided by the star's distance SC. The star's distance from Earth will be close to the value SC at all times.

Exercise 31.1

Given that the Earth's orbit has a radius of 150×10^6 km, at what distance from the Earth is a star for which the parallax angle is 2.5×10^{-5} rad?

Temperatures of stars and their spectra

When an object is so hot that it glows white hot, it is sending out a mixture of all the colours of the rainbow. This 'spectrum' of colour will range from red through to blue and can be displayed as a visible spectrum looking like a rainbow by using, for example, a diffraction grating.

Light is a wave with alternate crests and troughs. The number of crests per second is the frequency f and is equal to the frequency of the electron movements in the surface emitting the light. The wavelength is the distance between one crest of the wave and the next. For a stationary observer the wavelength decides the colour seen.

Wavelengths below the blue are invisible ultraviolet light and wavelengths greater than red are infrared (radiated heat).

How much of each colour is sent out (i.e. the intensity of each colour) will depend upon the nature and temperature of the emitting surface. A surface that is black when cold is the best possible radiator, a 'black-body radiator', and we assume a star to be the same. The effect of the temperature is shown in Fig. 31.12.

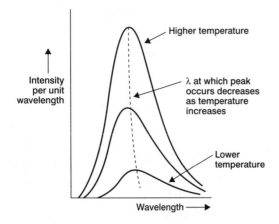

Fig. 31.12   Radiation from a hot object

The shape of the graph agrees with a formula discovered by Planck. The temperature of the object, e.g. a star, can be deduced from the graph shape and also from the Wien law, according to which

$$\lambda_{max} \times T = \text{Wien constant} \qquad (31.7)$$

where λ_{max} is the wavelength at which the peak of the graph occurs and T is the absolute temperature of the star. If the temperature is greater, then the peak is farther to the left.

It can also be deduced from the Planck equation that the radiated power P for a black-body radiator is given by

$$P = \sigma A T^4 \qquad (31.8)$$

where σ is the Stefan constant. A is the surface area of the star and T is its absolute temperature.

The spectrum of light from a star may show some wavelengths missing, so dark lines known as 'Fraunhofer lines' are seen when the spectrum is looked at. Light with these wavelengths has been stopped by absorption in the cooler layers of gas at the star's surface.

Example 7

The Wien constant is 2.90×10^{-3} m K and the Stefan constant σ is 5.67×10^{-8} W m^{-2} K^{-4}. Given that the Sun has a radius of approximately 7.0×10^8 m and its surface temperature is 5800 K, calculate:

(a) the wavelength at which the Sun's radiation peaks, and

(b) the energy emitted per second from the Sun's surface.

Method

(a) The Wien law states $\lambda_{max} T = $ Wien constant.

$$\therefore \quad \lambda_{max} \times 5800 = 2.90 \times 10^{-3}$$

$$\therefore \quad \lambda_{max} = \frac{2.90 \times 10^{-3}}{5800} = 5.0 \times 10^{-7} \text{ m}$$

(b) The energy per second is the power P, given $P = \sigma A T^4$, where A is the area of the radiating surface and T the temperature.

$$\therefore \quad P = 5.67 \times 10^{-8} \times 4\pi (7.0 \times 10^8)^2 \times 5800^4$$

$$= 3.95 \times 10^{26} \text{ watt}$$

Answers

(a) 5.0×10^{-7} m (b) 4.0×10^{26} W

Doppler effect, Hubble constant and red shift

For a light wave with frequency f one crest follows the next after a time T, the period, and this equals $1/f$ seconds. The distance between adjacent crests is the wavelength λ and is normally given by $\lambda f = c$ where c is the speed of light in the medium between source and observer (vacuum in the case of a star). So $\lambda = c/f$. However, if the source (e.g. a star) is moving away from the observer at a speed v, the wavelength is different. One crest leaving the star is followed by another at a time T later, but during this time the first crest has moved a distance c/f towards the observer and also the star has moved a distance v away from the observer, so that the second crest is emitted at a distance c/f plus vT behind the first. The wavelength is now

$$\lambda + vT \quad \text{or} \quad \lambda + \frac{v}{f} \quad \text{or} \quad \lambda + \frac{v\lambda}{c} \quad \text{or} \quad \lambda\left(1 + \frac{v}{c}\right)$$

Wavelength $(\lambda') = \lambda\left(1 + \frac{v}{c}\right)$

or $\lambda' - \lambda = \lambda v/c$, i.e.

$$\frac{\delta\lambda}{\lambda} = \frac{v}{c} \qquad\qquad (31.9)$$

where λ is the wavelength when the source is not moving away from the observer.

This change of wavelength due to movement is known as the 'Doppler Effect' and it affects the colour of light seen, in the case of a spectrum moving all wavelengths towards the red end. Similar reasoning shows a wavelength decreased to $\lambda\left(1 - \frac{v}{c}\right)$ if the speed v is towards the observer. Similar wavelength changes occur if the observer moves or both observer and source move.

The light from distant stars is moved to increased wavelengths on account of the stars moving away from the Earth. This effect is known as 'red shift'. The effect is greater for more distant stars because, according to the Hubble law the speed v of a galaxy away from the Earth increases in proportion to the distance d from the Earth.

$$v = H \times d \qquad\qquad (31.10)$$

where H is the Hubble constant.

When a star is spinning, one side of it is moving away and the opposite side is moving towards the observer, and a line in the expected spectrum will experience both an increase and a decrease due to Doppler effect: two lines will result.

Example 8

A cluster nebula in the Hydra galaxy is receding at a speed of about $6 \times 10^7 \, \text{m s}^{-1}$.

(a) What percentage increase in emission wavelengths does this speed cause, and

(b) What is the approximate distance of this nebula?

(Speed of light in vacuum $= 3.0 \times 10^8 \, \text{m s}^{-1}$, Hubble constant $= 1.7 \times 10^{-18} \, \text{s}^{-1}$.)

Method

(a) The wavelength increase is $\delta\lambda$, given by $\dfrac{\delta\lambda}{\lambda} = \dfrac{v}{c}$.

$\therefore$ The percentage increase in $\lambda = \dfrac{\delta\lambda}{\lambda} \times 100$

$$= \frac{v \times 100}{c} = \frac{6 \times 10^7 \times 100}{3.0 \times 10^8} = 2 \times 10^1$$

(b) The equation needed is $v = Hd$, where H is the Hubble constant, v the speed and d the distance.

$$\therefore \quad d = \frac{v}{H} = \frac{6 \times 10^7}{1.7 \times 10^{-18}} = 3.53 \times 10^{25} \, \text{m}$$

Answers

(a) 20%, (b) 3.5×10^{25} m approximately

Exercise 31.2

1 The Wien constant is 2.90×10^{-3} m K. What is the wavelength at which maximum radiation occurs from a star whose surface temperature is 5500 K?

2 In the spectrum of a certain star some wavelengths produced by hydrogen atoms can be distinguished. One of these wavelengths would be 4861×10^{-10} m, but on account of the star's moving away from the Earth the measured wavelength is greater by 0.7×10^{-10} m. What is the star's speed? (Speed of light in vacuum $= 3.0 \times 10^8 \, \text{m s}^{-1}$.)

The inverse square law

Consider light or other radiation emitted equally in all directions (i.e. uniformly) from the Sun, for example. If we are interested in both the light and invisible radiation, then the energy radiated per second (i.e. the power P) can be measured in watts as usual.

At a distance r large compared to the size of the source the radiation becomes spread evenly over an imaginary spherical surface having an area $4\pi r^2$. So the energy received per second by any square metre of surface at distance r, as in Fig. 31.13, is given by

$$I = \frac{P}{4\pi r^2} \qquad\qquad (31.11)$$

That is, I is proportional to $1/r^2$, inversely proportional to the square of the distance (the 'inverse square law'). Halving r, for example, quadruples I.

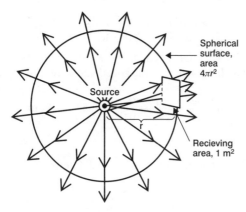

Fig. 31.13   The inverse square law

The I stands for 'intensity'.

For the same source (power P) the intensities I_1 and I_2 at distances r_1 and r_2 are related by

$$\frac{I_1}{I_2} = \frac{r_2^{\,2}}{r_1^{\,2}} \qquad\qquad (31.12)$$

If we are interested only in the light emitted and what the eye sees, it is usual for allowance to be made for the fact that some colours affect the eye more than others. The effective energy radiated is then called the 'luminous flux' and is measured in lumens. In this case the intensity I is properly called the 'luminous intensity' as opposed to 'radiant intensity'.

If a source does not radiate uniformly, a small receiving surface normal to the arriving radiation may have a small proportion of the total radiation coming in its direction, but halving its distance from the source will still quadruple the received intensity as long as it is not moved around the source at all. So Equation 31.12 can be applied to a non-uniform source.

Example 9

The total power radiated by the Sun is about 4.0×10^{26} watt. Calculate the energy received per second by the Earth from the Sun. Take the Earth's radius as 6.4×10^6 m and the distance of the Earth from the Sun as 1.5×10^{11} m.

Method

The intensity received is given by $I = \dfrac{P}{4\pi r^2}$ and is the energy received per metre^2 per second.

$$\therefore \quad I = \frac{4.0 \times 10^{26}}{4 \times 3.142 \times (1.5 \times 10^{11})^2}$$

$$= \frac{4.0 \times 10^{26}}{28.28 \times 10^{22}} = 1.4 \times 10^3 \,\mathrm{W\,m^{-2}}$$

The area of the Earth receiving this radiation, the Earth appearing as a disc to this radiation, is $4\pi r_{\mathrm{E}}^2$, which is $4 \times \pi \times (6.4 \times 10^6)^2$, and equals $5.148 \times 10^{14}\,\mathrm{m}^2$.

Hence the total energy received per second is

$$1.4 \times 10^3 \times 5.148 \times 10^{14} \quad \text{or} \quad 7.2 \times 10^{17}\,\mathrm{W}$$

Answer

$7.2 \times 10^{17}\,\mathrm{W}$

Magnitude of a star

In astronomy we usually compare the intensities and therefore the brightnesses of stars using a scale of 'magnitude'. This scale has magnitude 1 corresponding to an intensity 100 times greater than magnitude 6. For this range of 5, a difference of 1 in magnitude means a factor of $100^{1/5}$, which is 2.51, so that a magnitude 6 intensity will be multiplied by $2.51 \times 2.51 \times 2.51 \times 2.51 \times 2.51$ to change it to magnitude 1.

If the two intensities being compared are I_1 and I_2 then

$$\frac{I_1}{I_2} = 2.51^{m_2 - m_1} \quad \text{or} \quad \frac{I_1}{I_2} = 100^{(m_2 - m_1)/5}$$
$$\qquad\qquad (31.13)$$

(For example, the magnitude difference will be 1 if $I_1 = 2.51 \times I_2$.)

This equation can also be written as

$$\log \frac{I_1}{I_2} = 0.40\,(m_2 - m_1) \qquad (31.14)$$

Since we are dealing with a ratio of intensities, this formula will apply to both radiant intensities and luminous intensities.

For any star it is usual to specify the intensity it gives at a distance away from it of 10 parsecs and to use the magnitude scale to compare this intensity with the similar measurement of a standard star. We are then comparing their 'absolute magnitudes'.

A star would look brighter if you could get nearer to it, because the received intensity is greater and more light enters the eye from the star. At any distance from a star other than 10 parsecs its magnitude is called its 'apparent magnitude'. At 20 parsecs, for example, the intensity will be 4 times smaller due to the inverse square law, i.e. more than 2.5 times smaller, so that the magnitude will rise by more than 1.

Example 10

Given that the Sun has an apparent magnitude of −26.6 and is at 1.0 astronomical unit (AU) from the Earth, estimate the Sun's absolute magnitude (1 parsec = 206265 AU).

Method

At the greater distance of 10 parsec the intensity I_2 will be much smaller, the magnitude m_2 bigger.

Let I_1 be the intensity at distance d of 1.0 AU, which is $1/206\,265$ pc or 4.848×10^{-6} pc.

$$\log \frac{I_1}{I_2} = 0.4\,(m_2 - m_1) \quad \text{and} \quad \frac{I_1}{I_2} = \frac{10^2}{r^2}$$

$$= \frac{10^2}{(4.848 \times 10^{-6})^2} = 4.255 \times 10^{12}$$

so $\log \left(\dfrac{I_1}{I_2}\right) = 12.6289$ and $\dfrac{12.6289}{0.4} = 31.57$

$$= m_2 - m_1 = m_2 - -26.6 = m_2 + 26.6$$

$$\therefore \quad m_2 = 31.57 - 26.6 = 4.97$$

Answer

Absolute magnitude = 5.0

Cepheid variables

Certain stars, because of processes within them, show regular rises and falls in their radiated powers. The log of the period of the fluctuation rises in proportion to the mean absolute magnitude of the star.

Exercise 31.3

1 The star Altair has an absolute magnitude of +2.4. What is its apparent magnitude at 30 parsecs?

2 If the mean absolute magnitude m of a Cepheid variable is given by

$$m = -1.8 - \log\,(T/\text{day})$$

what is the value of m when $T = 8$ days?

3 Segin and Shedir are stars in the constellation of Cassiopeia and have absolute magnitudes of −2.9 and −0.9 respectively.

(a) If the intensity produced by Segin at a distance of 10 parsec is I_1 and for Shedir is I_2 what is the ratio I_1/I_2?

(b) Segin is 160 parsec from the Earth and Shedir is about 40 parsec from the Earth. Use the inverse square law to estimate the ratio of the intensities received at the Earth.

(c) What would the ratio be at equal distances, e.g. 20 parsec, from each star?

Binary stars

A 'binary star', although perhaps appearing as a single star, consists of two stars rotating around each other, i.e. around a common centre. A straight line drawn between the two stars passes through this centre at all times (If you like 'the stars keep opposite each other' (Fig. 31.14a).)

If during each rotation one star S_1, perhaps the smaller one, passes in front of the other star S_2, as seen by an observer on Earth (Fig. 31.14b),

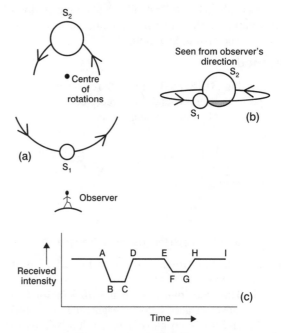

Fig. 31.14   Brightness of a binary star

then the brightness of the binary star system seen by the observer is reduced. Another reduction occurs half a revolution later when S_1 passes behind S_2.

In Fig. 31.14c, at A the star S_1 is about to cross S_2 as in Fig. 31.14b, at B all of it is in front of S_2, at C overlap begins to reduce and at D there is no overlap. At E we have S_1 beginning to pass behind S_2. At I the sequence starts again.

In a binary system one star is moving away from the observer when its companion star is moving towards the observer, and the Doppler effect will create two spectral lines in place of each expected line, as explained earlier for a single spinning star.

Example 11

(a) Make a sketch of the graph relating brightness to time for a binary star that consists of two identical stars which repeatedly overlap with their centres forming a straight line with the observer.

(b) How could you use such a graph to measure (i) the period of rotation, and (ii) the ratio $\dfrac{radius\ (r)\ of\ orbit}{radius\ (a)\ of\ star}$?

(c) Suppose that the above-mentioned ratio is 40 and that the period of rotation is measured as 5.0×10^6 s (about 58 days). Assuming that the density of the stars is $1.5 \times 10^3\ \mathrm{kg\,m^{-3}}$ and that the radius of the orbit is 2.0×10^7 km, (i) calculate the mass of each star, and (ii) calculate the period of rotation to show that the assumptions made agree with the measured period.

Method

(a) See Fig. 31.15: at B and E there is a momentary perfect overlap and the received intensity is produced by only the nearer star.

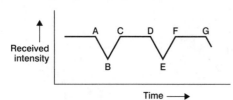

Fig. 31.15   Graph for answer to Example 10

(b) (i) The time from A to G in Fig. 31.15 is the period.
   (ii) The time from A to B is sufficient for S_1 to move distance a to the right in Fig. 31.14(b) and for S_2 to move a to the left, whereas in time A to G each star moves a distance $2\pi R$.

So $\dfrac{r}{a} = \dfrac{AG/2\pi}{AB}$.

(c) (i) Mass $m = \dfrac{4\pi a^3 \rho}{3}$ where ρ is density.

$$\therefore m = \frac{4 \times 3.142 \times (2.0 \times 10^{10}/40)^3 \times 1.5 \times 10^3}{3}$$

$$= 7.854 \times 10^{29}\ \mathrm{kg}$$

(ii) The period can be found if speed v is first obtained from the equations

$$F = \frac{GmM}{distance^2} \quad \text{and} \quad F = \frac{mv^2}{r}$$

where F is the force on either star.

Here $M = m$. This case is unlike that of the Earth orbiting the Sun (Example 5). The Sun's mass is so much greater than the Earth's that the force acting between the masses causes negligible movement of the Sun. Here both stars move. In this special and simple case of equal masses the stars must follow the same orbit whose centre will be midway between the two stars.

So $\dfrac{Gm^2}{(2r)^2} = \dfrac{mv^2}{r}$ where r is the orbit's radius,

And $v^2 = \dfrac{Gm}{4r} = \dfrac{6.67 \times 10^{-11} \times 7.854 \times 10^{29}}{4 \times 2.0 \times 10^{10}}$

$$= 6.55 \times 10^8\ \mathrm{m^2\,s^{-2}}$$

$$\therefore \quad v = 2.56 \times 10^4\ \mathrm{m\,s^{-1}}$$

$$\therefore \quad \text{period} = 2\pi r/v = \frac{2 \times 3.142 \times 2.0 \times 10^{10}}{2.56 \times 10^4}$$

$$= 4.91 \times 10^6\ \mathrm{s}$$

Answer

(c) (i) Mass $= 7.9 \times 10^{29}$ kg, (ii) period $= 4.9 \times 10^6$ s

Exercise 31.4

Capella is a binary star. Its two stars have masses not too different from 8×10^{26} kg each. If they are 3×10^9 m apart, what is their orbiting period? (Take the universal gravitational constant G as $6.7 \times 10^{-11}\ \mathrm{N\,m^2\,kg^{-2}}$.)

Exercise 31.5: Examination questions

1   For latitude $40°$ north what is the declination of a star

(a) if its altitude is $70°$

(b) if it is just visible on the northern horizon?

2 The right ascension for Vega is 18 h 36 min. At what GST will it culminate for a place 1 east of Greenwich?

3 Neptune orbits the Sun with a period of 165 years. What is the approximate radius of the orbit measured in AU?

4 Calculate the radius of a star with an absolute surface temperature of 6000 K and a power output of 3.5×10^{34} W, given that the Stefan constant is 5.67×10^{-8} W m^{-2} K^{-4}.

5 The absolute magnitude of the star Pollux is $+ 1.0$. What is its apparent magnitude at a distance of 35 light years? (1 light year $= 0.307$ parsec.)

6 A star of luminosity L is a distance d from the Earth. The intensity I of its radiation arriving at the Earth is given by the formula

$$I = L/(4\pi d^2)$$

Define the terms *luminosity* and *intensity*, and derive this formula. State *one* assumption which underlies the formula.

Deneb and Vega are two stars of similar colour. The table gives some information about them.

Star	Luminosity/W	Distance from Earth/m
Deneb	2.5×10^{31}	1.5×10^{19}
Vega	1.9×10^{28}	2.3×10^{17}

Which star would you expect to appear brighter as seen from Earth? Explain your answer.

[Edexcel 2001, part]

7 A certain galaxy G, visible in the constellation Ursa Major, is thought to be moving away from the Earth at a speed of 1.5×10^7 m s^{-1}. Use the formula

$$\frac{\Delta\lambda}{\lambda} = \frac{v}{c}$$

to calculate the apparent wavelength, measured using light from this galaxy, of a spectral line whose normal wavelength is 396.8 nm.

According to Hubble's law, the speed v at which a galaxy is moving away from the Earth is related to its distance d from the Earth by the formula

$$v = Hd$$

where H, the Hubble constant, has a value of approximately 1.7×10^{-18} s^{-1}. Estimate the distance of the galaxy G from the Earth.

(Take velocity of light in vacuum $c = 3.00 \times 10^8$ m s^{-1})

[Edexcel 2000, part]

32
Medical and health physics

Introduction

The medical profession has developed many techniques from physics. Thus medical and health physics is a very large topic embracing many branches of physics. We have already dealt with the following:

- Biomechanics of body forces (Chapter 4)
- The eye and correction of defective vision (Chapter 15)
- Fibre optics (Chapter 14)

This chapter deals with the physics of hearing, applications of ultrasonics including measurement of blood flow, the effects of ionising radiation and radiation protection (including absorption).

Physics of hearing

Intensity of sound

The eardrum vibrates according to the intensity of sound incident upon it. The intensity of a wave, be it a matter wave like sound, or electromagnetic wave like light, is given by

$$\text{Intensity } I = \frac{\text{Power}}{\text{Area}}$$

The unit of intensity is W m^{-2}.

A point source of sound emitting power P uniformly in all directions will result in a sound of intensity I at a distance r from the source given by:

$$I = P/4\pi r^2 \tag{32.1}$$

This is because the sound is spread uniformly over a spherical surface of radius r and of area $4\pi r^2$. It means that the inverse square law is obeyed (see also Chapters 28 and 31.).

Example 1

(a) Calculate the sound intensity at a distance of 20 m from a source of power 5.0 mW.

(b) If the ear of an observer can be assumed to be a circle of radius 0.8 cm, calculate the power of the sound entering the ear at 20 m from the source. Assume that the aperture of the ear is perpendicular to the arriving sound.

Method

(a) We use Equation 32.1 in which $P = 5.0 \times 10^{-3}$ and $r = 20$. Thus intensity I is given by

$$I = P/4\pi r^2 = 5.0 \times 10^{-3}/4\pi \times 20^2$$
$$= 0.995 \times 10^{-6} \text{ W m}^{-2}$$

(b) We know that 0.995×10^{-6} W of sound is incident on an area of 1.0 m^2 at a distance of 20 m from the source. A circle of radius 0.8 cm, or 0.8×10^{-2} m, has an area A given by:

$$A = \pi \times \text{radius}^2 = \pi \times (0.8 \times 10^{-2})^2$$
$$= 2.0 \times 10^{-4} \text{ m}^2$$

Thus, the power P_{ear} of sound entering the aperture of the ear is given by:

$$P_{\text{ear}} = \text{intensity} \times \text{area}$$
$$= 0.995 \times 10^{-6} \times 2.0 \times 10^{-4}$$
$$= 1.99 \times 10^{-10} \text{ W}$$

Answer

(a) $1.0 \,\mu\text{W m}^{-2}$, (b) 2.0×10^{-10} W

Intensity level

The ear can detect sounds over a vast range of intensities – its response is roughly logarithmic. For this reason a logarithmic scale called the decibel (dB) scale is used to record sound level. We define:

$$\text{Intensity level (in dB)} = 10 \log_{10}(I/I_0) \tag{32.2}$$

This is the intensity level of a sound of intensity I relative to a sound of intensity I_0. The average human ear can just detect a sound of intensity $I_0 = 1.0 \times 10^{-12} \, \text{W m}^{-2}$ This is called the threshold of hearing. If a sound is quoted as having an intensity level of, for example, 70 dB, it may be taken as being referred to this threshold of $1.0 \times 10^{-12} \, \text{W m}^{-2}$.

A doubling of sound intensity from I to $2I$ corresponds to a difference in intensity level of 3.0 dB. This can be seen from Equation 32.2 since

Increase in intensity level

$= 10[\log_{10}(2I/I_0) - \log_{10}(I/I_0)]$

$= 10\log_{10}(2I/I)$ (see Equation 2.12)

$= 10\log_{10} 2$

$= 3.01 \, \text{dB}$

Example 2

Calculate the intensity level of sounds having the following intensities:

(a) Loud music, $2.00 \times 10^{-2} \, \text{W m}^{-2}$;

(b) Noisy classroom, $5.00 \times 10^{-6} \, \text{W m}^{-2}$;

(c) Threshold of hearing, $1.00 \times 10^{-12} \, \text{W m}^{-2}$.

Method

We use Equation 32.2 in which $I_0 = 1.00 \times 10^{-12} \, \text{W m}^{-2}$.

(a) We have $I = 2.00 \times 10^{-2} \, \text{W m}^{-2}$. Thus

Intensity level

$= 10\log_{10}(I/I_0)$

$= 10\log_{10}(2.00 \times 10^{-2}/1.00 \times 10^{-12})$

$= 10\log_{10}(2.00 \times 10^{10})$

$= 103$

The intensity level is 103 dB.

(b) $I = 5.00 \times 10^{-6} \, \text{W m}^{-2}$. Thus

Intensity level

$= 10\log_{10}(5.00 \times 10^{-6}/1.00 \times 10^{-12})$

$= 10\log_{10}(5.00 \times 10^{6})$

$= 66.9$

The intensity level is 67 dB.

(c) $I = I_0$, hence

Intensity level $= 10\log(I_0/I_0)$

$= 10\log_{10} 1.00 = 0.00$

The intensity level for the threshold of hearing is zero.

Note that since intensity levels are taken relative to the threshold of hearing, it follows that this 'base' intensity level is zero, since we use a logarithmic scale.

Answer

(a) 103 dB, (b) 67 dB, (c) 0 dB.

Example 3

A music system can produce a sound of intensity $1.5 \times 10^{-5} \, \text{W m}^{-2}$. Replacing the amplifier with a more powerful one increases the intensity to $9.0 \times 10^{-4} \, \text{W m}^{-2}$. Express the increase in decibels.

Method

From Equation 32.2 we note that the difference in intensity level between two sounds of intensity I_2 and I_1 is given by

Intensity level difference

$$= 10[\log_{10}(I_2/I_0) - \log_{10}(I_1/I_0)] \qquad (32.3)$$
$$= 10\log_{10}(I_2/I_1)$$

We have $I_2 = 9.0 \times 10^{-4} \, \text{W m}^{-2}$ and $I_1 = 1.5 \times 10^{-5} \, \text{W m}^{-2}$. Equation 32.3 gives

Intensity level difference

$= 10\log_{10}(9.0 \times 10^{-4}/1.5 \times 10^{-5}) = 10\log_{10} 60$

$= 17.8 \, \text{dB}$

Answer

The intensity level difference is 18 dB.

Example 4

The sound intensity in a factory is $0.040 \, \text{W m}^{-2}$. The wearing of ear muffs by a worker results in a drop of 20 dB in perceived intensity level. Calculate:

(a) the sound intensity perceived by the worker when wearing ear muffs;

(b) the intensity level (i) without ear muffs and (ii) with ear muffs.

Method

(a) We use Equation 32.3 with $I_1 = 0.040$, Intensity level difference $= -20$ dB (note the negative sign, since intensity level has decreased). We require I_2. Thus

$-20 = 10\log_{10}(I_2/0.040)$

Rearranging gives

$I_2 = 0.040 \, \text{antilog}(-2.0) = 0.040/100$

$= 4.0 \times 10^{-4}$

263

The new sound intensity is $4.0 \times 10^{-4}\,\mathrm{W\,m^{-2}}$.

Note that a change of 20 dB is equivalent to a sound intensity change of 100 times (we can see this since $10 \log_{10} 100 = 20$).

(b) (i) We use Equation 32.2 in which $I = 0.040$ and $I_0 = 1.0 \times 10^{-12}$. Hence

Intensity level

$$= 10 \log_{10} (0.040/1.0 \times 10^{-12})$$

$$= 10 \log_{10} (4.0 \times 10^{10})$$

$$= 106\,\mathrm{dB}$$

(ii) The new intensity level is 20 dB less than 106 dB, which is 86 dB.

Answer

(a) $4.0 \times 10^{-4}\,\mathrm{W\,m^{-2}}$, (b) (i) 106 dB, (ii) 86 dB.

Exercise 32.1

(Take the threshold of hearing as $1.0 \times 10^{-12}\,\mathrm{W\,m^{-2}}$.)

1 Calculate the intensity level, in decibels, in the following circumstances:

(a) for a sound of intensity $4.0 \times 10^{-5}\,\mathrm{W\,m^{-2}}$

(b) at a distance of 10 m from a point source of sound of power 8.0 W. (Hint: first use Equation 32.1 to find the sound intensity).

2 The intensity level on a rocket launch pad is 170 dB. To what sound intensity does this correspond? (Note that this would rupture the ear drum.)

3 Calculate the difference in intensity level between two sounds of intensity $2.0 \times 10^{-4}\,\mathrm{W\,m^{-2}}$ and $5.0 \times 10^{-6}\,\mathrm{W\,m^{-2}}$.

4 Calculate the intensity of sound that has an intensity level (a) 5.0 dB above, (b) 5.0 dB below, sound of intensity $1.0 \times 10^{-8}\,\mathrm{W\,m^{-2}}$.

Applications of ultrasound

Reflection of ultrasound

The reflection of ultrasound is used to observe structures within the human body. An A scan can be used to measure distances in the body (e.g. biparietal diameter). A B scan can provide an outline image of the foetus.

Reflection occurs when an ultrasonic pulse passes across an interface between two media – for example tissue and bone. Some of the energy and intensity of the ultrasonic pulse is reflected as a result of the fact that the two media will have different 'characteristic acoustic impedances'. This is shown in Fig. 32.1.

The characteristic acoustic impedance Z of a medium is defined by:

$$Z = \text{density } \rho \times \text{velocity of ultrasound in the medium } c \tag{32.4}$$

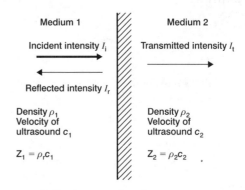

Fig. 32.1   Reflection at an interface

Values of density and velocity of ultrasound in different media are shown in Table 32.1.

Table 32.1   Density and velocity of ultrasound in certain materials

Material	Density/ $10^3\,\mathrm{kg\,m^{-3}}$	Velocity/ $\mathrm{km\,s^{-1}}$
Air	1.30×10^{-3}	0.330
Bone	1.91	4.08
Brain	1.03	1.54
Fat	0.952	1.45
Muscle	1.08	1.58
Soft tissue	1.06	1.54
Water	1.00	1.50

The fraction α_r of the intensity reflected is given by

$$\alpha_r = \frac{I_r}{I_i} = \left(\frac{Z_2 - Z_1}{Z_2 + Z_1} \right)^2 \tag{32.5}$$

where I_i = incident intensity from medium 1

I_r = intensity reflected at interface between medium 1 and 2

$Z_1 = \rho_1 c_1$ = characteristic impedance of medium 1

$Z_2 = \rho_2 c_2$ = characteristic impedance of medium 2

Note that if $Z_1 = Z_2$ there is no reflected intensity. If Z_1 and Z_2 are very different then most of the incident intensity is reflected at the interface.

Example 5

From the values given in Table 32.1, calculate the fraction of intensity reflected at the following interfaces: (a) air–soft tissue, (b) water–soft tissue, (c) soft tissue–bone.

Method

We use Equation 32.5 in each case.

(a) In this case air is medium 1 and soft tissue is medium 2. We have

$$Z_1 = \rho_1 c_1 = 1.30 \times 330 = 429 \, \text{kg m}^{-2} \text{s}^{-1}$$
$$Z_2 = \rho_2 c_2 = 1.06 \times 10^3 \times 1.54 \times 10^3$$
$$= 1.63 \times 10^6 \, \text{kg m}^{-2} \text{s}^{-1}$$

Substituting into Equation 32.5 gives

$$a_r = \left(\frac{Z_2 - Z_1}{Z_2 + Z_1}\right)^2$$
$$= \left(\frac{1.63 \times 10^6 - 0.429 \times 10^3}{1.63 \times 10^6 + 0.429 \times 10^3}\right)^2$$
$$= 1.00$$

Note that, to within three significant figures, all of the incident intensity is reflected. This is because the two media have very different characteristic acoustic impedances.

(b) In this case water is medium 1 and soft tissue is medium 2. We have

$$Z_1 = \rho_1 c_1 = 1.00 \times 10^3 \times 1.50 \times 10^3$$
$$= 1.50 \times 10^6 \, \text{kg m}^{-2} \text{s}^{-1}$$
$$Z_2 = \rho_2 c_2 = 1.06 \times 10^3 \times 1.54 \times 10^3$$
$$= 1.63 \times 10^6 \, \text{kg m}^{-2} \text{s}^{-1}$$

Substituting into Equation 32.5 gives

$$\alpha_r = \left(\frac{Z_2 - Z_1}{Z_2 + Z_1}\right)^2$$
$$= \left(\frac{1.63 \times 10^6 - 1.50 \times 10^6}{1.63 \times 10^6 + 1.50 \times 10^6}\right)^2$$
$$= 1.72 \times 10^{-3}$$

Thus, very little of the incident intensity is reflected and most of it is transmitted in to the soft tissue when ultrasound is incident from water. Comparison of (a) and (b) shows why it is essential to use an appropriate coupling medium to exclude air between the transducer which produces the ultrasound and the skin surface (soft tissue). If this were not so, the presence of an air gap would mean that very little ultrasound would pass from the transducer into the body.

(c) In this case soft tissue is medium 1 and bone is medium 2. We have:

$$Z_1 = \rho_1 c_1 = 1.06 \times 10^3 \times 1.54 \times 10^3$$
$$= 1.63 \times 10^6 \, \text{kg m}^{-2} \text{s}^{-1}$$
$$Z_2 = \rho_2 c_2 = 1.91 \times 10^3 \times 4.08 \times 10^3$$
$$= 7.79 \times 10^6 \, \text{kg m}^{-2} \text{s}^{-1}$$

Substituting into Equation 32.5 gives

$$\alpha_r = \left(\frac{Z_2 - Z_1}{Z_2 + Z_1}\right)^2$$
$$= \left(\frac{7.79 \times 10^6 - 1.63 \times 10^6}{7.79 \times 10^6 + 1.63 \times 10^6}\right)^2$$
$$= 0.428$$

Thus, 42.8% of the incident intensity is reflected at the boundary between soft tissue and bone. This is a typical value which enables structures to be seen within the human body.

Answer

(a) 1.00, (b) 1.72×10^{-3}, (c) 0.428.

Exercise 32.2

1 Use Table 32.1 to calculate the characteristic acoustic impedance of (a) brain, (b) muscle, (c) fat.

2 Calculate the fraction of intensity which will be reflected at the boundary between (a) brain and bone, (b) muscle and fat. Hence comment on the ability of ultrasonic reflection techniques to detect the boundaries in (a) and (b).

Blood flow measurement

The Doppler effect is a change in the observed frequency of waves as a result of movement of some kind*. In blood flow measurement use is made of the fact that when a beam of ultrasound is reflected from moving blood cells then the reflected waves have a different frequency to the incident waves. This Doppler shift Δf can be used to estimate the speed of blood flow.

*See also Chapter 31.

Fig. 32.2 shows an arrangement which is used to estimate the speed v of blood cells in a blood vessel using ultrasound.

It can be shown that the change in frequency Δf compared to the incident frequency f is given by

$$\Delta f = \frac{2fv\cos\theta}{c} \qquad (32.6)$$

where v is the blood flow velocity, c is the (average) speed of ultrasound and θ is the angle of incidence of the beam (see Fig. 32.2).

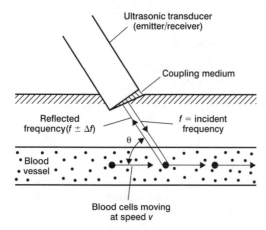

Fig. 32.2   Blood flow measurement

Example 6

In a measurement of blood flow in a patient, ultrasound of frequency 5.0 MHz is incident at an angle of 30° to the blood vessel and a Doppler shift in frequency of 4.4 kHz is observed. If the velocity of ultrasound can be taken as 1.5 km s^{-1} and the blood vessel is of diameter 1.0 mm calculate (a) the blood flow velocity, (b) the volume rate of blood flow.

Method

(a) To find the velocity v we refer to Equation 32.6 and Fig. 32.2. We have $\Delta f = 4.4 \times 10^3$ Hz, $f = 5.0 \times 10^6$ Hz, $\theta = 30°$ and $c = 1.5 \times 10^3$ m s^{-1}.

Rearranging Equation 32.6 gives

$$v = \frac{\Delta f \times c}{2f\cos\theta} = \frac{4.4 \times 10^3 \times 1.5 \times 10^3}{2 \times 5.0 \times 10^6 \times \cos 30°}$$

$$= 0.76 \text{ m s}^{-1}$$

(b) The volume rate of blood flow V_R is given by

$$V_R = A \times v$$

where A = area of cross section of the blood vessel and v = (average) blood flow velocity

We have

$$A = \pi \times (\text{radius})^2 = \pi \times (0.5 \times 10^{-3})^2 \text{ m}^2$$

and $v = 0.76 \text{ m s}^{-1}$

Thus $V_R = 0.60 \times 10^{-6} \text{ m}^3 \text{ s}^{-1}$

Answer

(a) 0.76 m s^{-1} (b) $0.60 \times 10^{-6} \text{ m}^3 \text{ s}^{-1}$

Exercise 32.3

1 Typical values present in ultrasound observation of blood flow velocity are as follows:

velocity of ultrasound = 1.5 km s^{-1},
blood velocity = 0.50 m s^{-1}
incident frequency = 10 MHz,
angle of incidence = 45°

Calculate the Doppler shift in frequency which would occur under these circumstances.

2 Ultrasound of frequency 4.0 MHz is incident at an angle of 30° to a blood vessel of diameter 1.6 mm. If a Doppler shift of 3.2 kHz is observed calculate:

(a) the blood flow velocity

(b) the volume rate of blood flow

Assume that the speed of ultrasound is 1.5 km s^{-1}.

Ionising radiation

The essentials of the following topics have already been dealt with:

- X-radiation in Chapter 28
- Radioactivity in Chapter 28 (including activity and half-life, absorption and half-thickness)

Those further aspects which occur in medical physics will be dealt with in this part.

Physical, biological and effective half-life

Substances containing radioactive isotopes may be introduced into the body, for example orally or by injection, for therapeutic purposes or to enable the body to be imaged. This radioactive material may be excreted from the body by biological means. Thus the decay of radioactivity in the body is governed by two factors:

1 'Normal' radioactive decay in which radioactive nuclei spontaneously decay and for which we have a *physical* or *radioactive half-life* T_R (this has been dealt with in Chapter 28 – see Equation 28.3).

2 Biological processes in which material is excreted from the body and for which we define a *biological half-life* T_B. The biological half-life is the time needed for one half of the original atoms present in an organ to be removed from the organ by biological processes alone (note this depends on the individual patient).

Taken together these two processes allow us to define an *effective half-life* T_E given by (since the two processes act 'in parallel')

$$\frac{1}{T_E} = \frac{1}{T_B} + \frac{1}{T_R} \tag{32.7}$$

or

$$T_E = \frac{T_B T_R}{(T_B + T_R)} \tag{32.8}$$

Note that T_E is less than T_B or T_R since the biological and physical processes are additive. Also if $T_B \gg T_R$ then $T_E \approx T_R$ and if $T_R \gg T_B$ then $T_E \approx T_B$.

Example 7

^{131}I has a physical (radioactive) half-life of 8.0 days. It is cleared from the thyroid with a half-life of 15 days.

(a) Calculate the effective half-life of ^{131}I in the thyroid.

(b) If the initial activity is 0.40 MBq, find the activity after 7.0 days.

Method

(a) We have $T_R = 8.0$ days and $T_B = 15$ days. From Equation 32.8 we can find T_E:

$$T_E = \frac{T_B T_R}{(T_B + T_R)} = \frac{15 \times 8.0}{15 + 8.0}$$

$$= 5.22$$

The effective half-life of ^{131}I in the thyroid is 5.2 days.

(b) The effective half-life governs the rate at which radioactivity actually decays in the organ, in this case the thyroid. The method by which activity can be calculated at some time has been dealt with in Chapter 28. We have

$$A = A_0 e^{-\lambda t} \tag{28.2b}$$

where A_0 = original activity at time $t = 0$

A = activity at time t

λ = radioactive decay constant (which can be calculated from the half-life T using Equation 28.3: $\lambda = \ln 2/T = 0.693/T$)

We have $A_0 = 0.40$ MBq, $T = T_E = 5.22$ days and require activity after 7.0 days. Equation 28.3 gives

$$\lambda = \frac{\ln 2}{T} = \frac{0.693}{5.22} = 0.133 \text{ days}^{-1}$$

Using Equation 28.2b with $\lambda t = 0.133 \times 7.0 = 0.931$ gives

$$A = 0.40 \, e^{-0.931}$$

$$= 0.158 \text{ MBq}.$$

The activity after 7.0 days is 0.16 MBq. (Note this can be checked using $A = A_0/2^Y$ where $Y = t/T$, as in Equation 28.4b.)

Answer

(a) 5.2 days, (b) 0.16 MBq.

Exposure and absorbed dose

Exposure of a material to radiation produces damage as a result of ionisation of the constituent atoms of the material. Exposure to radiation is defined in relation to the amount of charge Q, of one sign, produced in a mass m of *air* by the radiation. We define

$$\text{Exposure} = \frac{\text{Charge produced } Q}{\text{Mass of air } m} \tag{32.9}$$

Thus the unit of exposure is $C \, kg^{-1}$. The unit of exposure *rate* is $C \, kg^{-1} \, s^{-1}$.

Exposure does not describe the energy given to a material and thus a more useful measure of radiation damage is the *absorbed dose*. This effectively measures the deposition of energy E which causes radiation damage in a given mass m of material, such as tissue. This is defined as

$$\text{Absorbed dose} = \frac{\text{Energy deposited } E}{\text{Mass of material } m} \tag{32.10}$$

Thus the unit of absorbed dose is $J \, kg^{-1}$; known as the gray (Gy). Thus $1 \, J \, kg^{-1} = 1 \, Gy$. Absorbed *dose rate* is measured in $J \, kg^{-1} \, s^{-1}$ or $Gy \, s^{-1}$.

The relationship between exposure and the absorbed dose in a particular medium depends upon the radiation absorbing characteristics of the material – this is a function of the material and the energy spectrum of the absorbed radiation.

We say

Absorbed dose = Exposure × Energy required to produce unit charge in the material f

or

$$\text{Absorbed dose} = \text{Exposure} \times f \qquad (32.11)$$

Similarly

$$\text{Absorbed dose } rate = \text{exposure } rate \times f \qquad (32.12)$$

The units of f are $J\,C^{-1}$.

For x-radiation used in radiotherapy, f is about $34\,J\,C^{-1}$ for air and soft tissue, $37\,J\,C^{-1}$ for muscle and varies between $50\,J\,C^{-1}$ and $150\,J\,C^{-1}$ for bone, dependent upon the energy of the x-rays.

Example 8

The average ionisation energy for air is $55 \times 10^{-19}\,J$. Calculate (a) the absorbed dose in air at a point where the exposure is $50\,\mu C\,kg^{-1}$.

If this exposure occurs over a time of 20 minutes calculate (b) the exposure rate, (c) the absorbed dose rate. (Assume the charge on an electron $e = 1.6 \times 10^{-19}\,C$.)

Method

(a) We use Equation 32.11 with exposure $= 50 \times 10^{-6}\,C\,kg^{-1}$. First we must find f, the energy required to produce unit charge in the material. We are told that the ionisation energy is $55 \times 10^{-19}\,J$ – this is the energy needed to produce one ion pair, which has $1.6 \times 10^{-19}\,C$ of charge (of one sign). Thus the energy required to produce unit charge is given by

$$f = \frac{55 \times 10^{-19}}{1.6 \times 10^{-19}} = 34.4\,J\,C^{-1}.$$

From Equation 32.11:

$$\begin{aligned}\text{Absorbed dose} &= \text{Exposure} \times f \\ &= 50 \times 10^{-6} \times 34.4 \\ &= 1.72 \times 10^{-3}\,J\,kg^{-1} \\ &= 1.72\,mJ\,kg^{-1}\end{aligned}$$

(b) $\text{Exposure rate} = \dfrac{\text{Exposure}}{\text{Time } t}$

where $t = 20 \times 60 = 1200\,s$. Thus

$$\begin{aligned}\text{Exposure rate} &= \frac{50 \times 10^{-6}}{1200} \\ &= 4.17 \times 10^{-8}\,C\,kg^{-1}\,s^{-1}\end{aligned}$$

(c) $\text{Absorbed dose rate} = \dfrac{\text{Absorbed dose}}{\text{Time } t}$

$$\begin{aligned} &= \frac{1.72 \times 10^{-3}}{1200} \\ &= 1.43 \times 10^{-6}\,J\,kg^{-1}\,s^{-1}\end{aligned}$$

Answer

(a) $1.7\,mJ\,kg^{-1}$, (b) $4.2 \times 10^{-8}\,C\,kg^{-1}\,s^{-1}$, (c) $1.4 \times 10^{-6}\,J\,kg^{-1}\,s^{-1}$.

Dose equivalent and radiation levels

Exposure and absorbed dose are not adequate quantities to describe the radiation damage done by various types of radiation. We must recognise that certain types of ionising radiation (e.g. neutrons, α-particles) do more damage than other types (e.g. X-rays, γ-rays). In fact 1 Gy of densely ionising radiation does much more damage to a cell than 1 Gy of X-rays or γ-rays. We thus define the dose equivalent measured in sieverts (Sv) as

$$\text{Dose equivalent} = \text{Absorbed dose} \times \text{Quality factor} \qquad (32.13)$$

The quality factor is a pure number so that the sievert has the same physical dimensions as the gray ($J\,kg^{-1}$).

The magnitude of the quality factor accounts for the varying damage to cells by certain types of radiation. Values for quality factor (QF) for certain types of radiation are given in Table 32.2.

Table 32.2 Values for quality factor (QF) for various radiations

Radiation type	Quality factor (QF)
β-, X- and γ-rays	1.0
Slow neutrons	5.0
Fast neutrons, protons and α-particles	10

The following radiation doses apply for the UK:

- Average dose equivalent from background radiation = 1 mSv per year

- Maximum permissible dose equivalent for a radiation worker = 50 mSv per year
- Maximum permissible dose equivalent for a student = 5 mSv per year

Example 9

A radiation worker is exposed to α-radiation which results in a dose equivalent of 50 mSv over a year. If the technician works a 44 week year for 37 hours per week calculate the (average) absorbed dose rate at the technician's workplace. Neglect any background radiation dose.

Method

We use Equation 32.13 with dose equivalent = 50×10^{-3} Sv and from Table 32.2, QF = 10 (α-particles). Rearranging Equation 32.13 gives

$$\text{Absorbed dose} = \frac{\text{Dose equivalent}}{\text{QF}}$$

$$= \frac{50 \times 10^{-3}}{10}$$

$$= 50 \times 10^{-4} \text{ Gy}$$

This occurs in a time of $44 \times 37 \times 3600 = 5.86 \times 10^6$ s. Thus

$$\text{Absorbed dose rate} = \frac{\text{Absorbed dose}}{\text{Time}}$$

$$= \frac{50 \times 10^{-4}}{5.86 \times 10^6}$$

$$= 8.53 \times 10^{-10} \text{ Gy s}^{-1}$$

Answer

The (average) absorbed dose rate is 8.5×10^{-10} Gy s^{-1}.

Radiation protection

It is usually necessary to control the exposure to radiation of, for example, patients during radiotherapy treatment and/or to ensure that hospital personnel are not overexposed. This can be done in several ways:

1 By controlling the power of the source.
2 By controlling the time spent in the vicinity of the source.
3 By controlling the distance between the individual and the source – use is made of the inverse square law (see Chapter 28), since it is assumed that we have a point source.
4 By placing absorber between the source and exposed area – use is made of the notion of half-thickness to produce appropriate shielding (see Example 7, Chapter 28).

In the following two examples we refer to 3 and 4 only.

Example 10

The exposure rate at a distance of 0.50 m from a point source of radiation is 1.0 mC kg^{-1} h^{-1}. At what distance from the source will the exposure rate be 0.10 mC kg^{-1} h^{-1}?

Method

The inverse square law (see e.g. Equation 28.6) tells us that the exposure rate is proportional to $1/r^2$. Suppose that

$r_0 = 0.50$ m = Original distance, for which the original exposure rate is 1.0×10^{-3} C kg^{-1} h^{-1}

and

r_1 = new distance, for which the new exposure rate is 0.10×10^{-3} C kg^{-1} h^{-1}

Then we have

$$\frac{\text{Old exposure rate}}{\text{New exposure rate}} = \frac{r_1{}^2}{r_0{}^2}$$

Hence $\quad r_1{}^2 = r_0{}^2 \times \dfrac{\text{Old exposure rate}}{\text{New exposure rate}}$

and $\quad r_1 = r_0 \times \left(\dfrac{1.0 \times 10^{-3}}{0.1 \times 10^{-3}}\right)^{1/2}$

This gives $r_1 = 0.5 \times 10^{1/2} = 1.58$ m

Answer

1.6 m.

Absorption

X-rays and γ rays are both electromagnetic radiations and therefore show exponential attenuation by shielding. This type of absorption has been covered in Chapter 28 for γ rays.

We refer to Fig. 32.3 in which:

I_0 = incident intensity
I = emergent intensity
x = thickness (m)
μ = linear absorption coefficient (m^{-1})

Then

$$I = I_0 e^{-\mu x} \tag{32.14}$$

The *half value* thickness or half-thickness T is defined as that thickness of absorber which halves the intensity of the beam of radiation (see Chapter 28). Now:

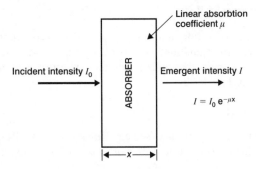

Fig. 32.3 Exponential absorption

$$I = I_0/2^Y \qquad (32.15)$$

where $Y = x/T$ and x is the thickness of the material, as above.

In addition

$$\mu = \ln 2/T \qquad (32.16)$$

Example 11

The *tenth value* thickness is that thickness of absorber which reduces the incident intensity by 10%, that is, leaves 90% of the original intensity. If the *half value* thickness of Aluminium for a given beam of X-rays is 2.00 mm, calculate:

(a) the tenth value thickness

(b) the linear absorption coefficient

(c) the fraction of the original intensity emerging through a 6.00 mm thickness of Aluminium.

Method

(a) Let the tenth value thickness be $x_{0.1}$. From Equation 32.15 we have:

$$I = I_0/2^Y \text{ with } Y = x_{0.1}/T \qquad (32.15)$$

In this case $I = 90\%$, $I_0 = 100\%$, $T = $ half value thickness $= 2.00$ mm and we require $x_{0.1}$.

Thus

$$90 = 100/2^Y \text{ so } 2^Y = 100/90$$

Taking logs to base 10 and noting that $Y = x_{0.1}/T$, gives

$$Y \log_{10} 2 = (x_{0.1}/T) \log_{10} 2 = \log_{10} (100/90)$$

Rearranging gives:

$$x_{0.1} = T \log_{10} (10/9)/ \log_{10} 2$$

Substituting $T = 2.00$ mm gives the tenth value thickness as 0.304 mm.

(b) To find μ we use Equation 32.16 in which $T = 2.00 \times 10^{-3}$ m. Thus:

$$\mu = \ln 2/T = 0.693/2.00 \times 10^{-3}$$
$$= 346.6 \, \text{m}^{-1}$$

(c) We can use Equation 32.14 in which we have $\mu = 346.6$, $x = 6.00$ mm and require the emerging fraction of intensity I/I_0. Since $\mu x = 346.6 \times 6.00 \times 10^{-3} = 2.080$, then rearranging Equation 32.14 gives:

$$I/I_0 = e^{-\mu x} = e^{-2.080} = 0.125.$$

Note that we could have obtained this answer more easily by using Equation 32.15 in which

$$Y = x/T = 6.00/2.00 = 3.00$$

(Note we have 3 half thicknesses of absorber)

Hence Equation 32.15 gives

$$I/I_0 = 1/2^{3.00} = 1/8 = 0.125.$$

Either approach is, of course, valid.

Answer

(a) 0.304 mm, (b) 347 m^{-1}, (c) 0.125 or 12.5%

Exercise 32.4

(Assume the following values: electron charge $e = 1.6 \times 10^{-19}$ C; energy required to produce unit charge in bone $= 140$ J; energy required to produce unit charge in air $= 34$ J. Refer to Table 32.2 for values of QF.)

1 The effective half-life of a radioactive isotope is 5.8 days. If it has a physical (radioactive) half-life of 8.0 days, calculate the biological half-life.

2 A certain radio-pharmaceutical has a physical half-life of 15 days and a biological half-life of 3.0 days. Calculate (a) the effective half-life, (b) the initial activity administered to the organ if the activity after 6.0 days is 50 kBq.

3 Calculate the exposure which gives rise to an absorbed dose of 5.0 mJ kg^{-1} in air.

4 A given exposure to radiation gives rise to an absorbed dose of 2.0 mJ kg^{-1} in air. Calculate the absorbed dose in bone for the same exposure.

5 An exposure rate is quoted as 75 μC kg^{-1} h^{-1}. Calculate the equivalent exposure rate in μC kg^{-1} s^{-1}.

6 A chest X-ray delivers a dose equivalent of 2.0 mSv in a time of 4.0 s. Calculate the average absorbed dose rate in this time.

7 A technician works in a hazardous environment in which the radiation dose can be considered to arise from slow neutrons. If the average absorbed dose in tissue can be taken as 1.5×10^{-4} Gy in one hour calculate:

(a) the absorbed dose rate in $Gy\,s^{-1}$:

(b) the maximum time that the technician can work in this environment in one year, assuming a maximum permissible whole body dose level of 50 mSv.

8 The exposure rate at a distance of 1.2 m from a source is $0.20\,mC\,kg^{-1}\,h^{-1}$. Calculate the exposure rate at a distance of 6.0 m from the source.

9 The tenth value thickness of lead for 50 kV X-rays is 0.180 mm. Calculate:

(a) the half value thickness

(b) the linear absorption coefficient

(c) the fraction of the original intensity emerging through a 6.00 mm thickness of lead.

Exercise 32.5: Examination questions

(Assume that the threshold of intensity of hearing is $1.0 \times 10^{-12}\,W\,m^{-2}$ unless otherwise stated.)

1 A point source of sound has a power of 12 mW. Calculate the maximum distance from the source at which it can just be heard by someone, given that the minimum sound intensity which that person's ear can detect is $3.0 \times 10^{-12}\,W\,m^{-2}$.

2 A student goes to a very loud disco at which the sound intensity level is 108 dB. Assuming that the student stays for two hours and that the sound may be assumed to be of constant intensity and to be collected by the eardrum over a surface area of $1.2\,cm^2$, calculate the total sound energy incident on the student's eardrum over this time.

3 The sound intensity next to a machine making cans is $3.6 \times 10^{-2}\,W\,m^{-2}$. It is decided that it is necessary to reduce this sound intensity to $3.6 \times 10^{-5}\,W\,m^{-2}$ in the interest of the health of the workers in that area. This is to be achieved by the wearing of ear muffs. Calculate:

(a) the intensity level of the sound next to the machine

(b) the reduction in intensity level which must be achieved by the ear muffs.

4 At a certain point P in a factory, noise can be heard from two different machines. The sound intensity level with both machines in operation is 86 dB. Calculate:

(a) the sound intensity at point P.

One of the machines is now switched off, which results in a drop in sound intensity level by 4 dB. Calculate:

(b) the sound intensity produced at P by the machine, prior to it being switched off.

5 (a) What is meant by the **intensity** of a sound?

(b) A person has normal hearing. For this person, the minimum audible intensity at a frequency of 1000 Hz is $1.0 \times 10^{-12}\,W\,m^{-2}$. The effective area of the entrance to the person's ear is $68\,mm^2$. Calculate the minimum acoustic (sound) power at 1000 Hz, incident on the entrance to the ear, which would cause the person to detect the sound.

(c) Sound entering a room through an open window produces a sound intensity level of 85 dB at a certain point in the room. When the window is closed, the sound intensity level at the point is reduced to 72 dB. Calculate the fraction of the sound energy which passes through the window glass.

[CCEA 2001]

6 The threshold of feeling is that sound intensity level at which the sensation of hearing changes to one of discomfort or pain. If this is taken to be 120 dB, calculate the sound intensity corresponding to the threshold of feeling.

7 (a) When ultrasound is used for the imaging of body structures, a coupling medium such as a water-based jelly is used between the ultrasonic transducer and the patient's skin. Explain why this is so. Be as quantitative as you can.

(b) The acoustic impedance of soft tissue is $1.63 \times 10^6\,kg\,m^{-2}\,s^{-1}$. A water-based jelly is formulated such that it acts as an ideal coupling medium to the skin (that is, there is no reflected intensity). If the velocity of sound in the jelly is $1.50 \times 10^3\,m\,s^{-1}$, calculate the density of the jelly.

8 The following information relates to ultrasound passing through body tissues:

Tissue type	Density/ $10^3\,kg\,m^{-3}$	Speed of ultrasound/ $10^3\,m\,s^{-1}$
Fat	0.9	1.5
Muscle	1.1	1.6

Calculate:

(a) the acoustice impedance of (i) fat, and (ii) muscle

(b) the fraction of intensity reflected when ultrasound passes from fat to muscle.

An ultrasonic pulse travels through fat and is reflected at a boundary between fat and tissue. A time interval of 0.018 ms is detected between the pulse entering the fat layer and being received back at the transducer. Calculate:

(c) the thickness of the layer of fat.

9 The following table shows values of acoustic impedance for various constituents of the body.

Constituent	Acoustic impedance/ $10^6 \, \text{kg m}^{-2} \, \text{s}^{-1}$
Blood	1.59
Brain	1.58
Muscle	1.70

Calculate the value of the reflection coefficient α at:

(a) brain–blood and

(b) muscle–blood interfaces.

Hence explain why ultrasound can be used to investigate blood flow within muscle but not within the brain.

10

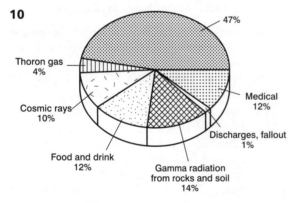

Fig. 32.4

(a) The pie chart in Fig. 32.4 shows the sources of background radiation in the United Kingdom. State the source of the 47% sector which is not labelled.

(b) Explain the relationship between an *absorbed dose* and a *dose equivalent*.

(c) The average annual dose equivalent for background radiation in the United Kingdom is 2.5 mSv. A typical chest X-ray has a dose equivalent of 30 μSv. Calculate how many chest X-rays would be needed in a year to give a patient a dose equivalent equal to the average annual value for background radiation.

(d) A patient involved in a serious road accident has 10 X-rays in 1 day. Suggest why this could be more dangerous than one year's background radiation, even though the total dose equivalent is much lower.   [OCR 2000]

11 A patient with cancer is treated using gamma rays from a cobalt-60 source. The source is surrounded by lead shielding, with an aperture through which the gamma rays pass.

The intensity received by the patient can be reduced by placing an absorber over the aperture. The gamma ray intensity I_t transmitted by the absorber is given by

$$I_t = I_0 e^{-\mu x}$$

where $I_0 = $ the gamma ray intensity incident on the absorber

$x = $ the absorber thickness (unit : mm)

$\mu = $ a constant (unit : mm^{-1})

(a) For the gamma rays emitted by a cobalt-60 source with a lead absorber 8.8 mm thick the fraction I_t/I_0 transmitted is 0.60. Calculate the value of μ.

(b) Using the information in (a) sketch a graph of I_t against x on Fig. 32.5 below.

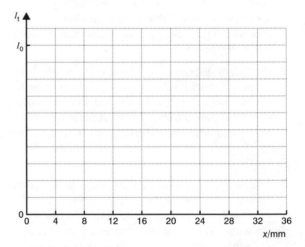

Fig. 32.5

[OCR 2001]

12 The intensity of (monochromatic) radiation from an X-ray source drops by 60% when a block of wood of thickness 30 mm is placed in its path. Calculate the half value thickness of wood for this radiation.

33 Rotational dynamics

Angular motion

A net force produces a linear acceleration such that (see equation 5.5)

$$\text{Force } F(\text{N}) = \text{Mass } m \, (\text{kg}) \times \begin{pmatrix} \text{Linear} \\ \text{acceleration } a \\ (\text{m s}^{-2}) \end{pmatrix}$$

Similarly a net torque, or moment, produces an angular acceleration such that

$$\begin{pmatrix} \text{Torque} \\ \Gamma(\text{N m}) \end{pmatrix} = \begin{pmatrix} \text{Moment of} \\ \text{inertia } I \\ (\text{kg m}^2) \end{pmatrix} \times \begin{pmatrix} \text{Angular} \\ \text{acceleration } \alpha \\ \text{rad s}^{-2} \end{pmatrix} \quad (33.1)$$

Comparing the two equations we see that Γ replaces F, I replaces m and α replaces a. Table 33.1 on page 275, lists a range of 'linear' quantities and gives their angular equivalents alongside.

Moment of inertia

The moment of inertia of various simple objects about an axis of rotation AA′ is shown in Fig. 33.1.

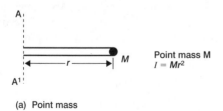

Point mass M
$I = Mr^2$

(a) Point mass

Fig 33.1   Moments of inertia of simple objects

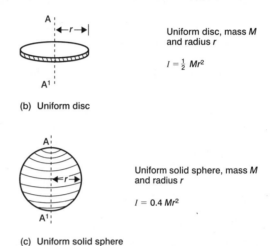

Uniform disc, mass M and radius r

$I = \tfrac{1}{2} Mr^2$

(b) Uniform disc

Uniform solid sphere, mass M and radius r

$I = 0.4 \, Mr^2$

(c) Uniform solid sphere

Fig 33.1 *continued*   Moments of inertia of simple objects

Note that the relationship $I = Mr^2$ can be applied to all objects for which the mass is effectively at a fixed distance from the axis of rotation – e.g. a hoop.

Note that the combined moment of inertia of two, or more, objects about a given axis is the sum of the separate moments of inertia.

Example 1

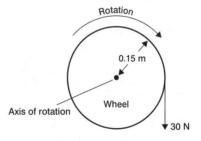

Fig. 33.2   Information for Example 1

Refer to Fig. 33.2. A constant tangential force of 30 N acts on a wheel of radius 0.15 m which rotates about its centre. Calculate (a) the torque acting on the wheel, (b) its angular acceleration if the moment of inertia of the wheel is 5.0 kg m². Neglect friction.

Method

(a)

$$\left(\begin{array}{c} \textbf{Torque} \\ \textbf{\Gamma (N\ m)} \end{array} \right) = \left(\begin{array}{c} \textbf{Force} \\ \textbf{F (N)} \end{array} \right) \times \left(\begin{array}{c} \textbf{Perpendicular} \\ \textbf{distance } d \textbf{ (m)} \\ \textbf{from axis of} \\ \textbf{rotation} \end{array} \right)$$

$$(33.2)$$

We have $F = 30$ and $d = 0.15$, so

$$\Gamma = 30 \times 0.15 = 4.5\,\text{N m}$$

This torque causes the angular velocity of the wheel to increase in the clockwise direction, i.e. it has an angular acceleration in the clockwise direction.

(b) Equation 33.1 gives $\Gamma = I\alpha$. We have $\Gamma = 4.5$ and $I = 5.0$, so

$$\alpha = \frac{\Gamma}{I} = \frac{4.5}{5.0} = 0.90\,\text{rad s}^{-2}$$

Every second the angular velocity of the wheel increases by $0.90\,\text{rad s}^{-1}$ in the clockwise direction.

Answer

(a) $4.5\,\text{N m}$, (b) $0.90\,\text{rad s}^{-2}$

Example 2

A flywheel on a motor increases its rate of rotation uniformly from $120\,\text{rev min}^{-1}$ to $300\,\text{rev min}^{-1}$ in $10\,\text{s}$. Calculate (a) its angular acceleration, (b) its angular displacement in this time.

Method

(a) We require initial angular velocity ω_0 and final angular velocity ω. In one revolution the angle swept out is 2π rad. So

$$\omega_0 = 120 \times 2\pi\,\text{rad min}^{-1} = 4\pi\,\text{rad s}^{-1}$$
$$\omega = 300 \times 2\pi\,\text{rad min}^{-1} = 10\pi\,\text{rad s}^{-1}$$

We have $t = 10$, so angular acceleration is

$$\alpha = \frac{\textbf{Change of angular velocity}}{\textbf{Time taken}}$$

$$= \frac{\omega - \omega_0}{t} \qquad (33.3)$$

So, $\alpha = \dfrac{10\pi - 4\pi}{10} = 0.6\pi\,\text{rad s}^{-2}$

(b) Angular displacement θ is given by

$$\theta = \textbf{Average angular velocity} \times \textbf{Time}$$

$$= \tfrac{1}{2}(\omega + \omega_0) \times t \qquad (33.4)$$

So, $\theta = \tfrac{1}{2}(4\pi + 10\pi) \times 10 = 70\pi$ rad

Note that we can readily solve this problem using the equations of uniform angular acceleration (see below).

Answer

(a) $0.6\pi\,\text{rad s}^{-2}$, (b) 70π rad.

Exercise 33.1

1 Calculate the required quantities:

	Γ/N m	I/kg m^2	α/rad s^{-2}
(a)	3.0	?	0.6
(b)	?	3.5	4.0
(c)	3.6	0.6	?

2 A torque of $15\,\text{N m}$ acts on a wheel of moment of inertia $6.0\,\text{kg m}^2$, initially at rest. Calculate (a) its angular acceleration, (b) its angular velocity after $20\,\text{s}$, (c) its angular displacement in this time.

3 A flywheel of moment of inertia $0.40\,\text{kg m}^2$ is initially rotating at $90\,\text{rev min}^{-1}$. It is brought to rest in $45\,\text{s}$ by a constant torque. Calculate (a) its initial angular velocity in rad s^{-1}, (b) its angular acceleration, (c) the magnitude of the torque, (d) its angular displacement in the first $15\,\text{s}$.

Equations of uniform angular acceleration

Table 33.1 on page 275 lists 'linear' quantities and their 'angular' equivalents. The equations of uniform angular acceleration are

$$\omega = \omega_0 + \alpha t \qquad (33.5)$$

$$\omega^2 = \omega_0^2 + 2\alpha\theta \qquad (33.6)$$

$$\theta = \omega_0 t + \tfrac{1}{2}\alpha t^2 \qquad (33.7)$$

They can be obtained by analogy with Equations 5.1, 5.3 and 5.4 or by combining Equations 33.3 and 33.4.

Example 3

A wheel is rotating initially at $90\,\text{rev min}^{-1}$. What torque is required to bring it to rest in 5.0 revolutions if the wheel has moment of inertia $0.80\,\text{kg m}^2$?

Method

We must first find α. We have

$$\omega_0 = 2\pi \times (90/60) = 3.0\pi$$

$\omega = 0$ and angular displacement $\theta = 5.0 \times 2\pi = 10\pi$.

So, from Equation 33.6

$$\omega^2 = \omega_0{}^2 + 2\alpha\theta$$

$$0^2 = (3\pi)^2 + 2 \times \alpha \times 10\pi$$

or $\quad \alpha = -0.45\pi \, \text{rad s}^{-2}$

Note the negative sign which indicates that the flywheel is slowing down (negative acceleration).

From Equation 33.1, with $I = 0.80$,

$$\Gamma = I\alpha = 0.80 \times (-0.45\pi)$$

$$= -0.36\pi \, \text{N m}$$

The negative sign indicates that the torque is applied in the opposite direction to that of the original rotation.

Answer

$-0.36\pi \, \text{N m}$.

Table 33.1 Rotational and translational quantities

Translational quantity	Rotational equivalent
Force F (N)	Torque Γ (N m)
Mass m (kg)	Moment of Inertia I (kg m^2)
Acceleration a (m s^{-2})	Angular acceleration α (rad s^{-2})
Time t (s)	Time t (s)
Initial velocity u (m s^{-1})	Initial angular velocity ω_0 (rad s^{-1})
Final velocity v (m s^{-1})	Final angular velocity ω (rad s^{-1})
Distance s (m)	Angular displacement θ (rad)
Momentum mv (kg m s^{-1})	Angular momentum $I\omega$ (kg m^2 rad s^{-1})
Kinetic energy $\frac{1}{2}mv^2$ (J)	Angular kinetic energy $\frac{1}{2}I\omega^2$ (J)

Example 4

A torque of 40 N m is applied to a wheel of moment of inertia 25 kg m^2, initially at rest. Calculate (a) the time it takes to make 10 revolutions and (b) its angular velocity at that time.

Method

(a) We have $\Gamma = 40$ and $I = 25$. So

$$\alpha = \frac{\Gamma}{I} = \frac{40}{25} = 1.6 \, \text{rad s}^{-2}$$

We use Equation 33.7, with $\omega_0 = 0$, $\alpha = 1.6$ and $\theta = 10$ revolutions $= 10 \times 2\pi$ rad, to find time t:

$$\theta = \omega_0 t + \tfrac{1}{2}\alpha t^2$$

$$20\pi = 0 \times t \times \tfrac{1}{2} \times 1.6 \times t^2$$

$$\therefore \quad t = \sqrt{(25\pi)} = 8.9 \, \text{s}$$

(b) We have $\omega_0 = 0$, $\alpha = 1.6$, $t = 8.9$ and require ω. So, using Equation 3.5:

$$\omega = \omega_0 + \alpha t = 0 + 1.6 \times 8.9 = 14$$

Answer

(a) 8.9 s, (b) 14 rad s^{-1}.

Example 5

A flywheel rotates on a bearing which exerts a constant frictional torque of 12 N m. An external torque of 36 N m acts on the flywheel for a time of 15 s, after which time it is removed. If the angular velocity of the flywheel increases from zero to 60 rad s^{-1} in the 15 s period, (a) calculate the moment of inertia of the flywheel, (b) find at what time the flywheel will come to rest.

Method

(a) To find I first find Γ and α. Net torque Γ is

$$\Gamma = \text{External torque} - \text{Frictional torque}$$

$$= 36 - 12 = 24 \, \text{N m}$$

We have $\omega_0 = 0$, $\omega = 60$ and $t = 15$. Rearranging Equation 33.5 gives

$$\alpha = \frac{\omega - \omega_0}{t} = \frac{60 - 0}{15} = 4.0 \, \text{rad s}^{-2}$$

So, since $I = \Gamma/\alpha = 24/4$, $I = 6.0 \, \text{kg m}^2$.

(b) When the external torque is removed, the flywheel slows down since a net torque Γ' of -12 N m now acts on it due to friction. Since $I = 6.0$, the angular acceleration α' is given by

$$\alpha' = \frac{\Gamma'}{I} = \frac{-12}{6} = -2.0 \, \text{rad s}^{-2}$$

Initial angular velocity $\omega_0' = 60$, final angular velocity (at rest) $\omega' = 0$ and $\alpha' = -2.0$. To find time t' to come to rest, we rearrange Equation 33.5:

$$t' = \frac{\omega' - \omega_0'}{\alpha} = \frac{0 - 60}{-2} = 30 \, \text{s}$$

Answer

(a) 6.0 kg m^2, (b) 30 s after the external torque is removed.

Exercise 33.2

1 Find α in each case, given the following:

(a) $\omega_0 = 10$, $\quad \omega = 25$, $\quad t = 5.0$

(b) $\omega_0 = 30$, $\quad \omega = 5.0$, $\quad t = 10$

(c) $\omega_0 = 30$, $\quad \omega = 10$, $\quad \theta = 200$

(d) $\omega_0 = 0$, $\quad t = 5.0$, $\quad \theta = 25$

Assume appropriate units in each case.

2 Calculate the torque which must be applied to a flywheel of moment of inertia $3.0\,\text{kg}\,\text{m}^2$ if it is to be accelerated uniformly from rest to $300\,\text{rev}\,\text{min}^{-1}$ in (a) $10\,\text{s}$, (b) 10 revolutions.

3 A flywheel of moment of inertia $20\,\text{kg}\,\text{m}^2$ is slowed down by a frictional torque of $8.0\,\text{N}\,\text{m}$. If it is initially rotating at $12\,\text{rad}\,\text{s}^{-1}$, calculate (a) the time it takes to stop, (b) the angular displacement in this time, (c) its angular velocity $15\,\text{s}$ after it starts to slow down.

4 A torque of $15\,\text{N}\,\text{m}$ is applied to a flywheel initially at rest, which then completes 5.0 revolutions in $2.0\,\text{s}$. Calculate (a) its angular acceleration, (b) its moment of inertia.

5 A wheel of moment of inertia $5.0\,\text{kg}\,\text{m}^2$ rotates on an axle which provides a constant frictional torque of $15\,\text{N}\,\text{m}$. (a) Calculate the external torque which must be supplied to increase its angular velocity from zero to $100\,\text{rad}\,\text{s}^{-1}$ in $20\,\text{s}$. (b) If the external torque is removed, calculate (i) the time before it comes to rest, (ii) the angular displacement in this time.

Rotational kinetic energy

The angular, or rotational kinetic energy of a wheel of moment of inertia I rotating with angular velocity ω is $\frac{1}{2}I\omega^2$.

Example 6

Calculate the rotational kinetic energy of a flywheel of moment of inertia $5.0\,\text{kg}\,\text{m}^2$ rotating at $120\,\text{rev}\,\text{min}^{-1}$.

Method

The rotation rate is $120\,\text{rev}\,\text{min}^{-1}$ or $2\,\text{rev}\,\text{s}^{-1}$. So $\omega = 2 \times 2\pi = 4\pi\,\text{rad}\,\text{s}^{-1}$, and since $I = 5.0$

$$\text{Rotational KE} = \tfrac{1}{2}I\omega^2 = \tfrac{1}{2} \times 5 \times (4\pi)^2 = 395\,\text{J}$$

Answer

$0.40\,\text{kJ}$.

Example 7

Calculate the total kinetic energy of a cylinder of mass $12\,\text{kg}$ and radius $0.20\,\text{m}$ if it is rolling along a plane with a translational velocity of $0.30\,\text{m}\,\text{s}^{-1}$. The moment of inertia of the cylinder is $0.24\,\text{kg}\,\text{m}^2$.

Method

Refer to Fig. 33.3. For there to be no sliding between point P on the wheel and the plane it touches, P must be stationary at the instant of contact. This means the translational velocity v (forwards) must be cancelled by speed $r\omega$ in the opposite direction due to rotation of the wheel, i.e. $v = r\omega$, so $\omega = v/r$. Now

$$\left(\begin{array}{c}\text{Total}\\\text{KE}\end{array}\right) = \left(\begin{array}{c}\text{Translational}\\\text{KE} \ (\frac{1}{2}mv^2)\end{array}\right) + \left(\begin{array}{c}\text{Rotational}\\\text{KE} \ (\frac{1}{2}I\omega^2)\end{array}\right)$$

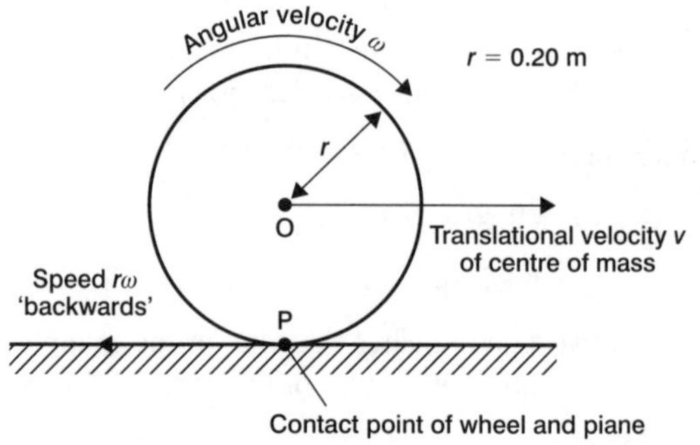

Fig. 33.3   Solution to Example 7

We have $m = 12$, $v = 0.30$, $I = 0.24$. Since $r = 0.20$ then $\omega = v/r = 0.3/0.2 = 1.5$. Hence

$$\text{Total KE} = \tfrac{1}{2} \times 12 \times (0.3)^2 + \tfrac{1}{2} \times 0.24 \times (1.5)^2$$
$$= 0.81\,\text{J}$$

Answer

$0.81\,\text{J}$.

Exercise 33.3

1 A wheel possesses $200\,\text{J}$ of rotational kinetic energy and has a moment of inertia of $0.80\,\text{kg}\,\text{m}^2$. Calculate its rate of rotation (a) in $\text{rad}\,\text{s}^{-1}$, (b) in $\text{rev}\,\text{min}^{-1}$.

2 The wheels of a car rotate 8.0 times each second. Each wheel has mass $15\,\text{kg}$, radius $0.30\,\text{m}$ and moment of inertia $0.27\,\text{kg}\,\text{m}^2$. Calculate (a) the translational speed of the car, (b) the total KE of the four wheels (combined).

Work, rotational energy and power

There are equivalent expressions for work and power, in linear and angular terms. Equation 6.1:

Work done (J) = Force (N) × Distance (m)

(6.1)

becomes, in angular terms

$$\text{Work done (J)} = \text{Torque } \Gamma \text{ (N m)} \times \text{Angular displacement } \theta \text{ (rad)} \quad \textbf{(33.8)}$$

Similarly equation (6.3):

$$\text{Power} \atop P \text{ (W)} = {\text{Force} \atop F \text{ (N)}} \times {\text{Velocity} \atop v \text{ (m s}^{-1}\text{)}} \qquad (6.3)$$

Becomes

$$\text{Power} \atop P \text{ (W)} = {\text{Torque} \atop \Gamma \text{ (N m)}} \times {\text{Angular velocity} \atop \omega \text{ (rad s}^{-1}\text{)}} \qquad (33.9)$$

Example 8

A torque of 8.0 N m is applied to a flywheel, initially at rest, for 15 s. If the flywheel has moment of inertia 2.0 kg m^2, calculate (a) the angular velocity acquired, (b) the kinetic energy acquired, (c) the work done by the torque. Comment on the magnitude of (b) and (c). Neglect friction.

Method

(a) We have $\Gamma = 8.0$ and $I = 2.0$.
Thus $\alpha = \Gamma/I = 4.0 \text{ rad s}^{-2}$. Now

$$\omega = \omega_0 + \alpha t$$

Since $\omega_0 = 0$, $\alpha = 4.0$ and $t = 15$, $\omega = 60 \text{ rad s}^{-1}$.

(b) Since $I = 2.0$ and $\omega = 60$,

$$\text{KE} = \tfrac{1}{2}I\omega^2 = 3600 \text{ J}$$

(c) Using Equation 33.7 with $\omega_0 = 0, t = 15$ and $\alpha = 4.0$

$$\theta = \omega_0 t + \tfrac{1}{2}\alpha t^2 = 450 \text{ rad}$$

To find the work done we use Equation (33.8) with $\Gamma = 8.0$ and $\theta = 450$. Hence:

$$\text{Work done} = \Gamma \times \theta = 8.0 \times 450$$
$$= 3600 \text{ J}$$

Note the answers to (b) and (c) are the same since, due to the absence of friction, all the work done by the torque becomes rotational KE. Note that there is no translational component of energy and no energy dissipated via friction.

Answer

(a) 60 rad s^{-1}, (b) 3.6 kJ, (c) 3.6 kJ

Example 9

A car engine is quoted as having an output power of 28.0 kW at a torque of 110 N m. Calculate the rate of rotation of the output shaft of the engine in revolutions per minute.

Method

We use Equation 33.9 in which $P = 28.0 \times 10^3$ and $\Gamma = 110$. Rearranging to find ω:

$$\omega = P/\Gamma = 28.0 \times 10^3/110$$
$$= 254.5 \text{ rad s}^{-1}$$

This equals $254.5 \times 60 = 15.27 \times 10^3 \text{ rad min}^{-1}$. We require the rate of rotation in revolutions per minute. Since $1 \text{ rev} = 2\pi \text{ rad}$ then:

$$\text{rate of rotation} = 15.27 \times 10^3/2\pi$$
$$= 2.43 \times 10^3 \text{ rev min}^{-1}$$

Answer

$2.43 \times 10^3 \text{ rev min}^{-1}$

Exercise 33.4

1

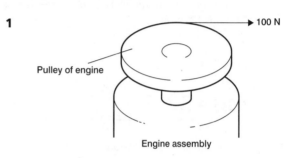

Fig. 33.4   Information for Question 1

Refer to Fig. 33.4. The engine of a lawn mower is turned over by applying a constant force of 100 N to a string wrapped round the pulley. If the moment of inertia of the lawn mower engine is 1.6 kg m^2, and the string is pulled out by 0.60 m, calculate (a) the work done by the force, (b) the angular velocity acquired by the pulley. Neglect friction.

2   (a) A 2 litre turbo diesel engine has a maximum torque of 250 N m at engine speed of $1.75 \times 10^3 \text{ rev min}^{-1}$. Calculate the output power at maximum torque.

   (b) The same engine develops maximum power, of 80.0 kW, at engine speed $4.00 \times 10^3 \text{ rev min}^{-1}$. Calculate the torque at maximum power.

Conservation of angular momentum

Provided that no external torque acts, angular momentum is conserved. This is the case, for example, when a spinning skater draws her arms in.

Example 10

A skater is turning at 3.0 rad s^{-1} with both arms outstretched, so that she has a moment of inertia of 4.0 kg m^2. Her arms are now drawn in, so that her new moment of inertia is $1.8\,\text{kg}\,\text{m}^2$. Calculate (a) her final angular velocity, (b) the increase in her rotational KE.

Method

(a) Angular momentum is conserved, so

$$I_0\omega_0 = I\omega \qquad (33.10)$$

We have $I_0 = 4.0$, $\omega_0 = 3.0$ and $I = 1.8$. Equation 33.10 gives $\omega = 6.67$.

(b) Original KE $= \frac{1}{2}I_0\omega_0^2 = \frac{1}{2} \times 4 \times 3^2 = 18\,\text{J}$

Final KE $= \frac{1}{2}I\omega^2 = \frac{1}{2} \times 1.8 \times 6.67^2 = 40.0\,\text{J}$

$\therefore$ Increase in rotational KE $= 22\,\text{J}$

This arises because of the work done by the skater as she pulls her arms in.

Answer

(a) $6.7\,\text{rad}\,\text{s}^{-1}$, (b) $22\,\text{J}$.

Exercise 33.5

1 A skater is rotating at $2.0\,\text{rad}\,\text{s}^{-1}$ with both arms outstretched, when her moment of inertia is $4.5\,\text{kg}\,\text{m}^2$. She now pulls in her arms and rotates at $8.0\,\text{rad}\,\text{s}^{-1}$. Calculate (a) her new moment of inertia, (b) the increase in her rotational KE.

2 A turntable is rotating freely about an axis with an angular velocity of $6.0\,\text{rad}\,\text{s}^{-1}$ and has a moment of inertia of $1.5\,\text{kg}\,\text{m}^2$. A rough disc is gently dropped on to the turntable so that the centres coincide. Eventually the combined turntable and disc rotate at $4.5\,\text{rad}\,\text{s}^{-1}$.* Calculate (a) the moment of inertia of the disc about the rotation axis, (b) the original rotational KE of the turntable, (c) the final rotational KE of the combination, (d) the rotational KE 'lost'. (Note: this is due to work done against friction as the angular velocity of the disc increases from zero to $4.5\,\text{rad}\,\text{s}^{-1}$.)

*Combined moment of inertia = Sum of separate moments of inertia.

Exercise 33.6: Examination questions

1 A flywheel has moment of inertia of $500\,\text{kg}\,\text{m}^2$. It is acted upon by an external driving torque of $750\,\text{Nm}$ and there is a constant frictional torque of $400\,\text{Nm}$. Calculate the angular acceleration.

2 A flywheel is rotating initially at $600\,\text{rad}\,\text{s}^{-1}$. A braking torque is applied which brings it to rest in $30.0\,\text{s}$. If the flywheel has moment of inertia $12.0\,\text{kg}\,\text{m}^2$, calculate:

(a) its angular acceleration as the brake was applied

(b) the number of revolutions it makes before it comes to rest

(c) the torque applied by the brake mechanism.

3 Calculate the moment of inertia of a flywheel which has rotational kinetic energy of $50.0\,\text{kJ}$ when it is rotating at $20.0\,\text{rad}\,\text{s}^{-1}$.

4 The drum of a spin drier has a moment of inertia of $0.24\,\text{kg}\,\text{m}^2$ when it is loaded with wet clothes. During operation it rotates with an angular velocity of $250\,\text{rad}\,\text{s}^{-1}$. Calculate:

(a) the rotational kinetic energy of the drum and wet clothes

(b) the net torque which must be supplied by the motor to accelerate the drum from rest to its operational angular velocity in $5.0\,\text{s}$.

5 A flywheel is initially at rest and a torque of $8.0\,\text{Nm}$ is applied to it. Calculate its rotational kinetic energy after it has completed 6.0 revolutions. Ignore the effects of friction.

6 A grinding wheel of radius $0.080\,\text{m}$ is driven by an electric motor at a constant speed of 50 revolutions per second. A piece of steel is pressed against the outer rim of the wheel, producing a tangential force on the wheel of $7.0\,\text{N}$.

(a) (i) Calculate the rate at which work is being done by the wheel.
    (ii) Identify **two** forms of energy change which are occurring, stating where the changes are taking place.

(b) A small fragment of the surface of the wheel breaks away from the wheel when it is in the position **P** shown in Fig. 33.5. The plane of the wheel is vertical.

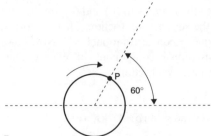

Fig. 33.5

(i) On Fig. 33.5 draw the path of the fragment as it leaves the wheel.
(ii) State the angle to the horizontal of this path. [OCR 2001]

7 In the petrol engine, illustrated in Fig. 33.6, the vertical motion of the piston is converted into rotational motion by the connecting rod and crankshaft.

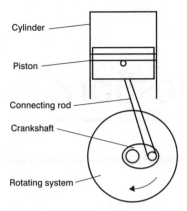

Fig. 33.6

As the crankshaft of the petrol engine rotates, the torque applied by the piston to the crankshaft varies. The variation, for two complete revolutions of the crankshaft, is shown in Fig. 33.7. The moment of inertia of the rotating system is 4.2×10^{-3} kg m^2. The crankshaft rotates at 3400 revolutions per minute.

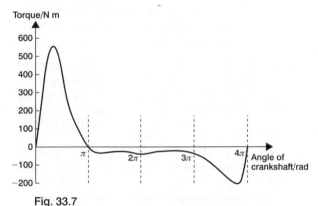

Fig. 33.7

(a) (i) Calculate the maximum angular acceleration during the cycle when there is no torque resisting the motion.

(ii) The average value of the torque during the cycle is 32 N m. Calculate the average power output of the engine.

(iii) Using data from Fig. 33.7 estimate the work done as the crankshaft rotates from π to 2π.

(iv) Calculate the rotational kinetic energy of the rotating system.

(b) A massive flywheel forms part of the rotating system of the engine. State and explain the purpose of the flywheel. [AQA 2000]

8 Assuming that the Earth moves in a circular orbit around the sun at a speed of 3.0×10^4 m s^{-1} and at a distance of 1.5×10^{11} m, calculate the angular momentum of the Earth around the sun, given that the mass of the Earth is 6.0×10^{24} kg.

9 (a) State the principle of conservation of angular momentum.

(b) It is thought that, in the past, comets have impacted the Earth. Fig. 33.8 shows a comet immediately before impact with the Earth at the equator. The comet is travelling in the same direction as the Earth's rotation and will impact horizontally on the Earth's surface. After the impact, the whole material of the comet remains on the surface of the Earth.

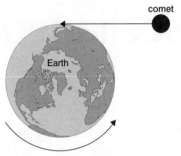

Earth's direction of rotation

Fig. 33.8

period of rotation of the Earth $= 8.6 \times 10^4$ s
radius of the Earth $\qquad = 6.4 \times 10^6$ m
moment of inertia of the Earth $= 9.8 \times 10^{37}$ kg m^2
velocity of the comet $\qquad = 2.1 \times 10^4$ m s^{-1}

(i) Show that the angular momentum of the Earth, before the comet's impact, is 7.2×10^{33} kg m^2 s^{-1}.

(ii) Consider the comet, just before impact, as though it is orbiting the Earth. Calculate the angular velocity of the comet about the Earth.

(iii) Calculate the angular momentum of the comet about the centre of the Earth, just before the impact.

moment of inertia of the comet about the Earth $= 3.8 \times 10^{35}$ kg m^2

(iv) Calculate the periodic time of the Earth's rotation after the comet's impact. In your calculation, you may assume that the moment of inertia of the Earth is unchanged by the impact.

(v) The impact of a larger comet, having the same angular momentum about the Earth, would significantly change the moment of inertia of the Earth. State and explain how the change in moment of inertia would affect the period of rotation of the Earth, compared with the value you calculated in part (iv).

[AQA 2000, part]

Section K
Further Revision Questions

34
Miscellaneous questions

Unless otherwise stated assume the following values:

$g = 9.8\,\mathrm{m\,s^{-2}}$

$R = 8.3\,\mathrm{J\,mol^{-1}\,K^{-1}}$

1 The aeroplane shown in Fig. 34.1 is travelling horizontally at $90\,\mathrm{m\,s^{-1}}$. It has to drop a 500 kg crate of emergency supplies to a village community following a disaster. To avoid damage to the crate the maximum vertical speed of the object on landing should be $36\,\mathrm{m\,s^{-1}}$.

the acceleration of free fall, $g = 9.8\,\mathrm{m\,s^{-2}}$

Fig. 34.1

(a) Assume that air resistance is negligible.
   (i) Calculate the maximum height from which the crate can be dropped.
   (ii) Calculate the time taken for the crate to reach the ground from this height.
   (iii) Explain why the mass of the crate has no effect on your answer to parts (i) and (ii).
   (iv) The crate has to land at a particular place, marked **X** on Fig. 34.1. Calculate the horizontal distance of the aeroplane from **X** when the crate is released from the maximum permitted height.

(b) The speed of the crate when it hits the ground is $97\,\mathrm{m\,s^{-1}}$. Calculate:

(i) the kinetic energy of the crate when it hits the ground;
(ii) the change in gravitational potential energy when the crate falls from the height calculated in (a)(i).

(c) In practice air resistance is not negligible. Suggest and explain how the quantities you have calculated in parts (a)(i) and (a)(ii) will compare with their actual value.

[AQA 2001]

2 The 'London Eye' is a large wheel which rotates at a slow steady speed in a vertical plane about a fixed horizontal axis. A total of 800 passengers can ride in 32 capsules equally spaced around the rim.

A simplified diagram is shown below.

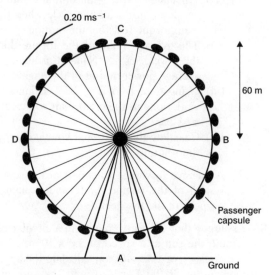

(a) On the wheel, the passengers travel at a speed of about $0.20\,\mathrm{m\,s^{-1}}$ round a circle of radius

60 m. Calculate how long the wheel takes to make one complete revolution.

(b) What is the change in the passenger's velocity when he travels from point B to point D?

(c) When one particular passenger ascends from point A to point C his gravitational potential energy increases by 80 kJ. Calculate his mass.

(d) Sketch a graph showing how the passenger's gravitational potential energy would vary with time as he ascended from A to C. Add a scale to each axis.

Discuss whether it is necessary for the motor driving the wheel to supply this gravitational potential energy. [Edexcel 2001]

3 In a dynamics experiment, a trolley is accelerated from rest along a horizontal runway as shown in Fig. 34.2.

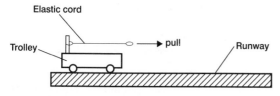

Fig. 34.2

The accelerating force is provided by an elastic cord. One end of the cord is attached to the trolley and the other end is pulled so that the extension of the cord remains constant as the trolley moves along the runway.

The acceleration a of the trolley varies with the extension x of the elastic cord as shown in Fig. 34.3.

The trolley experiences a constant frictional force when in motion.

(a) Use Fig. 34.3 to
   (i) determine the extension of the cord required to maintain constant speed of the trolley, giving a brief explanation for your answer,

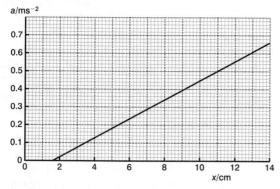

Fig 34.3

(ii) show that the increase in extension, beyond that found in (i), to produce an acceleration of $0.60 \, \text{m s}^{-2}$ is 11.2 cm.

(b) (i) Calculate the force required, in the absence of friction, to cause the trolley of mass 800 g to have an acceleration of $0.60 \, \text{m s}^{-2}$.
   (ii) Using your answers to (b)(i) and (a)(ii), determine the spring constant of the elastic cord. Assume that the cord obeys Hooke's law.
   (iii) Calculate the frictional force on the trolley.

(c) In one particular experiment, the extension of the cord is kept constant at 3.5 cm. Calculate
   (i) the speed of the trolley after it has travelled 1.2 m from rest along the runway,
   (ii) the time taken to travel a further 30 cm along the runway.

(d) By reference to Fig. 34.3, state and explain
   (i) whether the acceleration of the trolley is proportional to the extension of the cord,
   (ii) how it may be concluded that the Hooke's law limit of the cord has not been exceeded. [OCR 2000]

4 A gymnast does a hand-stand on a horizontal bar. The gymnast then rotates in a vertical circle with the bar as a pivot. The gymnast and bar remain rigid during the rotation and when friction and air resistance are negligible the gymnast returns to the original stationary position.

Fig. 34.4 shows the gymnast's position at the start and Fig. 34.5 shows the position after completing half the circle.

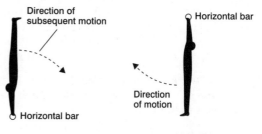

Fig. 34.4   **Fig. 34.5**

(a) The gymnast has a mass of 70 kg and the centre of mass of the gymnast is 1.20 m from the axis of rotation.

   acceleration of free fall, $g = 9.8 \, \text{m s}^{-2}$

   (i) Show clearly how the principle of conservation of energy predicts a speed of $6.9 \, \text{m s}^{-1}$ for the centre of mass when in the position shown in Fig. 34.5.

(ii) The maximum force on the arms of the gymnast occurs when in the position shown in Fig. 34.5

Calculate the centripetal force required to produce circular motion of the gymnast when the centre of mass is moving at $6.9\,\mathrm{m\,s^{-1}}$.

(iii) Determine the maximum tension in the arms of the gymnast when in the position shown in Fig. 34.5

(iv) Sketch a graph to show how the **vertical** component of the force **on the bar** varies with the angle rotated through by the gymnast during the manoeuvre. Assume that a downward force is positive.

Include the values for the initial force and the maximum force on the bar.

Only show the general shape between these values.

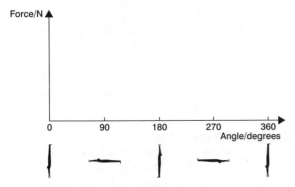

(b) The bones in each forearm have a length of 0.25 m. The total cross-sectional area of the bones in both forearms is $1.2 \times 10^{-3}\,\mathrm{m^2}$. The Young modulus of bone in compression is $1.6 \times 10^{10}\,\mathrm{Pa}$.

Assuming that the bones carry all the weight of the gymnast, calculate the reduction in length of the forearm bones when the gymnast is in the start position shown in Fig. 34.4.   [AQA 2000]

5 A toy frog has a spring which causes it to jump into the air. The force–compression graph for the spring is shown below.

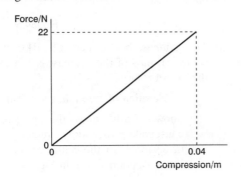

(a) Calculate the work done on the spring when it is compressed by 4.0 cm.

(b) The frog has a mass of 24 g and rises 0.60 m vertically into the air. Calculate the gravitational potential energy gained by the frog.

(c) Compare your two answers for energy and explain how they are consistent with the law of conservation of energy.   [Edexcel 2001]

6 This question is about lifts in tall buildings.

A typical lift, designed to carry a maximum of ten people, has a mass of 400 kg. The designers assume that the mass of the lift when full with passengers is unlikely to exceed 1500 kg.

(a) (i) Explain why this is a reasonable assumption.
   (ii) The lift is supported by a steel cable of cross-sectional area $3.2 \times 10^{-4}\,\mathrm{m^2}$. Show that the mass of 500 m of this cable is about 1300 kg.
      density of steel $= 8.0 \times 10^3\,\mathrm{kg\,m^{-3}}$

(b) (i) A single cable supports the lift. Suppose the lift starts from the bottom of a 500 m shaft. Show that the maximum tension in the cable must be greater than 33 kN to accelerate 1500 kg of loaded lift and 1300 kg of cable upwards at $2.0\,\mathrm{m\,s^{-2}}$. $g = 9.8\,\mathrm{N\,kg^{-1}}$.
   (ii) Calculate the stress that 33 kN produces in a steel cable with a cross-sectional area of $3.2 \times 10^{-4}\,\mathrm{m^2}$.

(c) (i) The yield stress of the steel cable is $2.9 \times 10^8\,\mathrm{N\,m^{-2}}$. For safety reasons the maximum stress in the cable should not exceed one quarter of the yield stress. Show that the addition of two more identical cables in parallel with the first will allow this condition to be met.
   (ii) Plans have been drawn up for a building 1 km high. Explain why a lift like this could not be used in a 1000 m tall shaft. Justify your answer.   [OCR Nuff 2000]

7 (a) State the *principle of moments*.

(b) To increase the extension of a stiff spring for a given load, a student set up the system shown in Fig. 34.6. The weight of the metal bar was 5.0 N and the tension the student achieved in the spring was 37 N.

the gravitational field strength, $g = 9.8\,\mathrm{N\,kg^{-1}}$

(i) Apply the principle of moments to calculate the mass of the load that the student used.

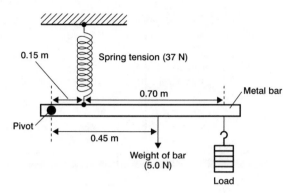

Fig. 34.6

(ii) Calculate the magnitude of the force exerted on the metal bar at the pivot.

(iii) Draw on Fig. 34.6 an arrow to show the direction of the force calculated in part (ii),

(c) The spring stiffness k of the spring was 550 N m^{-1}.

Calculate the energy stored in the spring.

[AQA 2001]

8 In 1628 BC there was a massive explosion of a volcano on the Mediterranean island of Thera. This coincided with the decline of Minoan civilisation on the nearby island of Crete. It has been proposed that this event may have given rise to the legend of the lost island of Atlantis.

It is possible to estimate the speed at which debris is ejected from an exploding volcano by studying how far blocks of rock are thrown.

The following passage is adapted from a description of a similar modern eruption:

'The largest blasts ejected blocks 0.5 m to 1.0 m in diameter up to 1 km from the vent, suggesting ejection speeds of up to 100 m s^{-1}. The average ejection speed was probably about 60 m s^{-1}.'

A block of debris is ejected at 100 m s^{-1} at an angle of 40° to the horizontal from a vent at ground level. Show that the vertical component of its velocity is about 65 m s^{-1}.

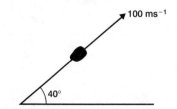

(a) Find the time for which the block is in flight.

(b) Show that the distance the block is thrown is about 1 km.

(c) Studies of the crater left behind suggest that the mass of material ejected was 6×10^{13} kg. Estimate the total kinetic energy of the ejected material.

(d) This eruption would have sent shock waves through the Earth's crust to Crete which is 110 km from Thera. If the waves had a frequency of 0.09 Hz and a wavelength of 40 km, find the time taken for the shock waves to reach Crete. [Edexcel S-H 2000]

9 (a) A wire of unextended length l and cross-sectional area A extends elastically by an amount Δl when the tension in the wire is increased by an amount F.

Write down expressions in terms of l, Δl, A and F for
(i) the spring constant k of the wire,
(ii) the Young modulus E of the material of the wire.

(b) In order to determine the Young modulus of the metal of a wire, a student set up the apparatus illustrated in Fig. 34.7.

The length of wire between the fixed end and the pulley is 1.4 m and the area of cross-section of the wire is 6.2×10^{-7} m^2. The wire passes over the pulley and is held taut by a mass attached to its free end.

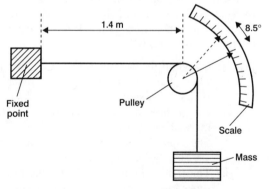

Fig. 34.7

When the mass attached to the end of the wire is increased by 7.0 kg, a pointer attached to the pulley rotates through an angle of 8.5°.
(i) The pulley has diameter 1.6 cm. The wire does not slip over the pulley as the pulley turns. Show that the extension of the wire resulting from the increased mass on its end is 1.2 mm.
(ii) Calculate
1. the increase in the stress in the wire,
2. the increase in the strain of the wire,
3. the Young modulus of the material of the wire.

283

(c) (i) Sketch a graph to show the variation with load F on the wire of its extension Δl. Assume that the elastic limit is not exceeded.

(ii) Use your graph to show that the strain energy E_s stored in the wire is given by the expression

$$E_s = \tfrac{1}{2}k(\Delta l)^2$$

where k is the spring constant of the wire.

(iii) For a total mass of 8.0 kg attached to the wire, the wire extends by 1.37 mm. Calculate the strain energy stored in the wire for this extension.

(d) (i) The specific heat capacity of the material of the wire in (b) is 420 J kg^{-1} K^{-1} and the mass of the wire is 6.2×10^{-3} kg. Calculate the change in temperature of the wire when the total mass of 8.0 kg is removed, assuming all the strain energy is converted into thermal energy in the wire.

(ii) Hence suggest why the steel head of a hammer can become warm when it is used repeatedly to hit nails into wood.

[OCR 2001]

10 A 1930s' racing car equipped with drum brakes is travelling on a horizontal track at 60 m s^{-1}. The brakes are applied to reduce the speed of the car to 10 m s^{-1}. The four brake drums are of iron, of **total** mass 30 kg. The total mass of the car is 1200 kg. The temperature of the drums before the brakes are applied is 0°C.

(a) (i) Calculate the temperature of the drums when the speed of the car reaches 10 m s^{-1}. Assume that all the kinetic energy lost by the car is converted to thermal energy in the drums.
The specific heat capacity of iron is 450 J kg^{-1} K^{-1}.

(ii) The friction lining material on the brake shoes loses all its braking effect when the drums reach 500°C.
Calculate how many successive stops of this type could be made before the brakes cease to function.

(b) State **two** reasons why modern racing cars use disc brakes. [OCR 2000]

11 (a) State the principle of conservation of linear momentum for two colliding bodies.

(b) A bullet of mass 0.010 kg travelling at a speed of 200 m s^{-1} strikes a block of wood of mass 0.390 kg hanging at rest from a long string. The bullet enters the block and lodges in the block.

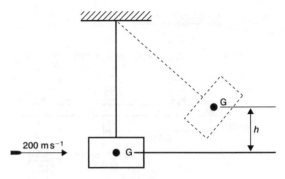

Calculate
(i) the linear momentum of the bullet before it strikes the block,
(ii) the speed with which the block first moves from rest after the bullet strikes it.

(c) During the collision of the bullet and block, kinetic energy is converted into internal energy which results in a temperature rise.
(i) Show that the kinetic energy of the bullet before it strikes the block is 200 J.
(ii) Show that the kinetic energy of the combined block and bullet immediately after the bullet has lodged in the block is 5.0 J.
(iii) The material from which the bullet is made has a specific heat capacity of 250 J kg^{-1} K^{-1}. Assuming that all the lost kinetic energy becomes internal energy in the bullet, calculate its temperature rise during the collision.

(d) The bullet lodges at the centre of mass G of the block. Calculate the vertical height h through which the block rises after the collision. [AQA 2001]

12 This question is about an electric hair-dryer. Fig. 34.8 shows the electrical circuit in the hairdryer.

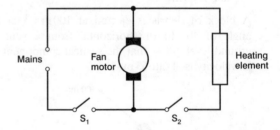

Fig. 34.8

When the fan motor is switched on, air is drawn over the heating element through the back of the hair-dryer. When the heating element is switched on, the air leaving the hairdryer from the nozzle is warmer than the air drawn in.

(a) Complete the table below, to show how the circuit works.

S_1	S_2	fan motor	heating element
open	open	off	off
closed	open		
open	closed		
closed	closed		

(b) The diameter of the fan is 5.0 cm. When the fan is switched on, air passes through the hairdryer at speed $v = 7.2\,\mathrm{m\,s^{-1}}$ (Fig. 34.9).

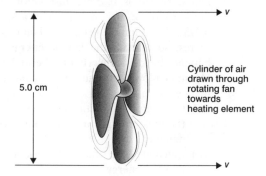

Cylinder of air drawn through rotating fan towards heating element

5.0 cm

Fig. 34.9

(i) Show that the volume of air drawn through the rotating fan every second is about $1.4 \times 10^{-2}\,\mathrm{m^3}$.

(ii) Calculate the mass of air drawn through the rotating fan every second.
density of air = $1.2\,\mathrm{kg\,m^{-3}}$.

(iii) The heater is rated at 300 W. Calculate the temperature of the air emerging from the nozzle. State any assumption that you make.
room temperature = 15 °C;
specific heat capacity of air = $1000\,\mathrm{J\,kg^{-1}\,^\circ C^{-1}}$.

(c) There are two fan speed settings. On the higher fan speed setting air is drawn through the hairdryer at a speed of $14.4\,\mathrm{m\,s^{-1}}$. Calculate the temperature at which the air now emerges from the nozzle.

[OCR Nuff 2000]

13 This question is about the feasibility of using pressurised air to propel a vehicle.

(a) A cylinder contains $0.10\,\mathrm{m^3}$ of air at a pressure of $50 \times 10^5\,\mathrm{N\,m^{-2}}$ at temperature 290 K.
(i) Show that there are about 210 moles of air in the cylinder.
molar gas constant $R = 8.3\,\mathrm{J\,mol^{-1}\,K^{-1}}$.
(ii) Calculate the mass of air in the cylinder.
molar mass of air = $0.030\,\mathrm{kg\,mol^{-1}}$.

(b) To obtain an estimate of the energy stored in the pressurised gas, consider the gas escaping into the atmosphere through a nozzle until the pressure in the cylinder has fallen to atmospheric pressure ($1 \times 10^5\,\mathrm{N\,m^{-2}}$). If we assume that the temperature remains constant, the pressure and volume change as shown in Fig. 34.10.

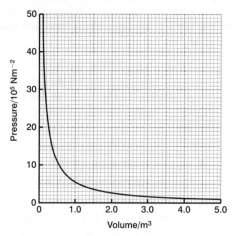

Fig. 34.10

(i) Estimate the energy released as the air expands into the atmosphere, using the area under the graph. Show clearly how you arrive at your answer.
(ii) In fact, the air cools as it expands. Explain why this means that the useful energy released is less than calculated in (b)(i), as the air expands into the atmosphere.

(c) It is proposed to utilise this energy to propel a vehicle.
(i) The aerodynamic drag force on the vehicle is 44 N, when travelling at $10\,\mathrm{m\,s^{-1}}$. Calculate the power required to propel the vehicle at this steady speed.
(ii) Calculate the maximum time the vehicle could travel at this steady speed on just one cylinder of pressurised air. Express your answer in hours.

(d) To increase the energy available from the pressurised air, it is proposed to increase the initial pressure in the cylinder from 50 atmospheres to 300 atmospheres. Suggest a possible disadvantage of this idea.

[OCR Nuff 2000]

14 In 1960 a brilliant physicist called Richard Feynman gave an interesting lecture entitled 'there's always room at the bottom'. He discussed the idea that we should eventually be able to stick individual atoms together to make useful new materials and build objects on a very small scale.

This has become known as *nanotechnology* and the paragraph below illustrates a possible application that may become a reality in the next twenty or thirty years.

If people suffer from very poor circulation, tissue can become damaged. A temporary solution would be to replicate one of the functions of red blood cells by manufacturing tiny spheres full of compressed oxygen and inject these directly into the blood stream. These spheres could then slowly release their oxygen.

Nanotechnology offers the promise of making extremely strong, diamond-like materials in any shape required. The spheres would be mass-produced very cheaply. Spheres of internal diameter 1.0×10^{-7} m could be filled with oxygen to a pressure of 1.0×10^8 Pa.

(a) State the meaning of the word *nano* when used as a prefix in front of a unit.

(b) The spheres are filled with oxygen at a body temperature of 310 K. Calculate the number of moles of oxygen in one sphere. Assume oxygen behaves as an ideal gas.

(c) The typical oxygen consumption of an adult is 2.5×10^{-6} m^3 per minute at atmospheric pressure $(1.0 \times 10^5$ Pa$)$. Calculate the volume in cm^3 of spheres required to sustain the oxygen requirement of an adult for one hour. Assume the volume of material used for the sphere is negligible compared with its internal volume.

(d) The strength of the material used to make these spheres must be extremely high. It would be comparable to diamond, with a breaking stress of 5.0×10^{10} Pa and a Young modulus of 1.0×10^{12} Pa. Calculate the strain in this material if taken to fracture, stating any assumption made.

Explain why large values of breaking stress and Young modulus would be such vital requirements for the material used in this application.

(e) State one other property that the materials used in this application should have.
[Edexcel S-H 2000]

15 (a) On your Data and Formulae Sheet the first law of thermodynamics is quoted in the form

$$\Delta Q = \Delta U + \Delta W$$

(i) For a system consisting of a fixed mass of gas, identify the terms ΔQ, ΔU and ΔW in this equation.

(ii) 50 J of thermal energy is supplied to a fixed mass of gas in a cylinder. The gas expands, doing 20 J of work.
  1. Use the first law of thermodynamics to calculate the change in internal energy of the gas. Indicate whether the change is an increase or a decrease.
  2. How could an experimenter detect that the internal energy of the gas had changed, and deduce the sign of the change?

(b) A sheet metal worker uses a hammer to beat out a thin piece of metal. The mass of the hammer-head is 0.45 kg. Just before it hits the work-piece, it is moving with a speed of 6.0 m s^{-1}; the impact brings the hammer-head to rest so that all the kinetic energy of the hammer-head is converted to thermal energy. Hammer blows continue at a regular rate of two per second.

(i) Calculate the kinetic energy converted to thermal energy in one blow of the hammer.

(ii) Calculate the rate of production of thermal energy.

(iii) The mass of the work-piece is 0.080 kg. The specific heat capacity of the metal of which it is made is 450 J K^{-1} kg^{-1}. Assuming that 70% of the thermal energy generated is transferred to the work-piece, calculate its initial rate of rise of temperature. [CCEA 2000]

16 The joule is the SI unit of energy. Express the joule in the base units of the SI system.

A candidate in a physics examination has worked out a formula for the kinetic energy E of a solid sphere spinning about its axis. His formula is

$$E = \frac{1}{2} \rho r^5 f^2$$

where ρ is the density of the sphere, r is its radius and f is the rotation frequency. Show that this formula is homogeneous with respect to base units.

Why might the formula still be incorrect?
[Edexcel 2001]

17 An electric kettle is marked '2.3 kW'.
(i) Calculate the current that it would take from a 230 V mains supply.
(ii) A householder runs a cable from his house to a shed at the bottom of his garden. He connects one end of the cable to the 230 V mains supply in his house and the other end to an electric socket into which he plugs the

kettle. In the cable the resistance of **each** of the two current-carrying wires is $0.75\,\Omega$.

(I) Assuming that the kettle still draws the current calculated in part (i), find the total power wasted in the cable.

(II) Explain why the assumption about the current is not quite correct.

[WJEC 2001, part]

18 The graph in Figure 34.11 shows how the resistance of a thermistor varies with temperature.

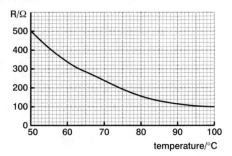

Fig. 34.11

(a) Explain why the resistance decreases at higher temperatures.

(b) The thermistor is included in the circuit shown in Fig. 34.12.

The thermistor has to be maintained at a temperature of $60\,°C$.

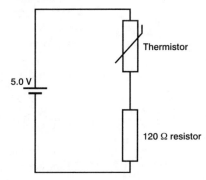

Fig. 34.12

Calculate:
(i) the potential difference across the thermistor;
(ii) the power that has to be removed from the thermistor to maintain the temperature at $60\,°C$. [AQA 2000, part]

19 A torch has three identical cells, each of e.m.f. $1.5\,V$, and a lamp which is labelled $3.5\,V$, $0.3\,A$.

(a) Draw a circuit diagram for the torch.

(b) Assume that the lamp is lit to normal brightness and that the connections have negligible resistance. Mark on your diagram

the voltage across each circuit component and the current flowing in the lamp.

(c) Calculate the internal resistance of one of these cells. [Edexcel 2000]

20

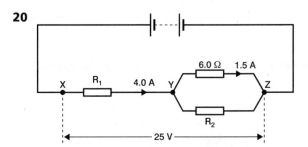

(a) In the circuit shown, the *potential difference* between X and Z is $25\,V$. Explain, in energy terms, what this statement means.

(b) Using the information in the diagram, calculate
(i) the potential difference between Y and Z,
(ii) the resistance of R_2,
(iii) the resistance of R_1,
(iv) the resistance of the combination of the three resistors, i.e. the effective resistance between X and Z.

[WJEC 2001]

21 (a) (i) Write down the equation linking the capacitance C of a capacitor to the charge Q on its plates and the potential difference V between them.
(ii) Define the farad, the unit of capacitance.

(b) Fig. 34.13 shows a capacitor of capacitance C, which is initially charged to a potential difference V_0. The capacitor is connected in series with a resistor of resistance R and a switch.

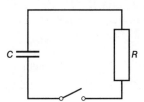

Fig. 34.13

When the switch is closed, the potential difference V across the capacitor decreases with time t. The rate at which V changes depends on a quantity called the **time constant** τ of the circuit.

(i) Your Data and Formulae Sheet shows that an expression for the time constant τ is $\tau = RC$. Show that the product of a resistance (unit : ohm) and a capacitance (unit : farad) has the unit second.

(ii) At a time equal to one time constant after the switch is closed, V is equal to xV_0, where x is a constant less than 1. Complete Table 34.1 giving the values of V at various times t.

Table 34.1

t	V
0	V_0
τ	xV_0
2τ	
3τ	
4τ	

(iii) Name the mathematical function describing the variation of V with t.

(iv) Obtain the numerical value of the constant x.

(v) The capacitor in the circuit of Fig. 34.13 has capacitance $22\,\mu F$. The resistor has resistance $47\,k\Omega$. Making use of your completed Table 34.1 and your answer to (iv), or otherwise, estimate the time after the switch is closed for the potential difference across the capacitor to fall to 5.0% of its initial value.
(e = 2.718) [CCEA 2000]

22 A coil of self-inductance 0.30 H and resistance 55 Ω is to be supplied from a 240 V (r.m.s.), 50 Hz a.c. source which has zero impedance. Find the values of the components that must be put in series with the coil if the current is to be 1.0 A (r.m.s.) and in phase with the applied voltage. [WJEC 2000, part]

23 The voltage output, V, of an a.c. source is given by the expression

$$V = 10 \sin (200\pi t)$$

where t is time, measured in seconds. The source is connected to a 0.50 H inductor in a circuit of negligible resistance.
(i) Calculate the peak current in the circuit.
(ii) State the phase relationship between the current and the supply voltage.
[OCR 2000, part]

24 (a) State Ohm's Law

(b)

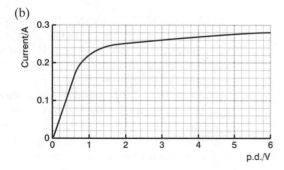

The graph shows how the current through the metal filament of a lamp depends on the potential difference applied across it.
(i) Calculate the resistance of the lamp for potential differences of
  (I) 0.60 V,
  (II) 6.0 V.
(ii) Discuss to what extent, if at all, Ohm's Law applies to the filament.
(iii) When the potential difference across the filament is 6.0 V its temperature is 2100°C. When the potential difference is 0.60 V the temperature of the filament may be approximated to 0°C.
  (I) Calculate the temperature coefficient of resistance of the filament.
  (II) Explain why the approximation is a reasonable one. [WJEC 2001]

25 Fig. 34.14 shows an X-ray tube.

(a) The cathode is a tungsten wire of length of 0.12 m and carries a current of 1.6 A. The voltage across the cathode is 6.3 V. Calculate the cross sectional area of the wire.

conductivity of tungsten $= 2.0 \times 10^7 \, \Omega^{-1} \, m^{-1}$

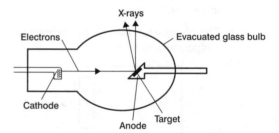

Fig. 34.14

(b) The X-ray power generated is 3.0 W, which is 1.0% of the power input to the tube. The remaining energy heats up the target in the anode. The target is a tungsten block of mass 0.045 kg. Calculate the initial rate of temperature rise of the tungsten block when the X-ray tube is turned on.

specific heat capacity of tungsten $= 140 \, J \, kg^{-1} \, K^{-1}$ [AQA 2000]

26 In the cathode ray tube illustrated in Fig. 34.15, electrons are accelerated by a potential difference of 1.8 kV between the cathode (**C**) and the anode (**A**).

(a) (i) Calculate the kinetic energy, in J, of the electrons after they have passed the anode.

charge on an electrode, $e = -1.6 \times 10^{-19} \, C$

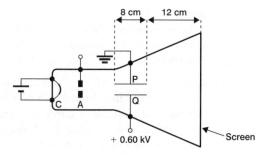

Fig. 34.15

(ii) Calculate the velocity of the electrons after they have passed the anode.

$$\text{mass of electron} = 9.1 \times 10^{-31}\,\text{kg}$$

(b) The plates **P** and **Q** are 8.0 cm long and are separated by a gap of 4.0 cm.
   (i) Define electric field strength.
   (ii) Calculate the force acting on an electron when it is between **P** and **Q** and state the direction of the force
   (iii) Calculate the time taken for an electron to pass between the plates.
   (iv) Calculate the vertical component of velocity at the time the electron leaves the electric field between **P** and **Q**.
   (v) Calculate the additional vertical displacement of the electron between the time it leaves the electric field between **P** and **Q** and when it reaches the screen.

[AQA 2000]

27 This question is about the radioactive material americium-241 used in domestic smoke detectors like that shown in Fig. 34.16.

Fig. 34.16

(a) An americium-241 nucleus decays by emitting an α-particle, and forms an isotope of neptunium. Complete the equation below that describes this decay by adding the missing nucleon and proton numbers.

$$^{241}_{95}\text{Am} \rightarrow {}^{237}\text{Np} + \text{He}$$

(b) The half-life of americium-241 is 430 years. Show that the decay constant, λ, for this isotope is $5.0 \times 10^{-11}\,\text{s}^{-1}$.
   1 year $= 3.2 \times 10^7\,\text{s}$

(c) (i) Calculate the number of americium-241 nuclei in a source with an activity of $4.6 \times 10^3\,\text{Bq}$.

(ii) Hence calculate the mass of americium-241 used in the source. Avogadro constant $= 6.0 \times 10^{23}\,\text{mol}^{-1}$

[OCR Nuff 2000]

28 Fig. 34.17 shows a series circuit containing a 2.0 V cell, a switch S, a 0.25 Ω resistor R, and an inductor L. The internal resistance of the cell and the resistance of L are negligible.

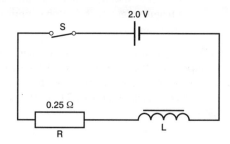

Fig. 34.17

(a) After closing S, the current in the circuit rises, eventually becoming steady. While the current is increasing from zero to 0.20 A, the rate of change of current can be assumed to be constant at $40\,\text{A s}^{-1}$.
   (i) Calculate, for the instant when the current is 0.20 A, the potential difference (p.d.)
      1. across R;
      2. across L.
   (ii) Use your result from (a)(i) 2 in calculating the inductance of L.

(b) The current in the circuit eventually becomes steady.
   (i) Calculate the magnitude of the steady current.
   (ii) Explain why the inductor L plays no part in determining the magnitude of this steady current. [OCR 2000]

29 (a) Uranium-238 decays by alpha emission to thorium-234. The table shows the masses in atomic mass units, u, of the nuclei of uranium-238 ($^{238}_{92}\text{U}$), thorium-234, and an alpha particle (helium-4).

Element	Nuclear mass/u
Uranium-238	238.0002
Thorium-234	233.9941
Helium-4, alpha particle	4.0015

1 atomic mass unit, u	$= 1.7 \times 10^{-27}\,\text{kg}$
speed of electromagnetic radiation, c	$= 3.0 \times 10^8\,\text{m s}^{-1}$
the Planck constant, h	$= 6.6 \times 10^{-34}\,\text{J s}$

(i) How many neutrons are there in a uranium-238 nucleus?

(ii) How many protons are there in a nucleus of thorium?

(b) (i) Determine the mass change in kg when a nucleus of uranium-238 decays by alpha emission to thorium-234.

(ii) Determine the increase in kinetic energy of the system when a uranium-238 nucleus decays by alpha emission to thorium-234.

(c) Wave particle duality suggests that a moving alpha particle (mass 6.8×10^{-27} kg) has a wavelength associated with it. One alpha particle has an energy of 7.0×10^{-13} J.

Calculate:

(i) the momentum of the alpha particle;

(ii) the wavelength associated with the alpha particle. [AQA 2000]

30 Assuming a value of 3×10^{-18} s^{-1} for the Hubble constant, find the distance from Earth of a galaxy for which the red shift for a particular spectral line is a tenth of the wavelength of the same spectral line observed from a stationary source.

(velocity of light in vacuum is 3.0×10^8 m s^{-1})

[Edexcel 2000]

Hints for examination questions

Chapter 3

Exercise 3.4

1 SI units are m for x, $\mathrm{m\,s^{-1}}$ for u and s for t, so m for ut, $\mathrm{m\,s^{-2}}$ for a, $\mathrm{s^2}$ for t^2

2 {force} is $\mathrm{M\,L\,T^{-2}}$ or $\mathrm{kg\,m\,s^{-2}}$. Dimensions of right-hand side of Equation?

3 (b) (i) See Equation 21.1 in Chapter 21
  (ii) Obtain value for π from calculator.
$$\frac{1}{\mathrm{F\,m^{-1}}} = \mathrm{m\,F^{-1}}$$

Chapter 4

Exercise 4.4

1 $2W \cos \theta$

2 (i) $T \sin 30° = 2000$ (ii) $T \cos 30°$

3 80 N tension throughout rope. For a section at $40°$ to horizontal:
  (i) horizontal force $80 \cos 40°$
  (ii) vertical force $= 80 \sin 40°$

4 Net resultant force and moment must be zero

5 (c) and (e) See Example 4

6

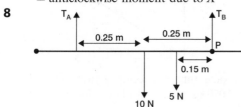

Principle of moments

7 (a) and (b) Net horizontal and vertical forces $= 0$
  (d) Clockwise moment due to W
    $=$ anticlockwise moment due to X

8

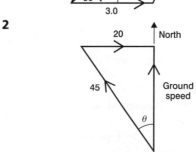

9

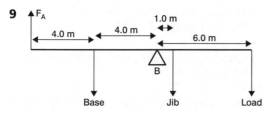

Principle of moments

10 $P = R + Q$ and $P \times 3d = Q \times 4d$

11 See Example 7

12 (b) I and II – take moments about P

13

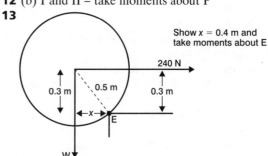

Show $x = 0.4$ m and take moments about E

Chapter 5

Exercise 5.9

1

2

3 Equation 5.1 to find a. Equation 5.4 with $t = 10$

4 Equation 5.4

5 (b) (ii) II and IV – see Example 5, Chapter 4
      III Net force $= ma$
    (iii) Equation 5.4

6 $s = \frac{1}{2}at^2$ with $s_1 = 30$ and $s_2 = 40$.
  Find t_1 and t_2

7 See Example 9

8 (a) (i) horizontal component is constant
      (ii) $18.8 = $ magnitude $\times \cos 40°$
    (b) Equation 5.3

(c) See Example 8(a) with u = vertical component $(15.8\,\mathrm{m\,s^{-1}})$. Add 1.80 to s.

(d) (i) Equation 5.3 with
s = maximum vertical height $(14.2\,\mathrm{m})$ and $a = g$

    (ii) Addition of horizontal and vertical components using Pythagoras

9 (i) Fig. 5.3, find $V = R \sin \alpha$, then Equation 5.3 with $v = 0$

  (ii) Fig. 5.3, find H

  (iii) See Example 8(b)

10 Net force = total mass $\times a$

11 (b) $F = ma$

  (c) Equation 5.1

  (d) Equation 5.4

12 (b) (i) Use Pythagoras   (ii) $F = ma$

13 Equation 5.1 then $F = ma$

14 (i) and (ii) See Example 13

  (iii) Equation 5.2 to find t; $(40 + t)$ = time when velocity $= 0$

15 (a) (ii) Example 13(b)

  (b) Example 8 and Equation 5.1.
For (iii) $s = -200\,\mathrm{m}$ or use 2 stages

  (c) (i) See Example 3 + calculate t from vertical component

    (ii) Fig. 5.3. H is constant. V decreases by $6.0\,\mathrm{m\,s^{-1}}$. Use Pythagoras.

16 (a) Equation 5.1   (c) Volume $\times$ density

17 (b) distance/time; assume falls approx 2 m

  (c) W = volume $\times$ density $\times g$

18 (a) (i) Example 15(a)   (ii) Equation 5.3

  (b) Stopping distances equal, so collide if thinking distance (at $30\,\mathrm{m\,s^{-1}}$) exceeds 15 m.

19 (a) Example 15(a)

  (b) (i) Equation 5.3   (ii) $F = ma$

Chapter 6

Exercise 6.5

1 Equation 6.1 (mass not needed)

2 (a) Fig. 5.3

  (b) (ii) Equation 6.1

    (iii) P = rate of doing work

  (c) P.E. gain

3 (a) Example 4   (b) Equation 5.3

  (c) mgh or use Fxd and $F = ma$

4 (a) (i) mgh

    (ii) P = rate of work done

    (iii) and (c) See Example 4(c)

5 PE becomes KE. KE = decrease in PE of B – increase in PE of A (no friction)

6 KE + PE, then KE

7 Energy dissipated = PE change – KE change

8 (a) Litres used $\times$ energy value

  (b) P = rate of transfer of energy

  (c) speed = distance/time

  (d) Thrust $\times$ speed = power

9 (a) (i) M = volume $\times$ density;
$PE = mgh$ with $h = 2.5\,\mathrm{m}$ (average height loss)

    (ii) Equation 6.2 and
P = rate of energy transfer

10 (a) (i) Volume = area $\times$ wind speed;
Mass = volume $\times$ density

    (ii) and (b) See Example 8

11 (a) Mgh; then Equation 6.2, then
power = rate of energy production

  (b) Walking energy = power $\times$ time; add body energy required for change in height

12 (b) (i) Fig. 6.1 to find h; then mgh

    (ii) power = work done/time taken

13 (a) Equation 6.1

  (b) $3.6\,\mathrm{m\,s^{-1}}$ along slope = $0.3\,\mathrm{m\,s^{-1}}$ vertically, then use mgh; half of mgh needed

14 (a) Equation 6.3   (b) Example 11

15 (b) Equation 6.3

  (c) (i) total = (air + friction) resistance

    (ii) friction $\propto$ speed^2

    (iii) Equation 6.3 using total resistance

16 (b) (ii) 60% of mgh is 12 MW, where
m = mass per second

    (iii) 12×10^6 is 84% of power reaching generator, all KE

    (iv) I. mgh becomes KE + friction losses
II. Estimate here; use Equation (6.3) with average velocity of (say) $20\,\mathrm{m\,s^{-1}}$

  (c) (i) Equation 6.2

    (ii) Overall efficiency = efficiency of pipe $\times$ efficiency of generator

Chapter 7

Exercise 7.6

1 Equation 7.1   (ii) Example 2

  (iii) $\frac{1}{2}mv^2$ (iv) KE becomes PE

2 (a) Example 1

  (b) Momentum conserved; positive to the right

3

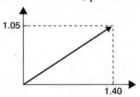

Use momentum as a vector quantity. On collision $m = 0.070\,\mathrm{kg}$, hence calculate velocities

4 (a) $F = ma$

  (b) (i) See Example 2, $u_2 = 0$

    (ii) See Example 3

5 (a) Equation 7.1

  (b) Calculate KE of 10 kg trolley; collision leads to KE reduction

6 (b) (i) Examples 2 and 3

    (ii) Example 4, use M and $(10/9)M$. Solve 2 simultaneous equations

(c) At each bounce, $v_R^2 = 0.9v^2$, where
v_R = rebound speed and v = impact speed
Use Equation 5.4 with $s = 0$ to find time between
each bounce. Remember v_R decreases at each bounce

7 (c) (i) See Example 5
  (ii) Power = rate of transfer of (kinetic) energy

8 See Example 5

9 $234v_1 = 4v_2$. Require ratio $(4v_2^2 : 234v_1^2)$

10 (b) (i) & (ii) See Example 1
  (iii) Remember (ii) is negative
  (iv) Equation 7.6
  (v) Calculate KE before and after contact

11 See Example 7

12 (b) (ii) Count squares; note t in ms and mass in g
  (iii) change in velocity = $2 \times$ incident velocity

13 See Example 8; for (ii) $a = F/m$

14 (a) (i) Equation 5.2   (ii) Equation 7.5
  (b) Example 8

15 See Example 8

16 (b) (i) Example 8(a); area = $\pi \times 18^2$
  (ii) Change in KE
  (c) Example 8(b), $v - u = 2$
  (e) Equation 6.2 for useful power output per wind
    turbine

Chapter 8

Exercise 8.4

1 (a) Equation 8.2   (b) Equation 8.1
  (c) Equation 8.3

2 See Example 1

3 See Fig. 8.4 and Equations 8.5 and 8.6.
  For T use P
  (d) $a = F/m$ (e) Equation 8.3

4 $mg + mv^2/r$

5 $T = mg + mv^2/r$ at bottom of swing. PE when
  pulled aside, at top of swing, becomes KE at
  bottom of swing; $mgh = \frac{1}{2}mv^2$. Calculate h from $5°$
  angle as in diagram

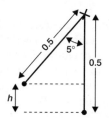

6 (a) (i) Equation 8.1   (ii) Equation 8.4
  (iii) Equation 8.4, $F = 180\,\text{N}$
  (b) See Example 3, $T_2 = 0$ at top of circle so
    $mg = mr\omega^2$

7 (a) (ii) Example 3; variation is $+$ or $-mg$
  (iii) $mv^2/r = 5.0$
  (iv) Equation 8.1
  (b) Equation 8.4; $r = L$

8 See Example 4

9 See Example 4. Force = $mg - mv^2/r$

10 (b) (i) Fig. 8.1   (ii) Equation 8.1 gives v in m s^{-1}
  (c) Example 4
  (d) Flash frequency = f in Hz, Example 1
  (e) See Example 5, for (iv) $t = 2\pi r/v$

11 Example 5. Note $r = (4.0 + 5.0 \sin 60°)$

Chapter 9

Exercise 9.5

1 Equation 9.1

2 Equate mass $\div$ distance^2

3 $F \propto 1/\text{distance}^2$

4 See Example 3

5 See Example 4

6 Equation 9.4

7 (a) Equation 9.1
  (b) Equation 9.4 and Example 4. Note U is per kg

8 (b) See Example 2(b)
    Quadratic equation gives two solutions:
    2.0×10^9, and 6.0×10^9 m

9 See Example 6

10 (a) $v = \sqrt{2gr}$ and $g = GM/r^2$ gives Equation 9.5
  (b) KE = $\frac{1}{2}mv^2$

11 (b) See Example 7
  (c) (i) See Example 5
    (ii) See Example 6, or find U from
      graph + Equations 9.4 and 9.5

12 (c) (i) and (ii) Equation 9.6
  (iii) Equation 9.3: G and M are constants

13 (b) (i) Equation 9.7   (ii) Equation 9.6
    (iii) KE = $\frac{1}{2}mv^2$
  (c) Compare KE of satellite to TNT explosive yield

14 (a) (i) Equation 9.7
    (ii) g provides v^2/r and $v = 2\pi r/T$

15 Geostationary orbit takes 24 hours. Use Kepler's
  3rd Law to find T. $N = 24/T$

16 (b) (i) Equation 9.7: height above surface
    (ii) Equation 9.3
  (c) Kepler's 3rd Law

17 (b) (iii) 1. Gradient = 0   2. g = slope of graph
  (c) (i) Note mass is ejected with velocity such that
    it just reaches peak of ϕ versus d curve.
    Then mass falls through $\Delta\phi$ towards
    surface: $\Delta\phi$ is greater for surface of Pluto
    than for surface of Charon.

Chapter 10

Exercise 10.5

1 Example 1

2 $F = ke$ with $k = 40$: Equation 10.2

3 (a) Find coordinates of A: use $F = ke$
  (b) Area under graph: 1 square equals 4 mJ

4 Extension $\propto l/\text{area}$: energy stored $\propto$ extension

5 (a) (i) $W = \sigma A$ at U.T.S.
    (ii) $\sigma = E\varepsilon$ then $W' = \sigma A$: $\varepsilon = 0.5\% = 0.005$

6 PE becomes stored energy. Wire of length l has extension e. So $mg(l+e) = \frac{1}{2}F_{UTS}\,e$.
Also $F_{UTS} = \sigma_{UTS}\,A$ and $\varepsilon = e/l = \sigma_{UTS}/E$.
Manipulate equations to find A. Note $e \ll l$

7 (b) (i) Equations 10.4 and 10.5
  (ii) Equations 10.2 and 10.3

8 (a) Example 7(a)   (b) Equation 10.2

9 (b) (i) πr^2
  (ii) Example 5 – read values from graph
  (iii) Equation 10.3 – read F from graph
  (iv) Example 2

10 (c) (i) Example 7
  (ii) Add load to shortest wire at base
  (iii) Equation 10.2: energy $\propto F$ for same extension

11 Equations 10.6 and 10.5

12 (b) (i) Equation 10.5
  (ii) Equation 10.6
  (iii) Equation 10.4

13 See Example 8
  (b) (i) and (ii) $\Delta l/l = 0.002 = \varepsilon$. Use Equations 10.7 and 10.5
  (iii) Equation 10.3

Chapter 11

Exercise 11.5

1 Example 1(a)

2 Equation 11.3

3 Resonant frequency at peak amplitude. Find T, use Equation 11.3 – or find ω, use Equation 11.4

4 (a) Find T, use Equation 11.3 – or find ω, use Equation 11.4
  (b) Equation 11.3, or use $f = 1/T \propto 1/\sqrt{m}$

5 (a) Treat as 100 kg (1000 N) over 1 spring – use Equation 11.5
  (b) Treat as 300 kg over 1 spring: Equation 11.3

6 (a) Example 3
  (b) Since $T \propto \sqrt{m}$ then $f \propto 1/\sqrt{m}$

7 (b) (i) Equation 11.5   (ii) Equation 11.3
  (c) (i) Equation 11.3 or $T \propto \sqrt{m}$

8 Example 3(b) and (c)

9 Equation 11.6 shows $T^2 \propto l$ and l becomes $(l+1.8)$

10 $T \propto 1/\sqrt{g}$ for pendulum, from Equation 11.6

11 Example 5

12 Takes $T/4$ where $T = 2\pi/\omega$ and $\omega = 20\pi$.
Or see Example 3(b)

13 (a) Equation 11.5
  (b) (i) Equation 11.3   (ii) Example 1(a)
  (iii) Example 7(c)   (iv) Example 6(b)

14 (a) & (b) – see Fig. 11.2
  (c) Example 6(b) but note this gives time after mid tide – also $y = (3.1 - 1.2)$ m

15 (a) Time between (for example) peaks
  (b) Equation 11.9: $\omega = 2\pi/T$ and $r = 0.080$ m
  (Compare also with PE $= mgh$)

(c) A $\sin^2$ graph: KE is zero at extremities of motion

16 $\frac{1}{2}mr^2\omega^2$ and $\omega = 2\pi/T$

Chapter 12

Exercise 12.8

1 Equation 12.1

2 Find λ and c, hence f

3 (b) Distance = speed × time: same distance but if time t for P wave, $t + 30$ for S wave. Solve simultaneous equations (Chapter 2)

4 Equation 12.3

5 See Example 2

6 (b) (i) I. f = number of cycles per second
  II. $\lambda/2$ or $3\lambda/2$ from $x = 0$ to $x = 0.10$ m
  (ii) Equation 12.1

7 (a) (i) Read off graph: $f = 1/T$
  (ii) Equation 12.1
  (b) (i) Distance = speed × time: same distance but if time t for P wave, $(t + 65)$ for S wave. Solve simultaneous equations (Chapter 2)

8 (b) Wave 1: Equation 12.1 to find f, then $T = 1/f$. Note 1 period $\equiv 360°$
  Wave 2: 36.0° is 1/10th of a period. Hence find T, then f and use Equation 12.1

9 Fig. 12.1(b): 2π equivalent to 1 wavelength. Then Equation 12.1

10 YP = 2.50 m

11 (b) (iv) 1. See Example 7
  2. Bright fringe means a whole number of fringe separations
  3. Half a fringe separation from C

12 $y \propto \lambda$ and $y \propto 1/a$

13 (b) Equation 12.7
  (c) (i) $y \propto \lambda$ (ii) $y \propto 1/a$
  (iii) More light passes through slits
  (d) (i) $_1n_2 = \lambda_1/\lambda_2$
  (ii) Calculate number of wavelengths in glass (N_g) and air (N_a). Then $(N_g - N_a)$ = extra wavelengths introduced by glass which causes the pattern shift. For direction – where does the zero fringe move to?

14 (a) 1.5 wavelengths in 0.87 m: $c = f\lambda$
  (b) 0.87 m is $\lambda/2$: $f = c/\lambda$

15 $\lambda = 2L$: $c = f\lambda$. Then find new λ, hence new f (or $f \propto 1/L$)

16 (b) (i) $\lambda = 2L$ (ii) $c = f\lambda$ (iii) Equation 12.4

17 (a) See Fig. 12.12(a)
  (b) 1. $x/d = \sin\theta$ 2. $d \propto 1/\sin\theta$

18 (a) (ii) fundamental, $\lambda = 4L$: $c = f\lambda$
  (b) $\lambda = c/f$. Number of wavelengths in tube = $0.46/\lambda$

19 (b) (i) Fig. 12.13(a)   (ii) $\lambda = c/f$ and $L = \lambda/4$
  (iii) Repeat above

Chapter 13

Exercise 13.3

1 Equation 13.1

2 $1/d$ given; use Equation 13.1

3 Equation 13.1. Set up 2 equations with same n and d value (d unknown). Divide equations to eliminate d

4 (b) Example 1 to find maximum n value. Note $d = (0.001/380)\,\text{m}$

5 $\theta = 22.3°$ for $n = 2$. Equation 13.1 to find d – convert to mm. Calculate $1/d$

6 (a) $\theta_{\text{red}} = 19.15°$; $\theta_{\text{blue}} = 12.55°$; Equation 13.1. Note $d = 0.001/500$.

   (b) See Example 1(c)

7 $d \sin 46° = 4\lambda_1 = 3\lambda_2$

8 (a) Equation 13.1 gives $d \sin \theta = 1360 \times 10^{-9}$. Find λ for $n = 1$ (infra red), $n = 2$ (red), $n = 3$ (violet) and $n = 4$ (ultra violet). For next common angle n becomes $2n$, then $3n$ and so on – determine if $2n$ or $3n$ etc are possible ($\sin \theta < 1$) and find θ

   (b) Example 1(c). Find maximum value of $n\lambda/d$ (less than 1)

9 (a) Equation 13.3   (b) $r = L\theta$

10 (a) Equation 13.1; $d = 2.1\,\mu\text{m}$

   (b) Equation 13.3; $W = 1.05\,\mu\text{m}$, $m = 1$

11 (a) Equation 13.1 with $d = 2W$ and $n = 2$; then Equation 13.3

   (b) Example 1(c)

   (d) Example 1(c); $n = 2$ also missing

12 Example 4

13 $r = L\theta$

14 (b) Example 5. Part (ii), ϕ decreases

15 See Example 5, r unknown $(= L\theta)$

Chapter 14

Exercise 14.4

1 (b) and (c) Equation 14.1

2 (a) Equation 14.1   (b) $\phi = (88.5 - \theta)°$

   (c) $360°$ equivalent to 24×60 minutes

3 Equation 14.4

4 (a) Equation 14.4

   (b) Equation 14.6 T.I.R. when $n_1 > n_2$ that is $c_1 < c_2$

5 (a) Equation 14.1

   (b) (i) Example 5

   (ii) Angle of incidence at wall $= (90 - \theta)°$; T.I.R.?

6 (a) Equation 14.1

   (b) Equation 14.4

   (c) Example 5(b) to find i_c

7 (b) (i) Equation 14.2   (ii) Figs. 14.6 and 14.7

8 See Example 6

9 Example 6. For (c) time = distance ÷ speed. Note Equation 14.4 to find speed in fibre.

Chapter 15

Exercise 15.4

1 (a) $u = v = 2f$ (b) $m = 2$, $v = 2u$

2 Example 1(b)

3 (c) Diverging lens, $f = -200$

   (i) Equations 15.1 and 15.3. See also Example 2

   (ii) 1. u becomes $(+)$ 360 mm

   2. Speed = distance moved ÷ time

   3. v decreases. See Fig. 15.2

4 (d) (i) Diverging lens $f = -150$ mm. Equations 15.1, 15.3 and 15.2 to find D

   (ii) Converging lens. Equation 15.2 gives $v = 0.2u$ or $u = 5v$. Substitute for u in Equation 15.1 to find v

   (iii) See Figs. 15.1(a) and 15.2

   (e) Equation 15.2 gives $m = 8.00$. So Equation 15.3 gives $v = 8u$
   Note $(v + u) = 567$ mm. Substitute in equations to find v and u. Then Equation 15.1 to find f

5 (a) (ii) Equation 15.5

   (b) (i) Objective (converging) lens forms image which acts as object for the diverging lens

   (ii) Equation 15.1 with $u = -21$ and $f = -20$

   (iii) Sign of v

   (c) Decrease u, increase v

6 Example 3 and Fig. 15.3

7 (a) (ii) Equations 15.1, 15.3 and 15.2

   (b) (ii) 1. Equation 15.2

   2. & 3. Equation 15.3 gives $u = -2.5v$: substitute in Equation 15.1 to find v, hence u. See also Fig. 15.2 and Example 2

8 (a) Equation 15.5

   (b) Equation 15.1; see also Example 3

9 Example 6

10 (b) (i) Equation 15.1 with $v = 1.7$ cm, f from Equation 15.5

   (ii) Find power for $u = \infty$

   (iii) Normal near point is 0.25 m

11 (a) Calculate f of lens.

   (b) Fig. 15.6(b)

12 (a) See Example 7

   (b) (i) Equation 15.6

   (ii) Calculate value of the new unaided far point

13 (d) Example 7

14 See Fig. 15.6 and Example 7

Chapter 16

Exercise 16.3

1 $D = 12.0$ cm; $v = -12.0$ cm

   (a) Equation 15.1   (b) Equation 16.4
   See also Example 1

2 (a)

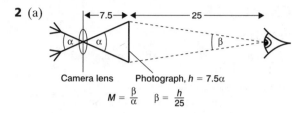

Camera lens   Photograph, $h = 7.5\alpha$

$$M = \frac{\beta}{\alpha} \qquad \beta = \frac{h}{25}$$

(b) For total magnification of 1 we require $\beta = \alpha$, so magnifying lens/eye must be placed 7.5 cm from the photograph, so $u = 7.5$ cm. From v ($= D$), find f using Equation 15.1

3 Example 3

4 Example 4

5 Find f_e, hence $M = f_o/f_e$. Then Equation 16.1

6 Find f_e, hence $M = f_o/f_e$. Equation 16.1 to find α. Then diameter = distance $\times \alpha$

Chapter 17

Exercise 17.4

1 (c) (i) Equation 17.1, $Q = P \times t$

2 (a) (i) Equation 17.4 and 17.2
    (ii) Equation 17.1

3 Equations 17.4 and 17.1

4 (b) (i) Equation 17.1
    (ii) Direct proportionality
    (iii) Equation 17.4; subtract from (i), then use direct proportionality

5 (b) (i) Mass = volume $\times$ density; per second
    (ii) Equation 17.1; per second

6 Example 5; water + calorimeter are equivalent to $(0.22 + 0.02) = 0.24$ kg water

7 (a) (i) 1 kg in 12 seconds
    (ii) Equations 17.4 and 17.2
  (b) (i) See Example 5
    (ii) Equation 17.1

8 $Q = P \times t$ = energy supplied by element. Equation 17.1 gives energy received by water. Difference is energy 'lost' to surroundings

9 Equation 17.6 and $Q = P \times t$

10 (a) Equation 17.4
  (b) Equation 17.7 with $VI = V^2/r$ and $H = 0$

11 In 1 minute $Q = mc \times 4.0$.
In 40 minutes $40\,Q = mL$
Divide equations to find c/L

12 $Q = mL$ and $Q = P \times t$

13 (a) (i) graph; 4 °C rise in 2 minutes
    (ii) $P = mc\,d\theta/dt$ and see Example 3
  (b) (ii) Equation 17.6 and $Q = P \times t$

14 Note Example 3. Let x = final temperature
Energy given out by warm water
$= 0.15 \times 4200 \times (18 - x)$
Energy absorbed by ice melting and the water formed warming $= (0.015 \times 3.4 \times 10^5)$
$+ (0.015 \times 4200 \times x)$
Equate the two terms to find x

15 (b) Calculate
    1. Energy given out if the lemonade cools to 0 °C
    2. Energy absorbed when 2 kg of ice melts
    Compare to see that only 1/2 of ice needs to melt
  (c) (i) $mc\,d\theta/dt = 0.25\,(X - \theta) = 0.25\,\Delta T$, where X is the external temperature
    Integrate and use $t = 0$, $\theta = 0$. Solve for ΔT. (Alternatively, consider the temperature difference in place of n in Equations 28.1 and 28.2a in Chapter 28.)
    (ii) Now $X = 30$ °C. Find t when $\theta = 6$ °C and $\theta = 18$ °C; find time difference (in hours). If ice molten, $m = 5.96$ kg and $c = 4200$.

16 (a) $(\theta_1 - \theta_2)/l$
  (b) Equation 17.10 and Example 7(a)

17 Equation 17.10; $\Delta Q/\Delta t \propto k/l$

18 Note 4 walls (0.9 m $\times$ 0.5 m) and 2 walls (0.5 m $\times$ 0.5 m). Equation 17.10

19 Equation 17.10

20 (a) (ii) and (iii) Equation 17.11
  (b) Equation 17.1; calculate Q per second

21 (b) (i) Equate $\Delta Q/\Delta t$; see Example 7(b)
    (ii) $\Delta\theta_s$ across steel is very small! Equation 17.10 with $\Delta\theta_p = 22$ K

Chapter 18

Exercise 18.4

1 Equation 18.4

2 Use total pressures. Equation 18.4

3 Example 4. Find n and M_g.
Density = mass $\div$ volume

4 (b) (i) Equation 18.2 with $V_1 = 5000$
    (iii) Use Equation 18.5 for initial and final conditions. Divide the 2 equations to find n_2/n_1.

5 Equation 18.5; set up two Equations for X and Y. Pressures equal. Divide equations

6 (a) Equation 18.4
  (b) Equation 18.5; $p \propto n$ and $n \propto$ mass

7 (a) and (b), see Example 4. Helium is monatomic
  (c) Force = net pressure $\times$ area (in m^2)

8 (b) (i) and (ii) Example 4
    (iii) Density = mass $\div$ volume
  (c) Equation 18.11

9 (b) (i) 1 mole contains N_A molecules
    (ii) Equation 18.5
    (iii) Example 8; density = mass $\div$ volume

10 See Example 7

11 (b) Equation 18.11 shows $m <c^2>$ is a constant and m depends on molar mass. Remember r.m.s. ratio is required
  (c) (i) Assume initially a fixed mass of gas (1.22 kg) with volume $V = 1$ m^3. Use Equation 18.2 to find new V. Hence density = mass $\div$ volume.

12 (b) (i) Mass = volume × density, which contains 1.03×10^{24} atoms

(ii) 1. Equation 18.9   2. Equation 18.11

13 (a) Equation 18.10

(b) Equation 18.5

(c) From Equation 18.10, $\frac{1}{2} M_g <c^2> = \frac{3}{2} pV$

(d) Example 2

14 (b) Equation 18.5; 1 mole contains N_A molecules

(c) (i) Equation 18.4 and force = pressure × area

(ii) $<c^2> \propto T$. See also Example 9

Chapter 19

Exercise 19.5

1 Equation 19.3

2 Equations 19.3 and 19.1. See Example 1

3 (a) $Q = mL$; $P = Q/t$

(b) (i) $pV = nRT$ (ii) Equation 19.3

(c) Example 2

4 Equation 18.2; $V_2 = V_1/3$, $p_2 = 5p_1$

5 Equation 18.2; $V_2 = V_1/2$, $p_2 = p_1$

6 See Example 4 to find T_2

7 Equation 19.8 and Example 6

8 Example 8

9 Example 8

10 (b) (i) work done =

energy supplied − energy expelled

(ii) efficiency = (work done)/(energy supplied) as a %

Chapter 20

Exercise 20.9

2 Convert mm^2 to m^2

3 Need Equation 20.7; use it twice if necessary

4 Combine series resistances first. Equations 20.6 and 20.7

5 (b) (i) Equation 20.5   (ii) Equation 20.10

6 From P to Q the shortest route is $1\,\Omega$ and other routes in parallel make R_{PQ} smaller still. x in parallel with x gives $x/2$. PS is $4\,\Omega$ in parallel with $6\,\Omega$

7 Equation 20.19

8 (a) See Equation 20.17

(b) (ii) Equation 20.5   (iii) Equation 20.16

9 (b) For each bulb $I = P/V$ (Equation 20.14). Could find bulb resistance ($P = V^2/R$). Then combine resistances in serial and parallel (Equations 20.6, 20.7)

10 Combine parallel resistances first (Equation 20.7). See Equation 20.6 for series. For power see Equation 20.14

11 Equation 20.5. PD across B is not 12 V. I through B? PD across B? Then ...

12 Need three Equations (use three loops of circuit)

13 Find PD across meter for full scale deflection; PD across shunt = PD across meter

Chapter 21

Exercise 21.2

1 (a) Kinetic energy = $\frac{1}{2}mv^2$ (see Chapter 6). For energy eV see Equation 21.5

(b) See Equations 8.4, 21.2 and 21.3

2 See Equations 21.4, 21.5

3 (a) Direction of force on a small + charge

(b) (iii) Equation 21.6

(iv) $V \propto 1/R$

(v) Potential × electron charge = PE of electron (Equations 20.3 and 21.5)

4 (a) Equations 8.4 and 21.1. Fm^{-1} is SI unit for ε

5 (a) (ii) Equation 21.1

(b) (ii) Equations 21.6 and 20.3

(iii) $V \propto 1/R$, $\dfrac{V_1}{V_2} = \dfrac{R_2}{R_1}$

Chapter 22

Exercise 22.3

1 (a) (i) Equation 22.4

(ii) Equation 22.1

(iii) Equation 22.2

(b) Equations 22.2 and 22.4. Note: need increase, not new energy

2 Equations 20.1 and 22.1
Equation 22.3

3 Exponential fall. Time constant, see Equation 22.7. $V = V_0 e^{-t/RC}$. $\ln V = ?$

4 (b) (i) Combine a pair of capacitors (Equation 22.6). (Can work in μF.) Then 3 pairs in parallel (Equation 22.5)

(c) (i) Equation 22.1

(ii)(3) $Q_z = 198\,\mu C - Q_{22}$

5 (a) Combine capacitors in series first (Equation 22.6). For parallel see Equation 22.5

(b) Q same for series capacitors

(c) PD shared by series capacitors

6 (a) Equation 22.4

(b) (i) Equation 22.6

(ii) Show charge on series combination is 1.0×10^{-4} C; then $V = Q/C$ for C_2

(c) Get half-life from graph (for half-life see text and Equation 22.8, also Chapter 28)

Chapter 23

Exercise 23.2

1 (i) Equation 23.1
   (ii) Need same value for $I_1 \times I_2$
2 (a) (ii) Torque or moment $= 2 \times F \times b/2$
   (b) (i) $i =$ net PD/resistance
3 (a) Convert mm to m. Equations 23.4 and 21.4
4 Equation 23.1 and $F = \mu/2\pi$ here

Chapter 24

Exercise 24.4

1 Equations 24.3 and 24.2 or 24.1
2 (a) Equation 24.7
   (b) (i) Equation 24.7, but need effective area
       perpendicular to B ($=A \cos \theta$) or effective
       B perpendicular to area A ($=B \cos \theta$)
       (ii) $\theta = \omega t$. Equations 24.5 or 24.6. Equation
       24.9
3 Equations 24.5 and 24.2
4 Equations 24.2 and 24.3. New flux is due to V_{vert}
   entering same face of window
5 See Example 4

Chapter 25

Exercise 25.2

1 Equation 25.1
2 (i) Equation 25.2
   (ii) $B \propto 1/L$
3 Equations 23.3, 25.1, 20.3, 20.1 and 24.8. Should
   find unit for μ is $\mathrm{N\,A^{-2}}$
4 (a) (i) Equation 25.1
       (ii) Corkscrew rule
   (b) (i) Opposing B values?
       (ii) Need Pythagoras and $\tan \theta$

Chapter 26

Exercise 26.4

1 Equations 24.9 and 26.1
2 Equation 26.7 and $\omega = 2\pi f$. $I = V/X_c$
3 (a) See text
   (b) (i) $f = 1/T$ (ii) Peak/$\sqrt{2}$
4 $Z^2 = R^2 + (\omega L - 1/\omega C)^2$ (Equation 26.8).
   Equation 26.5 and $\alpha = \tan^{-1}\dfrac{\omega L}{R}$ (Equation 26.9)
5 (a) Resonance
   (b) Equation 26.8, Equation 26.5
6 Equation 26.14 and Example 10(c)

Chapter 27

Exercise 27.2

1 (i) Left-hand rule
   (ii) Equation 23.4
   (iv) Equation 8.4

2 Equation 27.2. Note the 3 significant figures used
   for data in question. $KE = \dfrac{1}{2}mv^2$ (Chapter 6).
   De Broglie, see Equation 27.6
3 Equations 27.9 and 27.4, and look for energy
   change about 1.9 eV
4 (i) Equations 8.4 and 21.1
   (ii) (1) Substitute answer to (i) into given Equation
       (2) $n = 1$
       (4) Use equation 27.6 and substitute for mv
       using equation given. For other orbits
       deduce equation relating λ to n

Chapter 28

Exercise 28.5

1 (i) Equation 28.3
   (ii) Equation 28.1
   (iii) Equation 28.2b or 28.4b
2 (b) (i) Equation 28.4b and activity is $\propto$ mass (see
       Equation 28.5)
       (ii) Equation 28.2b with masses
3 For half-life read t from graph for 10^{20} atoms. Then
   use Equation 28.3. For rate of decay draw tangent to
   graph. For calculated λ use Equation 28.1
4 (b) Equations 28.1 and 28.3
   (c) 0.25 Bq is same as initial activity of bog wood.
       Use Equation 28.1. Simplify and take logs of
       both sides
5 Equation 28.10

Chapter 29

Exercise 29.2

1 (a) Rate of decay ($= \lambda N$) of Bi equals rate of
       decay of Tl. Equation 28.3
   (b) Nucleon numbers (or mass numbers) balance
       and proton numbers balance
2 (a) Neutron is $^1_0$n. See Example 3. Convert mass
       difference in u to kg, then use Equation 29.1 to
       get joules
   (b) 1.5×10^{26} pairs of nuclei
3 (a) (i) Nucleon numbers balance and proton
           numbers balance
       (ii) Equation 29.1
   (b) For mole see Chapter 3. 1 mole of U 235 atoms
       has mass 235 g or 0.235 kg
4 See Example 1. Convert kg to joules using
   Equation 29.1. 1 eV $= 1.6 \times 10^{-19}$ J

Chapter 30

Exercise 30.6

1 (c) Equation 21.6 in Chapter 21
   (e) Assume vacuum

2 (a) See Fig. 26.1 in Chapter 26. 230 V is r.m.s. Equation 26.4. Period = 1/frequency

(b) Equation 26.11 with cos α = 1 for resistance. Equation 26.4

3 See Fig. 28.1. One half-life divides activity by 2, 0.5 half-life divides activity by $\sqrt{2}$

4 (b) $y = hx$ – WFE? Gradient of graph? See Equations 30.1 and 27.2

5 Force = rate of change of momentum with time

6 (a) (i) k is slope of graph

(ii) Use $R = R_0 + k\theta$. $R_0 = 100$

(b) (iii) $\dfrac{\Delta R}{R_m} \times 100 = 2\%$ at 40 °C

7 (i) Equation 20.16

(ii) For graph $V = -Ir + E$. $y = mx + c$ (Equation 30.1) means a straight line. At $I = 0$, $V = E = 12$ V. At $I = 3$ A, $V = 3.0$ V. For current value, I must satisfy both curves

8 Photon energy is hf (see Chapter 27)

9 Straight line means charge $\propto$ PD. For area under graph see Equation 22.2 in Chapter 22

10 Charge = area under graph. Could count squares. Equation 20.1. Or find time constant (RC) (Equation 22.7) from graph directly or via half-life (Equation 22.8). Hence C.

11 Use log versus log graph. Log $T_{\text{on}} = \log(k\eta d^2) - \log V^2 = -2\log V + \text{constant}$ or T_{on} versus V^{-2}

12 (1) Amplitude means peak value

(2) Period = 1/frequency

Chapter 31

Exercise 31.5

1 Equation 31.2

2 See Example 2

3 See Equation 31.6

4 Equation 31.8. $A = 4\pi r^2$

5 Equations 31.11 to 31.14

6 Calculate ratio I_v/I_D

7 Equations 31.9 and 31.10

Chapter 32

Exercise 32.5

1 Equation 32.1

2 Equation 32.2 to find intensity. Then $P = I \times A$ to find power. For 2 hours. See Example 1

3 Equation 32.2 and Example 2. For (b) find reduction

4 (a) Equation 32.2

(b) See Example 4(a); intensities add

5 (b) $P = I \times A$

(c) Equation 32.3; see Example 4(a)

6 Equation 32.2

7 (b) $Z_1 = Z_2$

8 (a) and (b) Example 5

(c) Distance = speed × time; distance is twice thickness of fat layer

9 Equation 32.5

10 (c) Simple division

11 (a) Take natural logs; $\ln(I_t/I_o) = -\mu x$

(b) Transmitted intensity decreases by factor of 0.6 for each 8.8 mm of thickness

12 *Either* Equation 32.14 to find μ, then Equation 32.16 to find T

or Equation 32.15 to find Y and $Y = x/T$

Chapter 33

Exercise 33.6

1 Example 1(b); use net torque

2 (a) and (b) see Example 2

(c) Equation 33.1

3 KE $= \frac{1}{2} I \omega^2$

4 (a) Example 6

(b) Equations 33.5 and 33.3

5 Work done $= \Gamma \times \theta$

6 (a) (i) Equation 33.9

7 (a) (i) Equation 33.1; use maximum Γ

(ii) Equation 33.9; use average Γ

(iii) Equation 33.8; area under curve

(iv) KE $= \frac{1}{2} I \omega^2$

8 $I = mr^2$ and $\omega = v/r$

9 (b) (i) $\omega = 2\pi/T$ and $I\omega$

(ii) $\omega = v/r$

(iii) $I\omega$

(iv) Add angular momenta and sum $= I\omega_{\text{new}}$

(v) I increased now

Chapter 34

1 (a) (i) Equation 5.3. Vertical motion only

(ii) Equation 5.1

(iv) See Example 9(b) Chapter 5

(b) (i) KE $= \frac{1}{2} mv^2$

(ii) mgh

(c) Dissipation of energy lowers vertical speed

2 (a) Time = circumference ÷ speed

(b) See Example 1(c), Chapter 5

(c) mgh

(d) Other passengers descending

3 (a) (i) $a = 0$

(ii) $(x_{0.6} - 1.2)$ cm

(b) (i) $F = ma$

(ii) $k = \Delta F/\Delta x$ in N m^{-1}

(iii) Calculate ΔF when $\Delta x = 1.6$ cm

(c) (i) Equation 5.4

(ii) Equation 5.4 with $u = 0.49$, $s = 0.3$ and $a = 0.1$; solve quadratic (see Chapter 2) *or* use Equation 5.3 with $s = 1.2$ and $s = 1.5$ to find v in each case. Then time = distance ÷ average speed

4 (a) (i) PE becomes KE

(ii) $F = mv^2/r$

(iii) $mg + mv^2/r$

(iv) See (ii) and (iii) above

(b) Example 6, Chapter 10

5 (a) Work done $= \frac{1}{2}F \times e$
  (b) mgh
  (c) Energy 'lost'

6 (a) (i) Mass of average person?
    (ii) Mass = volume × density
  (b) (i) Tension $= mg + ma$
    (ii) $\sigma = F/A$
  (c) (i) m = mass of cables + mass of loaded lift. See (b). Maximum safe stress is yield stress ÷ 4
    (ii) Find mg for (say) cable alone, or 3 cables + static lift. Then $\sigma = mg/A$ for static situation. Compare to maximum safe stress. (Situation worse if lift is to accelerate upwards.)

7 (b) (i) Clockwise moments = anticlockwise moments
    (ii) and (iii) Total upwards force = total downwards force
  (c) Equations 10.1 and 10.2

8 Fig. 5.3
  (a) Example 8(b), Chapter 5
  (b) Example 3, Chapter 5
  (c) KE $= \frac{1}{2}mv^2$; use average speed
  (d) $c = f\lambda$ and time = distance ÷ speed

9 (b) (i) $\Delta l = r\theta$ and θ in rad
    (ii) 1. $\Delta\sigma = \Delta F/A$
      2. $\Delta\varepsilon = \Delta l/l$
      3. $E = \Delta\sigma/\Delta\varepsilon$
  (c) (iii) $\frac{1}{2}Fe$
  (d) Q = energy stored $= mc\Delta\theta$

10 (a) (i) $\Delta KE = mc\Delta\theta$ gives
      $\frac{1}{2}m_{car}(u^2 - v^2) = m_{brakes}c\Delta\theta$
    (ii) simple proportion

11 (b) (i) mass × velocity
    (ii) conservation of momentum; total mass = 0.40 kg
  (c) (i) & (ii) KE $= \frac{1}{2}mv^2$
    (iii) $\Delta KE = mc\Delta\theta$
  (d) $mgh = \frac{1}{2}mv^2$

12 (b) (i) Volume per second = area × speed
    (ii) Mass = volume × density (per second)
    (iii) $P = mc\Delta\theta$; m is mass per second
  (c) $\Delta\theta$ is half of (iii) since m doubles

13 (a) (i) $pV = nRT$ (ii) $M_g = nM_m$
  (b) (i) area under curve
    (ii) p depends on T
  (c) (i) Power = force × velocity
    (ii) Power = work done ÷ time taken

14 (b) $pV = nRT$
  (c) Simple proportion, $1\,m^3 = 10^6\,cm^3$; pV = constant
  (d) $E = \sigma/\varepsilon$

15 (a) See Equation 19.1 and explanation
  (b) (i) $\frac{1}{2}mv^2$
    (ii) Rate of energy transfer

(iii) 70% of (ii) $= mc$ × rate of temperature rise

16 For dimensions and units see Chapter 3. Joule is unit for work (work = force × distance, Equation 6.1). Acceleration = force/mass, Equation 7.5 $[f]$ is $[T]^{-1}$

17 (i) Equation 20.14
  (ii) Heat per second $= I^2R$

18 (b) (i) Equation 20.19
    (ii) Equation 20.14

19 For EMF and cells in series see Chapter 20. Need the three cells in series to get more than 3 V. PD across the battery of three cells is 3.5 V. Volt drop across the internal resistance of the cells is 1.0 V. See Equation 20.16

20 For resistances in series and parallel see Chapter 20, Equations 20.6 and 20.7
  (b) (i) Equation 20.5 can be applied to the 6.0 Ω resistor
    (ii) Current through R_2 is not 4 A nor 1.5 A
    (iii) PD between X and Y?

21 For capacitance see Chapter 22
  (b) (i) See Equations 20.5, 20.1 and 22.1
    (ii) Time constant is time for V to fall to V/e, see Chapter 22
    (iv) Consider $1/e$ as $1/3$ initially. Then try with $e = 2.718$, $1/e = 0.368$

22 For RMS values and for LCR series circuit see Chapter 26. For resonance see Equation 26.12. Current in phase with supply means resonance and current limited by resistance only.

23 (i) For inductance see Chapter 26. For similar Equation see Equation 26.1
    For peak values see definition of X_L preceding Equation 26.6
  (ii) For phase difference see text dealing with LCR circuit

24 (a) See Equation 20.4
  (b) (i) (1) $R = V/I$ (Equation 20.5) = 1/gradient
    (iii) See Equation 20.10

25 (a) Equation 20.8, $\sigma = 1/\rho$. Also Equation 20.5
  (b) For percentage see Chapter 2

26 (a) (i) Equation 21.5
    (ii) $KE = \frac{1}{2}mv^2$ (see Chapter 6)
  (b) (i) Equation 21.2
    (ii) $F = eE$ (Equation 21.2) and $E = V/d$ (Equation 21.4)
    (iii) Time = distance/velocity
    (iv) Equation 5.1

27 (a) See Example 29.2
  (b) Equation 28.3
  (c) (i) Equation 28.1
    (ii) 1 mole of particles = Avogadro number of particles and mass of 1 mole is A grams where A is mass number of particle (see Chapter 3)

28 (a) (i) (1) $V = IR$ (Equation 20.5)
          (2) $V_R + V_L$ must equal supply PD
     (ii) Equation 24.8
   (b) (i) $V = IR$ (Equation 20.5)

29 (a) Write equation with proton and neutron numbers including $_2^4He$. See Chapter 29 and Example 29.2

   (b) (i) See Example 29.1
     (ii) See Equation 29.1 and Example 29.1
   (c) (i) 7×10^{-13} J is kinetic energy and equals $\frac{1}{2}mv^2$. For De Broglie see Equation 27.6

30 See Equations 31.9 and 31.10 in Chapter 31

Answers

Chapter 3

Exercise 3.1

1 (a) ML^{-3} (b) L^2 (c) L^3T^{-1} (d) ML^2T^{-3}
2 (a) T (b) MLT^{-1} (c) T^{-1}
3 $MQ^{-1}T^{-1}$
4 $ML^2T^{-2}\theta^{-1}$ (θ is temperature)
5 MT^{-2}

Exercise 3.2

1 $\alpha = 1$, $\beta = 2$

Exercise 3.3

1 (a) $8.3\,\text{m s}^{-1}$ (b) $10^4\,\text{mm}^2$
  (c) $0.4\,\mu\text{m}$ (d) $2000\,\text{s}^{-1}$
2 $0.00001\,\text{m}\Omega^{-1}$

Exercise 3.4

1 (a) All terms have same units
  (c) All terms can have same units but numbers make equation incorrect, e.g. if $\frac{1}{2}$ were omitted in the equation given here.
3 (a) $\text{N m}^2\,\text{C}^{-2}$

  (b) (i) $k = \dfrac{1}{4\pi\varepsilon_0}$ (ii) $9.0 \times 10^9\,\text{m F}^{-1}$

Chapter 4

Exercise 4.1

1 (a) 24 N at 12° to 15 N (b) 36 N at 52° to 50 N
2 67.1 N at 63.4° to 30 N
3 79.4 N at 40.9° to 30 N

Exercise 4.2

1 (a) 4.1 N (OX), −1.4 N (OY)
  (b) 4.3 N at 19° to OX
2 (a) 30 N (b) 4.6 kg
3 (a) 5.3 N down plane (b) 9.7 N up plane

Exercise 4.3

1 (a) 0.27 kN (b) 1.5 kN
2 (a) 0.38 kN (b) 0.27 kN (c) 6.5 kg
3 (a) 1.2 kN (b) (i) 1.1 kN (ii) 0.31 kN
4 51°

Exercise 4.4: Examination questions

1 (i) 100 N (ii) 173 N (iii) Max value = 200 N
2 (i) 4.0 kN (ii) 3.5 kN (3.46 kN to 3 sig figs)
3 (b) (i) 0.14 kN (ii) 0.13 kN
4 B
5 (a) 200 N (c) 311 N
  (e) Tension increases markedly
6 D
7 (a) 100 N (b) Net force = 0
  (c) 15 Nm (d) 25 N
8 (ii) 6.5 N (iii) 8.5 N
9 (a) (i) 1. $(93 \times 10^4 + 8F_\text{A})\,\text{N m}$
       2. $112 \times 10^4\,\text{N m}$
  (ii) $F_\text{A} = 2.4 \times 10^4\,\text{N m}$
  (b) When $F_\text{A} = 0$
  (c) Increase angle beyond 60° until load just clears end of crane body
10 (b) (i) $P = 2500\,\text{N}$, $Q = 1875\,\text{N}$
  (ii) P, Q and R do not change
11 (ii) 0.21 kN (iii) 0.17(5) kN downwards
12 (b) (i) I 2.1(6) Nm II 36 N
  (ii) return (iii) Lower C of G, widen base
13 (c) (i) 180 N

Chapter 5

Exercise 5.1

1 (a) $\pi\,\text{s}$ (b) $8/\pi\,\text{m s}^{-1}$ to left
  (c) $8.0\,\text{m s}^{-1}$ upwards
2 $\sqrt{4.5}\,\text{m s}^{-1}$ at 135° to original direction
3 (a) $300\,\text{m s}^{-1}$ (b) 53°

Exercise 5.2

1 (a) $12\,\text{m s}^{-1}$ (b) 24 m
2 (a) $2.5\,\text{m s}^{-2}$ (b) 8.0 s, 128 m (c) 10 s, 125 m

Exercise 5.3

1 $360\,\text{m s}^{-2}$
2 (a) 3.0 s (b) 63 m
  (c) $t = 6.0\,\text{s}$ (for s increasing) and $t = 14\,\text{s}$ (for s decreasing)
3 $8.0 \times 10^{15}\,\text{m s}^{-2}$

Exercise 5.4

1 (a) 45 m (b) $30\,\text{m s}^{-1}$
2 $17\,\text{m s}^{-1}$
3 4.0 s

Exercise 5.5

1 (a) 4.5 s   (b) 18 m
(c) 45 m s^{-1}, 4.0 m s^{-1}
2 1.25 m
3 (a) 77 s   (b) 35 km   (c) 7.4 km

Exercise 5.6

1 (a) 7.5 N   (b) 0.50 m s^{-2} (c) 3.0 kg
2 (a) 4.0 m s^{-2} (b) 50 m
3 (a) 5.6 kN   (b) 3.2 kN   (c) 1.6 kN
4 (a) 42 m s^{-1} (b) 7.5 kN
5 3.6 kN

Exercise 5.7

1 (a) A – B, 0 m s^{-2}; B – C, 0.20 m s^{-2};
C – D, −0.6 m s^{-2}
(b) A – B, 40 m; B – C, 80 m; C – D, 30 m;
Total 150 m
2 3.9×10^{-5} m

Exercise 5.8

1 (a) 7.2 m   (b) 12 m   (c) 19 m
2 −5.0 m s^{-2}

Exercise 5.9: Examination questions

1 12 m s^{-1} at 39° to horizontal
2 (b) (i) 26°   (ii) 40 m s^{-1}
3 225 m
4 1.5 m s^{-2}
5 (b) (ii) I 0.50 kN;   II 0.25 kN;   III 0.15 kN;
     IV 0.43 kN
(iii) $t = 6.0$ s
6 A
7 (a) 30 s   (b) 7.5 km   (c) 0.39 km s^{-1}
8 (a) (i) 18.8 m s^{-1} (ii) 24.5 m s^{-1}
(b) (i) 10.0 m s^{-1} (c) 14.2 m
(d) (i) 16.9 m s^{-1} (ii) 25.3 m s^{-1}
9 (i) 55°   (ii) 4.6 m s^{-1} (iii) 1.3 s
10 A
11 (b) 3.68 m s^{-2} (c) 30.5 s   (d) 1.70(6) km
12 (b) (i) 300 N   (ii) 50 m s^{-2}
13 5.6 kN; Forces: 840 kN (weight); 16.8 kN (forward);
11.2 kN (backwards)
14 (i) 0.6 m s^{-2} (ii) I 120 m;   II 360 m
(iii) velocity = 0 at 55 s
15 (a) (i) v

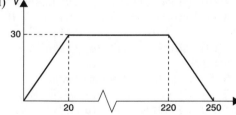

(ii) 6.7(5) km
(b) (i) 3.0 s   (ii) 245 m   (iii) 10 s
(c) (i) 59 m   (ii) 22 m s^{-1} at 22° to horizontal

16 (a) 2.0 m s^{-1} (c) 0.65×10^{-7} kg   (d) 7.7 m s^{-1}
17 (b) 4.2×10^{-3} m s^{-1}; time approx 5×10^2 s
(c) $W = (4/3)\pi r^3 \rho g$ (d) 6.1 μm
18 (a) (i) 0.67 s   (ii) −7.4 m s^{-2} (b) Collide
19 (a) 0.67 s   (b) (i) 6.6 m s^{-2} (ii) (−) 5.94 kN
(c) Deceleration reasonably constant, therefore
consistent
(d) Increase braking distance

Chapter 6

Exercise 6.1

1 (a) 1.8×10^5 J   (b) 4.0×10^8 J
(c) 1.8×10^{-16} J
2 (a) 15 m s^{-1} (b) 22.5 m
(c) 337.5 J   (d) 337.5 J
3 72 N
4 4.6 m s^{-1}

Exercise 6.2

1 (a) 24 J   (b) 13 m s^{-1}
2 (a) 5.0 mJ   (b) 0.20 m
3 6.0 J
4 1.6 m s^{-1}
5 (a) 0   (b) 1.5 m s^{-1} (c) 1.8 m s^{-1}

Exercise 6.3

1 (a) 0.42 kJ   (b) 28 m
2 18 J
3 (a) 18 J   (b) 4.4 N

Exercise 6.4

1 (a) 150 W   (b) 300 W   (c) 214 W
2 22.5 kW
3 (a) 10×10^3 kg s^{-1} (b) 20×10^3 kg s^{-1}
4 (a) 18 kW   (b) 36 kW   (c) 25 kW
5 (a) 23 kW   (b) 46 kW   (c) 33 kW
6 (a) 0.64 kN   (b) 1.0 kN
7 (a) 13.3 m s^{-1} (b) 8.0 m s^{-1}

Exercise 6.5: Examination questions

1 $1.0(5) \times 10^5$ J
2 (a) 319 N
(b) (i) Only the horizontal component does work
(ii) 2.4 MJ   (iii) 0.13 kW
(c) P.E. gain
3 (a) 41 m s^{-1} (b) (−) 7.1 m s^{-2} (c) 0.21 MJ
4 (a) (i) 1.02 MJ
(ii) 68.0 kW; PE becomes PE + KE
(b) 19.0 m s^{-1}
(c) Same speed since independent of mass
5 (a) 80 J   (b) 5.7 m s^{-1}
6 3.6 J
7 60 kJ
8 (a) 65×10^{11} J   (b) 38 MW   (c) 0.24 km s^{-1}

(d) 1.7×10^8 W per engine; significantly greater than (b)

9 (a) (i) 7.9×10^9 J   (ii) 2.9×10^5 W

10 (a) (i) $550\,\text{m}^3\,\text{s}^{-1}$; $660\,\text{kg}\,\text{s}^{-1}$ (ii) 33 kJ
    (b) 13 kW

11 (a) 0.40 kW   (b) 5.7 MJ

12 (b) (i) 32 J   (ii) 5.3 W

13 (a) 14 W
    (b) Valid claim (needs 105 W without resistive forces, 112 W with resistive forces)
    (c) Streamline flow produces less resistance to motion

14 (a) 30 kN   (b) 6.0 kN

15 (b) 1.8 kN
    (c) (i) 1.5(5) kN   (ii) 6.2 kN   (iii) 1.3×10^5 W

16 (b) (ii) $2.55 \times 10^4\,\text{kg}\,\text{s}^{-1}$
    (iii) I 14(.3) MW;   II 33(.5) $\text{m}\,\text{s}^{-1}$
    (iv) I 5.7 MW;   II 0.3 MN
    (c) (i) 71%   (ii) overall 59%

Chapter 7

Exercise 7.1

1 (a) $7.5\,\text{kg}\,\text{m}\,\text{s}^{-1}$ (b) $4.0\,\text{m}\,\text{s}^{-1}$
2 (a) $-10\,\text{kg}\,\text{m}\,\text{s}^{-1}$ (b) $10\,\text{kg}\,\text{m}\,\text{s}^{-1}$ (c) 0

Exercise 7.2

1 $3.3\,\text{m}\,\text{s}^{-1}$
2 $-0.67\,\text{m}\,\text{s}^{-1}$
3 $19\,\text{m}\,\text{s}^{-1}$

Exercise 7.3

1 0.33 kJ
2 16.3 kJ
3 2.4 J

Exercise 7.4

1 $0.40\,\text{m}\,\text{s}^{-1}$
2 $0.42\,\text{m}\,\text{s}^{-1}$
3 $807\,\text{km}\,\text{s}^{-1}$ forwards

Exercise 7.5

1 6 N
2 60 N
3 40 N s

Exercise 7.6: Examination questions

1 (i) 3.75 Ns   (ii) $1.24\,\text{m}\,\text{s}^{-1}$ (iii) 2.33 J
    (iv) 77.3×10^{-3} m
2 (a) 14 Ns   (b) 6.0 Ns to the right
3 $25\,\text{m}\,\text{s}^{-1}$
4 (a) 2.2×10^5 N   (b) (i) $0.36\,\text{m}\,\text{s}^{-1}$ (ii) 18 kJ
5 (a) **A**   (b) **B**
6 (b) (ii) Velocities; first ball v, second ball 0 (trivial solution) or; first ball $(-1/19)v$ and second ball $+(18/19)v$

(c) $t = 0.63 + \dfrac{(v\sqrt{0.9})}{5} \times (1/(1 - \sqrt{0.9}))$

7 (c) First bounce 0.63 s. To come to rest $t = 24$ s since where $v^2 = 40$
    (i) $5.8(5) \times 10^{-3}\,\text{m}\,\text{s}^{-1}$ (ii) 0.35 kW
    (iii) Velocities reversed – move towards each other

8 $0.24 \times 10^6\,\text{m}\,\text{s}^{-1}$

9 58.5

10 (b) (i) 2.2 Ns   (ii) -1.8 Ns
    (iii) -4.0 Ns   (iv) 28 N   (v) 5.9 J

11 (a) 5 Ns   (b) 7.5 Ns

12 (b) (i) Change in momentum of ball
    (ii) approx. 70 to $80\,\text{m}\,\text{s}^{-1}$
    (iii) approx. 35 to $40\,\text{m}\,\text{s}^{-1}$

13 (b) (i) 26.8×10^{-3} kg   (ii) $35.1 \times 10^{-3}\,\text{m}\,\text{s}^{-2}$

14 (a) (i) $0.60\,\text{km}\,\text{s}^{-1}$ (ii) 1.5 kN
    (b) 0.13(5) kN

15 (a) Area of cross section $\times$ speed $\times$ density $= 315v$
    (b) (i) $21\,\text{m}\,\text{s}^{-1}$ (ii) 1.4×10^5 Ns
    (iii) 1.4×10^5 N
    (c) e.g. friction, turbulence losses

16 (b) (i) 1.83×10^4 kg   (ii) $5.1(2) \times 10^5$ J
    (c) 37 kN
    (d) A large turning effect about base of tower
    (e) (approx.) 2450

Chapter 8

Exercise 8.1

1 (a) $4.7\,\text{rad}\,\text{s}^{-1}$ (b) $0.66\,\text{m}\,\text{s}^{-1}$ (c) 1.3 s
2 $0.033\,\text{rad}\,\text{s}^{-1}$

Exercise 8.2

1 1.5 kN, road friction
2 252 N, 132 N, $10\,\text{m}\,\text{s}^{-1}$ at bottom
3 90 m, leaves road

Exercise 8.3

1 (a) 1.8 s   (b) 5.8 N

Exercise 8.4: Examination questions

1 (a) $7.3 \times 10^{-5}\,\text{rad}\,\text{s}^{-1}$ (b) $0.47\,\text{km}\,\text{s}^{-1}$
    (c) $0.034\,\text{m}\,\text{s}^{-2}$
2 (b) (i) $94.2\,\text{rad}\,\text{s}^{-1}$ (ii) $18.8\,\text{m}\,\text{s}^{-1}$
3 (a) Vertical component $= P\cos 35° = W$
    (b) 4.79×10^5 N   (c) 2.74×10^5 N
    (d) $6.86\,\text{m}\,\text{s}^{-2}$ (e) 9.11 km
4 6.3 N
5 $T = 2.5(19)$ N at bottom of swing.
6 (a) (i) $1.2(5)\,\text{rad}\,\text{s}^{-1}$ (ii) 94 N   (iii) $3.5\,\text{m}\,\text{s}^{-1}$
    (b) (i) $5.0\,\text{rad}\,\text{s}^{-1}$ (ii) 40 N
7 (a) (ii) 0.30 kg   (iii) $2.6\,\text{m}\,\text{s}^{-1}$ (iv) $6.5\,\text{rad}\,\text{s}^{-1}$
    (b) $T \propto 1/L$ type graph
8 (a) 24(.5) $\text{m}\,\text{s}^{-1}$

(b) (i) Horizontal component, H, of reaction force produces part of centripetal force, F_C

(ii) $H = F_C$

9 5.5×10^3 N, downwards

10 (b) (i) $59.6 \, \text{rad s}^{-1}$ (ii) $16.69 \, \text{m s}^{-1}$ or $60 \, \text{km/hr}$

(c) (i) 28 m (ii) wheels leave ground

(d) (i) 3.0×10^3 r.p.m.

(ii) $0.31 \times 10^3 \, \text{rad s}^{-1}$ (iii) $47 \, \text{m s}^{-1}$

(e) (i) Vertical $T \cos \theta$; horizontal $T \sin \theta$

(ii) Vertical $= T \cos \theta$; $T = 17 \, \text{N}$ (16.8 N)

(iii) $2.6(0) \, \text{m s}^{-1}$ (iv) $2.2 \, \text{s}$ (2.18 s)

11 $1.44 \, \text{rad s}^{-1}$

Chapter 9

Exercise 9.1

1 27×10^{-9} N

2 49 N

3 36 kg

4 3.4×10^8 m from Earth

Exercise 9.2

1 (a) 7.3×10^{22} kg (b) $0.67 \, \text{N kg}^{-1}$

2 (a) $2.45 \, \text{m s}^{-2}$ (b) $19.6 \, \text{m s}^{-2}$ (c) $9.8 \, \text{m s}^{-2}$

3 $0.54 \, \text{m s}^{-2}$

Exercise 9.3

1 12×10^3 J

2 (a) $3.0 \times 10^6 \, \text{J kg}^{-1}$ (b) 4.5×10^9 J

3 $0.65 \, \text{km s}^{-1}$

4 $5.8 \times 10^7 \, \text{m s}^{-1}$

Exercise 9.4

1 (a) 5.1×10^3 s (b) 1.7×10^6 m

2 $R = 14 \times 10^{11}$, $T = 0.63$

Exercise 9.5: Examination questions

1 3.70 N

3 D

4 A

5 B

6 B

7 (a) 9.3×10^{-10} N, attractive

(b) Decrease of 3.7×10^{-11} J

8 (a) 2.0×10^9 m from centre of more massive star

(b) As (a). Note 6.0×10^9 m is also a solution

9 $11 \, \text{km s}^{-1}$

10 (a) $7.8 \, \text{km s}^{-1}$

(b) Energy required for launch ($\frac{1}{2} m v^2$) increases as m increases

11 (b) $4.2(4) \times 10^7$ m

(c) (i) 2.0×10^8 J (ii) $4.4 \, \text{km s}^{-1}$

12 (c) (i) 1.9×10^{27} kg (ii) $3.1 \times 10^4 \, \text{m s}^{-1}$

(iii) 3.4×10^4

13 (b) (i) $6.6(7) \times 10^6$ m (ii) $7.7(6) \times 10^3 \, \text{m s}^{-1}$

(iii) $3.0(2) \times 10^{10}$ J

(c) 1000 kg satellite has KE equivalent to approximately 7 tonnes of TNT

14 (a) (i) 7.8(9) hours (ii) $0.98 \, \text{N kg}^{-1}$

15 1.72 hours; 14 per day

16 (b) (i) 17.1×10^6 m (ii) $3.77 \, \text{m s}^{-2}$

(c) 1.26 days

17 (b) (iii) 1. 13.6×10^6 m from Pluto

2. $0.0536 \, \text{m s}^{-2}$

(c) (i) $0.935 \times 10^3 \, \text{m s}^{-1}$

(ii) $\Delta\phi$ between surface and peak of graph is less for Charon than Pluto

Chapter 10

Exercise 10.1

1 (a) $25 \, \text{N m}^{-1}$ (b) 26 mJ

2 (a) 4.0×10^{-2} J (b) 14×10^{-2} J

3 (a) 15 J (b) 30 J

Exercise 10.2

1 25 kN

2 0.50 mm

3 (a) 7.5×10^7 N (b) 3.0×10^{-4}

Exercise 10.3

1 (a) $57 \, \text{MN m}^{-2}$ (b) 0.45×10^{-3}

(c) $13 \times 10^{10} \, \text{N m}^{-2}$

2 7.5×10^5 N

3 (a) 1.4 mm (b) 42 N

4 (a) 0.19 mm (b) $0.94 \, \text{kJ m}^{-3}$

5 3.6 kJ

Exercise 10.4

1 0.58 kN

2 (a) 0.38 MN (b) 0.69 kJ

Exercise 10.5: Examination questions

1 (a) Straight line through origin to $x = 25$ mm, $F = 1.5$ N

(b) (i) 18 mJ

2 B

3 (a) $1.75 \times 10^3 \, \text{N m}^{-1}$ (b) Approximately 4.5 J

4 E

5 (a) (i) 72 N (ii) 8.0 N

6 $1.2 \times 10^{-5} \, \text{m}^2$

7 (b) (i) 12 mm (ii) 2.2 J

8 (a) 0.32 mm (b) 0.24×10^{-3} J

9 (b) (i) $0.10 \times 10^{-6} \, \text{m}^2$ (ii) 63×10^9 Pa

(iii) 1.4×10^8 Pa (iv) 0.32 J

10 (c) (i) Copper 4.6 mm; steel 7.9 mm

(ii) 54 N, to base of copper wire

(iii) Copper wire; larger force for same extension

11 $1.9 \times 10^5\,\mathrm{J\,m^{-3}}$; Nylon creeps, steel doesn't
12 (b) (i) $8.0 \times 10^{10}\,\mathrm{Pa}$ (ii) $48 \times 10^3\,\mathrm{J\,m^{-3}}$
  (iii) $0.55\,\mathrm{mm}$
13 (a) $6.9\,\mathrm{mm}$
  (b) (i) $87°\mathrm{C}$ (ii) $1.4 \times 10^8\,\mathrm{Pa}$ (iii) $0.28\,\mathrm{kN}$

Chapter 11

Exercise 11.1

1 (a) $4.0\,\mathrm{Hz}$ (b) $\pm1.3\pi^2\,\mathrm{m\,s^{-2}}, 0$
  (c) $-0.32\pi^2\,\mathrm{m\,s^{-2}}$
2 (a) $32 \times 10^3\,\mathrm{m\,s^{-2}}$ (b) $25\,\mathrm{kN}$
3 (a) $1.6\,\mathrm{Hz}$ (b) $0.63\,\mathrm{s}$
4 $2.9\,\mathrm{Hz}$

Exercise 11.2

1 (a) $0.20\,\mathrm{m}$ (b) $0.89\,\mathrm{s}$ (c) 68
2 (a) $1.5\,\mathrm{s}, 0.65\,\mathrm{Hz}$ (b) $0, \pm0.33\,\mathrm{m\,s^{-2}}$
3 (a) $0.25\,\mathrm{m}$ (b) $0.063\,\mathrm{m}$ (c) 60
4 $1.0\,\mathrm{m}, 0.84\,\mathrm{m}$

Exercise 11.3

1 (a)

t/s	0	0.25	0.50
$y/10^{-3}\,\mathrm{m}$	0	28	40
$v/\mathrm{m\,s^{-1}}$	0.13	0.089	0
$a/\mathrm{m\,s^{-2}}$	0	-0.28	-0.39

The remaining values follow by 'symmetry'
  (b) time = $0.10(3)\,\mathrm{s}$
2 (a) $0.30\,\mathrm{m\,s^{-2}}$ (b) $0.094\,\mathrm{m\,s^{-1}}$
  (c) $0.099\,\mathrm{m\,s^{-2}}, 0.089\,\mathrm{m\,s^{-1}}, r = 0.030, \omega = \pi$

Exercise 11.4

1 $0.87\,\mathrm{Hz}$
2 $18\,\mathrm{mJ}$
3 (a) $8.0\,\mathrm{mJ}$ (b) $0.5\,\mathrm{mJ}$ (c) $32\,\mathrm{mJ}$

Exercise 11.5: Examination questions

1 $1.4\,\mathrm{m\,s^{-2}}$
2 $28\,\mathrm{N\,m^{-1}}$
3 $1.15\,\mathrm{Hz}; 31\,\mathrm{N\,m^{-1}}$
4 (a) $2.5(3)\,\mathrm{N\,m^{-1}}$ (b) $8.0\,\mathrm{Hz}$
5 (a) $1.4 \times 10^4\,\mathrm{N\,m^{-1}}$ (b) $0.91\,\mathrm{s}$
  (c) (i) rolling and pitching (ii) bouncing
6 (a) $0.4\,\mathrm{s}$ (b) $0.5f$
7 (a) (ii) F in opposite direction to x
  (b) (i) $50\,\mathrm{N\,m^{-1}}$ (ii) $0.79\,\mathrm{s}$
  (c) (i) $0.56\,\mathrm{s}; T \propto \sqrt{m}$
    (ii) $0.79\,\mathrm{s}; T$ doesn't depend on amplitude
  (d) (i) Straight line starting at end of existing line, but with half the slope
    (ii) No, force non-linear
8 (a) $0.40\,\mathrm{s}$ (b) $2.25\,\mathrm{m\,s^{-2}}$
9 $0.6\,\mathrm{m}$
10 $\sqrt{2}T$

11 50.0
12 $25\,\mathrm{ms}$
13 (a) $80\,\mathrm{mm}$
  (b) (i) $0.56\,\mathrm{s}; 1.8\,\mathrm{Hz}$ (ii) $2.5\,\mathrm{m\,s^{-2}}$
    (iii) $0.22\,\mathrm{m\,s^{-1}}$ (iv) $23\,\mathrm{ms}$
14 (a) $3.1\,\mathrm{m}$
  (b) (i) 12.00 (12 noon) (ii) 15.00 (3.00 pm)
  (c) 13.26 (1.16 pm)
15 (a) $0.40\,\mathrm{s}$ (b) $0.20\,\mathrm{J}$ (0.197 J)
  (c)

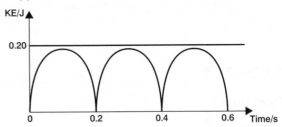

16 $3.9\,\mathrm{mJ}$

Chapter 12

Exercise 12.1

1 (a) $5.0 \times 10^{14}\,\mathrm{Hz}$ (b) $1.5\,\mathrm{km}$
2 $68 \times 10^9\,\mathrm{N\,m^{-2}}$
3 $3.6 \times 10^3\,\mathrm{m\,s^{-1}}$
4 $71\,\mathrm{m\,s^{-1}}$
5 (a) $0.10\,\mathrm{km\,s^{-1}}$ (b) $2.5 \times 10^{-4}\,\mathrm{kg\,m^{-1}}$

Exercise 12.2

1 (a) $+4.3\,\mathrm{cm}, -4.3\,\mathrm{cm}$ (b) $2\pi/3\,\mathrm{rad}, 10\pi/3\,\mathrm{rad}$
2 $3\frac{1}{3}\,\mathrm{cm}$
3 (a) $0.50\,\mathrm{Hz}$ (b) $12\,\mathrm{m}, 2.4\,\mathrm{m}$, for example

Exercise 12.3

1 (a) nothing (b) a double amplitude sound
2 $3.16\,\mathrm{cm}$
3 (a) $1.0\,\mathrm{m}, 340\,\mathrm{Hz}; 0.50\,\mathrm{m}, 680\,\mathrm{Hz}$
  (b) $2.0\,\mathrm{m}, 170\,\mathrm{Hz}; \frac{2}{3}\,\mathrm{m}, 510\,\mathrm{Hz}$
4 (a) $567\,\mathrm{Hz}$ (b) $283\,\mathrm{Hz}$

Exercise 12.4

1 $0.49\,\mathrm{mm}$
2 $0.42\,\mathrm{mm}$
3 (a) $0.51\,\mathrm{mm}$ (b) $0.80\,\mathrm{mm}$ (c) $1.6\,\mathrm{mm}$

Exercise 12.5

1 (a) $0.213\,\mathrm{m}$ (b) $1.1\,\mathrm{m\,s^{-1}}$
2 $2.8\,\mathrm{cm}, 11\,\mathrm{GHz}$

Exercise 12.6

1 (a) $1.8\,\mathrm{m}, 59\,\mathrm{Hz}; 0.90\,\mathrm{m}, 117\,\mathrm{Hz}; 0.60\,\mathrm{m}, 176\,\mathrm{Hz}$
  (b) $0.90\,\mathrm{m}, 117\,\mathrm{Hz}; 0.45\,\mathrm{m}, 234\,\mathrm{Hz}, 0.30\,\mathrm{m}, 351\,\mathrm{Hz}$
2 (a) $1.6 \times 10^{-3}\,\mathrm{kg\,m^{-1}}$ (b) $0.19\,\mathrm{km\,s^{-1}}$
  (c) $1.6\,\mathrm{m}$ (d) $0.81\,\mathrm{m}$
3 (a) $260\,\mathrm{Hz}$ (b) $300\,\mathrm{Hz}$ (c) $520\,\mathrm{Hz}$

Exercise 12.7

1 0.34 m
2 (a) 0.166 m
  (b) 0.498 m
3 170 Hz; 510 Hz
4 600 Hz; 1.00 kHz

Exercise 12.8: Examination questions

1 15 m s^{-1}
2 50 Hz
3 (b) 25 × 10^4 m
4 2.0 × 10^{11} N m^{-2}
5 (a) 5.0 × 10^3 m s^{-1}
  (b) Speed in metal track and in air differ; time
    interval 5.7 s.
6 (b) (i) I 2.5 Hz
      II 0.20 m; 0.067 m (0.5 λ or 1.5 λ in 0.10 m)
    (ii) 0.5 m s^{-1} (or 0.17 m s^{-1})
7 (a) (i) amplitude 8.7 (±0.2) mm; period
      30 × 10^{-3} s; frequency 33 Hz
    (ii) 0.15 m
  (b) (i) 1.1 × 10^6 m
8 (a) See Fig 11.2 (c)
  (b)

Wave	v	λ	f	t	ϕ_t	$\phi_{t+0.001}$
1	330	1.32	**250**	**0.0040**	0	**90.0°**
2	**340**	3.40	**100**	**0.010**	0	36.0°

9 0.15 km s^{-1}
10 (a) 120 mm
  (b) zero
11 (b) (ii) A – bright; B – dark
    (iii) λ = ay/d; d = fringe separation
    (iv) 1. 1.98 mm
      2. 9.9 mm corresponds to 5 whole fringe
        separations
      3. 0.99 mm
12 C
13 (b) 3.0 mm
  (c) (i) Fringe separation increases (to 4.0 mm)
    (ii) Fringe separation decreases (to 1.8 mm)
    (iii) Brighter pattern
  (d) (i) 300 nm
    (ii) 55(.6) fringes 'upwards' (direction BA)
14 (a) 70 m s^{-1} (69.6 m s^{-1})
  (b) 40 Hz
15 640 Hz
16 (b) (i) 20 × 10^{-6} m
    (ii) 0.20 km s^{-1}
    (iii) 16 × 10^{-8} N
17 (a) 225 nm
  (b) 1. 3.00 mm   2. Distance d decreases
18 (a) (i) loud sound   (ii) 0.33(1) km s^{-1}
  (b) (i) 0.368 m   (ii) See Fig. 12.13 (c)
19 (b) (i) as Fig. 12.13 (a)   (ii) 0.53 m
    (iii) Move 0.35 m downwards

Chapter 13

Exercise 13.1

1 638 nm
2 17.5°, 36.9°, 64.2°
3 1.0°, 2
4 7.4°

Exercise 13.2

1 (a) 12 × 10^{-3} rad or 0.69°   (b) 0.12 rad or 6.9°
2 9.2 mm
3 9.0 × 10^{-5} rad
4 0.20 m
5 4.0 m

Exercise 13.3: Examination questions

1 D
2 C
3 461 nm
4 (b) 9
5 395
6 (b) Red – 3: Blue – 4
7 (a) 600 × 10^{-9} m   (b) 4.0 × 10^5
8 (a) 42.9°
  (b) Violet – maximum order 8; angle 65.0°
9 (a) 0.047 rad   (b) 33 mm
10 (a) θ = 35° (34.8°)
  (b) For slit width = 1.05 μm, first order minimum
    for single slit occurs at θ = 35°
11 (a) d sin θ = nλ becomes (for d = 2W and n = 2)
    2W sin θ = 2λ. That is, sin θ = λ/W which is the
    minimum for a single slit, thus grating
    maximum not present.
  (b) nλ/d for n = 3 is greater than 1 (1.05)
  (c) Redistributed
  (d) 1st and 3rd orders (only) are present
12 1.5 m
13 2.0 m
14 (a) Resolving power defined as the minimum angle
    θ of separation of two sources which can just be
    distinguished as separate
  (b) (i) 37(.5) m
    (ii) Move away from the sources
15 D (7 × 10^{16} m)

Chapter 14

Exercise 14.1

1 50°
2 x = 35° y = 30°
3 40°

Exercise 14.2

1 (a) 33.4°   (b) 76°
2 16°
3 (a) 1.33   (b) 32.0°   (c) 540 nm

Exercise 14.3

1 1.3
2 28°
3 69.6°
4 (a) 80.5°   (b) 14.0°
5 (a) $2.00 \times 10^8 \, \mathrm{m \, s^{-1}}$
   (b) Min 30 m; Max 35 m (34.6 m)
   (c) Min 15×10^{-8} s; Max 17×10^{-8} s
6 Red – 19.3×10^{-8} s; Blue – 19.6×10^{-8} s

Exercise 14.4: Examination questions

1 (b) (i) 49° (49.5°)
   (c) (i) 35° (34.9°)
      (ii) 49° (49.5°) Same as (b)(i) since
         $n \sin i = \mathrm{constant}$
2 (a) 87.95°   (b) 0.55°   (c) 2.2 minutes
3 C
4 (a) 0.22   (b) 12.7° (air to water)
5 (a) $\theta = 22°$ (22.48°)
   (b) (i) 42° (41.8°)
      (ii)

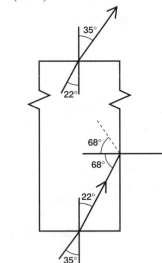

6 (a) 29°   (b) 2.42
   (c) Critical angle = 24.4° ∴ total internal
      reflection
7 (b) (i) 19° (19.47°)
      (ii)

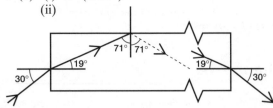

8 (a) (iii) 75.2°
   (b) (ii) Critical angle decreases so more light likely
      to stay in fibre
9 (a) 75.3°
   (b) 1. A = 9.8° & B = 80° (80.2°)
      2. Totally internally reflected
   (c) Shortest $7.6(0) \times 10^{-5}$ s; Longest 7.9×10^{-5} s
      $(7.86 \times 10^{-5}$ s)

Chapter 15

Exercise 15.1

1 +12 cm
2 (a) −30 cm   (b) +15 cm
3 (a) +5 cm   (b) −10 cm
4 −24 cm
5 (a) 11.1 cm   (b) 10.2 cm   (c) 0.9 cm

Exercise 15.2

1 −80/3 cm
2 (a) −0.50 m   (b) +3.0 D

Exercise 15.3

1 (a) 0.018 m   (b) +4.0 D
2 (a) −0.667 D   (b) 0.23 m to infinity
3 (a) 2.0 m
   (b) (i) −0.50 D   (ii) 2.0 D

Exercise 15.4: Examination questions

1 (a) Converging, 12.0 cm   (b) Converging, 10.7 cm
2 (a) 24 cm   (b) 18 mm   (c) Virtual
3 (c) (i) Position 400/3 mm from lens on same side
         as object: $m = (-) 1/3$
      (ii) 1. Position 129 mm (128.6 mm) from lens
         2. $2.4 \, \mathrm{mm \, s^{-1}}$ $(2.38 \, \mathrm{mm \, s^{-1}})$
         3. Towards lens
4 (d) (i) $D = 4.00$ mm
      (ii) Diverging lens – image is 120 mm from
         lens on same side as object
         Converging lens – image is 180 mm from
         lens on opposite side to object
      (iii) Diverging lens – virtual, upright,
         diminished
         Converged lens – real, inverted, diminished
   (e) $f = 56$ mm: converging lens
5 (a) (i) Distant object produces image at focal
         point
      (ii) 12.5 D
   (b) (i) A virtual object formed by the converging
         lens at distance $(80 - 59) = 21$ mm to
         right of eyepiece
      (ii) 420 mm from eyepiece and to the left of it
      (iii) Virtual: v is negative (−420 mm)
   (c) Move eyepiece away from objective lens so as
      to reduce object distance: u decreases so v
      increases for the diverging lens
6 (i) Away from film   (ii) 3.4 mm
7 (a) (ii) Distance 84 mm, linear magnification
         0.053, image height 21 mm
   (b) (ii) 1. (−) 0.4   2. (−) 120 mm   3. (+) 300 mm
8 (a) 20 D   (b) 66.7 mm
9 (a) 1.75 cm   (b) 1.68 cm
10 (b) (i) 0.85 m from eye lens
      (ii) Yes, lens power required is 59 D which is
         within the range quoted
      (iii) Hypermetropia

11 (a) (−) 0.800 m
    (b) Fig. 15.6 (b) shows combined effect of eye lens and correction (diverging) lens to produce a sharp image, on the retina, of a distant object
12 (a) (i) Diverging   (ii) (−) 2.0 m
       (iii) −0.50 D   (iv) Similar to Fig. 15.2
    (b) (i) −0.30 D
       (ii) Improved – lower power correction lens needed and new far point is 3.3 m from his eyes
13 (d) (i) −0.313 D   (ii) 12.5 cm
14 (a) 300 mm from eye
    (b) Moves from 200 mm from eye to 600 mm from eye

Chapter 16

Exercise 16.1

1 (a) 27 mm   (b) (−) 9.3 times
2 (a) 30 mm   (b) (−) 8.3 times
3 (a) 26 mm   (b) (−) 7.0 times

Exercise 16.2

1 (a) 20.0   (b) 63.0 cm   (c) 40.0×10^{-3} rad
2 (a) 18   (b) 5.6×10^{-3} rad
    (c) 2.1×10^3 km

Exercise 16.3: Examination questions

1 (a) 4.00 cm   (b) 3.00 times
2 (a) 0.30 times   (b) 10.7 cm
3 (a) 918 mm   (b) 50
4 (a) 125 mm   (b) 500 mm
5 20×10^{-2} rad
6 $2.1(5) \times 10^3$ km

Chapter 17

Exercise 17.1

1 5.0 K
2 7.1 V
3 1800 J kg^{-1} K^{-1}
4 39°C

Exercise 17.2

1 377 W
2 (a) 438 kJ kg^{-1} (b) 1.61 kJ   (c) 59.4 mg s^{-1}

Exercise 17.3

1 3.6 W
2 0.24 W m^{-1} K^{-1}
3 38 W, 36°C
4 (a) 40 kW   (b) 19(.4) °C   (c) 1.2 kW (1.17 kW)

Comment (a) 40 kW is very large and so a 20°C house temperature is unlikely to be maintained;
    (b) & (c) The insulation layer mainly determines the rate of energy transfer
5 (a) Thermal conductivity and thickness
    (b) 4 °C

Exercise 17.4: Examination questions

1 (c) (i) 65.3 °C
       (ii) Stop heat losses
    (d) (i) Conduction
       (ii) Free electrons (and atomic vibrations) transport KE
2 (a) (i) 0.90 MJ   (ii) 0.50(4) MJ
    (b) Heat losses; thermal capacity of dishwasher; evaporation of water
3 960 °C (957 °C)
    Discussion – • not all energy transferred to thermal energy
                • thermal energy losses to surroundings
                • thermal energy not uniformly distributed in bit
    First two points mean tip cooler, last point means tip is hotter
4 (b) (i) 66(.2) kJ
       (ii) 441 kJ per 100 g
       (iii) 432 kJ per 100 g
    (c) See Chapter 19; work done ΔW in pushing atmosphere back; $\Delta Q = \Delta U + \Delta W$ is greater if gas is allowed to expand
5 (b) (i) 0.060 kg s^{-1} (ii) 20 K (19.8 K)
6 42 °C
7 (a) (i) $\frac{1}{12}$ kg per second   (ii) 5.3 MJ
    (b) (i) 38(.3) °C. No energy transfer to surroundings, evaporation etc.
       (ii) 7.1 MJ
8 0.66 kW
9 4.0×10^3 s
10 (a) 10.7 kJ   (b) 97.0 kJ kg^{-1}
11 C
12 460 s
13 (a) (i) 0.033 K s^{-1} (0.03 to 0.04 acceptable)
       (ii) 56 W (50 to 70 acceptable)
    (b) (i) Ice melting at 0 °C
       (ii) 0.27 kg (0.24 to 0.34 acceptable)
14 9.0 °C
15 (b) 0 °C (when half of ice has melted) – assumes no energy transfer from surroundings
    (c) (i) $\Delta T = X e^{-0.25t/mc}$ (assumes external temperature X in °C, m = mass of liquid and c = specific heat capacity of liquid)
       (ii) 19 hours
16 (a) 100 K m^{-1} (b) 0.23 kW
17 25 W
18 460 W

19 1.50 K
20 (a) (i) Remove thermal energy at same rate as it flows into building from outside
   (ii) 0.22 kW
   (iii) 3 times larger
   (b) 22 kW
21 (b) (i) $1.3(3) \times 10^{-5}$ (ii) 73(.3) W

Chapter 18

Exercise 18.1

1 (a) 280   (b) 983   (c) 193   (d) 74
2 327 °C
3 $63 \times 10^{-3} \, \text{m}^3$
4 210 °C
5 170 cm³
6 $98 \times 10^{-3} \, \text{kg}$

Exercise 18.2

1 $22.5 \times 10^{-3} \, \text{m}$
2 (a) 22.7   (b) 1.37×10^{25}
3 (a) 0.100   (b) $28.0 \times 10^{-4} \, \text{kg}$
   (c) $2.19 \times 10^{-3} \, \text{m}^3$
4 2.30 kg, $73 \times 10^{-3} \, \text{m}^3$
5 $1.26 \times 10^5 \, \text{N m}^{-2}$

Exercise 18.3

1 $579 \, \text{m s}^{-1}$
2 (a) $30 \, \text{m s}^{-1}$ (b) $34 \, \text{m s}^{-1}$ (c) $35 \, \text{m s}^{-1}$
3 $0.179 \, \text{kg m}^{-3}$
4 $1.12 \, \text{km s}^{-1}$
5 $666 \, \text{m s}^{-1}$

Exercise 18.4: Examination questions

1 $1.13 \times 10^5 \, \text{Pa}$
2 456 kPa (355 kPa above atmospheric)
3 (a) $1.7(5) \times 10^{-3} \, \text{kg m}^{-3}$ (b) $3.2(5) \times 10^{19}$
4 (a) Silver absorbs less radiation (in daytime) and emits less radiation (at night-time)
   (b) (i) 43.5 kPa
   (ii) otherwise balloon would burst
   (iii) 19%
5 A
6 (a) 9.0% increase   (b) 8.3% released
7 (a) 2.48 mol   (b) 15.0×10^{23} atoms   (c) 0.90 N
8 (a) r.m.s. speed of molecules and pressure
   (b) (i) 1.2 mol (1.24)   (ii) 0.036 kg
   (iii) $3.6 \, \text{kg m}^{-3}$
   (c) (i) Same – temperature is same
   (ii) Different – mass of molecules is different
9 (b) (i) 38×10^{23} (ii) $5.2 \times 10^5 \, \text{Pa}$
   (iii) $0.48 \, \text{km s}^{-1}$
10 $424 \, \text{m s}^{-1}$
11 (b) 0.93(5)
   (c) (i) $0.412 \, \text{kg m}^{-3}$
   (ii) Composition of atmosphere changes with

height. More greenhouse gases at surface (e.g. carbon dioxide) and more oxygen and nitrogen at higher levels
12 (a) Equation 18.10
   (b) (i) $5.53 \times 10^{-26} \, \text{kg}$
   (ii) 1. $548 \, \text{m s}^{-1}$ 2. 401 K
13 (a) $0.58 \, \text{km s}^{-1}$ (b) 241 K
   (c) 6.0 kJ   (d) 361 K
14 (b) 5.0×10^{22}
   (c) (i) 55 N   (ii) 1.38

Chapter 19

Exercise 19.1

1 $-1.5 \times 10^2 \, \text{J}$
2 (a) $-50 \, \text{J}$ (b) $-70 \, \text{J}$
3 (a) $50 \times 10^{-3} \, \text{kg}$ (b) 113 kJ
   (c) 8.4 kJ   (d) 105 kJ

Exercise 19.2

1 (a) $6.73 \times 10^5 \, \text{N m}^{-2}$, 290 K
   (b) $14.4 \times 10^5 \, \text{N m}^{-2}$, 619 K
2 (a) $15.0 \times 10^{-4} \, \text{m}^3$, 280 K
   (b) $9.47 \times 10^{-4} \, \text{m}^3$, 177 K
3 232 K, 1.29 m Hg

Exercise 19.3

1 (a) 58%   (b) 3.0 MW   (c) 38%
2 5.0

Exercise 19.4

1 75 kJ
2 (a) 0.12 kJ   (b) 30 kW

Exercise 19.5: Examination questions

1 $3 \times 10^4 \, \text{J}$
2 (+) 150 J
3 (a) 458 s
   (b) (i) $0.85 \, \text{m}^3$ (ii) $0.86 \times 10^5 \, \text{J}$
   (c) $\Delta W = 0.86 \times 10^5 \, \text{J}$ $\Delta Q = 1.1 \, \text{MJ}$
   $\Delta U = 1.0(1) \, \text{MJ}$
4 500 K
5 150 K
6 134 K
7 35.0%
8 0.40 kJ
9 0.40 kJ
10 (a) AB   (b) (i) 200 J   (ii) 52.6%

Chapter 20

Exercise 20.1

1 12.5×10^{12}
2 1.25×10^{22}
3 0.75

Exercise 20.2

1 (a) 0.10 A   (b) 2.0 V
2 (a) 2.5 A   (b) 1.5 A
3 0.4 A, 80 V; 0.08 A, 160 V; 0.32 A, 160 V

Exercise 20.3

1 1.1 Ω
2 18×10^{-8} Ω m
3 2.0×10^{-2} K^{-1}

Exercise 20.4

1 2.4 kJ
2 (a) 2.4 kW h   (b) 8.6 MJ
3 240 J

Exercise 20.5

1 2.0 V
2 4.0 Ω
3 (a) 0.02 Ω   (b) 300 A
    (c) 1.8 kJ s^{-1} (1.8 kW)

Exercise 20.6

0.6 A

Exercise 20.7

1 (a) 10.0 mV   (b) 10 mA
2 (a) 0.040 Ω shunt   (b) 2.0 kΩ in series
    (c) 10 Ω in series

Exercise 20.8

1 (a) 1.2 V   (b) 1.0 V
2 (a) 4.0 V   (b) 3.4 V

Exercise 20.9

1 n = number of free electrons per unit volume
   v = drift velocity, i.e. mean speed of travel through length of wire

$\dfrac{n_Y}{n_X} = 1$ because both Y and X are copper

$\dfrac{I_Y}{I_X} = 1$ because conductors are in series

$\dfrac{v_Y}{v_X} = 2$ because area × v is same for Y and X

2 6.3 V
3 (i) 6 Ω   (ii) 0.9 A
4 2.0 Ω
5 (a) Increased vibrations of atoms hinders electron flow
    (b) (i) At 0 °C, 0.20 kΩ. At 100 °C, 0.29 kΩ
       (ii) 4.3×10^{-3} K^{-1}

6 B
7 Temperature increase releases more charge carriers. Resistance = 495 Ω

Increased p.d. produces increased current and so greater heating. This causes resistance to decrease and the 330 Ω gets higher proportion of the supply p.d.

8 (b) (i)

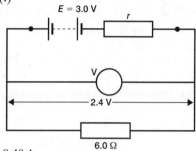

   (ii) 0.40 A
   (iii) 1.5 Ω
   (iv) 3.0 V ($V = E$)
   (v) Current increases so greater lost volts (Ir)
9 (a) 12 V   (b) 1.7 A
10 2.0 W
11 A   12 V, 3 A, 36 W, 4 Ω
    B   10 V, 2 A, 20 W, 5 Ω
    C   2 V, 2 A, 4 W, 1 Ω
    whole 12 V, 5 A, 60 W, 2.4 Ω
12 0.29 A
13

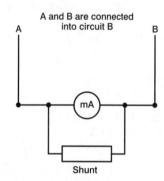

Shunt resistance needed = 0.16 Ω
14 (i) 80 Ω   (ii) 4.3 V
15 (i)

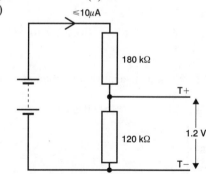

   (ii) 17 μA, 1.7×10^{-2} V

Chapter 21

Exercise 21.1

1 (a) 0.13 μN   (b) 0.40 nC
2 (a) (i) 1.8 kV m^{-1} (ii) zero   (b) 36 V
3 2.0 V

4 8.9 kV, 8.6 mJ
5 Zero, 2.8 kV m^{-1} to the right
6 8.8×10^{13} m s^{-2}, 3.0×10^6 m s^{-1}

Exercise 21.2

1 (b) (iv) 1.5×10^4 N C^{-1} (or V m^{-1})
    (v) 1.9×10^{-7} C
2 P.D. $= 90$ kV, Max K.E. $= 1.4 \times 10^{-14}$ J, max speed $= 1.8 \times 10^8$ m s^{-1}
3 (b) (iii) 1.6×10^{-19} C
    (iv) 7.2 and 4.8 J C^{-1}
    (v) -1.5×10^{-18} J

4 (a) $F = \dfrac{e^2}{4\pi\varepsilon r^2}$ (b) 2.1×10^{-10} m

5 (a) (ii) 1.8×10^4 N to the right (a large force but huge charges involved)
    (iii) Same size but to left
  (b) (ii) 36 J (Potential 9.0 MV)   (iii) 22 J
    (iv) zero

Chapter 22

Exercise 22.1

1 (a) 0.34×10^{-10} F   (b) 0.17×10^{-10} F
2 8.0×10^7 mm C^{-1}, 3.0 μF, 0.010 μF
3 $2.4 \times 10^{-1'0}$ F, 9.0×10^{-12} F m^{-1}
4 8.0 mA

Exercise 22.2

1 B
2 (a) and (b) 3.0 μC, 2.3 μJ, 1.5 V
    (c) 2.0 μC, 1.0 μJ, 1.0 V
3 (a) 0.80 mJ   (b) 14 V   (c) 0.53 mJ
4 6.7×10^{-10} F
5 A
6 (a) 20 V s^{-1} (b) 7.3 V s^{-1}

Exercise 22.3

1 (a) (i) 2.4×10^{-10} F   (ii) 1.2×10^{-7} C
    (iii) 30 μJ
  (b) (i) 0.9×10^{-4} J
    (ii) Work done against attraction of − and + plates
2 (a) 3.1×10^{-9} F   (b) 60 μA
3

Time constant $= 22$ s. Stays on for 15 s. R increase lengthens time on.
4 (b) (i) 33 μF   (ii) 100 V
  (c) (i) 2.0×10^2 μC
    (ii) (1) 64 μC   (2) 2.9 V   (3) 46 μF
5 (a) 26 μF   (b) 20 μF   (c) 20 μF
6 (b) (i) 8.3 μF
  (c) (i) 10 V at $t = 0$, $10 V \times \dfrac{1}{\sqrt{2}}$ at t = half-life/2

Chapter 23

Exercise 23.1

1 13 μN
2 4.0 mN
3 1.0 mA
4 (a) 3.2×10^{-5} N   (b) 6.4×10^{-6} N m
  (c) 6.0×10^{-6} N m
5 (a) 0.50×10^4 V m^{-1} (b) $0.010\,T$

Exercise 23.2

1 (i) 4.6 μN   (ii) 9.0 A
2 (a) (i) $F = BIL \times N$
  (b) (i) 0.33 A
    (ii) Small back EMF, large current, excessive heating
3 (a) 7.0×10^6 m s^{-1} (b) Negative
4 $4\pi \times 10^{-7}$ H m^{-1}

Chapter 24

Exercise 24.1

1 (a) 5.0 mV   (b) zero   (c) 4.3 mV
2 (a) 25 mV   (b) 2.5 mA
3 6.0×10^{-7} Wb

Exercise 24.2

1 (a) 1.6×10^{-4} Wb   (b) 80×10^{-4} Wb
  (c) 16 mV
2 1.0×10^{-2} H

Exercise 24.3

1 5.4 mA
2 (a) 13 mV   (b) 9.6 mV

Exercise 24.4

1 D
2 (a) 0.38 Wb
  (b) (i) $BAn\cos\theta$
3 B
4 Flux through closed window $= 18 \times 10^{-6}$ Wb. Induced e.m.f. $= 34$ μV. When sliding there is zero e.m.f.

5 (a) (i) (1) 0.050 V   (2) 1.95 V
(ii) 4.9×10^{-2} H
(b) (i) 8.0 A
(ii) Inductor only produces voltage when current is changing.

Chapter 25

Exercise 25.1

1 2.4×10^{-5} T
2 (a) 4 cm from 2.0 A wire, 6 cm from other wire
(b) 20 cm from 2.0 A wire, 30 cm from other wire
3 1.7×10^{-5} T
4 1.9×10^{-5} V

Exercise 25.2

1 0.75 mT
2 (i) 1.4×10^{-4} T   (ii) 1.8×10^{-4} T
4 (a) (i) 1.1×10^{-5} T
(ii) Clockwise in diagram, i.e. to right at P
(b) (i) 0.7×10^{-5} T north
(ii) 2.1×10^{-5} T at 31° east of north

Chapter 26

Exercise 26.1

1 (a) 2.8 V   (b) 0.14 A
2 (a) 2.5 ms   (b) 0.83 ms
3 (a) 0.10 kΩ   (b) 1.0 A   (c) 87 V   (d) 0.50 A
4 (a) 25 Ω   (b) 22 Ω
5 (a) 74 Ω   (b) 4.0 V
6 2.9 H

Exercise 26.2

1 0.25 kHz
2 5.1 μF, 2.4 A
3 (a) 0.36 kHz   (b) 0.20 A   (c) 89 V, 89 V, 20 V

Exercise 26.3

1 10 V RMS, 0.020 A RMS
2 50 kW, 0.56 W

Exercise 26.4

1 36
2 44 mA RMS
3 (a) A capacitor (current peaks a quarter cycle before its voltage peaks)
(b) (i) 1.0 Hz   (ii) 3.5 V   (iii) 1.0 Ω
4 Peak voltage = 0.25 kV. Peak supply p.d. leads the peak current by 32°
5 (a) 0.16 kHz   (b) (i) 30 mA RMS   (ii) 0.36 W
6 (a) Step up   (b) 1250   (c) 1.67 A

Chapter 27

Exercise 27.1

1 0.75×10^{15} Hz, 5.0×10^{-19} J
2 2.1 V
3 0.97×10^{15} Hz
4 6.6×10^{-34} J s
5 32×10^{14}
6 0.071 nm
7 1.9×10^{6} m s^{-1}

Exercise 27.2

1 (i) Direction is down page
(ii) 3.8×10^{-14} N
(iii) The force on the electron is perpendicular to the motion by left-hand rule, so the speed does not change but the electron is deflected, the force being the necessary inwards force for circular motion.
(iv) 1.5 mm
2 Max KE = 1.53×10^{-17} J. De Broglie wavelength = 1.3×10^{-10} m
3 For ionising transitions see figure

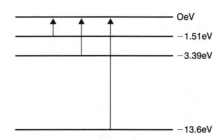

Ground state is lowest available energy level, normal level (unexcited state) for the hydrogen's electron
660 nm line is due to transition −3.39 to −1.51 eV

4 (i) $v = \dfrac{e}{\sqrt{4\pi\varepsilon_0 rm}}$

(ii) (1) $A = \dfrac{h^2 \varepsilon}{e^2 m_e}$ (2) 5.3×10^{-11} m

(3) radii in ratio 1, 4, 9, 16
(4) Deduce that wavelength = circumference divided by n, i.e. divided by 2, 3 and 4 for the next three orbits

Chapter 28

Exercise 28.1

1 3762 years or thereabouts
2 10 h, 0.069 h^{-1}
3 5.4×10^{10}

Exercise 28.2

1 1/3
2 (a) (i) 221 u   (ii) 3.7×10^{-22} g
(b) 27×10^{19} (c) 6.2×10^{18} Bq
3 2%

Exercise 28.3

1 218,84
2 0.50 m
3 12 thousand years

Exercise 28.4

1 4.1×10^{18} Hz, No
2 (a) 500 W   (b) 497.5 W   (c) 0.025 nm
3 (a) 1.4×10^{-15} J   (b) 1.4×10^{-16} J

Exercise 28.5

1 (i) $9.5 \times 10^{-3}\,\text{s}^{-1}$ (ii) 3.2×10^{6} (iii) 9.6 kBq
2 (a) (i) A is activity at time t, A_0 is initial activity, λ is radioactive decay constant
   (b) (i) $3.4\,\mu\text{g}$
       (ii) A further 49 years
3 Half life = 33 s. Decay constant = $0.021\,\text{s}^{-1}$. Rate of decay = $6 \times 10^{18}\,\text{s}^{-1}$. Calculated decay constant = $2 \times 10^{-2}\,\text{s}^{-1}$. More reliable to avoid drawing tangent method
4 (c) 1.8×10^{3} years
5 1.1×10^{-11} m

Chapter 29

Exercise 29.1

1 146
3 4.03 MeV

Exercise 29.2

1 (a) 2/5   (b) $^{207}_{81}\text{Tl} \rightarrow {}^{207}_{82}\text{Pb} + {}^{0}_{-1}\text{e}$
2 (a) 5.2×10^{-13} J   (b) 7.8×10^{13} J per kg
3 (a) (i) A = 141, Z = 56
       (ii) 3.24×10^{-11} J
   (b) 1.2×10^{19} W
4 7 MeV per nucleon

Chapter 30

Exercise 30.1

1
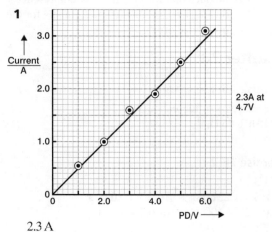

2.3 A

2
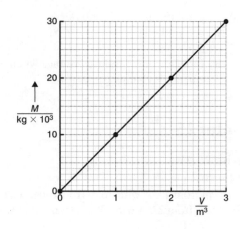

Exercise 30.2

1 $8\,\text{kg cm}^{-1}$
2 $L = 3.0 \times 10^{5}\,\text{J kg}^{-1}, h = 1.0\,\text{W}$

Exercise 30.3

1 $A = 3.5\,\text{A V}^{-2.5}, p = 2.5$
2 $0.010\,\text{day}^{-1}$

Exercise 30.4

1 7 m
2 96 mJ

Exercise 30.5

1 (a) 15 mV   (b) 5.0 kHz
2 $3.5\,\text{km s}^{-1}$
3 $40°$

Exercise 30.6

1 (a)
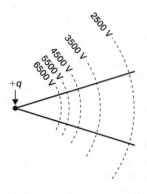

(b)

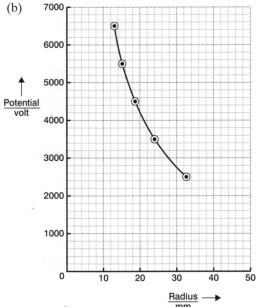

(c) $V = \dfrac{q}{4\pi\varepsilon r}$

(d) In figure for (b) $V \times r$ is constant approximately ($= 8.0 \times 10^4$ V mm) and so agrees with equation in part (c)

(e) 8.9×10^{-9} C

2 (a)

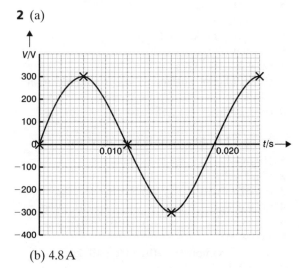

(b) 4.8 A

3 2800 disintegrations per second

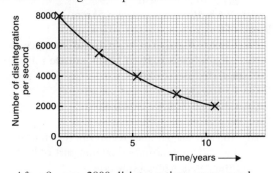

After 8 years 2800 disintegrations per second

4 (b) $hf = \text{WFE} + \text{KE}_{max}$ or $\text{KE}_{max} = hf - \text{WFE}$. Gradient of graph is h. $h = 6.8 \times 10^{-34}$ J s approximately

5 Momentum = mass $\times$ velocity. Slope = force on lorry. At $t = 20$ s slope is 1.5×10^4 N. Explanation of shape of graph is that friction and air resistance gradually increase the opposing force, i.e. net force decreases.

6 (a) (i) $38 \times 10^{-2}\,\Omega\,\text{K}^{-1}$
   (ii) $-263\,°\text{C}$. Close to absolute zero.
(b) (i) See columns 1 and 2 of table.

$\theta/°\text{C}$	R/Ω (= 100 + 0.380)	R_m/Ω	$\Delta R/\Omega$
0	100	100	0
100	138	138	0
200	176	175.5	0.5
300	214	212	2
400	252	247	5
500	290	281	9

(ii) See columns 2, 3 and 4 of table, and graph below.

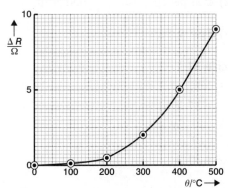

(iii) $300\,°\text{C}$

7 (i) 3.0 V   (ii) 2.1 A.

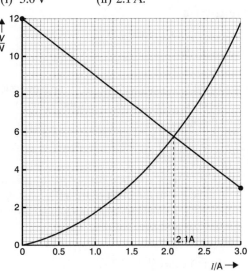

315

8 B

9 A

10 4 mC (800 squares each 5 s by 1 μA), Capacitance is between 4.4×10^{-4} F and 4.7×10^{-4} F.

11 (a)

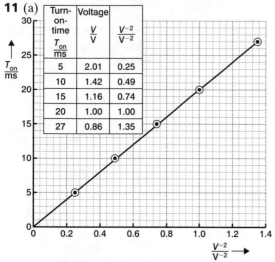

Turn-on-time T_{on} ms	Voltage $\frac{V}{V}$	$\frac{V^{-2}}{V^{-2}}$
5	2.01	0.25
10	1.42	0.49
15	1.16	0.74
20	1.00	1.00
27	0.86	1.35

(b) Straight line passing through $T_{on} = 0$, $V^{-2} = 0$ confirms that $T_{on} \propto 1/V^2$.

12 (1) 13 V

(2) 25 μs per cm (4 cm for period of 10^{-4} s)

Chapter 31

Exercise 31.1

6×10^{12} km

Exercise 31.2

1 5.3×10^{-7} m

2 4.3×10^{4} m s^{-1}

Exercise 31.3

1 4.8

2 -2.7

3 (a) 6.3   (b) 0.39   (c) 6.3

Exercise 31.4

3.1×10^{6} s (about 1 month)

Exercise 31.5

1 (a) 60°

(b) 50° (130° calculated means 50° up from celestial equator on other side of celestial sphere)

2 18 h 32 min

3 30 A.U.

4 6.2×10^{12} m

5 1.16

6 Intensity: energy received per second by unit area perpendicular to direction of travel

Luminosity: formula given shows luminosity is total energy radiated per second

Brighter star (greater intensity seen by observer on earth) is Vega $\left(\dfrac{I_V}{I_D} = 3.2\right)$

7 Apparent wavelength $= 4.2 \times 10^2$ nm, distance $= 8.8 \times 10^{24}$ m

Chapter 32

Exercise 32.1

1 (a) 76 dB   (b) 98 dB

2 1.0×10^5 W m^{-2}

3 16 dB

4 (a) 3.2×10^{-8} W m^{-2} (b) 3.2×10^{-9} W m^{-2}

Exercise 32.2

1 (a) 1.59×10^6 kg m^{-2} s^{-1} (b) 1.71×10^6 kg m^{-2} s^{-1}

(c) 1.38×10^6 kg m^{-2} s^{-1}

2 (a) 0.437, boundary easily detected

(b) 0.0114, boundary not easily detected

Exercise 32.3

1 4.7 kHz

2 (a) 0.69 m s^{-1} (b) 1.4×10^{-6} m^3 s^{-1}

Exercise 32.4

1 21 days

2 (a) 2.5 days   (b) 0.26 M Bq

3 1.5×10^{-4} C kg^{-1}

4 8.2 mJ kg^{-1}

5 2.1×10^{-2} μC kg^{-1} s^{-1}

6 0.5 mGy s^{-1}

7 (a) 4.2×10^{-8} Gy s^{-1}

(b) 66.7 hours, neglecting background radiation

8 8.0 μC kg^{-1} h^{-1}

9 (a) 1.2 mm   (b) 0.58×10^3 m^{-1} (c) 3.0%

Exercise 32.5: Examination questions

1 18 km

2 55 mJ

3 (a) 105(.6) dB   (b) 30 dB

4 (a) 4.0×10^{-4} W m^{-2} (b) 2.4×10^{-4} W m^{-2}

5 (b) 68×10^{-18} W   (c) 0.050 (1/20)

6 1.00 W m^{-2}

7 (a) See Example 5   (b) 1.09×10^3 kg m^{-3}

8 (a) (i) 1.35×10^6 kg m^{-2} s^{-1}

(ii) 1.76×10^6 kg m^{-2} s^{-1}

(b) 17.4×10^{-3} (c) 13(.5) mm

9 (a) 1.00×10^{-5} (b) 1.12×10^{-3}

Negligible reflection at brain–blood interface so this cannot be seen. Even blood–muscle reflection is small (but detectable).

10 (a) Radon gas

(b) Equation 32.13 + explanation

(c) 84 (83.3)

(d) No time for tissue to recover between doses and radiation absorbed by one part of body rather than whole body dose.

11 (a) 0.058 mm^{-1}

(b) Exponential decrease with, for example, the following points plotted:

$I =$	100	60	36	22	13
$x =$	0	8.8	17.6	26.4	35.2

12 23 mm

Chapter 33

Exercise 33.1

1 (a) 5.0   (b) 14   (c) 6.0
2 (a) $2.5\,\mathrm{rad\,s^{-1}}$ (b) $50\,\mathrm{rad\,s^{-1}}$ (c) 500 rad
3 (a) $9.4\,\mathrm{rad\,s^{-1}}$ (b) $-0.21\,\mathrm{rad\,s^{-2}}$
    (c) $-0.084\,\mathrm{N\,m}$ (d) 118 rad

Exercise 33.2

1 (a) +3.0   (b) −2.5
    (c) −2.0   (d) +2.0
2 (a) $3.0\pi\,\mathrm{N\,m}$ (b) $7.5\pi\,\mathrm{N\,m}$
3 (a) 30 s   (b) 180 rad   (c) $6\,\mathrm{rad\,s^{-1}}$
4 (a) $5.0\pi\,\mathrm{rad\,s^{-2}}$ (b) $0.95\,\mathrm{kg\,m^2}$
5 (a) $40\,\mathrm{N\,m}$ (b) (i) 33 s   (ii) 1.7×10^3 rad

Exercise 33.3

1 (a) $22\,\mathrm{rad\,s^{-1}}$ (b) $214\,\mathrm{rev\,min^{-1}}$
2 (a) $4.8\,\pi\,\mathrm{m\,s^{-1}}$ (b) 8.2 kJ

Exercise 33.4

1 (a) 60 J   (b) $8.7\,\mathrm{rad\,s^{-1}}$
2 (a) 45.8 kW   (b) $191\,\mathrm{N\,m}$

Exercise 33.5

1 (a) $9/8\,\mathrm{kg\,m^2}$ (b) 27 J
2 (a) $0.50\,\mathrm{kg\,m^2}$ (b) 27 J   (c) 20 J   (d) 7 J

Exercise 33.6: Examination questions

1 $0.7\,\mathrm{rad\,s^{-2}}$
2 (a) $-20.0\,\mathrm{rad\,s^{-2}}$ (b) 1.43×10^3 revolutions
    (c) (−) 240 N m
3 $250\,\mathrm{kg\,m^2}$
4 (a) 7.5 kJ   (b) 12 N m
5 0.30 kJ
6 (a) (i) 0.18 kW (176 W)
       (ii) –
   (b) (i) Perpendicular to radius
       (ii) 30°
7 (a) (i) $13\times10^4\,\mathrm{rad\,s^{-2}}$ (ii) 11 kW
       (iii) (−) 25π J   (iv) 0.27 kJ
   (b) To smooth out the motion
8 $2.7\times10^{40}\,\mathrm{kg\,m^2\,s^{-1}}$
9 (b) (i) $(7.16\times10^{33}\,\mathrm{kg\,m^2\,s^{-1}})$
       (ii) $3.3\times10^{-3}\,\mathrm{rad\,s^{-1}}$
       (iii) $1.2(5)\times10^{33}\,\mathrm{kg\,m^2\,s^{-1}}$
       (iv) 7.3×10^4 s (20 hours)

(v) Increased mass would increase total I thus decrease ω and increase T compared to part (iv)

Chapter 34:
Miscellaneous questions

1 (a) (i) 66 m   (ii) 3.7 s
      (iii) g independent of mass   (iv) 0.33 km
   (b) (i) 2.4 MJ   (ii) 0.32 MJ
   (c) Greater height possible. Longer time. Due to dissipative effects.
2 (a) 1.9×10^3 s (31 minutes)
   (b) $0.40\,\mathrm{m\,s^{-1}}$ (downwards)
   (c) 68 kg
   (d)

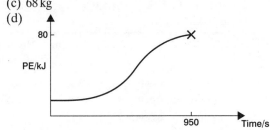

If wheel is equally loaded a passenger descending diametrically opposite means the motor does not have to supply the gravitational P.E. for the ascending passenger.
3 (a) (i) $x = 1.6$ cm when $a = 0$
       (ii) (12.8 − 1.6) cm
   (b) (i) 0.48 N   (ii) $4.3\,\mathrm{N\,m^{-1}}$
       (iii) 0.069 N
   (c) (i) $0.490\,\mathrm{m\,s^{-1}}$ (ii) 0.58 s
   (d) (i) Not directly proportional – graph does not pass through origin
       (ii) Acceleration, hence force, versus extension is a straight line
4 (a) (i) $(6.86\,\mathrm{m\,s^{-1}})$ (ii) 2.8 kN   (iii) 3.5 kN
       (iv) $F = +0.69$ kN at 0° and 360°; $F = +3.5$ kN at 180°
   (b) 8.9×10^{-6} m
5 (a) 0.44 J   (b) 0.14 J
   (c) Energy (of 0.30 J) dissipated via, for example, air resistance
6 (a) (i) Average person unlikely to have mass of 110 kg
       (ii) (1280 kg)
   (b) (i) 33(.04) kN   (ii) $1.0\times10^8\,\mathrm{N\,m^{-2}}$
   (c) (i) Maximum stress is 6.6×10^7 Pa which is less than the safe stress
       (ii) Greater length of cable leads to greater stress due to its own weight: e.g. cable alone gives $\sigma = 8.0\times10^7$ Pa
3 cables + static lift gives $\sigma = 9.5\times10^7$ Pa
Both situations lead to stresses greater than the maximum safe stress

7 (b) (i) 0.40 kg   (ii) 28 N   (iii) Downwards
   (c) 1.2(4) J
8 Vertical component 64 m s^{-1}
   (a) 13 s   (b) (1003 m)
   (c) 1.1 × 10^{17} J   (d) 31 s (30.6 s)
9 (b) (i) (1.19 mm)
      (ii) 1. 1.1 × 10^8 Pa   2. 0.85 × 10^{-3}
         3. 1.3 × 10^{11} Pa
   (c) (iii) 54 mJ
   (d) (i) 0.021 K
      (ii) Elastic strain energy 'stored' as thermal
         energy in hammer head
10 (a) (i) 156 K   (ii) 3 stops
   (b) Heat more rapidly dissipated and brakes
      lighter
11 (b) (i) 2.0 kg m s^{-1} (ii) 5.0 m s^{-1}
   (c) (i) 200 J   (ii) 5.0 J   (iii) 78 K
   (d) 1.3 m
12 (a)

S$_1$	S$_2$	Fan motor	Heating element
open	open	off	off
closed	open	on	off
open	closed	off	off
closed	closed	on	on

   (b) (i) 1.4(1) × 10^{-2} m^3 s^{-1}
      (ii) 1.7 × 10^{-2} kg s^{-1}
      (iii) 33 °C
   (c) 24 °C
13 (a) (i) 207(.5) mol   (ii) 6.2 kg
   (b) (i) 2.0 × 10^6 J (1.6 to 3.0 is acceptable)
      (ii) Pressure lower, so area under curve is
         smaller
   (c) (i) 0.44 kW
      (ii) 1.3 hours (1.0 to 2.1 is acceptable)
   (d) Need to have stronger, heavier, cylinder
14 (a) 10^{-9} (b) 2.0 × 10^{-17}
   (c) 0.15 cm^3 (at 1.0 × 10^8 Pa)   (d) 5.0 × 10^{-2}
   Large Young's modulus means spheres do not
   deform. Large breaking stress allows spheres to
   have thin shells.
   (e) Non toxic, bio-compatible, oxygen permeable
15 (a) (i) Equation 19.1 and explanation
      (ii) 1. $\Delta U = +30$ J
   (b) (i) 8.1 J   (ii) 16 W   (iii) 0.32 K s^{-1}
16 Might be incorrect number in the formula; the $\frac{1}{2}$
   might be incorrect.
17 (i) 10 A
   (ii) (I) 150 W
      (II) p.d. across kettle is less than 230 V by
         factor $\frac{R}{R+1.5}$
18 (a) Energy of thermal vibrations becomes greater
      and releases more current carriers

(b) (i) From the graph R is 340 Ω at 60°, PD is
      3.7 V
   (ii) 40 mW
19 (a) and (b)

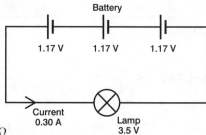

Battery
1.17 V   1.17 V   1.17 V
Current 0.30 A
Lamp 3.5 V

   (c) 1.1 Ω
20 (a) Work done in driving one coulomb of − charge
      from Z to X is 25 joule
   (b) (i) 9.0 V   (ii) 3.6 Ω
      (iii) 4.0 Ω   (iv) 6.25 or 6.2 Ω
21 (a) (i) $C = Q/V$
      (ii) $C = 1$ farad when $Q = 1$ coulomb and
         $V = 1$ volt
   (b) (i) –   (ii) x^2V_0, x^3V_0, x^4V_0
      (iii) $x = 1/e$ (iv) 0.37
      (v) 3 s
22 Need a capacitor in series. 34 μF, and extra
   resistance 185 Ω
23 (i) 32 mA
   (ii) Current lags on supply PD by 90° (1/4 cycle or
      $\frac{\pi}{2}$ radian). (Tan α in equation 26.9 is infinite.)
24 (a) $I \propto V$ for same temperature and conductor
   (b) (i) (I) 3.3 Ω   (II) 20 Ω
      (ii) Applies up to 0.18 A, 0.60 V
      (iii) (I) 2.4 × 10^{-3} K^{-1}
         (II) Resistance is constant at low
            temperatures and difference between
            room temperature and 0 °C is small
            compared with 2100 K
25 (a) 1.5 × 10^{-9} m^2 (b) 47 K s^{-1}
26 (a) (i) 2.9 × 10^{-16} J   (ii) 2.5 × 10^7 m s^{-1}
   (b) (i) $E = F/Q$
      (ii) 0.24 × 10^{-14} N, downwards
         towards + plate
      (iii) 3.2 × 10^{-9} s
      (iv) 8.4 × 10^6 m s^{-1}
      (v) 4.0 cm
27 (a) $^{237}_{93}$Np, $^4_2$He   (b) 5.0 × 10^{-11} s^{-1}
   (c) (i) 9.1 × 10^{13} (ii) 3.7 × 10^{-8} g
28 (a) (i) (1) 0.05 V   (2) 1.95 V
      (ii) 49 mH
   (b) (i) 8 A
      (ii) L produces no PD when current is
         constant, i.e. when $dI/dt = 0$
29 (a) (i) 146   (ii) 90
   (b) (i) 7.8 × 10^{-30} kg   (ii) 7.0 × 10^{-13} J
   (c) (i) 9.8 × 10^{-20} kg m s^{-1} (ii) 6.8 × 10^{-15} m
30 1.0 × 10^{25} m

Appendix

Table I
ABBREVIATIONS AND SYMBOLS

Abbreviations for SI units

kilogram (mass)	kg
metre (length)	m
second (time)	s
newton (force)	N
joule (work or energy)	J
watt (power)	W
kelvin (temperature)	K
volt (potential difference)	V
ampere (current)	A
ohm (resistance)	Ω
farad (capacitance)	F
henry (inductance)	H
tesla (magnetic flux density)	T
weber (magnetic flux)	Wb
coulomb (charge)	C
becquerel (activity)	Bq

Mathematical Symbols

equals	=
identical to	≡
proportional to	∝
square root	√
of the order of ('something like')	~
divided by	/
therefore	∴
approximately equal to	≏
greater than	>
less than	<

Table II
MULTIPLES AND SUBMULTIPLES FOR UNITS

The following prefixes are commonly used and their values are given below

Symbol	Prefix	Value
p	pico	10^{-12}
n	nano	10^{-9}
μ	micro	10^{-6}
m	milli	10^{-3}
c	centi	10^{-2}
k	kilo	10^{3}
M	mega	10^{6}
G	giga	10^{9}

ELECTRIC CIRCUIT SYMBOLS conforming to BS 3939

Component	Symbols	Notes
Conductor, wire or lead		Implies zero resistance
Crossing wires		No electrical connection between wires
Junction of two or more conductors		
Earth connection		
Switch (single way and two way)		Called 'single pole' since it 'breaks' only one conductor
Battery (one cell and multi cell)		
Resistor		
Variable resistor		
Potential or voltage divider		
Alternative method of showing voltage divider		Often referred to as potentiometer or a 'pot'
Capacitor		
Inductor (coil)		
Diode or Rectifier		Arrow indicates 'easy' direction of current flow
Bulb (lamp)		Second symbol is used for indicator lamps
Fuse link		Rating is usually shown by a number
Voltmeter and Ammeter		For millivoltmeter and milliammeter letters mV and mA are used
Galvanometer		
Thermistor		
Bipolar transistor		*pnp*
		npn
Light-emitting diode		*LED*
Zener diode		
Transformer		
Photoconductive cell		
Amplifier		

Index